# Pavement Design and Materials

**A.T. Papagiannakis PhD. P.E.**
Professor and Department Chair
Civil Engineering Department
University of Texas, San Antonio

**E.A. Masad PhD. P.E.**
Associate Professor
Zachry Department of Civil Engineering
Texas A&M University

John Wiley & Sons, Inc.

This book is printed on acid-free paper.♾



Published by John Wiley & Sons, Inc., Hoboken, New Jersey
Published simultaneously in Canada

Wiley Bicentennial Logo: Richard J. Pacifico



***Library of Congress Cataloging-in-Publication Data:***

Papagiannakis, A. T.
Pavement design and materials / A.T. Papagiannakis, E.A. Masad.
p. cm.
ISBN-13: 978-0-471-21461-8 (cloth)
1. Pavements—Design and construction. 2. Road materials. I. Masad, Eyad.
II. Title,
TE250.P25 2007
625.8–dc22

2007016255

Printed in the United States of America.

10 9 8

*Dedicated to my Father Stefanos, ATP*

*Dedicated to my wife Lina and to my family, EAM*

*"This was the only place in the world where pavements consist exclusively of holes with asphalt around them. They are the most economical because holes never go out of repair."*

*—Mark Twain Speech. October 15 and 17 1907*

# Contents

# Preface

We have embarked on this project with trepidation. Pavement engineering is a vast field covering a wide range of technical areas that are rapidly evolving. Our motivation was that only a handful of textbooks are in circulation addressing this topic. Furthermore, there have been several landmark technical developments in this area recently. These include the advent of *Superpave*™, the data generated by the Long-Term Pavement Performance (LTPP) project and the recent release of the Mechanistic-Empirical pavement design guide developed under NCHRP Study 1-37A. As a result, we felt that the time was right for recapturing the pavement engineering state of the art in a textbook.

This textbook covers pavement materials, analysis, design, evaluation, and economics of asphalt and portland concrete roadways. Its intended audience is engineering students at the undergraduate and junior graduate levels. In addition, practicing engineers may find it useful as a reference for practical design applications. Its structure focuses on the best established and currently applicable techniques for material characterization, analysis, and design, rather than offering a historical perspective of these techniques and the way they are applied by the multitude of jurisdictions dealing with roadway pavements. In compiling this textbook, our initial intention was to utilize metric (SI) units throughout. However, this was tempered by the number of empirical expressions still in use involving imperial units, including some adopted by the Mechanistic-Empirical design guide. In such cases, the use of dual units was unavoidable.

Implementing the various analytical techniques described in this textbook is facilitated through the use of software, a variety of which is available for pavement applications. Some of this software is proprietary, while other is free and can be downloaded from the Web. Since software evolves rapidly, we decided not to distribute software with this book. Instead, suggestions for software sources are given in a Web site associated with this book (www.wiley.com/go/pavement), and offered by application area following the chapter structure of the book (e.g., software for the structural analysis of rigid pavements, software for reducing pavement roughness profile measurements, and so on). This Web site is maintained to ensure that the most up-to-date software and its current source are recommended.

Finally, we want to express sincere thanks to all the agencies that authorized the use of some of their copyrighted material herein. These include: the Federal Highway Administration (FHWA), the American Association of State Highway and Transportation Officials (AASHTO), the Transportation Research Board (TRB), the Asphalt Institute, the Canadian Portland Cement Association (CPCA), and the American Society for Testing of Materials (ASTM). In addition, we would like to thank a number of companies that allowed use of photographs of their equipment in this book. We sincerely hope that this book will facilitate the instruction of pavement engineering and will serve as a reference to practicing pavement engineers worldwide.

A.T.Papagiannakis, PhD PE
San Antonio, Texas
and
E.A.Masad, PhD PE
College Station, Texas
November 2007

# 1 Introduction

## 1.1 Pavement Types

There are three general types of roadway pavements, namely flexible, rigid, and composite. Flexible pavements typically consist of asphalt concrete placed over granular base/subbase layers supported by the compacted soil, referred to as the *subgrade*. Some asphalt-paved surfaces consist of a simple bituminous surface treatment (BST), while other, lighter-duty asphalt-surfaced pavements are too thin, to be considered as flexible pavements, (i.e., combined layer thicknesses less than 15 cm). Rigid pavements typically consist of a portland concrete layer placed over the subgrade with or without a middle base layer. Composite pavements are typically the result of pavement rehabilitation, whereby portland concrete is used to cover damaged asphalt concrete or vice versa.

The terms *flexible* and *rigid* relate to the way asphalt and portland concrete pavements, respectively, transmit stress and deflection to the underlying layers. Ideally, a flexible layer transmits uniform stresses and nonuniform deflections, while the opposite is true for a rigid layer. In practice, the stress and deflection distributions in asphalt concrete and portland concrete pavements depend on the relative stiffness of these layers with respect to those of the underlying granular layers. This ratio is much lower for asphalt concrete than portland concrete, which justifies their generic designation as flexible and rigid, respectively. As described in later chapters, this affects significantly the way these two pavement types are analyzed and designed.

Figure 1.1 shows a typical cross section of a flexible pavement. The asphalt concrete layer, which may consist of two or more sublayers, or lifts, is placed on top of the granular base/subbase layers, which are placed on top of the subgrade. A tack coat layer may be applied to provide adhesion between layers, while a seal coat may provide a pavement surface barrier. A fabric or other geotextile placed between the base and the subgrade prevents migration of fines between them, and maintains their integrity. The base layers can be either compacted gravel, referred to as, simply, *granular*, or incorporate cement, referred to as *stabilized*. Typically, the asphalt concrete layer is designed with no interconnected voids (i.e., mix air voids 4–8%), and hence relies on surface runoff for precipitation drainage. Alternatively, asphalt concretes with interconnected voids, (i.e., mix air voids higher than 12%) allow drainage through the surface. This design requires a lower impermeable asphalt concrete layer, to prevent water from penetrating the base layer. Water runoff led to the edge of the pavement can be removed by surface evaporation, ditches, or drainage pipes.

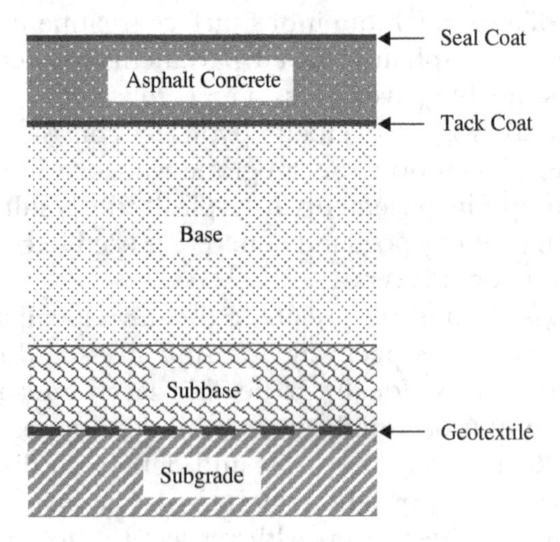


**Figure 1.1**
Typical Section of an Asphalt Concrete Pavement

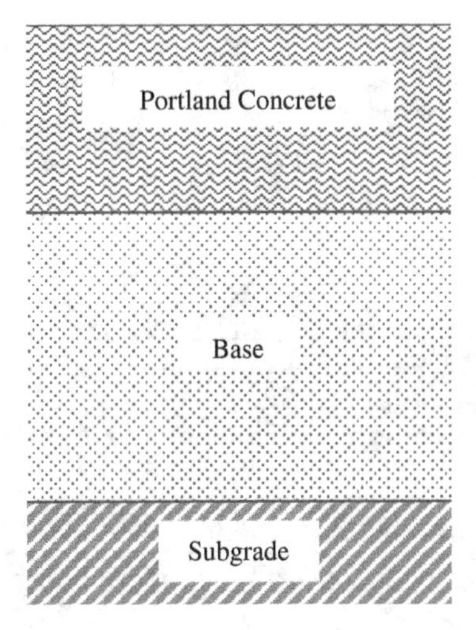


**Figure 1.2**
Typical Section of a Portland Concrete Pavement

Figure 1.2 shows a typical section of a rigid pavement. The portland concrete layer is placed either directly on top of the subgrade or on top of a granular base layer. Unreinforced portland concrete slabs, such as the one shown in Figure 1.2, tend to crack transversely where thermally induced tensile stresses exceed the tensile strength of the concrete. Hence, they require transverse joints at prescribed intervals. They are constructed by cutting a surface groove using a rotary saw before the concrete is fully cured. These joints, in addition to relieving thermal stresses, need to provide sufficient vertical load transfer between advancement slabs. Under a moving load, sufficient vertical load transfer provides a gradual buildup of stresses under the down stream slab, which controls the migration of moisture and fines under the joint and prevents downstream slab settlement, (i.e., faulting). Load transfer is accomplished either through aggregate interlock along the jagged edges of adjacent slabs (Figure 1.3a) or through dowel bars located at the neutral axis bridging the joint (Figure 1.3b). These dowel bars are smooth and epoxy-coated, to allow free horizontal movement while providing vertical displacement coupling between adjacent slabs. Collapsible end caps allow the expansion of the slabs without generating compressive stress in the dowel. These pavements are referred to as *jointed plain concrete pavements* (JPCP) and *jointed dowel reinforced concrete pavements*

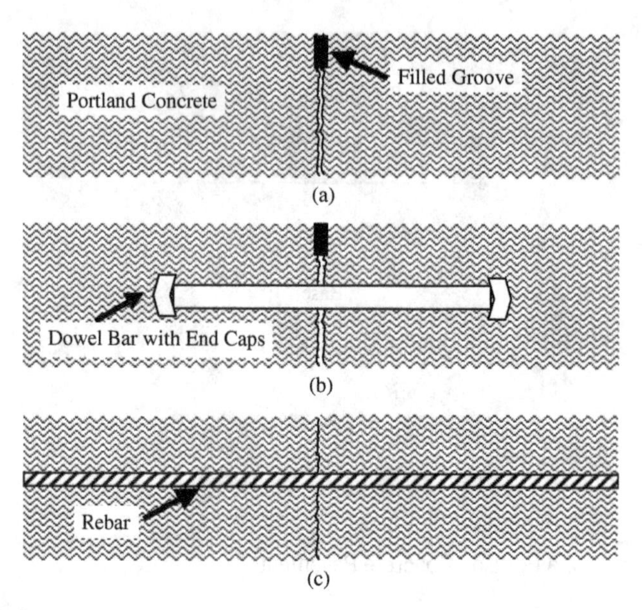


**Figure 1.3**
Typical Configuration of JPCPs, JDRCPs and CRCPs

(JDRCP), respectively. Continuously reinforced concrete slabs can withstand thermal stresses through the tensile strength of steel, hence they require no joints. The reinforcing steel is consists of deformed tiebars placed on the neutral axis of the slab. Thus, any thermal cracks in the concrete itself are not allowed to open, and the slab retains its structural integrity. These are called *continuously reinforced concrete pavements* (CRCP) (Figure 1.3c). It should be noted that hybrid rigid pavement structures have been developed, consisting of long CRCP slabs jointed through dowels in a JDRCP fashion.

A variety of other joint types are used in concrete pavements, including construction joints allowing continuity of the work between different days (Figure 1.4a) and expansion joints necessary where concrete pavements come against other rigid structures, such as bridge abutments (Figure 1.4b). Overall, the successful design and construction of joints and their reinforcement contributes significantly to the performance of concrete pavements. Reducing vehicle dynamics dictates the randomization of joint spacing, (e.g., 2.1, 2.7,

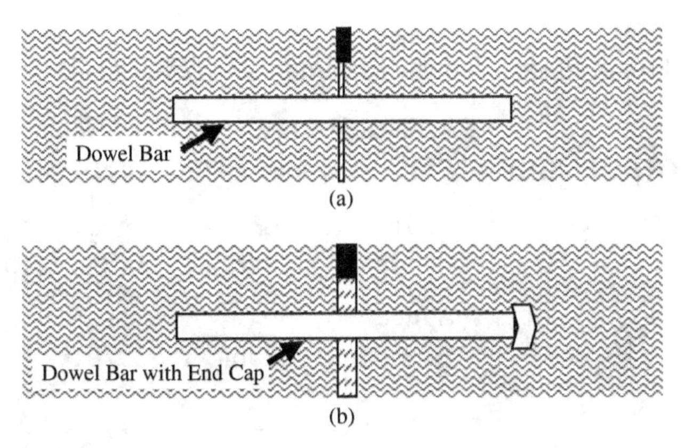


**Figure 1.4**
Special-Function Portland Concrete Pavement Joints

3.3, and 4.5 meters (m)), as well as their skewed arrangement with respect to the longitudinal axis of the pavement [5].

## 1.2 Pavement Infrastructure Overview

The staggering size of the roadway pavement infrastructure in the United States can be appreciated by the length inventory data shown in Tables 1.1 and 1.2 for rural and urban pavements, respectively[4]. These tables show centerline kilometers (km) length by roadway functional class, namely interstate, arterials, and collectors. These functional class designations relate to the geometric standards of the roadway, as well as the combination of access and mobility it affords. Minor collectors and local roads were excluded from these tables. Per the October 2006 count, they amounted to 1,584,764 and 1,094,428 centerline kilometers (km) in rural and urban areas, respectively, which brings the total length to 4.2 million km, or approximately 11 times the distance to the moon. The interstate system (74,000 km) and an additional 184,523 km of other freeways comprise the National Highway System (NHS), as designated in 1995 by Public Law 104–59 (6). The NHS represents about 4% of the total roadway pavement mileage but carries over 44% of the vehicle-kilometers traveled.

**Table 1.1**
Length Inventory (centerline km) of Rural Roadway Pavements (FHWA, 2004)

| OWNER/CLASS | PAVEMENT TYPE | | | | | |
|---|---|---|---|---|---|---|
| **State** | **BSTs** | **OTHER LIGHT** | **FLEXIBLE** | **COMPOSITE** | **RIGID** | **TOTAL** |
| Interstate | 455 | 1688 | 22911 | 9606 | 12590 | 47250 |
| Other Arterial | 1308 | 6531 | 104605 | 24282 | 14828 | 151555 |
| Minor Arterial | 5200 | 18865 | 153257 | 26145 | 8151 | 211619 |
| Major Collector | 51850 | 74984 | 202889 | 22531 | 3132 | 355386 |
| Subtotal | 58813 | 102068 | 483663 | 82564 | 38701 | 765809 |
| **Federal** | | | | | | |
| Interstate | — | — | — | — | — | — |
| Other Arterial | — | 13 | 143 | — | 2 | 158 |
| Minor Arterial | — | 800 | 447 | 3 | — | 1250 |
| Major Collector | 465 | 2791 | 1347 | — | — | 4603 |
| Subtotal | 465 | 3603 | 1938 | 3 | 2 | 6011 |
| **Other** | | | | | | |
| Interstate | — | — | 845 | 1147 | 401 | 2393 |
| Other Arterial | 58 | 77 | 328 | 166 | 604 | 1233 |
| Minor Arterial | 328 | 1622 | 2995 | 182 | 10 | 5137 |
| Major Collector | 46005 | 80166 | 107560 | 4521 | 11872 | 250124 |
| Subtotal | 46391 | 81866 | 111728 | 6016 | 12886 | 258886 |
| Total | 105669 | 187536 | 597329 | 88583 | 51589 | 1030707 |

## 1.3 Significance of Pavement Infrastructure to the Nation's Economic Activity

Roadway pavements play a very important role in the nation's economic activity. Approximately, 19% of average household expenditures is directly related to transportation (Figure 1.5). The predominant mode of personal transportation is by private motor vehicle, that is, 91.2% of the total vehicle-kilometers[6]. Furthermore, an average 89% of commercial freight transportation is carried by the highway system (Figure 1.6). The vehicle-miles traveled (VMT) is a very good indicator of the health of the economy, as suggested by its strong correlation to the gross domestic product (GDP), that is, the annual sum of goods and services transacted nationwide (Figure 1.7). These simple facts demonstrate the importance of the roadway infrastructure in the nation's economic well-being.

**Table 1.2**
Length Inventory (centerline km) of Urban Roadway Pavements (FHWA, 2004)

| MANAGER/CLASS | PAVEMENT TYPE | | | | | |
|---|---|---|---|---|---|---|
| **State** | **BSTs** | **OTHER LIGHT** | **FLEXIBLE** | **COMPOSITE** | **RIGID** | **TOTAL** |
| Interstate | 66 | 190 | 8618 | 6725 | 8087 | 23686 |
| Other Arterial | 499 | 2408 | 47817 | 23503 | 12517 | 86743 |
| Minor Arterial | 954 | 5412 | 34141 | 13118 | 2379 | 56003 |
| Major Collector | 3215 | 7546 | 15852 | 2535 | 853 | 30001 |
| Subtotal | 4735 | 15556 | 106427 | 45881 | 23836 | 196434 |
| **Federal** | | | | | | |
| Interstate | — | — | — | — | — | — |
| Other Arterial | — | 40 | 92 | — | 6 | 138 |
| Minor Arterial | — | 16 | 60 | 2 | 2 | 79 |
| Major Collector | 18 | 100 | 45 | 6 | — | 169 |
| Subtotal | 18 | 156 | 196 | 8 | 8 | 386 |
| **Other** | | | | | | |
| Interstate | — | 8 | 512 | 943 | 237 | 1699 |
| Other Arterial | 579 | 3858 | 18594 | 3555 | 3217 | 29803 |
| Minor Arterial | 5517 | 23669 | 60106 | 9991 | 8464 | 107745 |
| Major Collector | 10530 | 42728 | 70074 | 9999 | 6817 | 140148 |
| Subtotal | 16626 | 70262 | 149286 | 24488 | 18734 | 279396 |
| Total | 21378 | 85974 | 255909 | 70376 | 42578 | 476217 |

## 1.4 Funding Pavements

The value of this infrastructure is in the trillions of dollars. The ongoing annual expenditures for roadway preservation, capacity addition, and new route construction are in the billions, (e.g., the federal-only component of these expenditures in FY 2000 was $16.2 billion[6]). Although these figures include the cost of bridges, they demonstrate the extent of public investment in this vital piece of infrastructure.

Roadway pavements are financed through fuel taxes. Federal taxes on fuel date back to the 1930s. The Federal-Aid Highway Act of 1956 established the Highway Trust Fund and stipulated that 100% of the fuel tax be deposited into the fund. Between 1956 and 1982, the Highway Trust Fund was used solely to finance expenditures for the federal highway program. The Surface Transportation Act of 1982 legislated that approximately 20% of the federal fuel taxes revenues be allocated to a newly created mass transit account and be expended

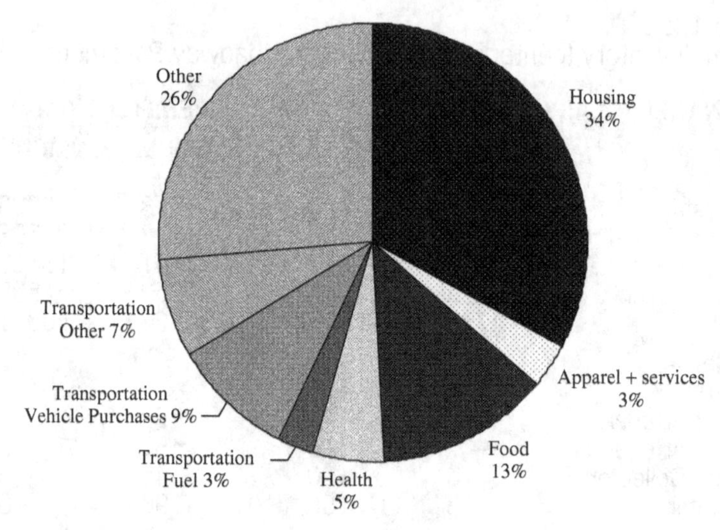


**Figure 1.5**
Distribution of Household Expenditures; 1999 Data (Ref. 6)

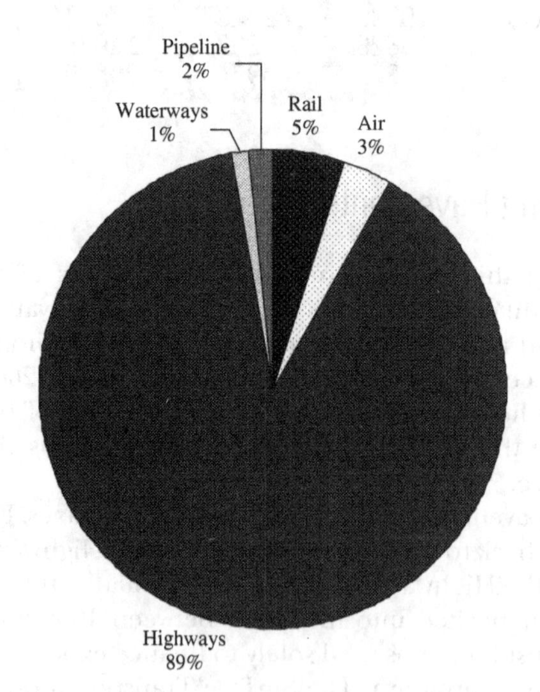


**Figure 1.6**
Distribution of Freight by Transport Mode; 1996 Data (Ref. 6)

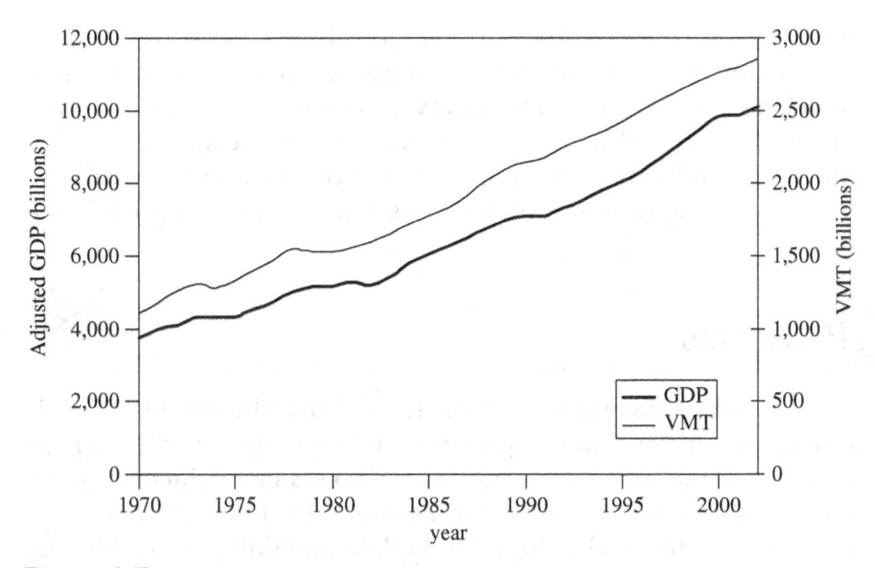


**Figure 1.7**
Correlation between VMT and GDP; 1970–2002 (Ref. 8)

to improve public transportation. The historic distribution of federal gasoline tax revenues is shown in Table 1.3.

In 2006, federal tax rates for gasoline and diesel were 18.4 cents per gallon and 24.4 cents per gallon, respectively; states impose their own taxes on fuel. As of 2004, the average state rates for gasoline and diesel were 19.1 cents per gallon and 24.4 cents per gallon, respectively[7]. The reason for higher diesel fuel tax rates is to compensate for the pavement damage caused by heavy trucks. For the

**Table 1.3**
Distribution of Gasoline Tax Revenues (1983–1997) (Ref. 1)

| Date | General Revenues | Highways | Mass Transit Account | Other Trust Funds |
|---|---|---|---|---|
| Before 1983 | | 100.0% | | |
| Apr. 1, 1983 | | 88.9% | 11.1% | |
| Dec 1, 1990 | 17.7% | 70.1% | 10.6% | 0.7% |
| Oct. 1, 1993 | 37.0% | 54.3% | 8.2% | 0.5% |
| Oct. 1, 1995 | 23.4% | 65.2% | 10.9% | 0.5% |
| Jan. 1, 1996 | 23.5% | 65.6% | 10.9% | |
| Oct. 1, 1997 | | 83.9% | 15.5% | 0.5% |

same reason, some states (e.g., Oregon, Idaho, New Mexico) are using a weigh-distance tax to replace part of the consumption-based diesel fuel taxes. This taxes heavy trucks in proportion to their weight and the distance they travel. It has been argued that this and similar taxation approaches provide a more equitable means of taxing various vehicle classes than fuel consumption-based taxes[2].

## 1.5 Engineering Pavements

The preceding discussion demonstrates clearly the extent of public investment in the roadway pavement infrastructure and its importance in the nation's economic vitality. As a result, engineering pavements requires the utmost care and use of state-of-the art technology, and involves the technology for both maintaining/rehabilitating existing pavements as well as designing/constructing new ones. This technology encompasses the characterization of the materials involved and the structural design of the layers selected to withstand prevailing traffic and environmental conditions. In addition, it necessitates the evaluation of their in-service performance with time as well as their economic implications to both the agency and the user. Although the last two topics relate to the broader subject of pavement management (e.g., Ref. 3), they are an integral part of pavement engineering, hence are included in this book.

## 1.6 Book Organization

This book contains information on roadway pavement materials, pavement structural analysis, pavement design, and pavement economic analysis. The remaining chapters are organized as follows:

- Chapter 2 describes with the characterization of traffic input.
- Chapters 3 and 4 deal with the characterization of pavement bases/subgrades and aggregates, respectively.
- Chapter 5 addresses asphalt binder and asphalt concrete characterization.

- Chapter 6 characterize, portland cement and portland concrete.
- Chapters 7 and 8 describe the analysis of flexible and rigid pavements, respectively.
- Chapter 9 discusses pavement evaluation.
- Chapter 10 addresses the environmental effects on pavements.
- Chapters 11 and 12 deal with the design of flexible and rigid pavements, respectively.
- Chapter 13 describes pavement rehabilitation.
- Chapter 14 deals with the economic analysis of alternative pavement designs.

## References


[1] Buechner, W. (2006). "History of the Gasoline Tax." American Roads and Transportation Building Association, (www.artba.org).

[2] FHWA (August 1997). Federal Highway Cost Allocation Study, Federal Highway Administration www.fhwa.dot.gov/policy/otps/costallocation.htm.

[3] Haas, R. C. G., W. R. Hudson, and J. Zaniewski, (1994). *Modern Pavement Management*, Krieger Publishing Co., Malabar, FL.

[4] FHWA (2005). "Highway Statistics; Section V": *Roadway Extent, Characteristics, and Performance*, Federal Highway Administration. Washington, DC.

[5] PCA (1980). *Joint Design for Concrete Highways and Street Pavements*, Portland Cement Association, Skokie IL.

[6] FHWA (2000). *Our Nation's Highways*, Federal Highway Administration, Washington, DC.

[7] FHWA (2007). State Motor Fuel Tax Rates, 1988–2003, Federal Highway Administration www.fhwa.dot.gov/policy/ohim/hs03/htm/mf205.htm.

[8] FHWA (2006). "Transportation Air Quality: Selected Facts and Figures," Federal Highway Administration, Washington, DC, Publication No. FHWA-HEP-05-045 HEP/12-05(8M)E.

## Problems

1.1 Find the length of roads in your state by functional class (i.e., interstate, other arterial, minor arterial, and major collector) and surface type, (i.e., BST, light-duty, flexible, composite, and rigid).

1.2 Find the rate of state tax levied on gasoline and diesel fuel in your state. What was the corresponding amount of total proceeds from the sale of gasoline and diesel fuel for road vehicles in the last year?

1.3 Compute the current annual amount of fuel tax, state and federal, paid for operating a typical privately owned vehicle in your state. Assume that the fuel consumption is 11.8 liters/100 km (20 miles/U.S. gallon) and that the vehicle is driven 24,000 km (15,000 miles) per year.

1.4 What was the percentage of the GNP expended on road transportation last year?

# 2 Pavement Traffic Loading

## 2.1 Introduction

Pavement deterioration is caused by the interacting damaging effects of traffic and the environment. Traffic loads, primarily those from heavy trucks, cause stresses/strains in pavement structures, whose effects accumulate over time, resulting in pavement deterioration, such as plastic deformation in asphalt concretes or fatigue cracking in portland concretes. Hence, truck traffic load data is an essential input to the pavement analysis and design process.

Truck traffic loads and their impact on pavements are quantified in terms of:

- Number of truck axles
- Configuration of these axles
- Their load magnitude

Axle configuration is defined by the number of axles sharing the same suspension system and the number of tires in each axle. Multiple axles involve two, three, or four axles spaced 1.2 to 2.0 meters apart, and are referred to as *tandem, triple,* or *quad,* respectively. They are treated differently from single axles because they impose pavement stresses/strains that overlap.

A number of additional traffic-related parameters are also important in analyzing pavements, namely:

- Timing of axle passes (i.e., time of the day and season within the year)
- Vehicle/axle speed
- Vehicle/axle lateral placement
- Tire inflation pressure

The timing of axle passes is important for both flexible and rigid pavements, mainly because of the seasonality in pavement layer properties and the time dependency of thermal stresses, respectively. The vehicle/axle speed is relevant mainly to flexible pavements, due to the viscoelastic behavior of the asphalt concrete. The lateral vehicle/axle weaving affects the lateral distribution of the accumulated damage, hence is also pertinent to pavement deterioration. Tire inflation pressure is relevant because it affects the contact pressure between tires and pavement. A common assumption is that the tire imprint has a circular shape and carries uniformly distributed vertical stress that is equal to the tire inflation pressure, (i.e., it is assumed that the tires are treadles and that the tire walls carry no load). As a result, its radius $a$ is given by:

$$a = \sqrt{\frac{P}{i\,\pi}} \tag{2.1}$$

where $P$ is the vertical load carried by the tire, and $i$ is the inflation/contact pressure.

As described next, state-of-the-art traffic monitoring equipment allows automated collection of all of these traffic data elements, with the exception of vehicle/axle lateral placement and tire inflation pressure. The following sections describe the basic features of traffic-monitoring technology, the methodologies used for summarizing the traffic load data for pavement analysis purposes, and the need for load limits and their enforcement.

## 2.2 Traffic-Monitoring Technology

Traffic load data is collected by a combination of traffic data monitoring systems, including automatic traffic recorders (ATR), automated vehicle classifiers (AVC), and weigh-in-motion (WIM) systems. These systems are typically installed in the driving lanes and record data at normal driving speeds. Static weigh scales, such as those installed in truck inspection stations, are used for load enforcement, rather than for data collection purposes.

### 2.2.1 Automated Traffic Recorder (ATR) Systems

ATRs are the least expensive traffic monitoring systems. They consist of a variety of sensors, ranging from pneumatic tubes to radar and a data acquisition system. The most common sensor in permanently installed ATRs is the inductive loop (Figure 2.1). Inductive loops are simple open-wire loops embedded near the pavement surface. They experience inductive currents in response to the movement of the metal mass of vehicles in their vicinity. The associated voltage lasts from the time the front of a vehicle crosses the leading edge of the loop until its rearmost crosses the downstream edge of the loop. The duration of the voltage is related to two variables, namely vehicle length and speed. Hence, single loops cannot be used to differentiate vehicle types and simply serve as vehicle counters. Vehicle counts are stored in a data acquisition box at the site. The data can be manually retrieved or uploaded electronically to a central data bank at regular intervals via telephone lines.

A number of alternative sensors have been introduced in ATR applications, using acoustic, radar, and ultrasonic technologies. These involve sensors placed overhead, rather than embedded into the pavement, and are preferable where traffic congestion results in

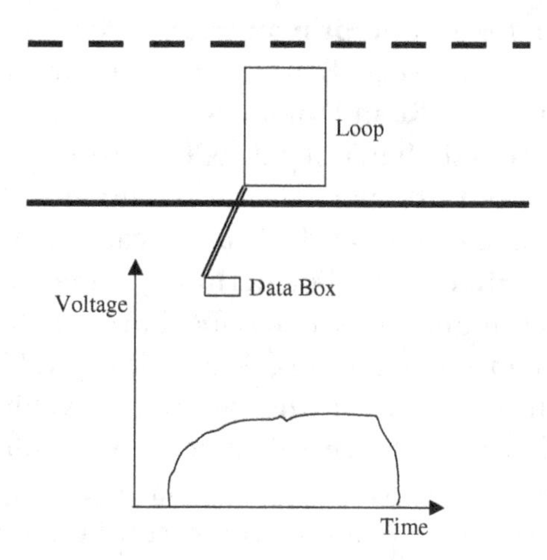


**Figure 2.1**
Schematic of an ATR System and the Voltage Output of Its Inductive Loop

variable vehicle speeds. Another advantage of such systems is that they can be moved between locations to provide short-term traffic count samples (e.g., 48 hours or 1 week) in a variety of locations without interrupting the traffic for sensor installation.

Clearly, ATRs collect only a subset of the traffic data elements needed for pavement analysis/design. Nevertheless, their low cost allows installation of a considerable number of these systems throughout a network of roads. This data, supplemented by AVC and WIM data, provides estimates of the axle loads experienced in multitude of locations throughout a road network.

### 2.2.2 Automated Vehicle Classifier (AVC) Systems

AVCs record vehicle volumes by vehicle classification. Vehicle classification is defined in terms of the number of axles by axle configuration. The majority of state departments of transportation (DOTs) use a FHWA-established vehicle classification system involving 13 vehicle classes (Table 2.1).

AVCs determine vehicle classification by detecting the number of axles and their spacing. This is done through a combination of vehicle and axle sensors. The most common configuration of AVC involves two inductive loops and a single-axle sensor (Figure 2.2). Two inductive loops allow calculation of vehicle speed as the ratio of their spacing divided by the difference in loop trigger timing. Given the vehicle speed, axle spacing is calculated from the trigger timing of the axle sensor. Axle sensor operation is based on mechanical, piezoelectric, or fiber-optic principles. A variety of manufacturers supply such axle sensors (e.g., Dynax Corp., Measurement Specialties Inc., and International Road Dynamics).

It should be noted that not all axle sensors can differentiate between two and four tires per axle. As a result, not all AVCs can distinguish Class 3 from Class 5 vehicles, nor can identify nonsteering truck axles on single tires. Furthermore, some unusual passenger car-trailer configurations could be hard to distinguish from some tractor-trailer combinations, (e.g., Class 8 vehicles). Another source of classification error is the variation in vehicle speed while passing an AVC system. Since this unavoidable under high-traffic volumes (e.g., level of service D or E), classifying vehicles through conventional AVC systems under these conditions is challenging. Camera-based sensors used for general traffic data collection purposes are emerging as potential AVC sensors. However, to date, no

**Table 2.1**
FHWA Vehicle Classes (Ref. 16)

| Class ID | Sketch | Description |
|---|---|---|
| 1 | | Motorcycles |
| 2 | | Passenger cars |
| 3 | | Two-axle, four-tire light trucks |
| 4 | | Buses |
| 5 | | Two-axle, six-tire, single-unit trucks |
| 6 | | Three-axle single-unit trucks |
| 7 | | Four or more axle single-unit trucks |
| 8 | | Four or fewer axle single-trailer trucks |
| 9 | | Five-axle single-trailer trucks |
| 10 | | Six or more axle single-trailer trucks |
| 11 | | Five or fewer axle multitrailer trucks |
| 12 | | Six-axle multitrailer trucks |
| 13 | | Seven or more axle multitrailer trucks |

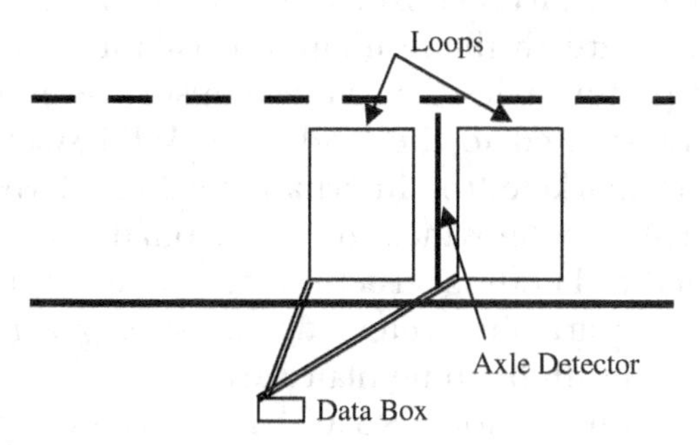


**Figure 2.2**
Schematic of an AVC System

**Table 2.2**
AVC System Accuracy Tolerances (Ref. 14)

| Element | Maximum Error |
|---|---|
| Unclassified Vehicles | 2% |
| Misclassified Trucks | 2% |

software exists that can automatically process vehicle images for the purpose of obtaining vehicle classes as defined in Table 2.1. Table 2.2 lists AVC system accuracy tolerances. Accuracy of AVC systems is evaluated on the basis of manual vehicle classification data obtained from several independent observers.

Clearly, although AVC data contains more information than ATR data, it still lacks a data element crucial to the pavement analysis/design process, namely the load of the axles.

### 2.2.3 Weigh-in-Motion (WIM) Systems

WIM technology expands on the information collected by AVCs by providing the load of each axle passing over the system. WIMs typically consist of a combination of inductive loops for detecting vehicle speed and one or several axle load sensors. Several technologies have been used for axle load sensing, including load-cell-equipped plates, strain-gauged plates, piezoelectric cables, quartz cables, fiber-optic cables, and capacitance mats. The first three are the most commonly used in permanently installed systems. An example of a load-cell-equipped plate WIM system is shown in Figures 2.3.

The common feature of WIM load sensors is their capability to respond/recover quickly, allowing multiple closely spaced axles to be weighed individually at highway speeds. The load-measuring principle, however, varies considerably between sensors. Load cell systems measure directly the resultant vertical force exerted on each wheel path by passing axles (i.e., a mechanism prevents shear forces from being transmitted to the load cell). WIM systems with load cell sensors are marketed by International Road Dynamics (IRD). Strain-gauged plate systems measure load from the response of strain gauges attached to bending structural components supporting the plates; hence, they are often referred to as *bending-plate systems*. WIM systems equipped with bending plates are marketed by a number of vendors, such as International Road Dynamics and Mettler/Toledo. Piezoelectric WIM sensor come from two manufacturers, namely Thermocoax and Measurement Specialties Inc. The Thermocax

**Figure 2.3**
Example of a Load Cell WIM system (Courtesy IRD Inc.)

sensors, marketed under the trade name Vibracoax, are formed by swaging/drawing a ceramic material sandwiched between a coaxial brass central core and a brass outer sheath, resulting in a coaxial circular section cable. The Measurement Specialties Inc. sensors are formed using a flat braided cable core enveloped by a piezoelectric copolymer extruded onto the outside of the braid. In both sensors, a polarization technique is used to produce piezoelectric sensitivity, whereby stress changes applied to the sensors generate a voltage differential between outer sheath and core. This voltage signal is electronically processed to yield the load of the axle that applied the stress. (More information on the mechanical properties of these two sensors and their laboratory and field performance can be found in references 6 and 7, respectively.)

One of the inherent features of all these WIM technologies is that they measure dynamic, rather than static, axle loads. Dynamic axle loads are substantially different from their static values—that is, those obtained with the vehicle stopped at a truck inspection station. This is due to the dynamic interaction between vehicles and pavement, which causes excitation of vehicle frame and axles. The resulting dynamic axle loads depend on pavement roughness, vehicle operating speed, and axle suspension type. Heavy truck suspensions include a variety of multiple-leaf springs, air springs, rubber springs, and torsion springs. Some of these suspensions are equipped with shock absorbers, others derive damping through frictional or torsional action, while others are not dampened at all. Quantifying these dynamic axle loads is important in understanding WIM accuracy. This can be done using either measurements onboard instrumented vehicles or simulation of vehicle dynamic behavior, as described in the voluminous literature on this subject, (e.g., see references 2, 15, 17). An example of dynamic axle load measurements from an instrumented Class 9 vehicle (i.e., five-axle semitrailer truck) is shown in Figure 2.4. These measurements were obtained from the lead drive and the lead trailer tandem axles, which were equipped with an air-spring and a rubber-spring suspension, respectively[9]. It can be seen that the dynamic loads are substantially different from the static loads of these axles, which amounted to about 100 Kilonewtons (kN). The importance of this difference on WIM measurements can be visualized by considering the width of the axle load sensors. Load cell or bending plate sensors have a width of about 0.6 meters, while piezoelectric cables have a width of a mere 0.03 meters. At highway speeds, it takes a fraction of a second to traverse such distances; hence, WIM sensors experience only a short section of the dynamic load waveform exerted along the length of the pavement. Figure 2.4 demonstrates this by including the width of a WIM sensor plate drawn to scale.

The magnitude of the dynamic load variation can be quantified by the coefficient of variation (*CV*) (i.e., standard deviation divided by the mean). Figure 2.5 shows the dependence of *CV* on vehicle

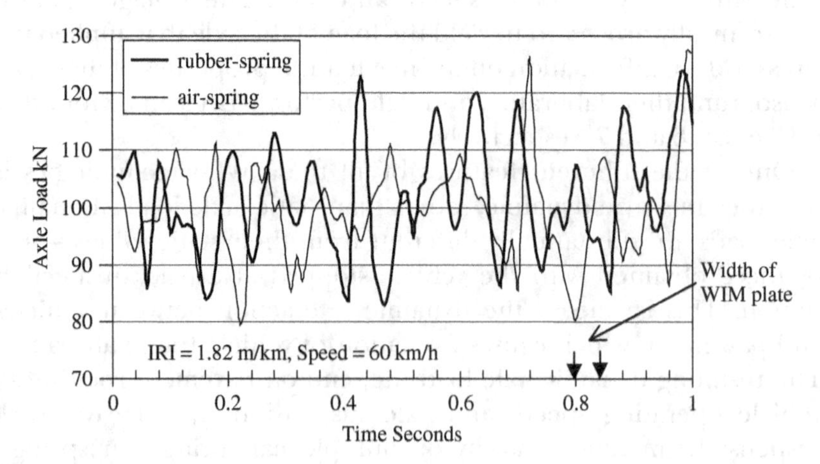


**Figure 2.4**
Example of Dynamic-Axle Load Measurements

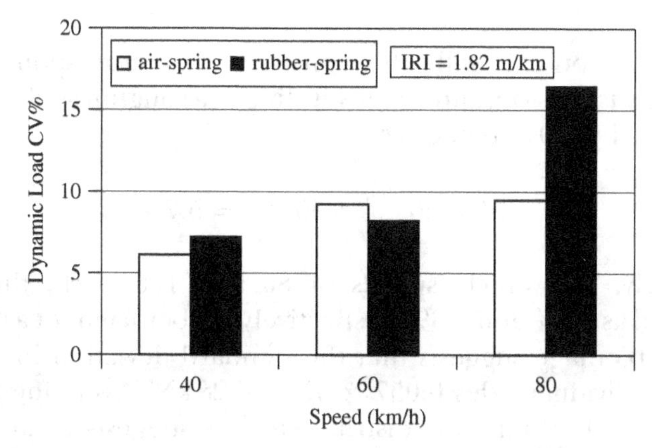


**Figure 2.5**
Example of the Effect of Vehicle Speed on the *CV* of Dynamic-Axle Load

speed and suspension type. The pavement roughness in this figure is indexed by the International Roughness Index (*IRI*), a unit that will be explained in detail in Chapter 9. In general, the dependence of *CV* in percent on roughness $R$ (*IRI* in m/km) and speed $V$ (km/hr) can be expressed as:

$$CV = V^a \, R^b \tag{2.2}$$

where the exponents $a$ and $b$ depend on suspension type. For the experimental data given in reference 9, the values of these regression constants are summarized in Table 2.3.

**Example 2.1**

A five-axle semitrailer truck is used for evaluating the accuracy of a WIM system. It is equipped with air-spring suspensions in both the drive and trailer axles. Each of these tandem axles carries a static load of 150 kN. The pavement roughness upstream from the WIM system is 1.5 m/km on the IRI scale. Compute the expected range in dynamic axle loads exerted on the pavement and, in turn, to the WIM sensors at vehicle speeds of 60, 80, and 100 km/h.

**Table 2.3**
Regression Constants for Equation 2.2 (Data from Ref. 9)

| Constant | Air-Spring Suspension | Rubber-Spring Suspension |
|---|---|---|
| a | 0.346 | 0.456 |
| b | 0.798 | 0.728 |

**ANSWER**

Using Equation 2.2 with the constants that correspond to the air-spring suspension, and by substituting in a roughness of 1.5 m/km and a speed of 60 km/h, gives:

$$CV = 60^{0.346}\,1.50^{0.798} = 5.7\%$$

Similarly, for vehicle speeds of 80 and 100 km/h, the *CV* is calculated as 6.3% and 6.8%, respectively. A coefficient of a variation of 5.7 percentage suggests that the standard deviation in dynamic load of individual axles $0.057 \times 75 = 4.28$ kN. Assuming that the dynamic load is normally distributed, this suggests a range from $75 - 2 \times 4.28 = 66.45$ kN to $75 + 2 \times 4.28 = 83.55$ kN at a 95% confidence. Consequently, this is the anticipated range in WIM measurements.

Several efforts have been made to establish WIM system accuracy with reference to dynamic rather than static loads. For this purpose, the dynamic behavior of test trucks was established either through onboard measurements or dynamic vehicle simulations, (e.g., references 10,11, respectively). Although these approaches give the true accuracy of WIM systems, they do not allow computing static axle loads from WIM measurements (This would require multiple WIM measurements and elaborate algorithms; e.g., reference 8). As a result, it is standard practice to establish WIM accuracy with reference to static loads instead[14]. Axle dynamics are reduced by specifying low pavement roughness in the 45-m approach to the WIM sensors, and, where possible, weigh vehicles at lower speeds. WIM accuracy is evaluated using a minimum of two test trucks performing several runs over the system at each of three vehicle speeds (i.e., minimum and maximum operating speeds at a site and an intermediate speed). The static axle loads of all these vehicles is established through static weighing using certified static scales. The percent error in individual measurements, *e*, is defined as:

$$e = \frac{WIM - static}{static}100 \tag{2.3}$$

Calibration consists of adjusting the WIM output to achieve a zero mean for the errors. WIM accuracy is defined in terms of the probability that individual axle load measurement errors are within prescribed limits (Table 2.4).

**Table 2.4**
WIM System Accuracy Tolerances (Ref. 14)

<table>
<tr><th rowspan="2">Element</th><th colspan="5">Tolerance for 95% Probability of Conformity</th></tr>
<tr><th>Type I</th><th>Type II</th><th>Type III</th><th colspan="2">Type IV</th></tr>
<tr><td>Wheel Load</td><td>±25%</td><td>-</td><td>±20%</td><td>≥2300 kg*</td><td>±100 kg</td></tr>
<tr><td>Axle Load</td><td>±20%</td><td>±30%</td><td>±15%</td><td>≥5400 kg</td><td>±200 kg</td></tr>
<tr><td>Axle Group Load</td><td>±15%</td><td>±20%</td><td>±10%</td><td>≥11300 kg</td><td>±500 kg</td></tr>
<tr><td>Gross Vehicle Weight</td><td>±10%</td><td>±15%</td><td>±6%</td><td>≥27200 kg</td><td>±1100 kg</td></tr>
<tr><td>Vehicle speed</td><td colspan="5">±1.6 km/h</td></tr>
<tr><td>Axle spacing</td><td colspan="5">±0.15 m</td></tr>
</table>

*Lighter masses and associated loads are of no interest in enforcement.

Four types of WIM systems are distinguished in terms of their operational and functional characteristics:

- Type I and Type II, which have the capability to collect wheel load and axle load data, respectively, at vehicle speeds ranging from 16 to 113 km/h.
- Type III, which has a traffic sorting function and operates at vehicle speeds from 24 to 113 km/h. Type III systems are installed on the approache ramps to truck inspection stations to single out trucks that are likely to be over the legal load limits, which in turn need to be weighed statically.
- Type IV is intended for load enforcement at vehicle speeds up to 16 km/h.

WIM systems that meet the pavement roughness requirement, but do not pass the evaluation test (i.e., their errors do not conform to the limits specified in Table 2.4 for the intended WIM type), are declared deficient. Given the variation in axle load dynamics discussed earlier, it is obvious that a significant part of the error ranges allowed in Table 2.4 is intended for accommodating axle dynamics.

**Example 2.2**

The load measurements shown in Table 2.5 were obtained from two Class 9 test vehicles running over a Type I WIM system. Determine whether this system meets ASTM accuracy tolerances in measuring axle loads[14]. Compute the corresponding calibration factor.

**Table 2.5**
Load Measurements (kN) for Evaluating a WIM System—Example 2.2

| | Static Load | WIM Pass 1 | WIM Pass 2 | WIM Pass 3 | WIM Pass 4 | WIM Pass 5 |
|---|---|---|---|---|---|---|
| **Test Vehicle 1** | | | | | | |
| Steering | 42.2 | 40.2 | 46.7 | 48.9 | 52.0 | 46.5 |
| Drive, Axle 1 | 69.5 | 74.0 | 69.6 | 72.0 | 69.8 | 72.0 |
| Drive, Axle 2 | 71.0 | 72.2 | 70.0 | 71.2 | 71.0 | 71.6 |
| Trailer, Axle 1 | 65.6 | 65.0 | 68.0 | 66.5 | 66.8 | 67.0 |
| Trailer, Axle 2 | 67.0 | 66.7 | 66.7 | 66.0 | 67.1 | 65.2 |
| GVW | 315.3 | 318.1 | 320.9 | 324.6 | 326.7 | 322.3 |
| **Test Vehicle 2** | | | | | | |
| Steering | 43.8 | 42.0 | 45.0 | 44.0 | 44.5 | 42.8 |
| Drive, Axle 1 | 50.0 | 51.1 | 53.3 | 49.2 | 50.3 | 51.5 |
| Drive, Axle 2 | 49.9 | 52.0 | 52.3 | 50.0 | 49.0 | 50.0 |
| Trailer, Axle 1 | 67.4 | 65.6 | 64.4 | 69.7 | 67.5 | 67.5 |
| Trailer, Axle 2 | 66.8 | 65.0 | 62.0 | 71.3 | 64.0 | 67.0 |
| GVW | 277.9 | 275.7 | 277.1 | 284.2 | 275.3 | 278.8 |

**ANSWER**

The errors in the WIM measurements of steering axles and tandem axles, calculated per Equation 2.3, is shown in Table 2.6. It can be seen that only one of the axle load measurements violates the limits given in Table 2.4 for a Type I WIM system (i.e., the steering axle measurement of the fourth pass of the first test vehicle has an error of +23.6%, which is larger than the prescribed +20%). This is one violation in 50 axle load measurements, which gives a conformity of 98%. The system exhibits no GVW measurement violations. In conclusion, this system passes the Type I requirements. The calibration factor is calculated, according to the ASTM standard[14] by averaging the errors in axle load measurements (Table 2.6). The result is 1.6%, which means that the output of the WIM system needs to be multiplied by 0.984 to produce zero average errors.

## 2.3 Summarizing Traffic Data for Pavement Design Input

In practice, traffic data is collected using a combination of traffic-monitoring technologies, including ATR, AVC, and WIM systems

**Table 2.6**
WIM Errors (Percent) Calculated for the Data Shown in Table 2.5

| | WIM Pass 1 | WIM Pass 2 | WIM Pass 3 | WIM Pass 4 | WIM Pass 5 | Average |
|---|---|---|---|---|---|---|
| **Test Vehicle 1** | | | | | | |
| Steering | −4.79 | 10.53 | 15.82 | 23.16 | 10.13 | 10.97 |
| Drive, Axle 1 | 6.47 | 0.14 | 3.60 | 0.36 | 3.60 | 2.83 |
| Drive, Axle 2 | 1.72 | −1.41 | 0.21 | 0.00 | 0.85 | 0.27 |
| Trailer, Axle 1 | −0.85 | 3.73 | 1.44 | 1.90 | 2.20 | 1.68 |
| Trailer, Axle 2 | −0.50 | −0.50 | −1.49 | 0.15 | −2.76 | −1.02 |
| GVW | 0.89 | 1.79 | 2.94 | 3.61 | 2.21 | — |
| **Test Vehicle 2** | | | | | | |
| Steering | −4.06 | 2.79 | 0.51 | 1.65 | −2.23 | −0.27 |
| Drive, Axle 1 | 2.22 | 6.67 | −1.60 | 0.50 | 3.00 | 2.16 |
| Drive, Axle 2 | 4.23 | 4.83 | 0.22 | −1.78 | 0.22 | 1.55 |
| Trailer, Axle 1 | −2.80 | −4.45 | 3.34 | 0.08 | 0.01 | −0.76 |
| Trailer, Axle 2 | −2.69 | −7.19 | 6.74 | −4.19 | 0.30 | −1.41 |
| GVW | −0.81 | −0.30 | 2.26 | −0.96 | 0.30 | — |
| | | | | | Overall Average 1.6 | |

distributed over the roadway network. Some of these systems are permanently installed, to record data continuously, while others are installed temporarily at a location, to record data over shorter periods of time and then arc moved to other locations. This allows expanding the area coverage of the traffic data collection with limited traffic-monitoring resources. Where data is collected over short periods of time (e.g., 48 hours or several weeks), appropriate factors are used to calculate the traffic volumes and axle loads over the desired interval (e.g., yearly)[16]. This data needs to be summarized in a format that can be readily input to the pavement design process. The methods used for this purpose are described next.

### 2.3.1 AASHTO 1986/1993 Pavement Design Approach

The 1986/1993 American Association of State Highway and Transportation Officials (AASHTO) Pavement Design Guide[1] utilizes an aggregate approach for handling traffic load input. It assigned dimensionless pavement damage units to each axle configuration and load magnitude, referred to as equivalent single-axle load (*ESAL*) factors. The origin of the *ESAL* concept is traced back to the AASHO Road Test conducted in the 1950s. As described in Chapters 11 and 12, this test involved observations of the performance of pavements subjected to accelerated (i.e., time-compressed) loading. The reference axle configuration/load for *ESAL* calculation was a single axle on dual tires inflated to 586 kPa (i.e., 85 lbs/in$^2$) carrying a load of 80 kN (i.e., 18,000 lbs). Pavement life was defined in terms of the number of load repetitions that cause pavement serviceability failure (i.e., a terminal value for the Present Serviceability Index (*PSI*) of either 2.0 or 2.5, as described in Chapter 9). Mathematically, the *ESAL* of an axle of load $x$ is defined as:

$$ESAL_x = \frac{\rho_{80}}{\rho_x} \tag{2.4}$$

where $\rho_{80}$ and $\rho_x$ are the observed number of repetitions to failure from the 80 kN reference axle and from the axle of load $x$, respectively. For a given axle configuration and load, the *ESAL* factors depend on the thickness of the pavement layers and the terminal serviceability selected. For flexible pavements, the thickness of the pavement layers is aggregated into the *S*tructural *N*umber (*SN*), defined as:

$$SN = a_1\, D_1 + a_2\, m_2\, D_2 + a_3\, m_3\, D_3 \tag{2.5}$$

where, $D_1$, $D_2$, and $D_3$ are the layer thicknesses of the asphalt concrete (inches), base, and subbase, respectively, and $m_2$, and $m_3$ are the drainage coefficients for the base and the subbase, respectively. The latter depend on the length of time to drain (see Table 10.1) and the percentage of time the pavement layer is exposed to moisture levels approaching saturation. The recommended values for the drainage coefficients are given in Table 2.7.

**Table 2.7**
Drainage Coefficients for Unbound Base and Subbase Layers (Ref. 1 Used by Permission)

| | **Percent of Time Layer Is Approaching Saturation** | | | |
|---|---|---|---|---|
| Drainage Quality* | < 1% | 1–5% | 5–25% | > 25% |
| Excellent | 1.40–1.35 | 1.35–1.30 | 1.30–1.20 | 1.20 |
| Good | 1.35–1.25 | 1.25–1.15 | 1.15–1.00 | 1.00 |
| Fair | 1.25–1.15 | 1.15–1.05 | 1.00–0.80 | 0.80 |
| Poor | 1.15–1.05 | 1.05–0.80 | 0.80–0.60 | 0.60 |
| Very Poor | 1.05–0.95 | 0.95–0.75 | 0.75–0.40 | 0.40 |

*See Table 10.1.

The *SN* layer coefficients $a_i$ are estimated from the remaining pavement life or the in situ pavement layer elastic moduli back-calculated from surface deflection measurements, as will be described in Chapters 13 and 9, respectively. Typically, layer coefficient values of 0.44, 0.14, and 0.11 are used for new asphalt concrete, unbound base and unbound subbase layers, respectively.

Tables 2.8 and 2.9 list flexible pavement *ESAL* factors for single and tandem axles, respectively. Tables 2.10 and 2.11 list rigid pavement *ESAL* factors for single and tandem axles, respectively. A complete list of *ESAL* factors for different axle loads/configurations can be found in Reference 1. These factors are for truck axles on dual tires. It is accepted that truck axles on single tires cause more pavement damage than identically loaded axles on dual tires. It is typically assumed that axles on single tires must carry about 10% lower load than axles on dual tires to cause the same pavement damage[12]. Accordingly, the *ESAL* factor of axles on single tires can be obtained from the dual-tire *ESAL* tables (Tables 2.8 to

**Table 2.8**
Flexible Pavement ESAL Factors for Single Axles on Dual Tires: Terminal Serviceability of 2.0 (Adapted from Ref. 1 Used by Permission)

| Axle Load (kN) | Pavement *SN* | | | | | |
|---|---|---|---|---|---|---|
| | 1 | 2 | 3 | 4 | 5 | 6 |
| 9 | .0002 | .0002 | .0002 | .0002 | .0002 | .0002 |
| 18 | .002 | .003 | .002 | .002 | .002 | .002 |
| 27 | .009 | .012 | .011 | .010 | .009 | .009 |
| 36 | .030 | .035 | .036 | .033 | .031 | .029 |
| 44 | .075 | .085 | .090 | .085 | .079 | .076 |
| 53 | .165 | .177 | .189 | .183 | .174 | .168 |
| 62 | .325 | .338 | .354 | .350 | .338 | .331 |
| 71 | .589 | .598 | .613 | .612 | .603 | .596 |
| 80 | 1.00 | 1.00 | 1.00 | 1.00 | 1.00 | 1.00 |
| 89 | 1.61 | 1.59 | 1.56 | 1.55 | 1.57 | 1.59 |
| 98 | 2.49 | 2.44 | 2.35 | 2.31 | 2.35 | 2.41 |
| 107 | 3.71 | 3.62 | 3.43 | 3.33 | 3.40 | 3.51 |
| 116 | 5.36 | 5.21 | 4.88 | 4.68 | 4.77 | 4.96 |
| 124 | 7.54 | 7.31 | 6.78 | 6.42 | 6.52 | 6.83 |
| 133 | 10.4 | 10.0 | 9.2 | 8.6 | 8.7 | 9.2 |
| 142 | 14.0 | 13.5 | 12.4 | 11.5 | 11.5 | 12.1 |

**Table 2.9**
Flexible Pavement ESAL Factors for Tandem Axles on Dual Tires: Terminal Serviceability of 2.0 (Adapted from Ref. 1 Used by Permission)

| Axle Load (kN) | Pavement *SN* | | | | | |
|---|---|---|---|---|---|---|
| | 1 | 2 | 3 | 4 | 5 | 6 |
| 44 | .007 | .008 | .008 | .007 | .006 | .006 |
| 53 | .013 | .016 | .016 | .014 | .013 | .012 |
| 62 | .024 | .029 | .029 | .026 | .024 | .023 |
| 71 | .041 | .048 | .050 | .046 | .042 | .040 |
| 80 | .066 | .077 | .081 | .075 | .069 | .066 |
| 89 | .103 | .117 | .124 | .117 | .109 | .105 |
| 98 | .156 | .171 | .183 | .174 | .164 | .158 |
| 107 | .227 | .244 | .260 | .252 | .239 | .231 |
| 116 | .322 | .340 | .360 | .353 | .338 | .329 |
| 124 | .447 | .465 | .487 | .481 | .466 | .455 |
| 133 | .607 | .623 | .646 | .643 | .627 | .617 |
| 142 | .810 | .823 | .843 | .842 | .829 | .819 |
| 151 | 1.06 | 1.07 | 1.08 | 1.08 | 1.08 | 1.07 |
| 160 | 1.38 | 1.38 | 1.38 | 1.38 | 1.38 | 1.38 |
| 169 | 1.76 | 1.75 | 1.73 | 1.72 | 1.73 | 1.74 |
| 178 | 2.22 | 2.19 | 2.15 | 2.13 | 2.16 | 2.18 |

2.11) after multiplying the load by a factor of 1.1. The accumulated damaging effect of the variety of axles passing over a pavement section is calculated simply by adding the *ESAL* factors of these axles.

**Example 2.3**

The following axle load measurements were obtained from weighing a Class 9 vehicle empty and full (see Figure 2.6). Compute the total *ESALs* caused by one pass of this vehicle, and the pavement-related efficiency of this vehicle in terms of kN of cargo carried per *ESAL*. Given a flexible pavement with an *SN* of 4 and a terminal *PSI* of 2.0.

**ANSWERS**

The *ESAL* values corresponding to this vehicle are tabulated in Table 2.12. The kN of cargo carried per *ESAL* is 204.6/2.39 = 85.6 kN/ESAL. This is a good indicator of the relative cargo-carrying efficiency of various truck configurations with reference to the pavement damage caused.

**Table 2.10**
Rigid Pavement ESAL Factors for Single Axles on Dual Tires: Terminal Serviceability of 2.0 (Adapted from Ref. 1 Used by Permission)

| Axle Load (kN) | Slab Thickness (mm) | | | | | | | | |
|---|---|---|---|---|---|---|---|---|---|
| | 152 | 178 | 203 | 229 | 254 | 279 | 305 | 330 | 356 |
| 9 | .0002 | .0002 | .0002 | .0002 | .0002 | .0002 | .0002 | .0002 | .0002 |
| 18 | .002 | .002 | .002 | .002 | .002 | .002 | .002 | .002 | .002 |
| 27 | .011 | .010 | .010 | .010 | .010 | .010 | .010 | .010 | .010 |
| 36 | .035 | .033 | .032 | .032 | .032 | .032 | .032 | .032 | .032 |
| 44 | .087 | .084 | .082 | .081 | .080 | .080 | .080 | .080 | .080 |
| 53 | .186 | .180 | .176 | .175 | .174 | .174 | .173 | .173 | .173 |
| 62 | .353 | .346 | .341 | .338 | .337 | .336 | .336 | .336 | .336 |
| 71 | .614 | .609 | .604 | .601 | .599 | .599 | .598 | .598 | .598 |
| 80 | 1.00 | 1.00 | 1.00 | 1.00 | 1.00 | 1.00 | 1.00 | 1.00 | 1.00 |
| 89 | 1.55 | 1.56 | 1.57 | 1.58 | 1.58 | 1.59 | 1.59 | 1.59 | 1.59 |
| 98 | 2.32 | 2.32 | 2.35 | 2.38 | 2.40 | 2.41 | 2.41 | 2.41 | 2.42 |
| 107 | 3.37 | 3.34 | 3.40 | 3.47 | 3.51 | 3.53 | 3.54 | 3.55 | 3.55 |
| 116 | 4.76 | 4.69 | 4.77 | 4.88 | 4.97 | 5.02 | 5.04 | 5.06 | 5.06 |
| 124 | 6.58 | 6.44 | 6.52 | 6.70 | 6.85 | 6.94 | 7.00 | 7.02 | 7.04 |
| 133 | 8.92 | 8.68 | 8.74 | 8.98 | 9.23 | 9.39 | 9.48 | 9.54 | 9.56 |
| 142 | 11.9 | 11.5 | 11.5 | 11.8 | 12.2 | 12.4 | 12.6 | 12.7 | 12.7 |

### 2.3.2 NCHRP 1-37A Pavement Design Approach

The load input to the NCHRP 1-37A pavement design approach[5] is in terms of axle load distributions (i.e., load spectra) by axle configuration. This approach is a significant improvement over the aggregate *ESAL*-based method described previously, because it allows a mechanistic pavement design approach. As described extensively in Chapters 11 and 12, this involves computing the pavement structural responses to load, (i.e., stresses/strains), translating them into damage, and accumulating the damage into distresses, which reduce pavement performance over time. An example of load spectra is shown in Figure 2.7.

In general, this type of traffic data is assembled by combining data from WIM, AVC, and ATR systems distributed throughout a roadway network. Table 2.13 outlines the four traffic input levels distinguished by the NCHRP 1-37A approach.

- Level 1 input requires project/lane-specific data on volume/classification/axle load distribution, which can be collected only with a WIM system operated at the design site over extended periods of time.

**Table 2.11**
Rigid Pavement ESAL Factors for Tandem Axles on Dual Tires: Terminal Serviceability of 2.0 (Adapted from Ref. 1 Used by Permission)

| Axle Load (kN) | Slab Thickness (mm) | | | | | | | | |
|---|---|---|---|---|---|---|---|---|---|
| | 152 | 178 | 203 | 229 | 254 | 279 | 305 | 330 | 356 |
| 44 | .014 | .013 | .013 | .012 | .012 | .012 | .012 | .012 | .012 |
| 53 | .028 | .026 | .026 | .025 | .025 | .025 | .025 | .025 | .025 |
| 62 | .051 | .049 | .048 | .047 | .047 | .047 | .047 | .047 | .047 |
| 71 | .087 | .084 | .082 | .081 | .081 | .080 | .080 | .080 | .080 |
| 80 | .141 | .136 | .133 | .132 | .131 | .131 | .131 | .131 | .131 |
| 89 | .216 | .210 | .206 | .204 | .203 | .203 | .203 | .203 | .203 |
| 98 | .319 | .313 | .307 | .305 | .304 | .303 | .303 | .303 | .303 |
| 107 | .454 | .449 | .444 | .441 | .440 | .439 | .439 | .439 | .439 |
| 116 | .629 | .626 | .622 | .620 | .618 | .618 | .618 | .618 | .618 |
| 124 | .852 | .851 | .850 | .850 | .850 | .849 | .849 | .849 | .849 |
| 133 | 1.13 | 1.13 | 1.14 | 1.14 | 1.14 | 1.14 | 1.14 | 1.14 | 1.14 |
| 142 | 1.48 | 1.48 | 1.49 | 1.50 | 1.51 | 1.51 | 1.51 | 1.51 | 1.51 |
| 151 | 1.90 | 1.90 | 1.93 | 1.95 | 1.96 | 1.97 | 1.97 | 1.97 | 1.97 |
| 160 | 2.42 | 2.41 | 2.45 | 2.49 | 2.51 | 2.52 | 2.53 | 2.53 | 2.53 |
| 169 | 3.04 | 3.02 | 3.07 | 3.13 | 3.17 | 3.19 | 3.20 | 3.20 | 3.21 |
| 178 | 3.79 | 3.74 | 3.80 | 3.89 | 3.95 | 3.98 | 4.00 | 4.01 | 4.01 |

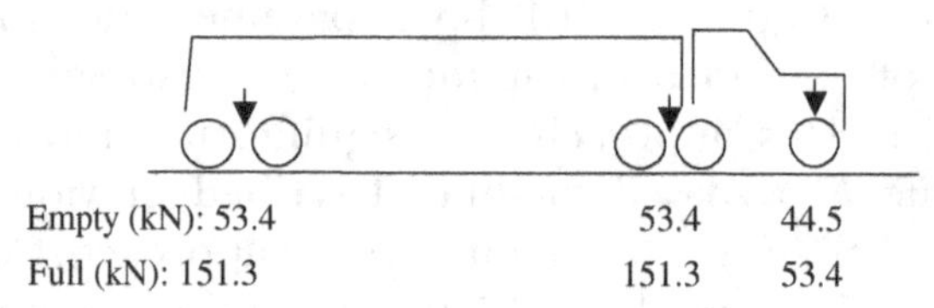


**Figure 2.6**
Example 2.3; *ESAL* Factor Calculations

- Level 2 input requires project/lane-specific data on traffic volumes by vehicle class, combined with representative/regional axle load distribution data—although it is possible with a site/lane-specific installed AVC system, combined with WIM installations on roads of similar truck traffic composition as the design site.
- Level 3 input requires site/lane-specific data on traffic volumes and an estimate of the percentage of trucks, which is also

**Table 2.12**
ESAL Calculations for Example 2.3

| ESAL Factor | Empty | Full |
|---|---|---|
| Steering (from Table 2.8) | 0.140 | 0.23 |
| Drive Tandem (from Table 2.9) | 0.014 | 1.08 |
| Trailer Tandem (from Table 2.9) | 0.014 | 1.08 |
| Total | 0.168 | 2.39 |

possible with a site/lane-specific ATR and manual truck percentage observations, combined with regional AVC and WIM data.

- Level 4 input is similar to the Level 3 input, except that national representative or default AVC and WIM data is utilized, were instead of regional data.

The accuracy and the time coverage of the data in each level defines the confidence in the computed axle load spectra, which in turn defines the reliability in pavement design. Table 2.14 summarizes the traffic data elements required as input to the NCHRP 1-37A approach and the flow of calculations carried out by the software in assembling the axle load spectra information by axle configuration and month of the year. The major variables utilized are defined next.

The average annual daily truck traffic for vehicle class $c$ ($AADTT_c$) is computed from AVC data as[16]:

$$AADTT_c = \frac{1}{7}\sum_{i=1}^{7}\left[\frac{1}{12}\sum_{j=1}^{12}\left(\frac{1}{n}\sum_{k=1}^{n} AADTT_{ijkc}\right)\right] \tag{2.6}$$

where:

$AADTT_{ijkc}$ = average daily traffic volume for vehicle class $c$, for day $k$ of day of the week (DOW) $i$, and month $j$.

$i$ = DOW, ranging from 1 to 7 for Monday to Sunday, respectively.

$j$ = month of the year ranging from 1 to 12 for January to December, respectively.

$n$ = number of times data from a particular DOW is available for computing the average in a given month (i.e., 1, 2, 3, 4, or 5).

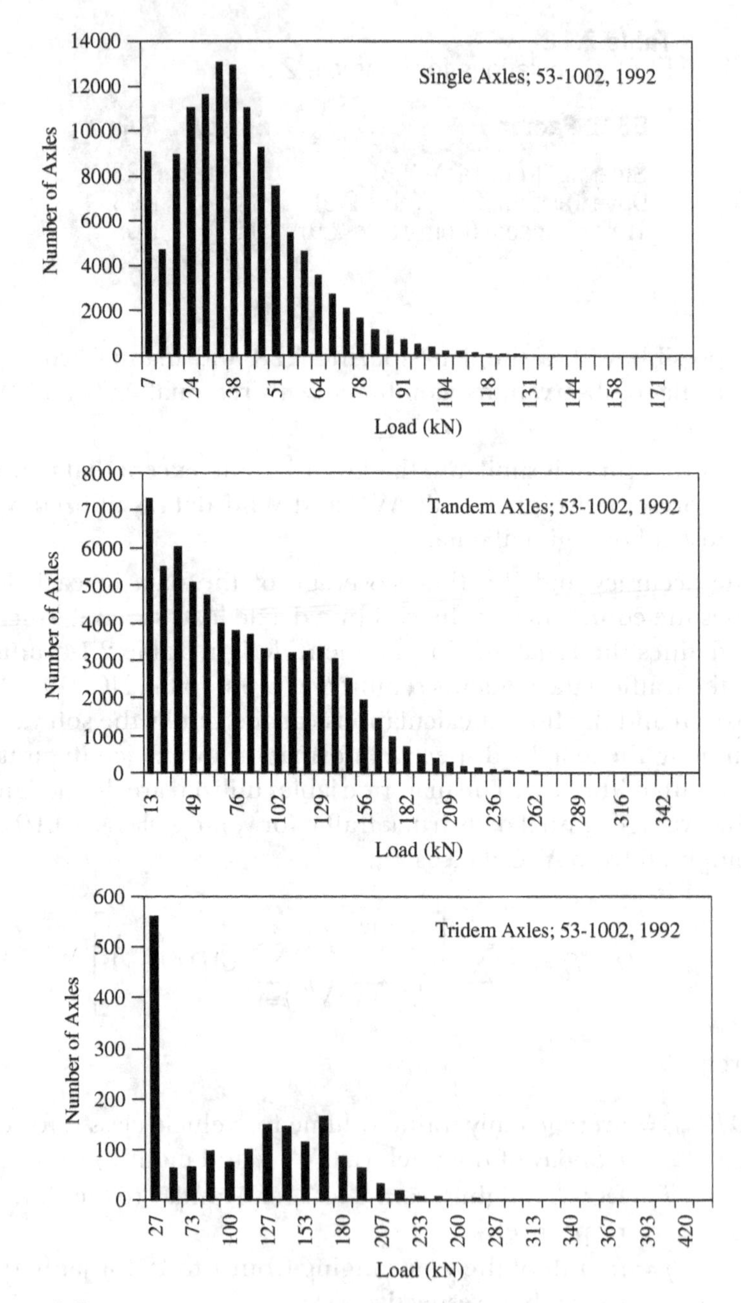


**Figure 2.7**
Annual Axle Load Spectra Example (Ref. 4)

**Table 2.13**
Traffic Input Levels in the NCHRP 1-37A Approach (Ref. 5)

| Data Element/Input Variables | Traffic Input Level 1 | 2 | 3 | 4 |
|---|---|---|---|---|
| WIM Data–Site/Segment-Specific | x | | | |
| WIM Data—Regional Representative Weight Data | | x | x | |
| AVC Data–Site/Segment-Specific | x | x | | |
| AVC Data—Regional Represent. Truck Volume Data | | | x | |
| ATR—Site-Specific | | | x | x |

**Table 2.14**
Traffic Input Elements and Flow of Calculations in Assembling Axle Load Spectra

| Traffic Input Component | Main Data Element | Input Array Size | Calculation and Result |
|---|---|---|---|
| 1 | Average annual daily truck traffic (AADTT) in the design lane | 1 | — |
| 2 | Distribution of trucks by class (i.e., FHWA classes 4–13). | 1 × 10 | 1 × 2 = annual average daily number of trucks by class |
| 3 | Monthly adjustment factors (MAF) by truck class | 12 × 10 | 1 × 2 × 3 = adjusted average daily number of trucks by class, by month |
| 4 | Number of axles by axle configuration, (single, tandem, triple, quad) by truck class | 4 × 10 | 1 × 2 × 3 × 4 = average number of axles by axle configuration, by month |
| 5 | Load frequency distribution (%) by axle configuration, month, and truck class | 4 × 12 × 10 × 41 | 1 × 2 × 3 × 4 × 5 = number of axles by load range, by axle configuration, by month |

The monthly adjustment factor for month $j$ ($MAF_j$) is computed using[16]:

$$MAF_j = \frac{AADTT_c}{VOL_{cj}} \tag{2.7}$$

where, $AADT_c$ = average annual daily truck traffic volume for vehicle class $c$, and $VOL_{cj}$ = average annual daily truck traffic volume for vehicle class $c$ and month $j$ that can be obtained from AVC data.

It should be noted that the NCHRP 1-37A approach assumes a uniform traffic distribution within each month and a constant distribution of the hourly traffic within each day. It also accepts a single tire inflation as input and utilizes it for all axle configurations.

## 2.4 Load Limits and Enforcement

The need for load enforcement rises from the highly nonlinear relationship between axle load and pavement damage. This can be illustrated by observing the change in *ESAL* factors as a function of axle load, as shown for example in Figure 2.8 for tandem axles on a flexible pavement with a *SN* of 4.

In addition to this empirical evidence, there is indisputable mechanistic proof that pavement damage is a highly nonlinear function of axle load. As discussed in Chapters 11 and 12, asphalt concrete fatigue, for example, depends in a highly nonlinear function on strain level. Similar considerations apply not only to pavements but also to bridges. Obviously, there is a fundamental need for protecting the roadway infrastructure. This is done by imposing load limits on commercial vehicles, and enforcing them. On the interstate system, minimum load limits are set by the federal government[13] and are 89 kN (20,000 lbs), 151 kN (34,000 lbs) for single and tandem axles, respectively, and 356 kN (80,000 lbs) for gross vehicle weight (GVW). In addition, the total load limit, $W$ (kN), allowed on $N$

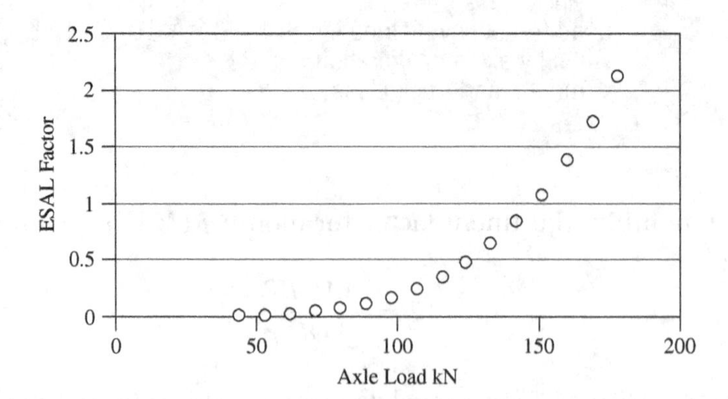


**Figure 2.8**
Relationship between ESAL Factors and Axle Load (Ref. 1)

consecutive axles is given by:

$$W = 2.224\left(\frac{0.3048\ LN}{N-1} + 12\ N + 36\right) \tag{2.8}$$

where $L$ is the distance (m) between the extreme axles in this group of $N$. This formula makes axle spacing allowances for load and is designed to limit the number of heavy axles that can be simultaneously carried by bridges—hence its name "bridge formula." Note that state governments may impose in their jurisdictions different axle loads than the federal limits given here. States are also responsible for administrating a system of special permits and fees for commercial loads that exceed those legal limits but cannot be subdivided. It is important to understand that the need to protect the public investment in the roadway infrastructure is coupled with the need for cost-effective transportation of goods on the highway network. The latter is directly related to the amount of cargo that can be carried by commercial vehicles. Load limit regulations and the associated legislation are continuously evolving in response to these two needs.

**Example 2.4**

The weight and dimension measurements shown in Figure 2.9 were obtained for two commercial vehicles at a truck inspection station. Determine whether they are legally loaded with reference to axle load limits, GVW limits, and the bridge formula.

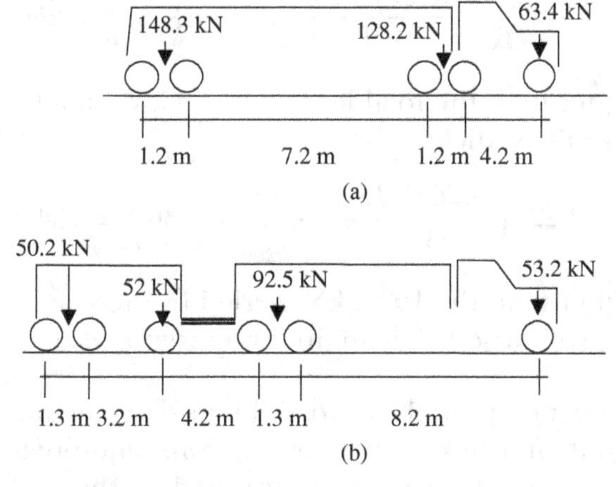


**Figure 2.9**
Weight and Dimension Measurements for Example 2.4

**ANSWER**

Both vehicles meet the axle group limits—that is, their single axles carry less than 89 kN and their tandem axles carry less than 151 kN. They also satisfy the GVW requirement, (i.e., their GVW is 339.9 kN and 247.9 kN, respectively). Test the compliance of vehicle (a) with respect to the bridge formula, (i.e., Equation 2.8), starting with the three axles of the tractor:

$$W = 2.224\left(\frac{0.3048\ 5.4\ 3}{3-1} + 12\ 3 + 36\right) = 165.6\ kN$$

which is lower than actual 191.6 kN carried by the tractor. As a result, vehicle (a) is in violation of the bridge formula, hence is overloaded, (i.e., tractor is simply too short for the load it carries).

Test the compliance of vehicle (b) to the bridge formula, beginning with the load of the three tractor axles:

$$W = 2.224\left(\frac{0.3048\ 9.5\ 3}{3-1} + 12\ 3 + 36\right) = 169.8\ \mathrm{kN}$$

which is higher than the 145.7 kN carried. Test next the first four axles:

$$W = 2.224\left(\frac{0.3048\ 13.7\ 4}{4-1} + 12\ 4 + 36\right) = 199.2\ \mathrm{kN}$$

which is higher than the 197.7 kN carried. Test next all six axles:

$$W = 2.224\left(\frac{0.3048\ 18.2\ 6}{6-1} + 12\ 6 + 36\right) = 255\ \mathrm{kN}$$

which is higher than the total load of 247.9 kN. Finally, test the last three axles of the vehicle:

$$W = 2.224\left(\frac{0.3048\ 4.5\ 3}{3-1} + 12\ 3 + 36\right) = 164.7\ \mathrm{kN}$$

which is higher than the 102.2 kN carried by these axles. In conclusion, vehicle (b) passed all load limit requirements.

The enforcement of these load limits is state jurisdiction. It is carried out at truck inspection stations equipped with static weigh scales, and is typically administrated by the highway patrol. High-speed WIM systems have not been used for load enforcement purposes in North America, because in-motion axles can substantially deviate from static load due to dynamics, as discussed earlier. Their role in load enforcement is, therefore, solely for sorting trucks as they approach truck inspection stations. Sorting WIM systems identify trucks that are potentially violating axle load limits or the bridge formula. Only these trucks need to be pulled off the traffic stream for static weighing; the rest can proceed without stopping.

Truck inspection stations can be augmented by incorporating technology for the automatic identification of commercial vehicles. This facilitates checking the credentials of the carrier and the type of cargo being carried. Such systems consist of transponders onboard the vehicles and roadside antennas that record the unique numbers being transmitted. They allow the automated clearance of commercial vehicles through truck inspection stations without stopping. A number of such systems are in operation across the United States (e.g., Oregon's Green Light).

## References


1. AASHTO (1993) *AASHTO Guide for the Design of Pavement Structures*, "American Association of State Highway and Transportation Officials", Washington, DC.
2. Cebon, D. (1999). *Handbook of Vehicle-Road Interaction*, Swets & Zeitlinger, Rotterdam, NL.
3. FHWA (August 2001). Data Collection Guide for SPS Sites, Version 1.0, Federal Highway Administration, Washington, DC.
4. FHWA (2001). DataPave Release 3.0, Federal Highway Administration, FHWA-RD-01-148, Washington, DC.
5. NCHRP (July 2004). "*2002 Design Guide: Design of New and Rehabilitated Pavement Structures*", Draft Final Report, National Cooperative Highway Research Program NCHRP Study 1-37A, Washington, DC.
6. CERF (2001). "Evaluation of Thermocoax Piezoelectric WIM Sensors," Technical Evaluation Report, Highway Innovative Technology Evaluation Center, Civil Engineering Research Foundation, Report No, CERF 40586, Washington, DC.
7. CERF (2001). "Evaluation of Measurement Specialties Inc. Piezoelectric WIM Sensors," Technical Evaluation Report, Highway

Innovative Technology Evaluation Center, Civil Engineering Research Foundation, Report No, CERF 40587, Washington, DC.

[8] Gonzalez, A., A. T. Papagiannakis, and E. O'Brien (2003). ''Evaluation of an Artificial Neural Network Technique Applied to Multiple-Sensor Weigh-in-Motion Systems'' *Journal of the Transportation Research Board*, Record No. 1855, pp. 151–159.

[9] Papagiannakis, A. T., R. C. G. Haas, J. H. F. Woodrooffe, and P. LeBlanc, ''Effects of Dynamic Load on Flexible Pavements,'' Transportation Research Record No. 1207, pp. 187–196, Washington, DC.

[10] Papagiannakis, A. T., W. A. Phang, J. H. F. Woodrooffe, A. T. Bergan, and R. C. G. Haas (1989). ''Accuracy of Weigh-in-Motion Scales and Piezoelectric Cables,'' Transportation Research Record No. 1215, pp. 189–196, Washington DC.

[11] Papagiannakis, A. T. (March 1997). ''Calibration of WIM Systems through Vehicle Simulations,'' *ASTM Journal of Testing and Evaluation*, Vol. 25, No. 2, pp. 207–214.

[12] Papagiannakis, A. T. and R. C. Haas (December 1986). ''Wide-Base Tires: Industry Trends and State of Knowledge of Their Impact on Pavements,'' Ministry of Transportation and Communications of Ontario, Ottawa, Ontario, Canada Report Under Contract, No. 21160.

[13] TRBNRC (2002). ''Regulation of Weights, Lengths, and Widths of Commercial Motor Vehicles,'' Special Report 267, Transportation Research Board National Research Council, Washington, DC.

[14] ASTM (2002). ''Standard Specification for Highway Weigh-in-Motion (WIM) Systems with User Requirements and Test Method,'' American Society of Testing of Materials, ASTM E 1318–02, Vol. 04.03, West, Conshohocken, PA.

[15] Sweatman, P. F. (1983). ''A Study of the Dynamic Wheel Forces in Axle Group Suspensions of Heavy Vehicles'' SR No. 27, Australian Road Research Board, Sydney, Australia.

[16] FHWA (2001). ''Traffic Monitoring Guide,'' U.S. Department of Transportation, Federal Highway Administration, Office of Highway Policy Information, Washington, DC.

[17] Woodrooffe, J. H. F., P. A. Le Blanc, and K. R. Le Piane, (1886). ''Effect of Suspension Variations on the Dynamic Loads of Heavy Articulated Vehicles, Vehicle Weights, and Dimensions Study,'' Technical Report No. 11, Road and Transportation Association of Canada, Washington, DC.

## Problems

2.1 Compute the *ESAL* factor for the two vehicles shown in Figure 2.9, given a flexible pavement with an *SN* of 4 and a terminal serviceability of 2.0.

2.2 Compute the expected range in the dynamic axle loads of the tandem axles of a five-axle semitrailer truck running at 70 km/h on a road with an IRI roughness of 2.2 m/km at 90% confidence level. Its tractor and trailer axles weigh 150 and 140 kN, and are equipped with air and rubber suspensions, respectively.

**Table 2.15**
Load Measurements (kN) for Evaluating a WIM System—Problem 2.4

| | Static Load | WIM Pass 1 | WIM Pass 2 | WIM Pass 3 | WIM Pass 4 | WIM Pass 5 |
|---|---|---|---|---|---|---|
| **Test Vehicle 1** | | | | | | |
| Steering | 45.9 | 38.2 | 51.3 | 43.3 | 50.1 | 46.5 |
| Drive, axle 1 | 70.2 | 74.0 | 60.3 | 75.3 | 58.2 | 48.9 |
| Drive, axle 2 | 75.3 | 68.3 | 78.2 | 81.3 | 60.2 | 80.9 |
| Trailer, axle 1 | 69.7 | 60.0 | 58.2 | 66.5 | 48.3 | 67.0 |
| Trailer, axle 2 | 65.4 | 55.3 | 45.3 | 42.3 | 67.1 | 65.2 |
| GVW | 326.5 | 295.8 | 293.3 | 308.7 | 283.9 | 308.5 |
| **Test Vehicle 2** | | | | | | |
| Steering | 43.2 | 38.0 | 45.0 | 44.0 | 44.5 | 42.8 |
| Drive, axle 1 | 62.3 | 51.1 | 68.9 | 70.3 | 48.3 | 55.8 |
| Drive, axle 2 | 63.2 | 49.3 | 52.3 | 78.3 | 55.0 | 49.9 |
| Trailer, axle 1 | 69.3 | 65.6 | 64.4 | 69.7 | 67.5 | 67.5 |
| Trailer, axle 2 | 70.2 | 65.0 | 62.0 | 71.3 | 64.0 | 67.0 |
| GVW | 308.2 | 269.0 | 292.6 | 333.6 | 279.3 | 283.0 |

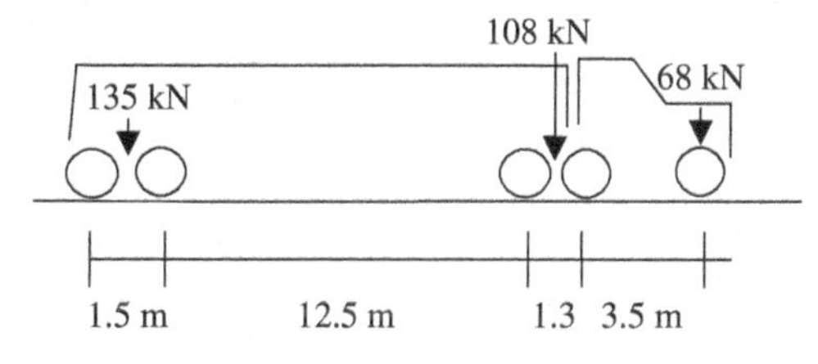


**Figure 2.10**
Weight and Dimension Measurements for Problem 2.5

2.3 For the vehicle in question 2 operating on the same road, estimate the maximum operating speed that can be allowed if the largest range in the axle dynamics desired is ± 20 kN at a confidence of 90%.

2.4 The load measurements shown in Table 2.15 were obtained from two Class 9 test vehicles running over a Type II WIM system. Determine whether this system meets ASTM accuracy tolerances in measuring axle loads, and compute the corresponding calibration factor.

2.5 Determine if the vehicle shown in Figure 2.10 is legally loaded.

# 3 Characterization of Pavement Subgrades and Bases

## 3.1 Introduction

The properties of the base/subbase and subgrade layers play a vital role in the structural integrity and performance of pavements. In flexible pavements, the base and subbase layers are structural components that need to provide sufficient strength, while reducing stresses to levels that can be sustained by the subgrade. In rigid pavements, the base layer is used for leveling and structural strengthening of weak subgrades. Furthermore, properly constructed base/subbase layers can provide internal drainage, while preventing water ingress into the subgrade. The properties of the subgrade and base layers can be improved through compaction or chemical stabilization under controlled moisture conditions.

## 3.2 Mechanical Behavior

Granular base/subbase layers exhibit an elastoplastic behavior in response to the loading and unloading conditions imposed by traffic loads. Upon unloading, this entails an elastic (i.e., recoverable) and a plastic (i.e., permanent) deformation components. This behavior can be described with the aid of the "shakedown" theory, as modified by Werkmeister et al. and illustrated in Figure 3.1.[28] At small stresses, the behavior can be purely elastic, whereby no plastic strain develops upon unloading. In this purely elastic response, the loading and unloading paths are the same, and there is no shift in the horizontal direction, indicating that the energy input in deforming the solid grains is released upon unloading. However, if the applied load increases, the material begins to develop small levels of permanent strain over a few cycles. Nevertheless, subsequent cycles at the same strain level yield no additional plastic deformation (Figure 3.1). This response is referred to as *elastic shakedown.* The permanent strain under these conditions is small, which is attributed to limited slipping of particles and changes in density, while the material adjusts to the applied loads. In a laboratory experiment, this permanent strain can take place as a result of speciman conditioning and adjusting under applied loads. The elastic shakedown is shown in Figure 3.1 by a linear stress-strain relationship in which the loading and unloading paths coincide.

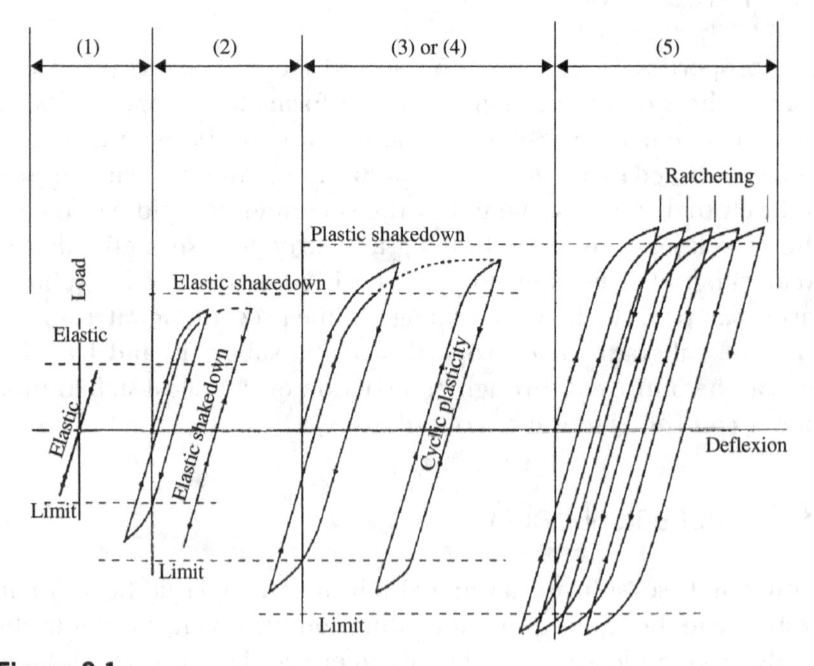


**Figure 3.1**
Illustration of the Granular Material Response under Cyclic Loading, Based on Shakedown Theory (Ref. 28)

Further increases in the applied loads cause a *plastic shakedown* behavior. In this behavior, the aggregate develops plastic strain higher than that in the elastic shakedown region. After a certain number of cycles, the plastic strain deformation development ceases. The stress level at which this condition is achieved is referred to as the *plastic shakedown limit.* In many aggregate materials, plastic strain does not stop at the plastic shakedown limit, but continues to develop at a constant rate. This region has been referred to by Werkmeister et al. as the "plastic creep region."[28] It is attributed to the gradual and low-level abrasion of the aggregates. Soils in the plastic shakedown and plastic creep regions experience a constant level of resilient (elastic) strain. The last region in Figure 3.1 illustrates the response of soil that experiences plastic strain at an increasing rate until complete failure. In this range, aggregates experience significant crushing, abrasion and breakdown.[28] Figure 3.2 summarizes the plastic strain development as a function of loading cycles that correspond to regions 2 to 5 identified in Figure 3.1.

## 3.3 Resilient Response

Pavement design methods commonly assume that the response of granular unbound bases/subbases and subgrades involves strains sufficiently small to correspond to the regions 2 and 3 identified in Figures 3.1 and 3.2. Accordingly, their modulus is assumed adequately described by the elastic-only component of the response, referred to as the *resilient modulus.* Some of the design methods

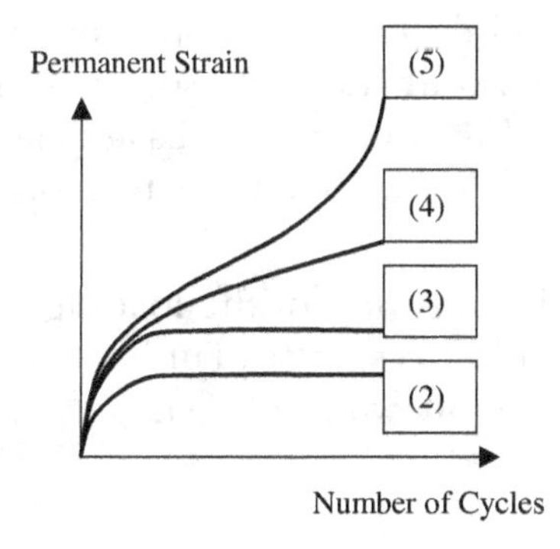


**Figure 3.2**
Plastic Strain Development for the Different Regions Shown in Figure 3.1

assume the resilient modulus to be constant (linear stress-strain behavior), while others express it as a function of stress level. It is typically assumed that the mechanical properties of the unbound layers are isotropic (i.e., direction-independent). In such a case, the elastic response is described by only the elastic modulus and the Poisson's ratio.

The resilient or elastic properties of unbound layers are determined using repeated load triaxial tests. Consider a cylindrical specimen that is subjected to a confined triaxial stress state under a constant confinement and a dynamic axial compressive load. In this case, the confining stress in the radial direction represents the minor and intermediate principal stresses, while the vertical stress is the major principal stress. The measured modulus is defined as the ratio of the applied deviatoric stress (axial stress minus the radial stress) divided by the resilient strain. The resilient modulus $M_r$ and the Poisson's ratio $\mu$ are defined by Equations 3.1 and 3.2, respectively:

$$M_r = \frac{\Delta(\sigma_1 - \sigma_3)}{\varepsilon_{1,r}} \tag{3.1}$$

$$\mu = -\frac{\varepsilon_{3,r}}{\varepsilon_{1,r}} \tag{3.2}$$

where, $\sigma_1$ is the major principal stress, $\sigma_3$ is the minor principal stress, $\varepsilon_{1,r}$ is the major principal resilient strain, and $\varepsilon_{3,r}$ is the minor principal resilient strain. The resilient modulus can be used as input to layer elastic analysis models to calculate the pavement structural response to wheel loads. As a result, it is an essential input to the structural design of pavement structures. The response of a granular material under one cycle of loading is shown in Figure 3.3. Tables 3.1 and 3.2 list resilient moduli and Poisson's ratio values, respectively, for unbound granular and subgrade soil materials recommended by NCHRP study 1–37A.[30] Subgrade soils are classified according to the AASHTO and the Unified Soil Classification system (USCs).

### 3.3.1 Factors Affecting the Resilient Properties

Lekarp et al.[12] provide a review of the loading and material factors that influence the resilient response of unbound layers. These factors are summarized in the following subsect.

#### STRESS LEVEL

The resilient modulus of granular materials used in base and subbase layers is typically assumed to be a function of the confining stress $\sigma_3$,

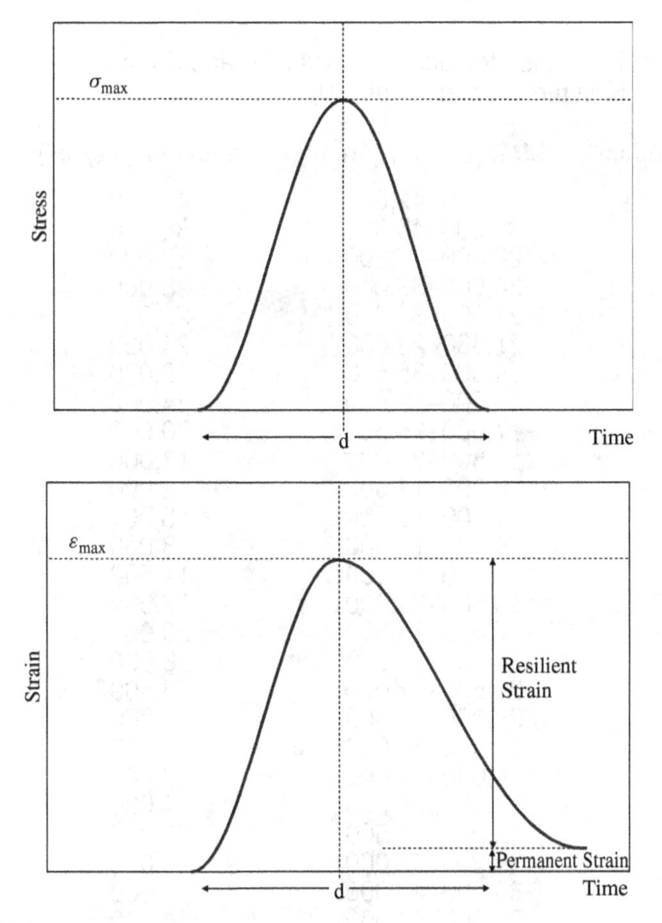


**Figure 3.3**
Response of Granular Materials under an Applied Load

or the sum of the principal stresses or bulk stress $\theta$, as expressed in Equations 3.3 and 3.4.

$$M_r = k_1 P_a \left( \frac{\sigma_3}{P_a} \right)^{k_2} \tag{3.3}$$

$$M_r = k_1 P_a \left( \frac{\theta}{P_a} \right)^{k_2} \tag{3.4}$$

The atmospheric pressure constant is used to eliminate the influence of the units of pressure on the calculated resilient modulus.

**Table 3.1**
Typical Resilient Modulus Values for Unbound Granular and Subgrade Materials at Optimum Moisture Content (Ref. 30)

| Material Classification | $M_r$ Range (lbs/in$^2$) | Typical $M_r$(lbs/in$^2$) |
|---|---|---|
| A-1-a | 38,500–42,000 | 40,000 |
| A-1-b | 35,500–40,000 | 38,000 |
| A-2–4 | 28,000–37,500 | 32,000 |
| A-2–5 | 24,000–33,000 | 28,000 |
| A-2–6 | 21,500–31,000 | 26,000 |
| A-2–7 | 21,500–28,000 | 24,000 |
| A-3 | 24,500–35,500 | 29,000 |
| A-4 | 21,500–29,000 | 24,000 |
| A-5 | 17,000–25,500 | 20,000 |
| A-6 | 13,500–24,000 | 17,000 |
| A-7–5 | 8,000–17,500 | 12,000 |
| A-7–6 | 5,000–13,500 | 8,000 |
| CH | 5,000–13,500 | 8,000 |
| MH | 8,000–17,500 | 11,500 |
| CL | 13,500–24,000 | 17,000 |
| ML | 17,000–25,500 | 20,000 |
| SW | 24,000–33,000 | 28,000 |
| SW-SC | 21,500–31,000 | 25,500 |
| SW-SM | 24,000–33,000 | 28,000 |
| SP-SC | 21,500–31,000 | 25,500 |
| SP-SM | 24,000–33,000 | 28,000 |
| SC | 21,500–28,000 | 24,000 |
| SM | 28,000–37,500 | 32,000 |
| GW | 39,500–42,000 | 41,000 |
| GP | 35,500–40,000 | 38,000 |
| GW-GC | 28,000–40,000 | 34,500 |
| GW-GM | 35,500–40,500 | 38,500 |
| GP-GM | 31,000–40,000 | 36,000 |
| GC | 24,000–37,500 | 31,000 |
| GM | 33,000–42,000 | 38,500 |

Hicks and Monismith[6] found that the resilient modulus is highly influenced by confining pressure, but to a lesser extent by the deviatoric stress. Figure 3.4 shows the relation between the bulk stress and resilient modulus at different values of confinement.

The resilient modulus has also been described as a function of both the confining pressure and the deviatoric stress, as shown

**Table 3.2**
Typical Poisson's Ratio Values for Unbound Granular and Subgrade Materials at Optimum Moisture Content (Ref. 30)

| Material Description | $\mu$ (Range) | $\mu$ (Typical) |
|---|---|---|
| Clay (saturated) | 0.4–0.5 | 0.45 |
| Clay (unsaturated) | 0.1–0.3 | 0.2 |
| Sandy clay | 0.2–0.3 | 0.25 |
| Silt | 0.3–0.35 | 0.325 |
| Dense sand | 0.2–0.4 | 0.3 |
| Coarse-grained sand | 0.15 | 0.15 |
| Fine-grained sand | 0.25 | 0.25 |
| Bedrock | 0.1–0.4 | 0.25 |

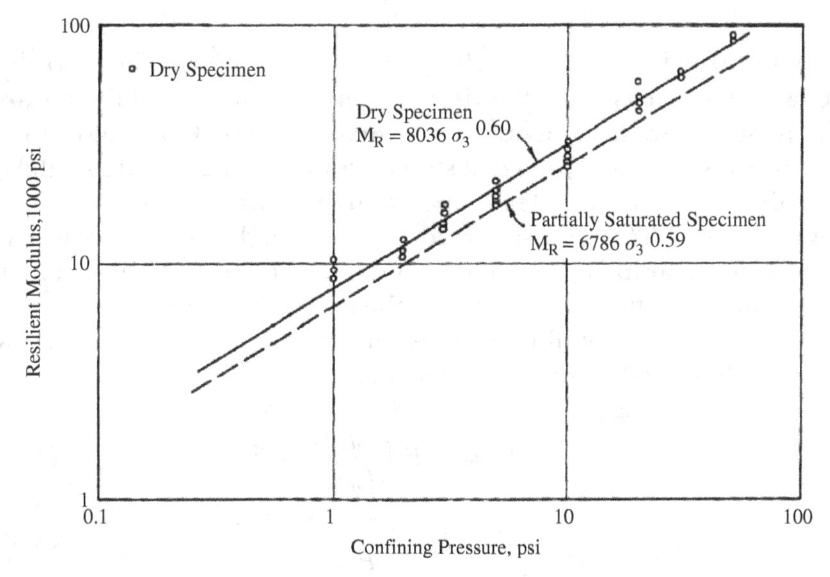


**Figure 3.4**
Example of the Correlation between the Resilient Modulus ($M_r$) and Confining Stress ($\sigma_3$) (Ref. 6)

in Equation 3.5.[17,26]

$$M_r = k_1 P_a \left(\frac{\theta}{P_a}\right)^{k_2} \left(\frac{\sigma_d}{P_a}\right)^{k_3} \tag{3.5}$$

where $\sigma_d = \sigma_1 - \sigma_3$ is the deviatoric stress. The coefficient $k_2$ is positive, indicating that an increase in confinement causes an increase in the modulus, while the coefficient $k_3$ is negative, indicating that an increase in the deviatoric stress causes a reduction in the resilient modulus. Equation 3.5 is applicable to the triaxial test, where $\sigma_2$ is equal to $\sigma_3$. For the three-dimensional case in a pavement structure, $\sigma_2$ is not necessarily equal to $\sigma_3$; the deviatoric stress is replaced with the octahedral stress, as follows:

$$M_r = k_1 P_a \left(\frac{\theta}{P_a}\right)^{k_2} \left(\frac{\tau_{oct}}{P_a}\right)^{k_3} \tag{3.6}$$

where,

$$\tau_{oct} = \frac{1}{3}\sqrt{(\sigma_1 - \sigma_3)^2 + (\sigma_2 - \sigma_3)^2 + (\sigma_1 - \sigma_2)^2} \tag{3.7}$$

For the triaxial case in which $\sigma_2 = \sigma_3$, $\tau_{oct} = \frac{\sqrt{2}}{3}\sigma_d$. The work by Uzan[26] has shown that the decrease in resilient modulus with an increase in deviatoric stress occurs when the ratio of the major principal stress to minor principal stress is lower than 2 or 3 depending on the material type. The experimental measurements by Hicks and Monismith[6] were conducted at a ratio higher than 2, where a dense granular material would experience dilation and the resilient modulus is a function primarily of the confining pressure ($\theta$).

The resilient modulus of fine-grained soils or cohesive soils is usually described by Equation 3.8 or 3.9.

$$M_r = k_1 P_a \left(\frac{\sigma_d}{P_a}\right)^{k_2} \tag{3.8}$$

$$M_r = k_1 P_a \left(\frac{\tau_{oct}}{P_a}\right)^{k_2} \tag{3.9}$$

In these Equations, the $k_2$ value is negative, indicating that an increase in deviatoric stress causes a reduction in the resilient modulus.

**COMPACTION AND AGGREGATE STRUCTURE**

Several experimental studies have shown that the unbound granular layers exhibit cross-anisotropic properties. This behavior is caused by the preferred orientation of aggregates in the unbound layers, due partially to aggregate shape and compaction forces. This results in base and subbase layers that are stiffer in the vertical direction than in the horizontal direction. The main advantages of using anisotropic properties are to describe the dilative elastic behaviour of unbound layers and to reduce the unrealistically large tensile stresses predicted in the granular bases using isotropic models[11,16,25].

In the anisotropic model, expressions similar to Equation 3.5 are used to describe the vertical resilient modulus $M_r^y$, the horizontal resilient modulus $M_r^x$, and the shear modulus $G_r^{xy}$. The corresponding expressions are:[1]

$$M_r^y = k_1 P_a \left(\frac{\theta}{P_a}\right)^{k_2} \left(\frac{\tau_{oct}}{P_a}\right)^{k_3} \tag{3.10}$$

$$M_r^x = k_4 P_a \left(\frac{\theta}{P_a}\right)^{k_5} \left(\frac{\tau_{oct}}{P_a}\right)^{k_6} \tag{3.11}$$

$$G_r^{xy} = k_7 P_a \left(\frac{\theta}{P_a}\right)^{k_8} \left(\frac{\tau_{oct}}{P_a}\right)^{k_9} \tag{3.12}$$

The anisotropic analysis of granular materials requires the $k_i$ coefficients, as well as the properties $n$, $m$, $\mu'$, defined as:

$$n = \frac{M_r^x}{M_r^y}, \quad m = \frac{G_r^{xy}}{M_r^y}, \quad \mu' = \frac{\mu_{xx}}{\mu_{xy}} \tag{3.13}$$

where $\mu_{xx}$ and $\mu_{xy}$ are the directional Poisson's ratios for the horizontal and vertical directions, respectively.

Tutumluer and Thompson[25] and Adu-Osei et al.[1] clearly demonstrated that the unbound aggregate base material should be modeled as nonlinear and anisotropic to account for stress sensitivity and the significant differences between vertical and horizontal moduli and Poisson's ratios. Figure 3.5 shows examples of the vertical, horizontal, and shear moduli, which illustrate that the vertical modulus is typically higher than the horizontal modulus.

#### MATERIAL FACTORS

The material factors that influence the resilient modulus include density, gradation, fines content, and moisture content. As reported by Lekarp et al.[12], some studies concluded that density has a significant influence on resilient modulus, while others found the influence of density to depend primarily on aggregate shape and confinement level. The influence of density was found to be more significant for partially crushed aggregates, but almost negligible for

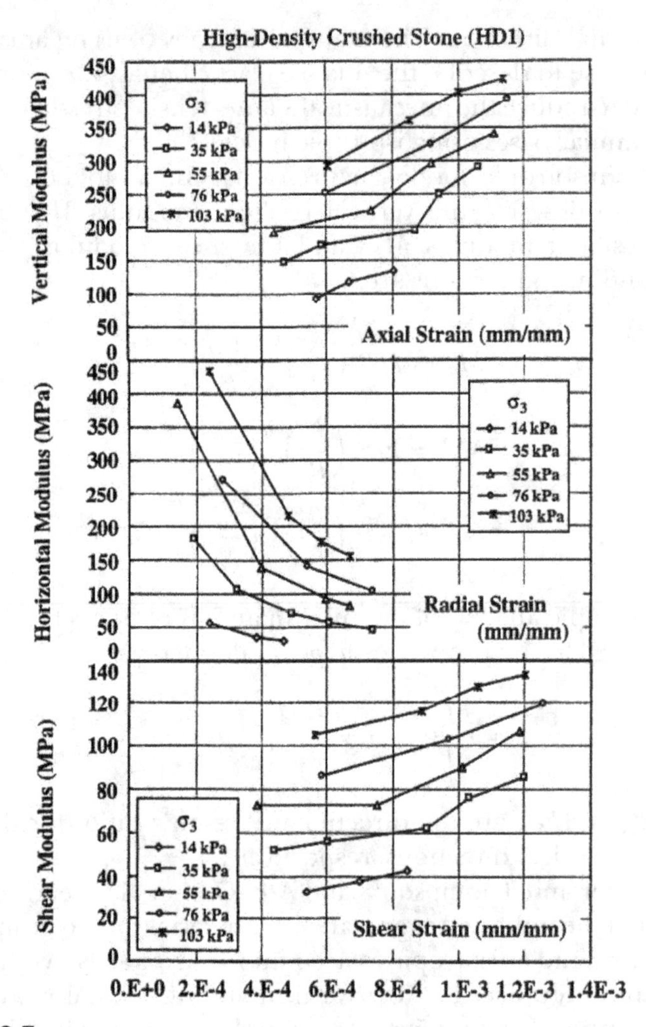


**Figure 3.5**
Examples of the Measurements of Vertical, Horizontal, and Shear Moduli (after Ref. 25)

fully crushed aggregates, as reported by Hicks and Monismith[6]. The resilient modulus was found to increase with an increasing density at low-confining stress levels, while it was less sensitive to density at high-confining stress levels.

There seems to be no consensus in the literature on the effect of gradation on the resilient modulus. Hicks and Monismith[6] found that a variation in fines content between 2% to 10% has a slight effect on resilient modulus. Barksdale and Itani[3] however, reported a detrimental effect of the increase in fines on resilient modulus. A possible explanation for this discrepancy was offered by Jorenby and Hicks[7], who suggested that, up to a point, an increase in fines could cause an increase in the contacts and a filling of the voids between large particles. Beyond that point, excess fines could displace the coarse particles leading to an aggregate matrix of fines that has a reduced load-carrying capability. Some studies showed that well-graded aggregates have a higher resilient modulus than uniformly graded aggregates[27], while other studies concluded the opposite[21].

The moisture content or degree of saturation is a critical factor influencing the resilient modulus. Dawson et al.[4] found that below the optimum moisture content, an increase in moisture level causes an increase in resilient modulus. On the other hand, above the optimum moisture content, increases in moisture cause a decrease in the resilient modulus. This behavior can be explained by three mechanisms:

- At low levels of moisture content, suction may increase the apparent cohesion between particles and result in an increase in the resilient modulus.[4]
- At high levels of moisture content, pore pressures can cause a decrease in effective stress, hence a reduction in the resilient modulus.[3,6]
- Fines may have a "lubricating" effect, even without the development of pore pressure, and can cause a reduction in the resilient modulus.[20]

In terms of aggregate physical characteristics, angular and rough-textured particles have been found to have a higher resilient modulus than uncrushed or partially crushed particles.[3,22]

### 3.3.2 Experimental Measurements

The method for measuring the resilient modulus of soils and aggregate materials is described in the AASHTO T 307–99 standard. It includes a procedure for measuring the resilient modulus of untreated granular base/subbase materials, which are defined as soil-aggregate mixtures and naturally occurring materials that do not include a stabilizing agent. It also includes a procedure for measuring the resilient modulus of subgrades, which are defined as soil compacted before the placement of subbase and/or base layers.

The resilient modulus is measured using repeated load triaxial compression tests using a setup similar to that shown in Figure 3.6. A repeated axial cyclic stress of fixed magnitude with a load duration of 0.1 second within a cycle duration between 1.0 and 3.1 seconds is applied to a cylindrical test specimen subjected to a confining stress. The axial stress is applied in a haversine format, which is mathematically expressed as $(1 - \cos \omega)/2$ with

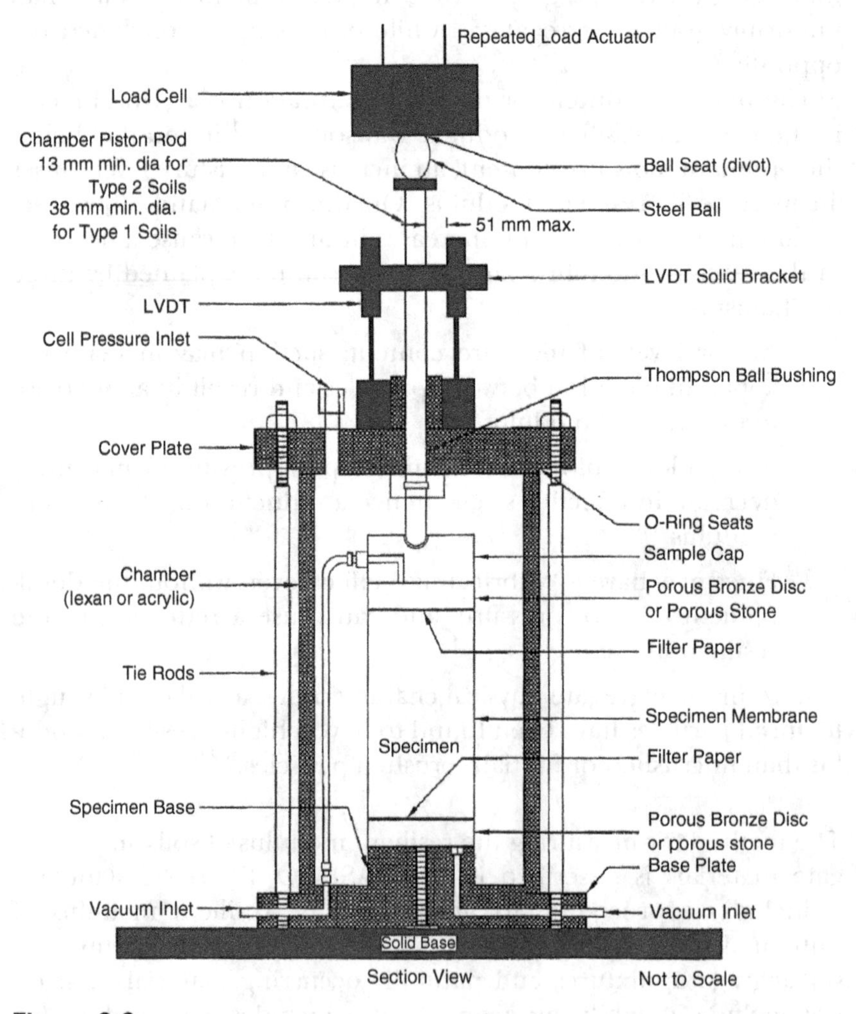


**Figure 3.6**
Schematic of the Triaxial Setup Used for Measuring the Resilient Modulus (AASHTO T 307–99)

$\omega$ ranging from 0 to $2\pi$. The total resilient (i.e., recoverable) axial deformation response of the specimen is measured and used to calculate the resilient modulus. This test is intended for determining the $k_i$ coefficients for the base/subbase materials and the subgrades involved in Equation 3.6 and Equation 3.9, respectively.

Undisturbed subgrade soil specimens are tested at the natural moisture content. Reconstituted test specimens are prepared to approximate the in-situ wet density and moisture content. If either the in-situ moisture content or the in-place density is not available, it is recommend to use a certain percentage of the maximum dry density and the corresponding optimum moisture content according to AASHTO T 99 or AASHTO T 180. The testing program for the subgrade soil in shown in Table 3.3; the testing program for the base and subbase materials is shown in Table 3.4.

NCHRP study 1–28A[29] made a number of recommendations for enhancing the AASHTO T 307–99 testing procedure. These include a sample size that depends on the maximum aggregate size being tested, a longer loading time of 0.2 second for subgrade soils, and a shorter loading period of 0.8 second for base aggregates, and a different loading sequence. The resilient modulus model adopted

**Table 3.3**
Testing Sequence for Subgrade Soil (after AASHTO T 307–99)

| Sequence No. | Confining $\sigma_3$(kPa) | Max Axial $\sigma_{max}$(kPa) | Cyclic (kPa) | Constant $0.1\sigma_{max}$ (kPa) | No. of Cycles |
|---|---|---|---|---|---|
| 0 | 41.4 | 27.6 | 24.8 | 2.8 | 500–1000 |
| 1 | 41.4 | 13.8 | 12.4 | 1.4 | 100 |
| 2 | 41.4 | 27.6 | 24.8 | 2.8 | 100 |
| 3 | 41.4 | 41.4 | 37.3 | 4.1 | 100 |
| 4 | 41.4 | 55.2 | 49.7 | 5.5 | 100 |
| 5 | 41.4 | 68.9 | 62 | 6.9 | 100 |
| 6 | 27.6 | 13.8 | 12.4 | 1.4 | 100 |
| 7 | 27.6 | 27.6 | 24.8 | 2.8 | 100 |
| 8 | 27.6 | 41.4 | 37.3 | 4.1 | 100 |
| 9 | 27.6 | 55.2 | 49.7 | 5.5 | 100 |
| 10 | 27.6 | 68.9 | 62 | 6.9 | 100 |
| 11 | 13.8 | 13.8 | 12.4 | 1.4 | 100 |
| 12 | 13.8 | 27.6 | 24.8 | 2.8 | 100 |
| 13 | 13.8 | 41.4 | 37.3 | 4.1 | 100 |
| 14 | 13.8 | 55.2 | 49.7 | 5.5 | 100 |
| 15 | 13.8 | 68.9 | 62 | 6.9 | 100 |

**Table 3.4**
Testing Sequence for Base/Subbase Materials (after AASHTO T 307–99)

| Sequence No. | Confining $\sigma_3$ (kPa) | Max Axial $\sigma_{max}$ (kPa) | Cyclic (kPa) | Constant $0.1\sigma_{max}$ (kPa) | No. of Cycles |
|---|---|---|---|---|---|
| 0 | 103.4 | 103.4 | 93.1 | 10.3 | 500–1000 |
| 1 | 20.7 | 20.7 | 18.6 | 2.1 | 100 |
| 2 | 20.7 | 41.4 | 37.3 | 4.1 | 100 |
| 3 | 20.7 | 62.1 | 55.9 | 6.25 | 100 |
| 4 | 34.5 | 34.5 | 31.0 | 3.5 | 100 |
| 5 | 34.5 | 68.9 | 62.0 | 6.9 | 100 |
| 6 | 34.5 | 103.4 | 93.1 | 10.3 | 100 |
| 7 | 68.9 | 68.9 | 62.0 | 6.9 | 100 |
| 8 | 68.9 | 137.9 | 124.1 | 13.8 | 100 |
| 9 | 68.9 | 206.8 | 186.1 | 20.7 | 100 |
| 10 | 103.4 | 68.9 | 62.0 | 6.9 | 100 |
| 11 | 103.4 | 103.4 | 93.1 | 10.3 | 100 |
| 12 | 103.4 | 206.8 | 186.1 | 20.7 | 100 |
| 13 | 137.9 | 103.4 | 93.1 | 10.3 | 100 |
| 14 | 137.9 | 137.9 | 124.1 | 13.8 | 100 |
| 15 | 137.9 | 275.8 | 248.2 | 27.6 | 100 |

by the NCRHP 1–28A study is:

$$M_r = k_1 P_a \left(\frac{\theta - 3\,k_6}{P_a}\right)^{k_2} \left(\frac{\tau_{oct}}{P_a} + k_7\right)^{k_3} \tag{3.14}$$

where $k_1$, $k_2 \geq 0$, $k_3$, $k_6 \leq 0$, and $k_7 \geq 1$.

**Example 3.1** A series of repeated triaxial tests were conducted to determine the resilient modulus of an unbound granular material that is intended as a base layer for an asphalt concrete pavement (Table 3.5). The tests were carried out at different confining and axial stresses, as shown in Table 3.5. Based on these results:

- Plot the resilient modulus $M_r$ versus bulk stress $\theta$ and versus octahedral stress $\tau_{oct}$ and discuss these relationships.
- Determine the appropriate constants for these relationships, (i.e., $k_1$, $k_2$, $k_3$) by fitting Equations 3.4 and 3.6 to this data.

**ANSWER**

For the triaxial stress condition of this test, the bulk stress and the octahedral shear stress are computed using $\theta = \sigma_1 + 2\sigma_3$

**Table 3.5**
Data for Example 3.1

| Sequence No. | Confining $\sigma_3$ (kPa) | Max Axial $\sigma_{max}$ (kPa) | Cyclic (kPa) | Constant $0.1\sigma_{max}$ (kPa) | $M_r$ (kPa) |
|---|---|---|---|---|---|
| 1 | 20.7 | 20.7 | 18.6 | 2.1 | 69,527.30 |
| 2 | 20.7 | 41.4 | 37.3 | 4.1 | 68,981.57 |
| 3 | 20.7 | 62.1 | 55.9 | 6.25 | 73,893.20 |
| 4 | 34.5 | 34.5 | 31.0 | 3.5 | 112,849.34 |
| 5 | 34.5 | 68.9 | 62.0 | 6.9 | 111,939.53 |
| 6 | 34.5 | 103.4 | 93.1 | 10.3 | 119,866.90 |
| 7 | 68.9 | 68.9 | 62.0 | 6.9 | 217,409.41 |
| 8 | 68.9 | 137.9 | 124.1 | 13.8 | 215,703.02 |
| 9 | 68.9 | 206.8 | 186.1 | 20.7 | 231,037.44 |
| 10 | 103.4 | 68.9 | 62.0 | 6.9 | 341,371.40 |
| 11 | 103.4 | 103.4 | 93.1 | 10.3 | 319,456.68 |
| 12 | 103.4 | 206.8 | 186.1 | 20.7 | 316,926.56 |
| 13 | 137.9 | 103.4 | 93.1 | 10.3 | 438,251.20 |
| 14 | 137.9 | 137.9 | 124.1 | 13.8 | 419,713.98 |
| 15 | 137.9 | 275.8 | 248.2 | 27.6 | 416,389.81 |

and $\tau_{oct} = \frac{\sqrt{2}}{3}\sigma_d$, respectively. The calculated values are shown in Table 3.6. The relationships between the resilient modulus and the bulk stress is plotted in Figure 3.7, which shows clearly that the resilient modulus increases with increasing bulk stress. The relationship between resilient modulus and the octahedral shear stress is shown in Figure 3.8. It is noted that at a given bulk modulus, an increase in octahedral shear stress causes a decrease in the resilient modulus (i.e., it is anticipated that the coefficient $k_3$ in Equation 3.6 is negative). However, it is difficult to conduct the test at a constant bulk stress. For a given stress combination, the increase in octahedral shear stress can be combined with an increase in the bulk stress. Therefore, depending on the values of the exponents $k_2$ *and* $k_3$ in Equation 3.6, the resilient modulus can increase or decrease with increasing octahedral shear stress, for a given confining pressure ($\sigma_3$). For this data, the modulus tended to increase with increasing octahedral stress at low-confining pressures, while it decreased with increasing octahedral stress at high-confining pressures. Hence, a granular base resilient modulus model needs to account for the effects of both bulk stress and octahedral stress changes.

**Table 3.6**
Computing Bulk and Octahedral Stresses for Example 3.3

| $\theta$ (kPa) | $\tau_{oct}$ (kPa) |
|---|---|
| 82.8 | 9.758074 |
| 103.5 | 19.51615 |
| 124.2 | 29.27422 |
| 138 | 16.26346 |
| 172.4 | 32.47977 |
| 206.9 | 48.74323 |
| 275.6 | 32.47977 |
| 344.6 | 65.00668 |
| 413.5 | 97.48645 |
| 379.1 | 32.47977 |
| 413.6 | 48.74323 |
| 517 | 97.48645 |
| 517.1 | 48.74323 |
| 551.6 | 65.00668 |
| 689.5 | 130.0134 |

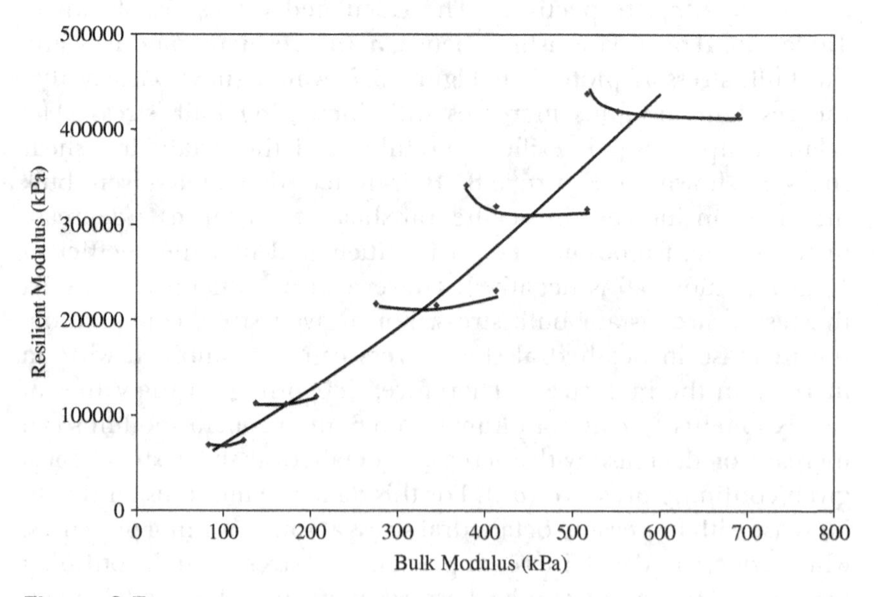


**Figure 3.7**
Relationship between Resilient Modulus and Bulk Stress for Example 3.1

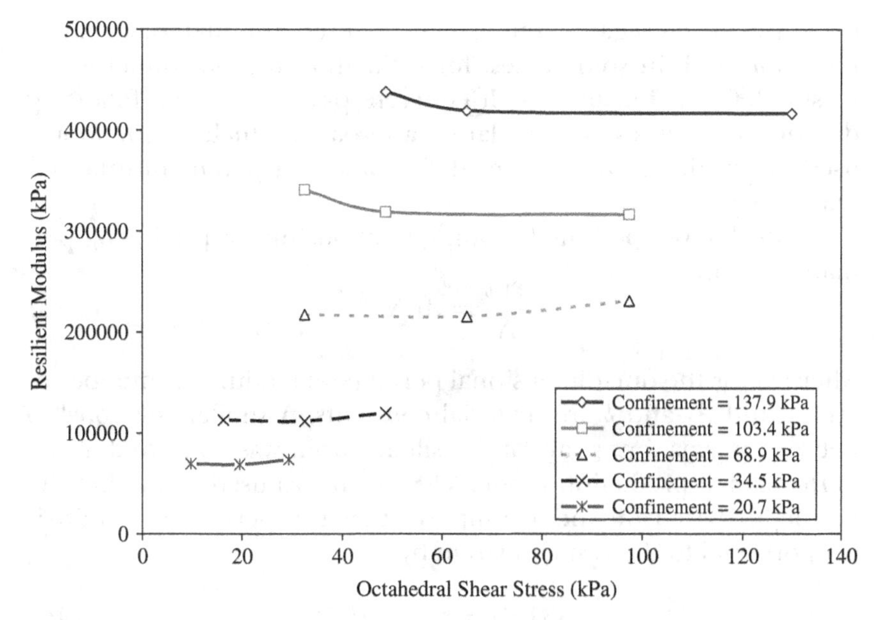


**Figure 3.8**
Relationship between Octahedral Shear Stress and Resilient Modulus for Example 3.1

Finally, the requested coefficients are computed through regression on the given data. It was carried out with a spreadsheet after logarithmic transformation of the data. They are as follows:

- Equation 3.4: $k_1 = 709.8$ and $k_2 = 0.985$
- Equation 3.6: $k_1 = 306.06$, $k_2 = 1.415$, and $k_3 = -.467$

## 3.4 Plastic Response

Estimating the plastic component of the strain in granular materials is important for quantifying their permanent deformation. The plastic response of aggregates can be modeled on the basis of two distinct approaches. The first approach derives the three-dimensional stress-strain behavior of the aggregates based on plasticity theory. The second approach is based on using laboratory results to develop a one-dimensional relationship between stress level, number of cycles, and accumulation of permanent strain. Obviously, the latter is not capable of determining the three-dimensional plastic response of aggregates; but it can be used for material characterization and, in some cases, for estimating the one-dimensional plastic deformation in asphalt concrete pavements. The following description focuses on the latter approach, which is commonly used in predicting permanent deformation in pavement granular materials[13].

Khedr[10] developed the following relationship for predicting permanent strain:

$$\frac{\varepsilon_{1,p}}{N} = A_1\, N^{-b} \tag{3.15}$$

where $\varepsilon_{1,p}$ is the one-dimensional permanent strain, $N$ is number of cycles, and $A_1$ and $b$ are material constants. A similar relationship between permanent strain and resilient strain was employed in the performance prediction model VESYS.[9] It was used to predict the rut depth based on the assumption that the permanent strain is proportional to the resilient strain by:

$$\varepsilon_{1,p}(N) = \varepsilon_{r,200}\ \upsilon N^{-\alpha} \tag{3.16}$$

where $\varepsilon_{1,p}(N)$ is permanent strain due to a single-load application at the $N$th cycle, $\varepsilon_{r,200}$ is the resilient strain at 200 cycles, $\upsilon$ represents the constant of proportionality between permanent and resilient strain, and $\alpha$ is a material representing the rate of decrease in permanent strain with the number of load applications.

Tseng and Lytton[24] developed the following relationship to predict the permanent strain involving three parameters, namely $\varepsilon_0$, $\rho$ and $\beta$:

$$\varepsilon_{1,p} = \varepsilon_0\ e^{-\left(\frac{\rho}{N}\right)^{\beta}} \tag{3.17}$$

where, $\varepsilon_0$ is the maximum permanent strain at a very high number of loading cycles. As described in Chapter 11, this expression can be extended to predict permanent deformation in an aggregate layer of thickness, $h$, using:

$$\delta_{1,p} = \left(\frac{\varepsilon_0}{\varepsilon_r}\right) e^{-\left(\frac{\rho}{N}\right)^{\beta}}\ \varepsilon_v\, h \tag{3.18}$$

where $\varepsilon_r$ is the resilient strain measured in the laboratory and $\varepsilon_v$ is the vertical elastic strain in the pavement layer computed from layer elastic analysis (Chapter 7). According to Tseng and Lytton[24], the three parameters $\varepsilon_0$, $\rho$, and $\beta$ can be expressed as functions of the

**Table 3.7**
Data for Example 3.2

| Loading Cycles $N$ | Axial Permanent Strain ($\varepsilon_{1,p}$) $10^{-3}$ |
|---|---|
| 10 | 0.5 |
| 100 | 0.7 |
| 1,000 | 1.2 |
| 2,000 | 1.5 |
| 4,000 | 1.7 |
| 8,000 | 2.0 |
| 16,000 | 2.3 |
| 32,000 | 2.4 |
| 64,000 | 2.6 |
| 128,000 | 2.9 |
| 256,000 | 3.1 |

resilient modulus, the confining pressure, and the water content. In addition, for subgrade soils, these three parameters depend on deviatoric stress. This model was incorporated into the NCHRP 1–37A pavement design guide[30] for computing the plastic deformation of granular layers, as described in Chapter 11.

**Example 3.2**

A triaxial permanent deformation test was performed on a granular base material. The data obtained from this test is given in Table 3.7. Analyze this data to determine the parameters of Equation 3.17.

**ANSWER**

Taking natural logarithms, translate Equation 3.17 into:

$$\ln(\varepsilon_{1,p}) = \ln(\varepsilon_o) - \left(\frac{\rho}{N}\right)^{\beta}$$

The derivative of this expression with respect to $N$ is:

$$\frac{d(\ln(\varepsilon_{1,p}))}{dN} = -(\rho)^{\beta}(-\beta)N^{-\beta-1}$$

In the preceding equation, substitute $\ln N$ for $N$, using $\frac{d(\ln N)}{dN} = \frac{1}{N}$:

$$\frac{d(\ln(\varepsilon_{1,p}))}{d(\ln N)} = -(\rho)^{\beta}(-\beta)N^{-\beta}$$

**Table 3.8**

Computing $\frac{\Delta(\ln(\varepsilon_{1,p}))}{\Delta(\ln N)}$ and $\varepsilon_0$ for Example 3.2

| Loading Cycles N | $\frac{\Delta(\ln(\varepsilon_{1,p}))}{\Delta(\ln N)}$ | $\varepsilon_0(10^{-3})$ |
|---|---|---|
| 10 | | 0.01349 |
| 100 | 0.544 | 0.00539 |
| 1,000 | 0.234 | 0.00322 |
| 2,000 | 0.322 | 0.00332 |
| 4,000 | 0.181 | 0.00322 |
| 8,000 | 0.234 | 0.00334 |
| 16,000 | 0.202 | 0.00348 |
| 32,000 | 0.120 | 0.00348 |
| 64,000 | 0.057 | 0.00340 |
| 128,000 | 0.054 | 0.00335 |
| 256,000 | 0.052 | 0.00333 |

Taking logarithms, translate the preceding equation into:

$$\log\frac{d(\ln(\varepsilon_{1,p}))}{d(\ln N)} = \log(\beta\rho^{\beta}) - \beta\log N$$

The left-hand-side derivative is approximated using the following finite difference expression:

$$\frac{d(\ln(\varepsilon_{1,p}))}{d(\ln N)} = \frac{\Delta(\ln(\varepsilon_{1,p}))}{\Delta(\ln N)}$$

The values calculated for these finite differences are shown in the second column of Table 3.8. Subsequently, a regression equation is fitted, as shown in Figure 3.9, to obtain the slope and intercept of the preceding equation.

Figure 3.9 gives a slope $\beta = 0.3147$. The intercept, which is expressed as $\log(\beta\rho^{\beta})$, is equal to 0.437, which gives a $\rho$ value of 946.1. Finally, the parameter $\varepsilon_0$ is calculated for each number of cycles using Equation 3.17. The results are shown in the third column of Table 3.8. Typically, a representative $\varepsilon_0$ is obtained as the average of all the $\varepsilon_0$ values computed. In this case, however, the best fit between predicted and observed data was found by averaging all but the first two $\varepsilon_0$ values, yielding a value of 0.00335 (Figure 3.10).

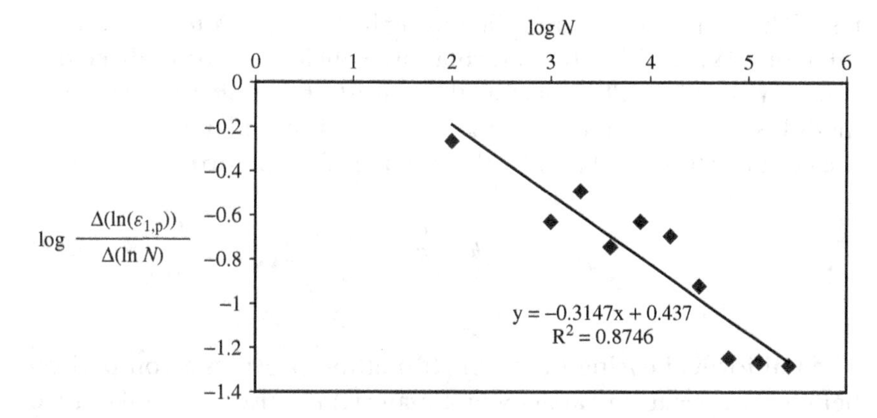


**Figure 3.9**
Relationship $\log\frac{d(\ln(\varepsilon_{1,p}))}{d(\ln N)} = \log(\beta\rho^{\beta}) - \beta \log N$ for Example 3.2

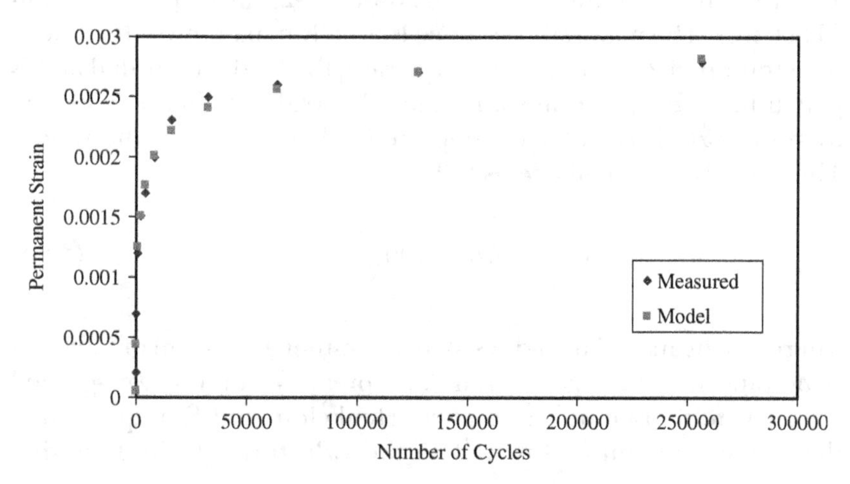


**Figure 3.10**
Observed versus Predicted Plastic Strain Data for Example 3.2

## 3.5 Other Aggregate Layer Indices

### 3.5.1 Modulus of Subgrade Reaction

The modulus of subgrade reaction is a material input to the analysis of portland concrete pavements, as will be discussed in Chapter 8. It represents the elastic constant of a series of springs supporting the portland concrete slabs under the ''liquid'', or Winkler slab foundation assumption (Figure 8.2). It is measured through plate-loading tests. The vertical stress is applied using hydraulic jack to a maximum value of 69 kPa, while the deformation is measured using three-dial gauges placed at 120° apart at the outside edge of the plate. The modulus of subgrade reaction $k$ (MPa/m) is calculated by dividing the contact stress $\sigma$ (MPa) by the circular plate deformation $\delta$ (m).

$$k = \frac{\sigma}{\delta} \tag{3.19}$$

### 3.5.2 California Bearing Ratio

The California bearing ratio (*CBR*) method provides a comparison between the bearing capacity of a material relative to a well-graded crushed stone, which has a reference *CBR* value of 100%. The test involves applying load to a small penetration piston at a rate of 1.3 mm (0.05 in.) per minute and recording the total load at penetrations ranging from 0.64 mm (0.025 in.) up to 7.62 mm (0.300 in.). The material resistance is usually maximum at the lowest penetration of 2.54 mm. If this is the case, the load is recorded at this penetration. However, in some cases, the load is higher at 5.08 mm, then the *CBR* is calculated using the load at 5.08 mm penetration. The *CBR* (%) is calculated as follows:

$$CBR = 100\left(\frac{x}{y}\right) \tag{3.20}$$

where $x$ is the material load resistance at either a 2.54 mm or 5.08 mm penetrations, and $y$ is the standard pressure for the well-graded crushed stone used as a reference. Heukelom and Klomp[5] related the resilient modulus (lbs/in$^2$) to the *CBR* through the following empirical expression:

$$M_r = 1500\ CBR \tag{3.21}$$

### 3.5.3 *R*-Value

This test procedure infers the stiffness of unbound granular materials from their resistance to deformation. The experimental setup applies a vertical pressure on a cylindrical sample and measures the lateral pressure transmitted in response. A vertical stress of 1.1 MPa is applied to a sample with a diameter of 102 mm and a height of about 114 mm. The system measures the resulting horizontal pressure induced in response to this vertical force. The *R*-value is calculated as follows:

$$R = 100 - \frac{100}{\frac{2.5}{D}\left(\frac{p_v}{p_h} - 1\right) + 1} \tag{3.22}$$

where $R$ is the resistance value, $p_v$ is the applied vertical pressure, and $p_h$ is the measured horizontal pressure, and $D$ is the displacement in the fluid needed to increase the horizontal pressure from 35 to 690 kPa under an applied vertical pressure of 1.1 MPa. Note that $R = 0$ for a liquid with $p_v = p_h$, while $R = 100$ for the case of $p_h = 0$.

The Asphalt Institute[2] developed the following empirical relationship to relate the resilient modulus in (lbs/in$^2$) to the *R*-value:

$$M_r = 1155 + 555\,R \tag{3.23}$$

Table 3.9 gives typical values of *CBR*, *R*, and $M_r$, based on the work of the Asphalt Institute.[2]

### 3.5.4 Coefficient of Lateral Pressure

The coefficient of lateral pressure $k_a$ is defined as the ratio of the lateral stress divided by the vertical stress. For unbound granular, subgrade, and bedrock materials, the in-situ typical $k_a$ ranges from 0.4 to 0.6. The coefficient of lateral pressure can be estimated for cohesionless and cohesive soils using Equations 3.24 and 3.25, respectively.

$$k_a = \frac{\mu}{1-\mu} \tag{3.24}$$

$$k_\alpha = 1 - \sin\phi \tag{3.25}$$

**Table 3.9**
Values of *CBR*, *R*, and Resilient Modulus (Ref. 2)

| Soil Description | CBR | R-Value | $M_r$(lbs/in$^2$) |
|---|---|---|---|
| Sand | 31 | 60 | 16,900 |
| Silt | 20 | 59 | 11,200 |
| Sandy Loam | 25 | 21 | 11,600 |
| Silt-Clay Loam | 25 | 21 | 17,600 |
| Silty Clay | 7.6 | 18 | 8,200 |
| Heavy Clay | 5.2 | <5 | 1,600 |

**Table 3.10**
Typical Effective Angle of Internal Friction for Unbound Granular, Subgrade and Bedrock Materials (Ref. 30)

| Material Description | Angle of Internal Friction $\phi$ | Coefficient of Lateral Pressure, $k_a$ |
|---|---|---|
| Clean, sound bedrock | 35 | 0.495 |
| Clean gravel, gravel-sand mixtures, and coarse sand | 29 to 31 | 0.548 to 0.575 |
| Clean fine to medium sand, silty medium to coarse sand, silty or clayey gravel | 24 to 29 | 0.575 to 0.645 |
| Clean fine sand, silty or clayey fine to medium sand | 19 to 24 | 0.645 to 0.717 |
| Fine sandy silt, nonplastic silt | 17 to 19 | 0.717 to 0.746 |
| Very stiff and hard residual clay | 22 to 26 | 0.617 to 0.673 |
| Medium stiff and stiff clay and silty clay | 19 | 0.717 |

where $\mu$ is the Poisson's ratio and $\phi$ is the soil angle of internal friction. Typical ranges of $\phi$ and $k_a$ are presented in Table 3.10.

### 3.5.5 The Atterberg Limits

The Atterberg limits are index properties used to determine the consistency of soils at different moisture contents. The liquid limit (LL) is determined by placing a soil sample in a standard device that consists of a cup and a crank to repeatedly drop the cup. A groove is cut in the device using a standard tool. The cup is dropped repeatedly using the crank. The number of drops to close the groove by 1/2 inch (12.5 mm) is recorded. The test is repeated at different moisture contents. The liquid limit is defined by the moisture content at which 25 drops are needed to close the groove by 1/2 inch (12.5 mm). In essence, this test standardizes the amount of energy that goes into the deformation of the soil and the amount of plastic deformation the soil undergoes, while using the moisture content as the index of the associated shear strength.

The plastic limit (PL) is determined by hand-rolling a soil sample into threads. The thread becomes thinner and, eventually, breaks as the rolling process continues. The soil is at the plastic limit when it breaks at a diameter of 0.125 inches (3.2 mm). In essence, this tests dries up the sample gradually through hand-rolling, until

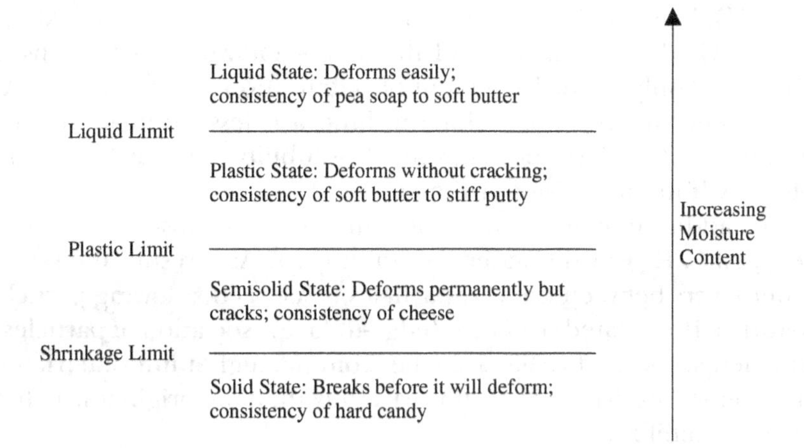


**Figure 3.11**
Consistency of Fine-Grained Soils at Different Moisture Contents (Ref. 19)

soil suction exceeds its tensile strength. Figure 3.11 illustrates the consistency of soils at different moisture contents.

The soil with a moisture content between the liquid limit and plastic limit is considered to be in a plastic state. The plasticity index (PI = LL−PL) is a measure of the range of moisture contents that define this plastic state. Soils with a large clay content retain this plastic state over a wide range of moisture contents, and thus have a large PI value.

## 3.6 Aggregate and Soil Stabilization

Expansive clay subgrades exhibit significant changes in volume in response to changes in water content. These volume changes result in severe pavement distresses, (e.g., cracking and permanent deformation). Stabilization is used to reduce the volume change potential and increase the strength and stiffness of subgrades[18]. Similarly, marginal aggregates in base layers are stabilized in order to reduce plasticity and increase strength and stiffness. The primary traditional stabilizers are hydrated lime, portland cement, and fly ash.

Lime has been used in different forms in soil stabilization. These forms are hydrated high-calcium lime, $Ca(OH)_2$; dihydrated dolomitic lime, $Ca(OH)_2Mg(OH)_2$; monohydrated dolomitic lime,

$Ca(OH)_2MgO$; Calcitic quicklime, $CaO$; and dolomite quicklime, $CaOMgO$[14,15]. The majority of the lime stabilization is done using $Ca(OH)_2$; only about 10 percent of stabilization is done using $CaO$. The different forms of dolomitic lime are less popular in stabilization because $Mg(OH)_2$ has much solubility than $Ca(OH)_2$, and $MgO$ hydrates much slower than $CaO$.

The addition of lime causes calcium to replace most of the cations (e.g., $Na^+$, $K^+$) in the water system in clays. As a result, the size of water layers between the clay particles is reduced, allowing particles to form a flocculated structure (edge-to-face association of particles). The new soil structure has a smaller volume, higher internal friction (greater strength), and better workability than the original structure prior to stabilization.

Lime stabilization promotes long-term gain of strength through reactions with soil silica and soil alumina. When sufficient lime is added to a soil, the pH of the soil-lime mixture increases to about 12.4. The pH elevation increases the solubilities of silica and alumina.[8] The soil-lime reaction can be described qualitatively as follows:

$$\begin{aligned}
&Ca[OH]_2(\textit{Calcium Hydrate}) \rightarrow Ca^{++} + 2[OH]^- \\
&Ca^{++} + 2[OH]^- + SiO_2(\textit{Silica}) \rightarrow CSH(\textit{Calcium Silicate Hydrate}) \\
&Ca^{++} + 2[OH]^- + Al_2O_3(\textit{Alumina}) \rightarrow CAH(\textit{Calcium Aluminate Hydrate})
\end{aligned} \tag{3.26}$$

The properties that influence the soil-lime reactivity are soil pH; organic carbon content; drainage; presence of excessive quantities of exchangeable sodium; clay mineralogy; degree of weathering; presence of carbonates, sulfates, or both; extractable iron; silica-sesquioxide ratio; and silica-alumina ratio.[23] Soil-lime reactions are time- and temperature-dependent and continue for long periods of time. Little[14,15] has developed a protocol for the design and testing of soil-lime mixtures that consists of the following steps:

1. Determine the optimum lime content using ASTM D-6274.
2. Simulate field conditions through the use of AASHTO T-180 compaction and seven-day curing at 40° C. Subject the samples to 24 to 47 hours of moisture conditioning.

3. Evaluate stiffness, moisture sensitivity, and compressive strength using ASTM D-5102.
4. Measure the resilient modulus using AASHTO T-307.

Little[14,15] presented data showing that the benefits of lime stabilization include higher strength, better resilient properties, greater resistance to fracture and fatigue, and lower sensitivity to changes in moisture content.

The addition of cement is another method for improving the properties of base and subgrade soils. The soil-cement mixtures are tested using durability and strength tests. Durability is assessed by measuring the weight loss under wet-dry (ASTM D559) and freeze-thaw (ASTM D560) tests. Strength is typically measured using the unconfined strength test using the ASTM D1633 procedure.

Cement stabilization of soil is used to achieve one or more of the following objectives:[18]

- Reduce the PI.
- Increase the shrinkage limit.
- Reduce the volume change.
- Reduce the clay/silt-sized particles.
- Improve strength.
- Increase the resilient modulus.

The third traditional method of stabilization is the use of coal fly ash. Fly ash is a synthetic pozzolan that is produced as a result of the combustion of coal. Although self-cementing fly ash (Class C) contains lime, or CaO, only a small percentage of CaO is available as "free" lime that can be used in pozzolanic reaction. Therefore, the use of ash for the stabilization of clay soils can be effective only with the addition of either lime or cement. The lime is produced by the hydration of cement in soil. The addition of lime or cement modifies the clay surface and provides the necessary lime for the pozzolanic reaction of fly ash. When fly ash is used, it is important to consider the rate of the hydration process, the influence of moisture content, the percentage of free lime, and the percentage of sulfates. It is especially important to avoid the reaction of soil sulfates with the calcium or the carbonate present in the stabilizing agents, because they form ettringite and thaumasite, which are very expansive, and produce swelling.

## References


[1] Adu-Osei, A., Little, D., and Lytton, R. (2001). Cross-Anisotropic Characterization of Unbound Granular Materials, *Transportation Research Record 1757, Journal of the Transportation Research Board*, pp. 82–91.

[2] Asphalt Institute, (1982). *Research and Development of the Asphalt Institute's Thickness Design Manual (MS-1)*, 9th ed. Research Report 82–2; Asphalt Institute, Lexington, KY.

[3] Barksdale, R. D., and Itani, S. Y. (1989). "Influence of Aggregate Shape on Base Behaviour," in Transportation Research Record 1227, Transportation Research Board, Washington, DC, pp. 173–182.

[4] Dawson, A. R., Thom, N. H., and Paute, J. L. (1996). "Mechanical Characteristics of Unbound Granular Materials as a Function of Condition". *Flexible Pavements, Proc., Eur. Symp. Euroflex 1993*, A. G. Correia, ed., Balkema, Rotterdam, the Netherlands, pp. 35–44.

[5] Heukelom, W., and Klomp, A. J. G. (1962). "Dynamic Testing as a Means of Controlling Pavements During and After Construction," *Proceedings of the 1st International Conference on the Structural Design of Asphalt Pavements*, Ann, Arbor, MI pp. 667–685.

[6] Hicks, R. G., and Monismith, C. L. (1971). "Factors Influencing the Resilient Properties of Granular Materials," in Transportation Research Record 345, Transportation Research Board, National Research Council, Washington DC, pp. 15–31.

[7] Jorenby, B. N., and Hicks, R. G. (1986). "Base Course Contamination Limits," in Transportation Research Record 1095, Transportation Research Board, Washington, DC, pp. 86–101.

[8] Keller, W. D. (1957). *The Principles of Chemical Weathering*. Lucas Brothers Publishers, Columbia, MO.

[9] Kenis, W. J. (1977). "Predictive Design Procedures: A Design Method for Flexible Pavements Using the VESYS Structural Subsystem," *Proceedings of the 4th International Conference on the Structural Design of Asphalt Pavements*, Ann, Arbor, MI, Vol. 1, pp. 101–147.

[10] Khedr, S. (1985). ''Deformation Characteristics of Granular Base Course in Flexible Pavement,'' in Transportation Research Record 1043, Transportation Research Board, Washington, DC, pp. 131–138.

[11] Kim. S. H., Little, D., Masad, E., and Lytton, R. (2005). Estimation of Level of Anisotropy in Unbound Granular Layers Considering Aggregate Physical Properties, *International Journal of Pavement Engineering*, Vol. 6, No. 4, pp. 217–227.

[12] Lekarp, F., Isaacson, U., and Dawson. A. (2000). State of the Art, Part I: Resilient Response of Unbound Aggregate, *Journal of Transportation Engineering*, American Society of Civil Engineers, Vol. 126, No. 1, pp. 66–75.

[13] Lekarp, F., Isaacson, U., and Dawson, A. (2000). State of the Art, Part II: Permanent Strain Response of Unbound Aggregate, *Journal of Transportation Engineering*, American Society of Civil Engineers, Vol. 126, No. 1, pp. 76–83.

[14] Little, D. N. (1999). *Evaluation of Structural Properties of Lime-Stablized Soils and Aggregates*, Vol. 1: Summary of Findings, National Lime Association, Arlington, VA.

[15] Little, D. N. (1999). *Evaluation of Structural Properties of Lime-Stablized Soils and Aggregates*, Vol. 2: Documentation of Findings, National Lime Association, Arlington, VA.

[16] Masad, S., Little, D., and Masad, E. (2006). Analysis of Flexible Pavement Response and Performance Using Isotropic and Anisotropic Material Properties, *Journal of Transportation Engineering*, American Society of Civil Engineers, Vol. 132, No. 4, pp. 342–349.

[17] May, R. W., and Witczak, M. W. (1981). ''Effective Granular Modulus to Model Pavement Responses,'' in Transportation Research Record 810, Transportation Research Board, Washington, DC, pp. 1–9.

[18] Petry, T. M., and Little, D. (2002). ''Review of Stabilization of Clays and Expansive Soils in Pavements and Lightly Loaded Structures—History, Practice, and Future,'' *Journal of Materials in Civil Engineering*, American Society of Civil Engineering, Vol. 14, No. 6, pp. 447–460.

[19] Sowers, G. F. (1992). *Introductory Soil Mechanics and Foundations: Geotechnical Engineering*, 4th Ed., Macmillan, New York.

[20] Thom, N. H., and Brown, S. F. (1987). "Effect of Moisture on the Structural Performance of a Crushed-Limestone Road Base," in Transportation Research Record 1121, Transportation Research Board, Washington, DC, pp. 50–56.

[21] Thom, N. H., and Brown, S. F. (1988). The Effect of Grading and Density on the Mechanical Properties of a Crushed Dolomitic Limestone. *Proc., 14th Australian Road Research Board, Conf. Canberra, Australia*, Vol. 14, Part 7, pp. 94–100.

[22] Tongyan, P., Tutumluer, E., and Anocie-Boateng, J. (2006). "Aggregate Morphology Affecting Resilient Behavior of Unbound Granular Materials," in Transportation Research Record 1952, Journal of the Transportation Research Board, pp. 12–20.

[23] TRB (1987). Lime Stabilization: Reactions, Properties, Design and Construction, State of the Art, Report 5, Transportation Research Board, National Research Council, Washington DC.

[24] Tseng, K. H. and Lytton, R. L. (1989). "Prediction of Permanent Deformation in Flexible Pavements Materials," in *Implication of Aggregates in the Design, Construction, and Performance of Flexible Pavements,* ASTM STP 1016, H. G Schreuders and C. R. Marek, eds., American Society for Testing and Materials, Philadelphia, PA, pp. 154–172.

[25] Tutumluer, E., and Thompson, M. (1997). "Anisotropic Modeling of Granular Bases in Flexible Pavements," in Transportation Research Record 1577, National Research Council, Washington DC, pp. 18–26.

[26] Uzan, J. (1985). "Characterization of Granular Material," in Transportation Research Record 1022, Transportation Research Board, National Research Council, Washington DC, pp. 52–59.

[27] Van Van Niekerk, A. A., Houben, L. J. M., and Molenaar, A. A. (1998). "Estimation of Mechanical Behaviour of Unbound Road-Building Materials from Physical Material Properties". *Proceeding, 5th International Conference Trondheim, Norway on the Bearing Capacity of Roads and Airfields,* R. S. Nordal and G. Rafsdal, eds., Vol. 3, pp. 1221–1233.

[28] Werkmeister, S., Dawson, A., and Wellner, F. (2001). "Permanent Deformation Behavior of Granular Materials and the Shakedown Concept," *Transportation Research Record 1757, Journal of the Transportation Research Board*, pp. 75–81.

[29] Witczak, M. (2003). Harmonized Test Methods for Laboratory Determination of Resilient Modulus for Flexible Pavement Design, NCHRP Research Results Digest 285, National Cooperative Highway Research Program, Washington, DC.

[30] NCHRP (July 2004). *2002 Design Guide: Design of New and Rehabilitated Pavement Structures*, Draft Final Report, NCHRP Study 1–37A, National Cooperative Highway Research Program, Washington, DC.

## Problems

3.1 Provide a summary of the subgrade, subbase, and base properties that are needed as input to the proposed design guide from the NCHRP 1–37A design approach.

3.2 Plot the relationships between resilient modulus and CBR given in Equations 3.21 and 3.22 for a range of *CBR* values from 5 to 50. Comment on the predictions of these two equations.

3.3 Discuss the benefits of subgrade and base stabilization using lime.

3.4 Based on your background in geotechnical engineering, describe the influence of moisture content on the resilient modulus

3.5 Using the data given in Example 3.1, calculate the difference in the estimates of the resilient modulus obtained from Equations 3.4 and 3.6. Plot the difference versus the ratio of $\tau_{oct}$ to $\theta$. What do you conclude from this plot?

# 4 Aggregates

## 4.1 Introduction

Aggregates refer to the material derived from natural rocks, or are the by-product of the manufacturing process of other materials, (e.g., the manufacturing of steel generates slag as a by-product that has been used as an aggregate). Aggregates are an important ingredient of the materials used in highway construction. They constitute 70% to 85% by weight of portland cement concrete (PCC) and hot-mix asphalt (HMA). By volume, the corresponding ratios are 60% to 75% for PCC and 75% to 85% for HMA, respectively. The physical, mechanical, and chemical properties of aggregates play an important role in the performance of both rigid and flexible pavements. This chapter discusses the different classifications of aggregates, aggregate properties, and the influence of these properties on the performance of pavements.

## 4.2 Aggregate Types and Classifications

### 4.2.1 Classification Based on Source

Aggregates derived from natural rocks can be classified on the basis of size as crushed stone, sand, or gravel. Crushed stone refers to the different rock types and sizes that are produced by blasting and then crushing. Sand and gravel comprise any clean mixture of aggregate sizes found in natural deposits, such as stream channels. The word "natural" in reference to sand is sometimes used to indicate that this aggregate is available in natural deposits and not produced through crushing processes. In some cases, the word "manufactured" in reference to sand is used to refer to the small sizes of crushed stone. Based on information from the Bureau of Mines, the *Aggregate Handbook* states that the about 2.1 billion tons of aggregates derived from rocks are produced annually, of which 897 million tons are sand and gravel; the remaining 1.2 billion tons are crushed stones[13]. The percentages of crushed stones produced by geological origin are given in Figure 4.1.

Aggregates can also be classified on the basis of the geological origin of their parent rock, as igneous, sedimentary, and metamorphic. Igneous rocks are formed by cooling of the molten liquid silicate

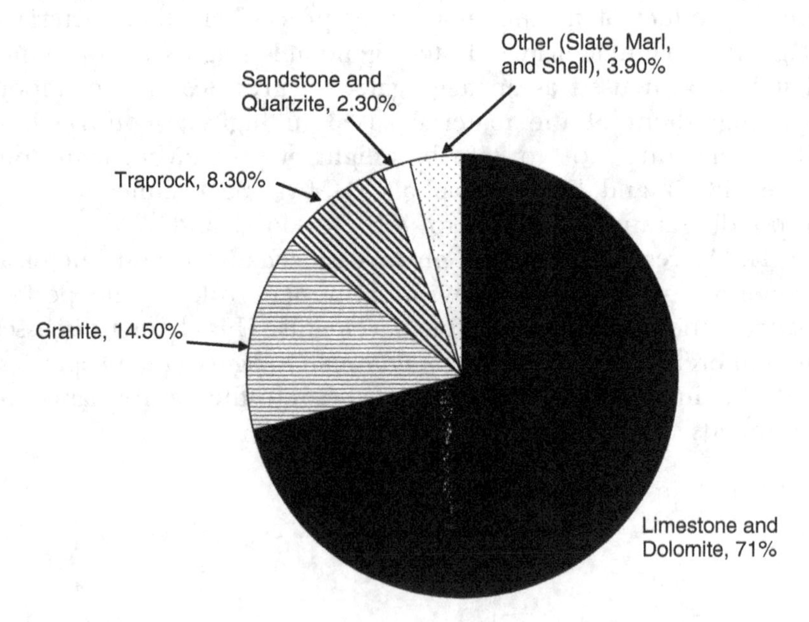


**Figure 4.1**
Percentages of Different Types of Crushed Stone Produced Annually

referred to as *magma*. Igneous rocks are either crystalline in structure, with fine or coarse grains, or have a noncrystalline or glassy structure. Their texture size and type depend on the geological process that created them. Extrusive igneous rocks are formed where the magma reached the earth's surface as ash or lava and cooled rapidly at the surface to form fine-grained or glassy rock. Intrusive igneous rocks are formed where the magma cooled slower under the earth's crust and formed larger crystals. Basalts and granites are examples of extrusive and intrusive rocks, respectively.

Sedimentary rocks are formed at the earth's surface or under water, due to consolidation of sedimentary materials or chemical precipitates. The sedimentary materials are the result of the disintegration of existing rocks under the effect of clastic processes such as weathering and abrasion by wind, water, ice, or gravity. The sediments harden due to the cementation by silica and carbonate minerals and the pressure under the weight of overlying deposits. The rocks formed by sediments are referred to as *clastic rocks*. Examples of clastic rocks are sandstone, siltstone, and shale. The sedimentary rocks formed by chemical precipitates are called *carbonate rocks*. These rocks are formed by the deposition and cementation of the shells of marine animals, shells of marine plants, and fine carbonate mud that precipitates from marine water. Examples of carbonate rocks are limestone and dolomite.

Metamorphic rocks are formed by the recrystallization of sedimentary and igneous rocks under the influence of pressure and temperature. Examples of metamorphic rocks are gneiss, quartzite, and marble.

Aggregates that are by-products of the manufacturing of another material are referred to, in some cases, as *man-made aggregates* or *artificial aggregates*. An example of by-product aggregate is slag, which is produced during the metallurgical processing of steel, iron, tin, and copper. The most widely used variety is blast-furnace slag, which is a nonmetallic product that is developed in a molten condition simultaneously with iron in a blast furnace. Expanded slag is produced by expanding blast-furnace slag by mixing it with water while it is still molten.

### 4.2.2 Classification Based on Size

Aggregates are classified in terms of size, as fine and coarse. The size that separates fine from coarse aggregates differs, based on the application and the intended use of the aggregates. According to ASTM C125, which relates to PCC, fine aggregate sizes are defined as those passing the No. 4 sieve (4.75 mm) and predominantly retained on the No. 200 (75 $\mu$m) sieve. Coarse aggregates are defined as those predominantly retained on the No. 4 (4.75 mm) sieve. For HMAs, the No. 4 (4.75 mm) sieve or the No. 8 (2.36 mm) sieve are typically used to separate the fine aggregate from the coarse aggregate sizes.

## 4.3 Aggregate Properties

There are many AASHTO and ASTM specifications and tests for aggregates, as summarized in Tables 4.1 and 4.2, respectively. Some of these tests were adapted for use in Superpave,™ a contraction of Superior Performing Pavements (a trademark of the U.S. Department of Transportation), which was developed under the Strategic Highway Research Program (SHRP). The objective of the discussion in the following sections is not to describe the details of these tests but rather to explain the impact of the properties of aggregates on the behavior and performance of the pavement layers, where these aggregates are used.

### 4.3.1 Physical Properties

**GRADATION AND SIZE DISTRIBUTION**

Aggregate gradation gives the percentage of each of the sizes in a blend. It is typically expressed as the percentage of the aggregate blend passing sieves with standard openings. The size distribution of aggregate particles is directly related to the performance of the pavement layers. In general, aggregate size distributions are classified as *gap graded*, *uniform*, *well-graded*, or *open graded*. These distributions are shown in a semi-long scale in Figure 4.2. The sieves that are typically used in determining the gradation are 2 in, 1–½ in., 1 in., ¾ in., ½ in., 3/8 in., No. 4, No. 8, No. 16, No. 30, No. 50, No. 100, and No. 200 (50.8 mm, 37.5 mm, 25.4 mm, 19.0 mm, 12.5 mm, 9.5 mm, 4.75 mm, 2.36 mm, 1.18 mm, 0.6 mm, 0.3 mm, 0.15 mm and 0.075 mm, respectively).

Aggregate gradation is typically presented in a graphical form in which the percent of aggregate passing a sieve size is plotted on the ordinate in an arithmetic scale, and the particle size is plotted on the abscissa in a logarithmic scale. Alternatively, it can be plotted using the Fuller and Thompson method, whereby the percent passing is plotted versus the particle size, raised to an exponent $n$. Fuller

**Table 4.1**
ASTM and AASHTO Aggregate Specifications (Ref. 13)

| AASHTO Specifications | ASTM Specifications | Title |
|---|---|---|
| M-43 | C 448 | Standard Sizes of Coarse Aggregate |
| M-283 | | Coarse Aggregates for Highway and Airport Construction |
| | D 2940 | Graded Aggregate for Bases or Subbases |
| M-147 | | Materials for Aggregate and Soil-Aggregate Subbase, Base, and Surface Courses |
| M-155 | | Granular Material to Control Pumping Under Concrete Pavement |
| M-29 | D 1073 | Fine Aggregate for Bituminous Paving Mixtures |
| | D 692 | Aggregate for Bituminous Paving Mixtures |
| M-17 | D 242 | Mineral Filler for Bituminous Paving Mixtures |
| R-12 | | Bituminous Mixture Design Using Marshall and Hveem Procedures |
| | D 3515 | Hot-Mixed, Hot-Laid Bituminous Paving Mixtures |
| | D 693 | Crushed Aggregate for Macadam Pavements |
| | D 1139 | Crushed Stone, Crushed Slag, and Gravel for Bituminous Surface Treatments |
| M-6 | | Fine Aggregate for PC Concrete |
| M-80 | | Coarse Aggregate for PC Concrete |
| | C 38 | Concrete Aggregates |
| M-195 | C 330 | Lightweight Aggregates for Structural Concrete |
| R-1 | E 380 | Metric Practice Guide |
| R-10 | | Definitions of Terms for Specifications and Procedures |
| R-11 | E 29 | Practice for Indicating Which Places of Figures Are To Be Considered Significant in Specified Limiting Values |
| M-145 | | Classification of Soils and Soil-Aggregate, Fill Materials, and Base Materials |
| M-146 | | Terms Related to Subgrade, Soil-Aggregate, and Fill Materials |
| | D 8 | Definitions of Terms Relating to Materials for Roads and Pavements |
| | C 125 | Terminology Relating to Concrete and Concrete Aggregates |
| | D 3665 | Random Sampling of Construction Materials |

and Thompson observed that the aggregate reaches its maximum possible density (i.e., densest packing of particles) when its gradation matches the following expression:

$$P = 100(d/D)^n \tag{4.1}$$

**Table 4.2**
ASTM and AASHTO Aggregate Tests (Ref. 13)

| AASHTO Procedures | ASTM Procedures | Title |
|---|---|---|
| M-92 | E 11 | Wire Cloth Sieves for Testing Purposes |
| M-132 | D 12 | Terms Relating to Density and Specific Gravity |
| M-231 | — | Weights and Balances Used in Testing |
| — | D 3665 | Evaluation of Inspecting and Testing Agencies for Bituminous Paving Materials |
| — | C 1077 | Practice for Laboratories Testing Concrete and Concrete Aggregates |
| T-2 | D 75 | Sampling Aggregates |
| T-248 | C 702 | Reducing Field Samples of Aggregate to Testing Size |
| T-87 | D 421 | Dry Preparation of Disturbed Soil and Soil Aggregate Samples for Tests |
| T-146 | D 2217 | Wet Preparation of Disturbed Soil Samples for Tests |
| T-27 | C 136 | Sieve Analysis of Fine and Coarse Aggregates |
| T-11 | C 117 | Amount of Material Finer Than the No. 200 Sieve |
| T-30 | | Mechanical Analysis of Extracted Aggregates |
| T-88 | D 422 | Particle Size Analysis of Soils |
| T-37 | D 546 | Sieve Analysis of Mineral Filler |
| T-176 | D 2419 | Sand Equivalent Test for Plastic Fines in Graded Aggregates and Soils |
| — | D 4318 | Liquid Limit, Plastic Limit, and Plasticity Index of Soils |
| T-210 | D 3744 | Aggregate Durability Index |
| T-104 | C 88 | Soundness of Aggregate by Use of Sodium Sulfate or Magnesium Sulfate |
| T-103 | — | Soundness of Aggregates by Freezing and Thawing |
| — | D 4792 | Potential Expansion of Aggregates from Hydration Reactions |
| T-161 | C 666 | Resistance of Concrete to Rapid Freezing and Thawing |
| — | C 671 | Critical Dilation of Concrete Specimens Subjected to Freezing |
| — | C 682 | Evaluation of Frost Resistance of Coarse Aggregates in Air-Entrained Concrete by Critical Dilation Procedures |
| T-96 | C 131 or C 535 | Resistance to Abrasion of Small- or Large-Size Coarse Aggregate by Use of the Los Angeles Machine |
| T-21 | C40 | Organic Impurities in Sands for Concrete |
| T-71 | C 87 | Effect of Organic Impurities in Fine Aggregate on Strength of Mortar |
| T-112 | C 142 | Clay Lumps and Friable Particles in Aggregate |
| T-113 | C 123 | Lightweight Pieces in Aggregate |
| — | C 294 | Nomenclature of Constituents of Natural Mineral Aggregate |
| — | C 295 | Practice for Petrographic Examination of Aggregates for Concrete |

**Table 4.2**
*(Continued)*

| *AASHTO Procedures* | *ASTM Procedures* | *Title* |
|---|---|---|
| — | C 227 | Alkali Reactivity Potential of Cement-Aggregate Combinations |
| — | C 289 | Potential Reactivity of Aggregates |
| — | C 586 | Potential Alkali Reactivity of Carbonate Rocks for Concrete Aggregate |
| — | D 4791 | Flat or Elongated Particles in Coarse Aggregate |
| — | C 342 | Volume Change Potential of Cement-Aggregate Combinations |
| — | C 441 | Mineral Admixture Effectiveness in Preventing Excessive Expansion Due to Alkali Aggregate Reaction |
| T-165 | D 1075 | Effect of Water on Cohesion of Compacted Bituminous Mixtures |
| T-182 | D 1664 | Coating and Stripping of Bitumen—Aggregate Mixtures |
| T-195 | D 2489 | Determining Degree of Particle Coating of Bituminous Aggregate Mixtures |
| T-270 | — | Centrifuge Kerosene Equivalent and Approximate Bitumen Ratio (ABR) |
| T-283 | — | Resistance of Compacted Bituminous Mixture to Moisture-Induced Damage |
| — | D 4469 | Calculating Percent Absorption by the Aggregate in an Asphalt Pavement Mixture |
| — | D 1559 | Resistance to Plastic Flow—Marshall Apparatus |
| — | D 1560 | Deformation and Cohesion–Hveem Apparatus |
| T-99 | D 689 | Moisture-Density Relationship Using a 5.5-Pound Rammer and a 12-Inch Drop |
| T-180 | D 1557 | Moisture-Density Relationship Using a 10-Pound Rammer and an 18-Inch Drop |
| T-215 | D 2434 | Permeability of Granular Soils |
| T-224 | D 4718 | Correction for Coarse Particles in Soil Compaction Tests |
| T-238 | D 2922 | Density of Soil and Soil Aggregate In-Place by Nuclear Methods |
| T-239 | D 3017 | Moisture Content of Soil and Soil Aggregate In-Place by Nuclear Methods |
| — | D 4253 | Index Density of Soils Using a Vibratory Table |
| T-191 | D 1556 | Density of Soil In-Place by the Sand Cone Method |
| T-205 | D 2167 | Density of Soil In-Place by the Rubber Balloon Method |

*(continued overleaf)*

**Table 4.2**
*(Continued)*

| *AASHTO Procedures* | *ASTM Procedures* | *Title* |
|---|---|---|
| T-190 | D 2844 | Resistance R-Value and Expansion Pressure of Compacted Soils |
| T-193 | C 1883 | California Bearing Ratio |
| T-234 | D 2850 | Strength Parameters of Soils by Triaxial Compression |
| T-274 | | Resilient Modulus of Subgrade Soils |
| T-212 | D 3397 | Triaxial Classification of Base Materials, Soils, and Soil Mixtures |
| T-84 | C 128 | Specific Gravity and Absorption of Fine Aggregate |
| T-85 | C 127 | Specific Gravity and Absorption of Coarse Aggregate |
| T-19 | C 29 | Unit Weight and Voids in Aggregate |
| T-242 | E 374 | Frictional Properties of Paved Surfaces Using a Full-Scale Tire |
| T-279 | D 3319 | Accelerated Polishing of Aggregates Using the British Wheel |
| T-278 | E 303 | Measuring Surface Frictional Properties Using the British Pendulum Tester (BPT) |
| — | D 3042 | Insoluble Residue in Carbonate Aggregates |
| — | E 707 | Skid Resistance of Paved Surfaces Using the NC State Variable-Speed Friction Tester |
| — | E 660 | Accelerated Polishing of Aggregates or Pavement Surfaces Using a Small-Wheel Circular Polishing Machine |
| — | D 4791 | Flat or Elongated Particles in Coarse Aggregate |
| — | D 3398 | Index of Aggregate Particle Shape and Texture |
| TP-58 | D6928 | Resistance of Coarse Aggregate to Degradation by Abrasion in the Micro-Deval Apparatus |

where $P$ is the percentage of aggregates passing the sieve size $d$, $D$ is the maximum aggregate size in the gradation, and the exponent $n$ has a value of 0.5. Note that the FHWA has recommended a slightly smaller value for this exponent of 0.45. An example of Fuller's maximum density line for an exponent for a maximum aggregate size of 25.0 is shown in Figure 4.3.

Aggregate blends are designated by their maximum aggregate size or their nominal maximum aggregate. According to ASTM C 125, the maximum size refers to the smallest sieve through which 100% of the aggregate sample particles pass, and the nominal maximum size as the largest sieve that retains some (i.e., less than 10% by

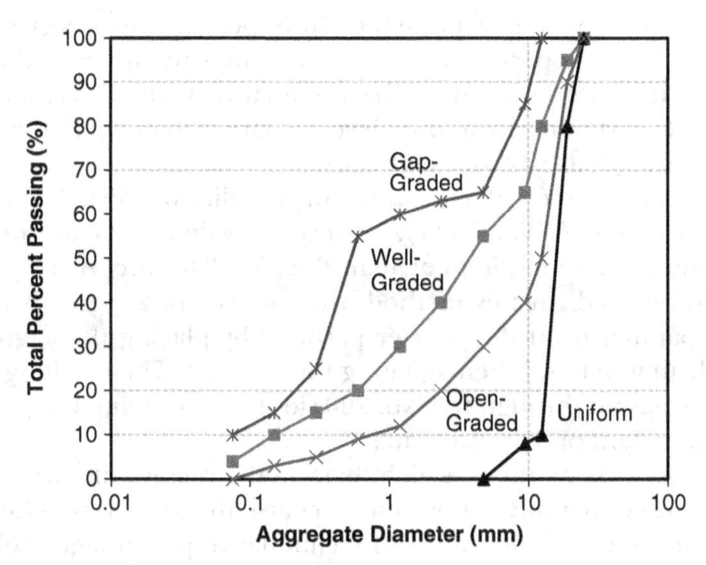


**Figure 4.2**
Examples of Different Aggregate Size Gradations

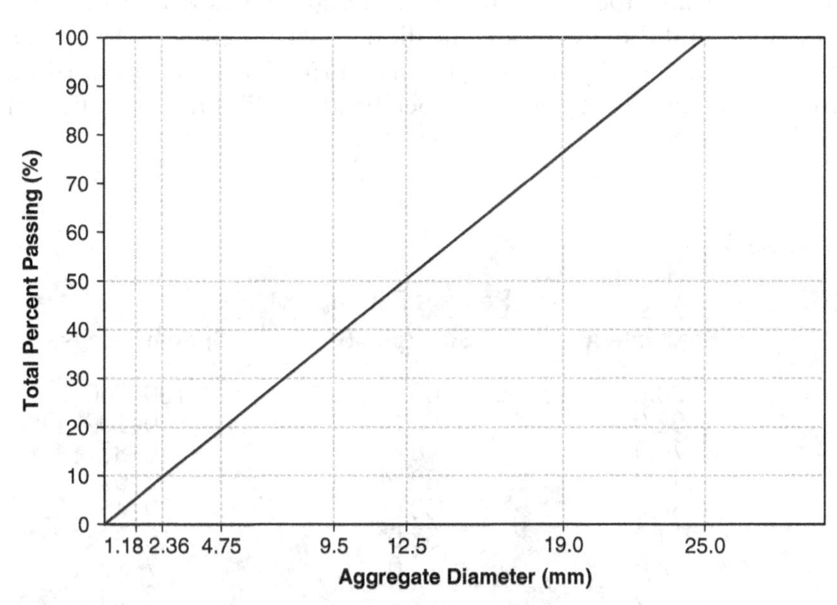


**Figure 4.3**
Illustration of Fuller's Maximum-density Line in a 0.45 Power Chart

weight) of the aggregate particles. The Superpave mix design system defines these properties differently. The maximum size is defined as one sieve size larger than the nominal maximum size, and the nominal maximum size as one sieve size larger than the first sieve to retain more than 10% by weight.

Sieve analysis is conducted under dry conditions (ASTM C 136) or wet conditions (ASTM C 117). To accurately determine the amount of material in a sample finer than the No. 200 sieve, ASTM C 117 should be used. In this method, the minus No. 200 size particles are separated from the coarser particles by placing the aggregate sample in water and then agitating the solution. The resulting wash water is passed through the No. 200 sieve to determine the percent by total weight of fines accurately.

The fineness modulus, which denotes the relative fineness of the sand, is used to indicate how fine or coarse the sand. It is defined as one-hundredth of the sum of the cumulative percentages held on the standard sieves (Nos. 4, 8, 16, 30, 50, and 100) in a sieve test of sand. The smaller the value of the fineness modulus, the finer the sand.

**Example 4.1** Two aggregate stockpiles are given, designated as A and B. Their gradation and the ASTM specification limits are given in Table 4.3. Examine whether, by blending them in a 75/25 percent proportion, they can meet ASTM gradation specifications. Plot the gradation of

**Table 4.3**
Gradation data for Example 4.1

| | Percent Passing (%) | | |
|---|---|---|---|
| **Seive Size** | **Stockpile A** | **Stockpile B** | **Specifications** |
| 1/2 in. (12.5 mm) | 100.0 | 100.0 | 100–100 |
| 3/8 in. (9.5 mm) | 92.0 | 100.0 | 90–100 |
| No. 4 (4.75 mm) | 70.0 | 98.0 | 50–85 |
| No. 8 (2.36 mm) | 43.0 | 89.0 | 32–67 |
| No. 16 (1.18 mm) | 34.0 | 60.0 | 20–45 |
| No. 30 (600 $\mu$m) | 14.0 | 53.0 | 15–32 |
| No. 100 (150 $\mu$m) | 2.0 | 27.0 | 7–20 |
| No. 200 (75 $\mu$m) | 1.0 | 9.0 | 2–10 |
| Bulk-Specific Gravity | 2.750 | 2.600 | |
| Absorption | 1.5% | 2.5% | |

**Table 4.4**
Gradation of Blended Aggregate for Example 4.1

| Sieve Size | Percent Passing (%) | | | | |
|---|---|---|---|---|---|
| | Stockpile A | Stockpile B | Specifications | Blend | Status |
| 1/2 in. (12.5 mm) | 100 | 100 | 100–100 | 100 | Ok |
| 3/8 in. (9.5 mm) | 92 | 100 | 90–100 | 94 | Ok |
| No. 4 (4.75 mm) | 70 | 98 | 50–85 | 77 | Ok |
| No. 8 (2.36 mm) | 43 | 89 | 32–67 | 54.5 | Ok |
| No. 16 (1.18 mm) | 34 | 60 | 20–45 | 40.5 | Ok |
| No. 30 (600 μm) | 14 | 53 | 15–32 | 23.8 | Ok |
| No. 100 (150 μm) | 2 | 27 | 7–20 | 8.3 | Ok |
| No. 200 (75 μm) | 1 | 9 | 2–10 | 3.0 | Ok |
| Bulk-Specific Gravity | 2.750 | 2.600 | | 2.711 | |
| Absorption | 1.50% | 2.50% | | 1.75% | |

the blended aggregate and determine its maximum aggregate size and its nominal maximum aggregate size per ASTM specification C 125.

**ANSWER**

The data on the blended gradation is shown in Table 4.4 and plotted along with the ASTM specification limits in Figure 4.4, which suggests that the blend meets gradation specifications. According to ASTM C 125, the maximum size is designated as the smallest sieve through which 100% of the aggregate sample particles pass; and nominal maximum size is designated as the largest sieve that retains some (i.e., less than 10%) of the aggregate particles. Therefore, the maximum aggregate size and nominal maximum sizes are 12.5 mm and 9.5 mm, respectively.

**SPECIFIC GRAVITY AND ABSORPTION**

The specific gravity and absorption of coarse and fine aggregate are determined by ASTM C 127 and C 128 procedures, respectively. These tests are based on the Archimedes principle, which states that a solid immersed in water is subjected to a vertical buoyant force equal to the weight of the water it displaces. The following are the definitions of the different measurements of specific gravity:

- *Apparent specific gravity:* The apparent specific gravity is the ratio of the weight of dry aggregate to the weight of water having a

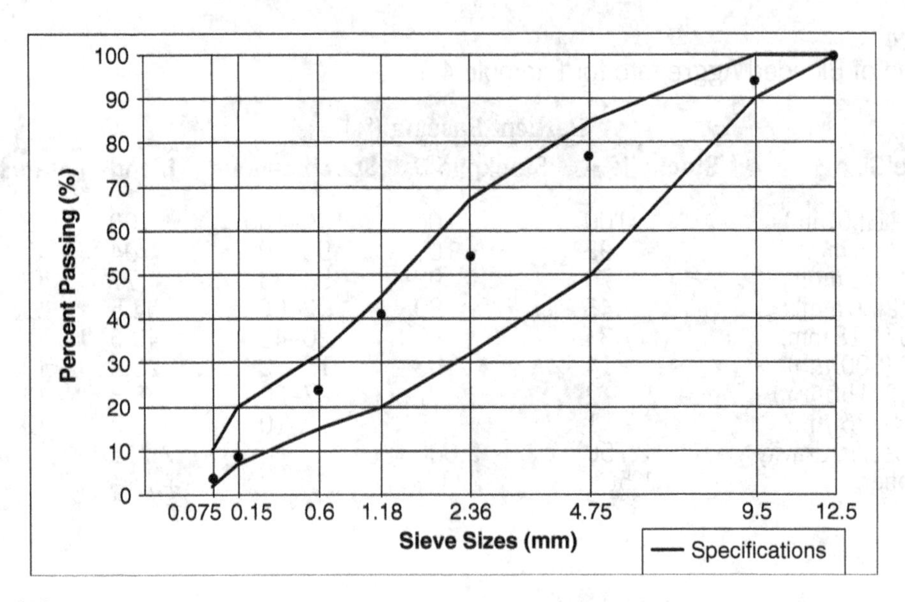


**Figure 4.4**
Gradation Results for Example 4.1

volume equal to the solid volume of the aggregate, excluding its permeable pores.

- *Bulk-specific gravity:* This specific gravity is the ratio of the weight of dry aggregate to the weight of water having a volume equal to the volume of the aggregate, including both its permeable and impermeable pores.
- *Bulk-specific gravity—saturated, surface dry* (SSD)*:* The SSD-specific gravity is the ratio of the weight of the aggregate, including the weight of water in its permeable voids, to the weight of an equal volume of water.

These specific gravity values and absorption are determined by the following equations:

$$\text{Apparent specific gravity} = \frac{A}{A - C} \tag{4.2}$$

$$\text{Bulk-specific gravity} = \frac{A}{B - C} \tag{4.3}$$

$$\text{Bulk-specific gravity, SSD} = \frac{B}{B - C} \tag{4.4}$$

$$\text{Water-absorption (\%)} = \frac{B - A}{A} 100 \tag{4.5}$$

where:

$A$ = Weight of oven-dry sample of aggregate in air
$B$ = Weight of saturated, surface-dry sample in air
$C$ = Weight of saturated sample in water

The specific gravity of fine aggregate is measured by immersing a saturated and surface-dried aggregate sample in a pycnometer that is filled with water. The pycnometer is rolled, inverted, agitated, and subjected to suction to eliminate air bubbles. The total weight of the pycnometer, sample, and water is determined. The fine aggregate is removed, dried to a constant weight, and weighed. The weight of the pycnometer is determined, and the bulk-specific gravity, bulk-saturated surface-dry-specific gravity, apparent specific gravity, and the absorption are calculated as follows:

$$\text{Apparent specific gravity} = \frac{A}{B + A - C} \tag{4.6}$$

$$\text{Bulk-specific gravity} = \frac{A}{B + D - C} \tag{4.7}$$

$$\text{Bulk-specific gravity, SSD} = \frac{D}{B + D - C} \tag{4.8}$$

$$\text{Absorption} = \frac{D - A}{A} 100 \tag{4.9}$$

where:

$A$ = weight of oven-dry specimen in air
$B$ = weight of pycnometer filled with water
$C$ = weight of pycnometer with specimen and water to calibration mark
$D$ = weight of saturated surface-dry specimen

It is necessary in some cases to blend different stockpiles of aggregates with known gradations. The percentage of the combined aggregate passing a given sieve size ($P$) is calculated using Equation 4.10.

$$P = A\,a + B\,b + C\,c + \cdots \tag{4.10}$$

where $A$, $B$, $C \cdots$ are the percentages of each aggregate that passes a given sieve size, and $a$, $b$, $c \ldots$ are the proportions of each aggregate needed to meet the requirements for material passing the given sieve where $a + b + c + \cdots = 1.00$. The blended specific gravity, $G$, and absorption are calculated using Equations 4.11 and 4.12, respectively.

$$\text{Combined specific gravity } G = \frac{1}{\dfrac{a}{G_A} + \dfrac{b}{G_B} + \cdots} \tag{4.11}$$

$$\text{Combined absorption} = a\,\text{Absorption}_A + b\,\text{Absortpion}_B + \cdots \tag{4.12}$$

**Example 4.2** The following information was obtained on a fine aggregate sample:

Mass of wet sand = 627.3 g
Mass of dry sand = 590.1 g
Absorption = 1.5%.

Calculate the total moisture content of the wet sand, and its saturated surface dry weight.

**ANSWER**

The moisture content is computed as:
(Mass of wet sand − Mass of dry sand)/Mass of dry sand = (627.3 − 590.1)/590.1 = 6.30%
The saturated surface-dry weight of the sand is computed as:
Absorption × Dry Weight + Dry Weight = 1.5/100 × 590.1 + 590.1 = 599.0 g

**Example 4.3** Given the data presented in Example 4.1, compute the bulk-specific gravity and the absorption of the blended aggregate.

**ANSWER**

The bulk-specific gravity of the blend is computed from Equation 4.11 as:

$$G = \frac{1}{\dfrac{0.75}{100 \times 2.750} + \dfrac{0.25}{100 \times 2.600}} = 2.771$$

The absorption of the blend can be found using Equation 4.12.

$$\text{Absorption}_{BLEND} = 0.75 \times 1.5 + 0.25 \times 2.5 = 1.75\%$$

**PORE STRUCTURE**

Pore structure refers to the size, volume, and shape of the void spaces within an aggregate particle.[13] It is not desirable for aggregates to have large volumes of permeable pores. Large volumes of pores make the aggregate more susceptible to degradation or breakage under the repeated cycles of freezing/thawing and/or wetting/drying. In HMAs, a large volume of permeable pores increases the absorption of binder. Also, an aggregate that is porous increases the possibility of selective absorption taking place. This phenomenon refers to the absorption of oily constituents of the asphalt into the aggregate, leaving the harder residue on the surface. As a result, it is possible for selective absorption to lead to raveling and stripping of asphalt binder from aggregate.[13]

**GEOMETRY**

A particle geometry can be fully expressed in terms of three independent properties, namely form (or shape), angularity (or roundness), and surface texture.[2] A schematic diagram that illustrates the differences between these properties is shown in Figure 4.5. Shape reflects variations in the proportions of a particle. Angularity reflects variations at the corners. Surface texture is used to describe the surface irregularity or asperities at a very small scale. Texture is a function primarily of aggregate minearlogy, while angularity is influenced by crushing techniques. As discussed in Chapter 9, these shape characteristics affect the texture and frictional characteristics of a pavement surface.

The fine aggregate angularity can be measured using the ASTM C 1252 method. This method is often referred to as the Fine Aggregate Angularity (FAA) test. It measures the loose uncompacted

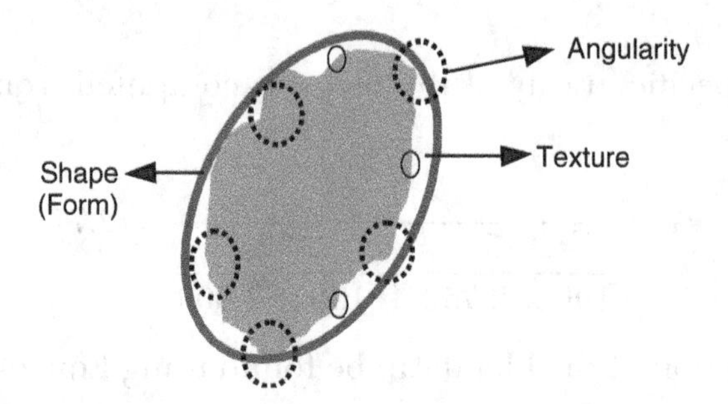


**Figure 4.5**
Components of Aggregate Shape Properties: Shape, Angularity, and Texture

porosity, (i.e., volume of voids divided by the total volume) of a sample of fine aggregate that falls from a fixed distance through a given-sized orifice. Low porosity is associated with more rounded and smooth-surfaced fine aggregates that allow closer packing of particles. This procedure is used by Superpave to determine aggregate angularity to ensure that fine aggregates have sufficient internal friction to produce rut-resistant HMAs. The apparatus used in this test method is shown in Figure 4.6.

A similar method is described in AASHTO TP 56 to measure the loose uncompacted porosity of coarse aggregates. The apparatus used in this method is shown in Figure 4.7.

The ASTM D 3398 method is used to obtain an index of overall aggregate particle shape and texture. The test is based on measuring the change in voids as the aggregate sample is compacted in a standard mold. It accounts for the combined effect of shape, angularity, and texture, as well as aggregate uniformity. The index is calculated using the following equation:

$$I_a = 1.25V_{10} - 0.25V_{50} - 32.0 \tag{4.13}$$

where $I_a$ is particle index value, $V_{10}$ is the percent of voids in the aggregate compacted with 10 blows per layer, and $V_{50}$ is the percent of voids in the aggregate compacted with 50 blows per layer.

Coarse aggregate angularity is measured using ASTM D 5821. This method is based on evaluating the angularity of an aggregate sample by visually examining each particle and counting the number of crushed faces, as illustrated in Figure 4.8. It is also the method currently used in the Superpave system for evaluating the angularity of

**Figure 4.6**
Uncompacted Void Content of Fine Aggregate Apparatus. Courtesy of FHWA

coarse aggregates used in HMAs. The percent of aggregate crushed faces (one face, two or more faces) is associated with angularity. The shape of coarse aggregates is measured by determining the percentage by number or weight of flat, elongated, or both flat and elongated particles in a given sample of coarse aggregate using the ASTM D 4791 procedure. This procedure uses a proportional caliper device, shown in Figure 4.9, to measure the dimensional ratio of aggregates. The aggregates are classified according to the undesirable ratios of width to thickness or length to width, respectively. Superpave specifications characterize an aggregate particle by comparing its length to its thickness or the maximum dimension to the minimum one.

**Figure 4.7**
Uncompacted Void Content of Coarse Aggregate Apparatus. Image courtesy of Gilson, Inc.

**Figure 4.8**
Illustration of Counting Percent of Fractured Faces. Image courtesy of Gilson, Inc.

Recently, image analysis methods have been used to quantify the shape, angularity, and texture of aggregates. These methods rely on capturing images of particles and then using mathematical functions to describe the geometry captured in these images. A system that was developed and used by different research centers

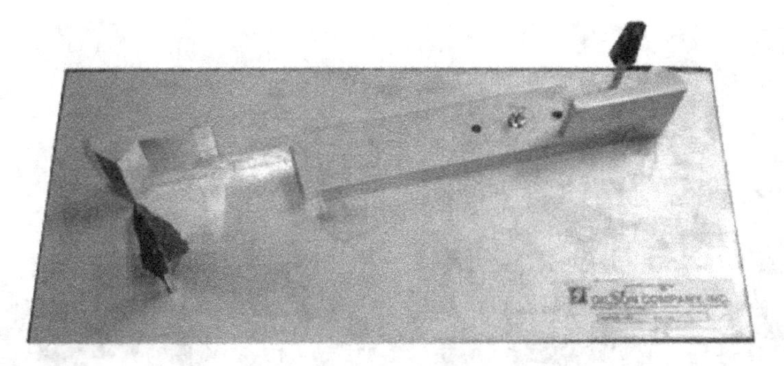

**Figure 4.9**
Flat and Elongated Coarse Aggregate Caliper

is called the Aggregate Imaging System (AIMS).[10] The first module of this system is for the analysis of fine aggregates; black-and-white images are captured using a video camera and a microscope. The second module is devoted to the analysis of coarse aggregate; gray images as well as black-and-white images are captured. Fine aggregates are analyzed for shape and angularity, while coarse aggregates are analyzed for shape, angularity, and texture. The video microscope is used to determine the depth of particles, while the images of two-dimensional projections provide the other two dimensions. These three dimensions quantify shape. Angularity is determined by analyzing the black-and-white images, while texture is determined by analyzing the gray images. AIMS is shown in Figure 4.10.

Many studies have shown that an increase in aggregate angularity and texture increase the strength and stability of HMAs.[10] In general, open-graded mixtures were found to be influenced more by the angularity and texture of aggregates than dense-graded mixtures. Also, fine aggregate angularity plays a more critical role than coarse aggregate angularity in influencing the properties of HMAs. The presence of excessive flat and elongated aggregate particles is undesirable in asphalt mixtures because they tend to break down (especially in open-graded mixtures) during production and construction, thus affecting the durability of HMAs.

Performance of PCCs is influenced by aggregate properties. The properties of the aggregate used in the concrete affect the performance of both fresh and hardened PCC. Aggregate characteristics affect the proportioning of concrete mixtures, the rheological

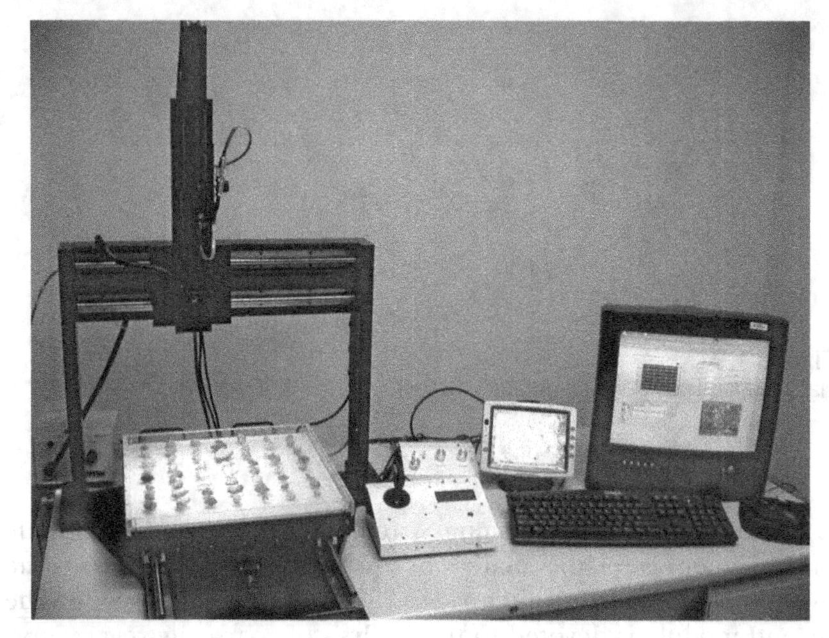

**Figure 4.10**
Aggregate Imaging System (AIMS).

properties of the mixtures, the aggregate-mortar bond, and the interlocking strength (load transfer) of the concrete joint/crack.

Meininger[11] indicated that fine aggregate content and properties mostly affect the water content needed in the concrete mix. Thus, selecting the proper fine aggregate content and proper particle shape and texture ensures a workable and easily handled mix. Coarse aggregate particle shape and angularity are related to critical performance parameters such as transverse cracking, faulting of joints, and cracks. Using a high percentage of flat and/or elongated particles might cause problems when placing the concrete, since it may result in voids and incomplete consolidation, which in the long run may result to spalling. Coarse aggregate shape, angularity, and surface texture are believed to have a significant effect on the bond strength between aggregate particles and the cement paste. Weak bonding between aggregates and mortar leads to distresses in concrete pavements, including longitudinal and transverse cracking, joint cracks, spalling, and punchouts. The increase in bond strength is a consideration in selecting aggregate for concrete, where flexural strength is important or where high compressive strength is needed.

Kosmatka et al.[7] indicated that aggregate properties (shape and surface texture) affect freshly mixed concrete more than hardened concrete. Rough-textured, angular, and elongated particles require more water to yield workable concrete than smooth and rounded aggregates. Angular particles require more cement to maintain the same water-to-cement ratio. However, with satisfactory gradation, both crushed and noncrushed aggregate of similar rock origin generally give the same strength, given the same cement amount. Another detrimental effect of angular and poorly graded aggregates is that they may be difficult to pump.

The performance of unbound granular pavement base and subbase layers is greatly affected by the properties of the aggregates used. Poor performance of unbound granular base layers can result in premature pavement surface failure in both HMAs and PCCs. Distresses in an asphalt pavement due to poor unbound layers include rutting, fatigue cracking, longitudinal cracking, depressions, corrugations, and frost heave. There is significant correlation between aggregate shape properties and the resilient modulus and shear strength properties of unbound aggregates used in base layers (e.g., reference [1]), as discussed earlier (Chapter 3).

Figure 4.10 shows the correlation between the shear strength of unbound aggregates and aggregate angularity measured using image analysis methods. The trend in this data suggests that as the angularity values increase, the angle of internal friction increases exponentially. The correlation between the deviator stress at failure and the angularity value is plotted in Figure 4.11 for the three confining pressures. As the angularity value of the unbound aggregate material increases, the deviator stress at failure also increases for each of the three confining pressures.

#### DURABILITY AND SOUNDNESS

Durability and soundness of aggregates refer to their resistance to temperature and moisture, as well as to the presence of deleterious substances such as soft particles, clay, and organic materials.[13] Clay particles can adversely affect the bond between the aggregates and asphalt, which can lead to stripping and raveling in HMAs. Also, the presence of these particles weakens the bond between aggregates and paste in PCCs, leading to cracking. The sand-equivalent test (ASTM D 2419) is used to determine the relative proportions of clay and dust in fine aggregates. In this test, a sample of fine aggregates

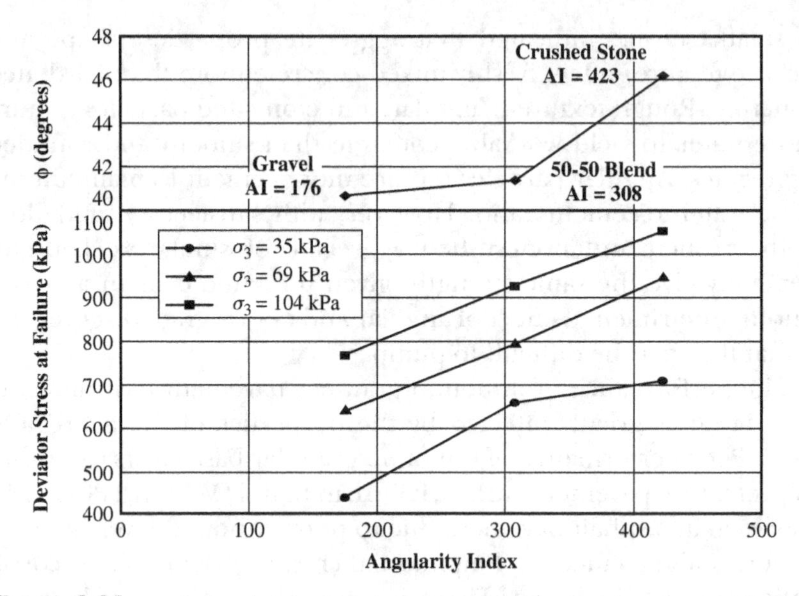


**Figure 4.11**
Correlation between Coarse Aggregate Angularity (AI) Angle of Internal Friction $\phi$ and Deviatoric Stress Failure at Different Confinement Stresses $\sigma_3$ (Ref. 14).

is agitated in water and allowed to settle. The sand particles settle and separate from the flocculated clay, and the heights of clay and sand in the cylinder are measured. The sand equivalent is the ratio of the height of sand to the height of clay times 100. The ASTM C 142 procedure (clay lumps and friable particles in aggregates) is another method that can be used to measure the cleanliness of fine aggregates. The ASTM C 40 procedure is used to determine the organic impurities in sands for concrete. This procedure is used to determine if organic impurities are present to a level that requires further tests before they are approved for use. ASTM D 3744 (aggregate durability index) is another procedure that can be used to determine the relative resistance of an aggregate in producing detrimental clay like fines when subjected to mechanical methods of degradation in the presence of water.

The coefficient of thermal expansion is defined as the change in the volume of aggregate produced by a unit change in temperature. It is desirable to have a small difference in the coefficient of thermal expansion between the aggregate parent rock minerals to reduce the differential expansion within the aggregate that can cause internal fracture. The coefficient of thermal expansion of aggregates ranges from $0.5 \times 10^{-6}$ to $9 \times 10^{-6}$ in/in/°F.[13]

Thermal conductivity is the capability of an aggregate to transmit heat. The advantage of low thermal conductivity is to decrease the depth of frost penetration through a pavement. However, there is an advantage of higher thermal conductivity in minimizing the differential temperatures between the top and bottom of PCC slabs: a uniform temperature through the slab decreases curling and cracking. Thermal compatibility between aggregate and cement paste in a PCC pavement is also an important consideration. An increase in the difference in the coefficient of thermal expansion of the aggregate and that of the cement binder increases the chances of PCC cracking.

The wetting and drying process influences the volume change of aggregates and the pavement layer in which the aggregates are used. Aggregates should exhibit little or no volume change with variations in moisture content. The freezing and thawing process is another phenomenon that could cause fracture of aggregates due to the buildup of internal stresses as a result of the increase of water volume inside the aggregate. The sulfate soundness test (ASTM C 88) is used to measure the aggregate resistance to freezing and thawing. The aggregate sample is immersed into a solution of sodium or magnesium sulfate of specified concentration for a period of time. The sample is then removed and permitted to drain, after which it is placed in an oven to dry to constant weight. The process of immersion and drying is typically repeated for five cycles. During the immersion cycle, the sulfate salt solution penetrates the aggregate. Oven-drying dehydrates the sulfate salt precipitated in the aggregate pores. The internal expansive force, due to the formation of the sulfate salt crystals, is intended to simulate the expansion of water upon freezing. The aggregate sample is washed and sieved, and the reduction in aggregate sizes due to breakage is used as the indication of durability of the aggregate.

The influence of freezing and thawing and the presence of deleterious materials cause major distresses in PCC, such as pitting, D-line cracking, and map cracking. Pitting is the result the disintegration of weak aggregates near the surface under the freezing and thawing action. D-line cracking appears initially as fine cracks near transverse joints and cracks in pavements, and progresses to become cracks near longitudinal joints and free edges. This distress is caused by the change in volume and breakage of coarse aggregates as a result of water freezing in aggregate pores. Map cracking refers to random, well-distributed cracks over the pavement surface. This distress is attributed to the expansion of aggregate particles along with shrinkage of the concrete mortar. A complete description of the pavement distresses is given in Chapter 9.

### 4.3.2 Chemical Properties

The chemical properties of aggregates influence their adhesion to asphalt. Poor adhesion of the asphalt to the aggregate in the presence of moisture leads to stripping and raveling. There are several theories that explain the asphalt-aggregate adhesion mechanism.[5] Aggregates that are susceptible to adhesion loss in the presence of moisture are typically called *hydrophilic* (exhibiting water affinity) or *acidic*.[15] Aggregates that have good adhesion with asphalt and exhibit good resistance to moisture damage are called *hydrophobic* (exhibiting water aversion) or *basic*. The nature of electric charges on the aggregate surface when in contact with water also influences the adhesion between the aggregate and the asphalt. Most siliceous aggregates (e.g., sandstone, granite, quartz, and siliceous gravel) are negatively charged in the presence of water. Other aggregates, such as limestone, exhibit a positive charge in the presence of water. A classification of aggregates based on their silica or alkaline content is shown in Figure 4.12.

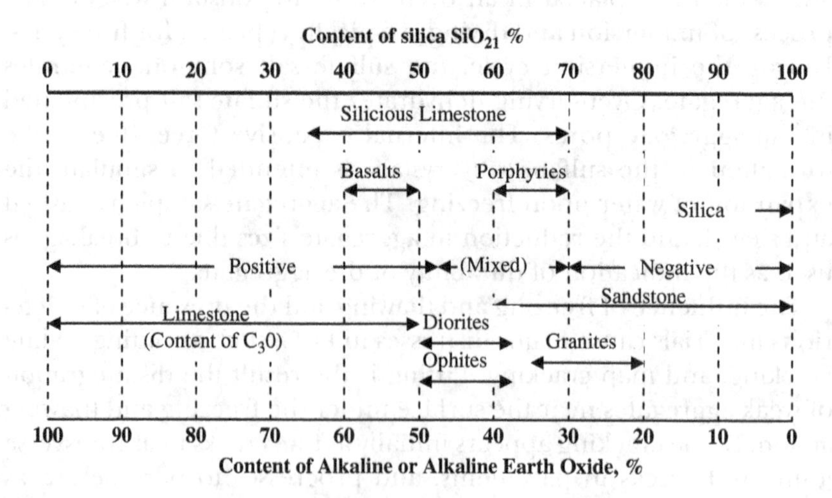


**Figure 4.12**
Classification of Aggregates by Surface Charges (Ref. 15).

Certain forms of silica and siliceous material react with the alkali released during the hydration of portland cement, which leads to the formation of a gel-like material around aggregate particles. This gel-like material expands, leading to random cracking at the concrete surface. The ASTM C 289 procedure, entitled potential reactivity of aggregates, or chemical method, is used to quickly determine the potential reactivity of an aggregate with the alkali in PCCs.

Aggregate surface energy measurements have been shown to be good indicators of the adhesion potential between aggregates and asphalt binders[3,8]. The surface-free energy of a material is defined as the amount of work required to create a unit area of a new surface of that specific material in a vacuum. The surface-free energy of a material is divided into three separate components. These components are the monopolar acidic $\gamma^+$, the monopolar basic $\gamma^-$, and the apolar, or Lifshitz-van der Waals $\gamma^{LW}$. The total surface-free energy of a material, $\gamma^{Total}$, is obtained from the three components, as shown in Equation (4.14).

$$\gamma^{Total} = \gamma^{LW} + 2\sqrt{\gamma^+\gamma^-} \tag{4.14}$$

The bond energy between the aggregate and asphalt binder without ($W_{AS}$) and with moisture ($W_{ASW}$) is calculated using Equations 4.15 and 4.16, respectively.[3,8]

$$W_{AS} = 2\sqrt{\gamma_A^{LW}\gamma_S^{LW}} + 2\sqrt{\gamma_A^+\gamma_S^-} + 2\sqrt{\gamma_A^-\gamma_S^+} \tag{4.15}$$

$$W_{ASW} = \gamma_{AW} + \gamma_{SW} - \gamma_{AS} \tag{4.16}$$

where, the subscripts $A$, $S$, and $W$ represent the asphalt binder, aggregate (stone), and water, respectively. The surface energy with two subscripts, $\gamma_{ij}$ represents the energy of the interface between any two materials, $i$ and $j$, which is computed using the surface-free energy components as follows:

$$\gamma_{ij} = \gamma_i + \gamma_j - 2\sqrt{\gamma_i^{LW}\gamma_j^{LW}} - 2\sqrt{\gamma_i^+\gamma_j^-} - 2\sqrt{\gamma_i^-\gamma_j^+} \tag{4.17}$$

### 4.3.3 Mechanical Properties

The mechanical properties of aggregates are related to their resistance to degradation due to abrasion, polishing, impact, or loading stress in the pavement layers. The nature and mineralogy of the aggregates determine the aggregate resistance to these factors. Aggregate strength and stiffness have been used as indicators of aggregate resistance to degradation. Strength is quantified by the maximum tensile or compressive stress that an aggregate sample can sustain prior to failure. Stiffness is the resistance of an aggregate particle to deformation, and it is quantified by the modulus of elasticity. Strength and stiffness are measured on cylinders or cubes cored from the parent rock. It is difficult to measure these properties on individual aggregate particles, as their irregular shape makes it impractical. An aggregate with a high degree of stiffness is desired for most construction applications. However, overly stiff aggregates can cause microcracking of the cement paste surrounding the particle during the plastic shrinkage stage of the hydration process.

Aggregates are exposed to impact and/or abrasion forces during plant operations and compaction. These forces might cause changes in aggregate size distribution, leading to field-produced mixes different from the laboratory-designed ones. Aggregate breakage can also occur under traffic loads, especially in pavement layers that rely on stone-on-stone contacts such as the open-graded friction course (OGFC), stone matrix asphalt (SMA), and unbound layers. These materials experience high stresses at the aggregate contact points, which might cause aggregate fracture and reduction in load-carrying capacity.

The Los Angeles Degradation Test (ASTM C 131) is the most widely specified test for evaluating the resistance of coarse aggregates to abrasion and impact forces. In this test, a sample of coarse aggregates with a specified gradation is placed in a steel drum along with steel balls. The drum is rotated, and the tumbling action causes abrasion between particles and between steel balls and particles. Following the completion of the revolutions, the sample is removed and sieved. The percent passing the No. 12 sieve is used as a measure of the Los Angeles degradation value for the sample.

Aggregate resistance to breakage due to impact can also be measured using the Page Impact Test (ASTM D 3). This test measures the resistance of a cylindrical core specimen to the impact of a hammer dropped freely from different heights. The height of drop in centimeters causing fracture of the specimen is reported as the toughness value.

The aggregate impact value (AIV) (British Standard: BS 812-Part-112) is also used to measure aggregate resistance to impact. A standard sample of coarse aggregates is placed in a cylindrical mold. The sample is subjected to blows from a hammer. Degradation is measured by the aggregate weight passing a British Standard 2.40-mm sieve, which corresponds to approximately the U.S. Standard No. 8 sieve size. The aggregate crushing value (ACV) (British Standard: BS 812-Part-110) is another test in which a coarse aggregate sample in a cylindrical mold is subjected to a continuous load transmitted through a piston in a compaction test machine. A load is applied gradually over a specified time. The ACV is equal to the percentage of fines created that pass the British Standard (BS) 2.40-mm sieve, expressed as a percentage of the initial percent passing value.

The Micro-Deval test (ASTM D 6928) was developed in the 1960s in France for measuring aggregate resistance to abrasion. Coarse aggregate abrasion takes place in this test through the interaction among aggregate particles and between aggregate particles and steel balls in the presence of water. Sieve analysis is conducted after the Micro-Deval test to determine the weight loss in the coarse aggregate sample as the material passing sieve No. 16 (1.18 mm). Figure 4.13 shows the components of the Micro-Deval test, while Figure 4.14 shows aggregate particles before and after abrasion in the Micro-Deval.

The resistance of aggregates to polishing influences asphalt pavement frictional resistance, also known as skid resistance. The fine aggregate fraction in PCCs is important to the friction characteristics of the surface. The friction characteristics of HMAs are influenced mostly by the coarse aggregate exposed at the surface.

The most widely used method for measuring aggregate resistance to polishing employs the British Wheel device (ASTM D 3319). Aggregate friction can be measured by the British Portable Pendulum Tester (ASTM E 303) device (Figure 4.15). This device involves a rubber slider at the end of a pendulum. The friction, hence the degree of polishing, is measured by the height reached by the slider as it swings past the point of contact with the test specimen. Figure 4.16 shows an example of aggregate coupons before and after polishing.

Many studies have shown that the measurements of the British pendulum are a function of many other factors besides aggregate

**Figure 4.13**
Micro-Deval Test Components. Image courtesy of Gilson, Inc.

**Figure 4.14**
Aggregate Particles (a) before Abrasion in the Micro-Deval, and (b) after Abrasion in the Micro-Deval

texture. These factors include coupon curvature, aggregate arrangement, and aggregate size. Kandhal et al.[6] and Mahmoud[9] indicated that the measurements are within a narrow range, which makes it difficult to distinguish among aggregates based on their polishing resistance.

Crouch and Dunn[4] developed the Micro-Deval Voids at nine hours (MDV9) to evaluate aggregate resistance to polishing. In this

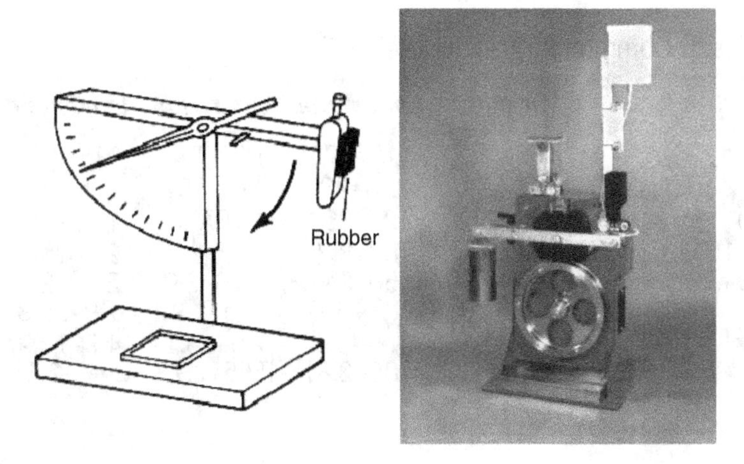


**Figure 4.15**
British Pendulum Tester and British Polishing Wheel

**Figure 4.16**
Aggregate Coupons (a) before Polishing and (b) after Polishing Using the British Wheel.

method, the Micro-Deval apparatus is used to polish an aggregate sample for nine hours. Then the voids in an uncompacted sample of aggregates are used to measure their packing, as reflected by their void ratio. The lower the void ratio, the smoother the polished aggregates, hence the lower their polishing resistance. Mahmoud[9] used the Micro-Deval to polish an aggregate sample, and measured the resulting loss in texture using the Aggregate Imaging System (AIMS).

**Table 4.5**
Typical Physical Properties of Common Aggregate (Ref. 13)

| *Property* | *Granite* | *Limestone* | *Quartzite* | *Sandstone* |
|---|---|---|---|---|
| Unit Weight (lbs/in$^3$) | 162–172 | 117–175 | 165–170 | 119–168 |
| Compressive Strength ( $\times$ 10$^3$ lbs/in$^2$) | 5-67 | 2.6–28 | 16–45 | 5-20 |
| Tensile Strength (lbs/in$^2$) | 427–711 | 427–853 | NA* | 142–427 |
| Modulus of Elasticity ( $\times$ 10$^6$ lbs/in$^2$) | 4.5–8.7 | 4.3–8.7 | NA* | 2.3–10.8 |
| Water Absorption (% by weight) | 0.07–0.30 | 0.50–24.0 | 0.10–2.0 | 2.0–12.0 |
| Average. Porosity (%) | 0.4–3.8 | 1.1–31.0 | 1.5–1.9 | 1.9–27.3 |
| Linear Expansion ( $\times$ 10$^{-6}$ in/in/°C) | 1.8–11.9 | 0.9–12.2 | 7.0–13.1 | 4.3–13.9 |
| Specific Gravity | 2.60–2.76 | 1.88–2.81 | 2.65–2.73 | 2.44–2.61 |

*NA = Data not available.

The ASTM D 3042 procedure is used to determine the percentage of insoluble residue in carbonate aggregates (e.g., limestone, dolomite). The test is primarily used to identify carbonate aggregates that are prone to polishing, hence should be avoided in constructing HMA friction courses. In this test, a hydrochloric acid is added to an aggregate sample and agitated. Then the diluted solution and residue are washed from the aggregates. The diluted solution is discarded, and the insoluble residue is dried, sieved, and weighted. The higher the weight of the insoluble residue, the higher the resistance of aggregates to polishing. Shown in Table 4.5 are typical values for some of the aggregate properties discussed in this chapter.

# References


1. Barksdale, R. D., and Itani, S. Y. (1994). ''Influence of Aggregate Shape on Base Behavior,'' Transportation Research Record 1227, Transportation Research Board, National Research Council, Washington, DC, pp. 171–182.
2. Barrett, P. J. (1980). The Shape of Rock Particles: A Critical Review, *Sedimentology*, Vol. 27, pp. 291–303.
3. Bhasin, A., Masad, E., Little, D., and Lytton, R. (2006). ''Limits on Adhesive Bond Energy for Improved Resistance of Hot-Mix Asphalt to Moisture Damage,'' in Transportation Research Record 1970, *Journal of the Transportation Research Board*, pp. 3–13.

[4] Crouch, L., and Dunn, T. (2005). "Identification of Aggregates for Tennessee Bituminous Surface courses," Tennessee Department of Transportation TDOT Project Number TNSPR-RES1149, *Final Report*, Cookeville, TN.

[5] Hefer, A. W., Little, D. N., and Lytton, R. L. (2005). "A Synthesis of Theories and Mechanisms of Bitumen-Aggregate Adhesion, Including Recent Advances in Quantifying the Effects of water," *Journal of the Association of Asphalt Paving Technologists*, Vol. 74, pp. 139–192.

[6] Kandhal, P., Parker, F. Jr., and Bishara, E. (1993). "Evaluation of Alabama Limestone Aggregates for Asphalt-Wearing Courses," Transportation Research Record 1418, Transportation Research Board, Washington, DC, pp. 12–21.

[7] Kosmatka, S. H., Kerkhoff, B., and Panarese, W. C. (2002). *Design and Control of Concrete Mixtures*, 14th edition, Portland Cement Association, Skokie, IL.

[8] Little, D. N., and Bhasin, A. (2007). *Using Surface Energy Measurements to Select Materials for Asphalt Pavement*, Texas Transportation Institute, Draft Final Report Under Project 9-37, submitted to National Cooperative Highway Research Program, Washington, DC.

[9] Mahmoud, E. (2005). "Development of Experimental Methods for the Evaluation of Aggregate Resistance to Polishing, Abrasion, and Breakage," "MS Thesis", Texas A&M University, College Station, TX.

[10] Masad, E., Al-Rousan, T., Button, J., Little, D., and Tutumluer, E. (2007). *Test Methods for Characterizing Aggregate Shape, Texture, and Angularity*, Report Number 555, National Cooperative Highway Research Program, National Research Council, Washington, DC.

[11] Meininger, R. C. (1998). "Aggregate Test Related to Performance of Portland Cement Concrete Pavement," National Cooperative Highway Research Program Project 4-20A Final Report. Transportation Research Board, National Research Council, Washington, DC.

[12] Mertens, E. W., and Borgfeldt, J. J. (1965). "Cationic Asphalt Emulsions," *Bituminous Materials: Asphalts, Tars, and Pitches*, A. J. Hoiberg, ed., Vol. 2, No. 1, Interscience Publishers, John Wiley & Sons, Inc., New York.

[13] NSSGA (2001). *The Aggregate Handbook*, R. D. Barksdale, ed., National Stone, Sand, and Gravel Association, Arlington, VA.

[14] Rao, C., Tutumluer, E., and Kim, I. T. (2002). "Quantification of Coarse Aggregate Angularity Based on Image Analysis," Transportation Research Record 1787, Transportation Research Board, National Research Council, Washington, DC, pp. 117–124.

[15] Roberts, F. L., Kandhal, P. S., Brown, E. R., Lee, D., and Kennedy, T. (1996). *Hot-Mix Asphalt Materials, Mixture Design, and Construction*, NAPA Education Foundation, Lanham, MD.

## Problems

4.1 Name three common minerals and identify their geological origin. Which one is more porous, and why?

4.2 Name three common types of igneous rocks used as aggregates.

4.3 Given the following measurements on a sample of fine aggregate, calculate the bulk dry, bulk SSD, and apparent specific gravities:

Aggregate saturated surface dry (SSD) weight = 459.34 g
Weight of pyenometer and water = 2345.67 g
Weight of pyenometer, water, and sample = 2640.35 g
Aggregate weight after being dried in oven = 454.12 g

4.4 Plot typical gradation curves for two samples, one with a fineness modulus of 2.2 and the other of 3.2.

4.5 Calculate and plot the gradation of the sieve analysis data shown in Table 4.6 on a semilog plot.

4.6 Repeat the solution to problem 4.5, using the Fuller-Thompson approach. Also plot the Fuller line using an exponent of 0.45. How dense do you think this aggregate packs?

4.7 Table 4.7 shows the grain size distribution for two aggregates and the specification limits for an asphalt concrete. Determine the blend proportion required to meet the specification and the gradations of the blend. On a semilog gradation graph, plot the gradations of aggregate A, aggregate B, the selected blend, and the specification limits.

**Table 4.6**
Data for Problem 4.5

| Sieve Size | Amount Retained, g | Cumulative Amount Retained, g | Cumulative Percent Retained | Percent Passing |
|---|---|---|---|---|
| 25 mm (1 in.) | 0 | | | |
| 9.5 mm (3/8 in.) | 35.2 | | | |
| 4.75 mm (No. 4) | 299.6 | | | |
| 2.00 mm (No. 10) | 149.7 | | | |
| 0.425 (No. 40) | 125.8 | | | |
| 0.075 mm (No. 200) | 60.4 | | | |
| Pan | 7.3 | | | |

**Table 4.7**
Data for Problem 4.7

| | 19 mm | 12.5 mm | 9.5 mm | 4.75 mm | 2.36 mm | 0.6 mm | 0.30 mm | 0.15 mm | 0.075 mm |
|---|---|---|---|---|---|---|---|---|---|
| Specification Limits | 100 | 80–100 | 70–90 | 50–70 | 35–50 | 18–29 | 13–23 | 8–16 | 4–10 |
| Aggregate A | 100 | 85 | 55 | 20 | 2 | 0 | 0 | 0 | 0 |
| Aggregate B | 100 | 100 | 100 | 85 | 67 | 45 | 32 | 19 | 11 |

4.8 Laboratory measurements of the specific gravity and absorption of two coarse aggregate sizes are:

Aggregate A: Bulk dry-specific gravity = 2.81; absorption = 0.4%

Aggregate B: Bulk dry-specific gravity = 2.44; absorption = 5.2%

What is the average specific gravity of a blend of 50% aggregate A and 50% aggregate B by weight, and what is its average absorption?

# 5 Asphalt Materials

## 5.1 Introduction

The hot-mixed asphalt (HMA) used in constructing the surface layer of asphalt concrete pavements consists of asphalt binder, aggregates, and, in some cases, chemical additives. Asphalt binder is a blend of hydrocarbons of different molecular weights. It is the product of the distillation of crude oil. The chemical additives are usually used to enhance the mixture resistance to some pavement distresses, such as moisture susceptibility, rutting, or fatigue cracking. Around 30 million tons of asphalt are used in highway construction in the United States every year.

Aggregate properties have been discussed in Chapter 4. This chapter discusses the chemical, physical, and mechanical properties of asphalt binders and asphalt mixtures that influence the design and performance of asphalt concrete pavements. Included here is a brief description of the state-of-the-art tests used for measuring these properties. More details on these tests can be found in the pertinent ASTM and AASHTO standards.

## 5.2 Chemical Composition of Asphalt Binders

The purpose of this section is twofold: first, to introduce the chemistry of asphalt binders at the molecular level and the chemistry of the interactions among these molecules, and, second, to present a model for the chemical composition of asphalt binders. The main sources for the information presented are Robertson[18] and the Western Research Institute (WRI).[20]

### 5.2.1 Asphalt Chemistry at the Molecular and Intermolecular Levels

At the molecular level, asphalt consists of compounds called hydrocarbons, made of hydrogen and carbon atoms. Asphalt molecules have: (1) an aliphatic structure of straight or branched chains; (2) an unsaturated ring or aromatic structure; or (3) saturated rings or branches, which have the highest hydrogen-to-carbon ratio. Examples of these structures are given in Figure 5.1. The atoms within asphalt molecules are held together by strong covalent bonds.

In addition to hydrocarbons, asphalt includes heteroatoms such as nitrogen, sulfur, oxygen, and metals. Although these heteroatoms exist in small percentages compared to the hydrocarbons, they influence the interactions among molecules and asphalt properties. Distribution of metals such as vanadium, nickel, and iron depends on the crude oil source. As such, these metals can be used to identify

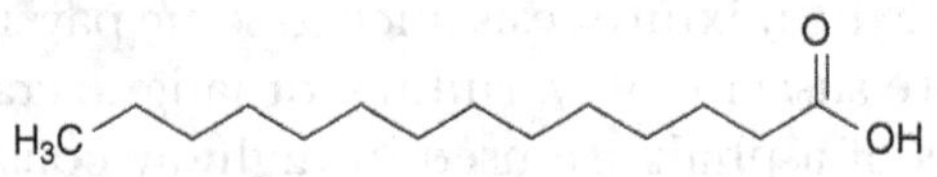

(a) Aliphatic Molecule

(b) Saturated Molecule with Branched Structure

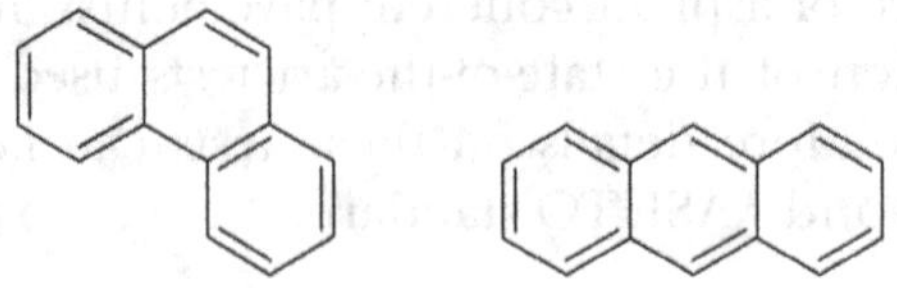

(c) Aromatic Molecules

**Figure 5.1**
Examples of Asphalt Molecules

**Table 5.1**
Elemental Analyses of Representative Petroleum Asphalts (Ref. 15)

| Elements | B-2959 Mexican Blend | B-3036 Arkansas-Louisiana | B-3051 Boscan | B-3602 California |
|---|---|---|---|---|
| Carbon, percent | 83.7 | 85.78 | 82.90 | 86.77 |
| Hydrogen, percent | 9.91 | 10.19 | 10.45 | 10.93 |
| Nitrogen, percent | 0.28 | 0.26 | 0.78 | 1.10 |
| Sulfur, percent | 5.25 | 3.41 | 5.43 | 0.99 |
| Oxygen, percent | 0.77 | 0.36 | 0.29 | 0.20 |
| Vanadium, ppm | 180 | 7 | 1380 | 4 |
| Nickel, ppm | 22 | 0.4 | 109 | 6 |

the binder source. Table 5.1 shows elemental compositions of a number of asphalts.

The distribution of heteroatoms causes an asymmetric charge distribution or polarity in the molecules. The molecules remain neutral in terms of the overall charge, but the center of electron density is different from the center of the positive charge. This polarity increases the interaction and association among the molecules to form large groups of connected or associated molecules. This interaction can occur in the form of moderately strong electrostatic forces among the polar part of the molecules or through weak forces among the nonpolar part of the molecules called Vander Waal's forces. Therefore, an increase in temperature or the application of stress influence more the weakening of these intermolecular bonds than the intramolecular covalent bonds.

Oxidation has a significant effect on the asphalt structure. It changes the chemical structure of the molecules (i.e., changes the covalent bonds) and increases the polarity of the molecules. Therefore, it promotes greater association among molecules, leading to a more brittle asphalt structure. The level of association of molecules affect the binder sensitivity to changes in temperature (temperature susceptibility) and its sensitivity to changes in shear stress (stress susceptibility).

### 5.2.2 Asphalt Chemical Model

Several models are available for describing the asphalt binder structure. The model presented by WRI[20] consists of polar and nonpolar molecules. The polar molecules tend to associate strongly and are dispersed in the continuous phase of the relatively nonpolar portion

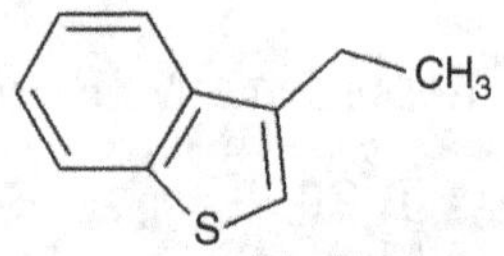

**Figure 5.2**
Example of Resin Molecule

of the asphalt. As the temperature rises, associations of the least polar molecules decrease, and the material becomes less viscous. As the temperature rises higher, the higher-polarity molecules disassociate. A decrease in temperature causes greater association among asphalt molecules, leading to higher resistance to flow.

The polar component of the model is matrix-structured. Each matrix consists of a nucleus that has mostly an aromatic structure of highly polar and associated molecules (asphaltene), surrounded by aromatic molecules with branches (adsorbed resin), as shown in Figure 5.2, and an immobilized solvent. This matrix structure follows the model presented by Pfeiffer and Saal,[16] as shown in Figure 5.3. The matrices are dispersed in a relatively nonpolar solvent, which is referred to as *maltene*. The maltene consists of the free resin (not adsorbed to the asphaltene) and oils. Oils consist primarily of nonpolar aliphatic and saturated molecules.

Asphaltenes are generally dark brown solids; resins are generally dark and semisolid; oils are usually colorless or white liquids. Several methods have been used to separate asphalt into these different chemical fractions. However, the proportions of these fractions have been found to depend on the method and the chemical used to separate the groups. For example, when n-heptane is used to separate the asphalt, the highly aromatic nucleus precipitate is referred to as *n-heptane asphaltene*; when iso-octance is used, the nucleus and the adsorbed resin with the matrix precipitate is referred to as *iso-octane asphaltene*.

The amounts of adsorbed resin and immobilized solvent depend on temperature and applied stresses. At mix-plant temperatures, the asphaltene molecular associations are stabilized by fairly small amounts of adsorbed resins. As the temperature is lowered, more resin adheres to the asphaltene, causing an increase in its volume and in the amount of immobilized solvent.[20] In addition to the increase of the volume of the matrix, a floc formation occurs in which more solvent is immobilized between the matrices, according

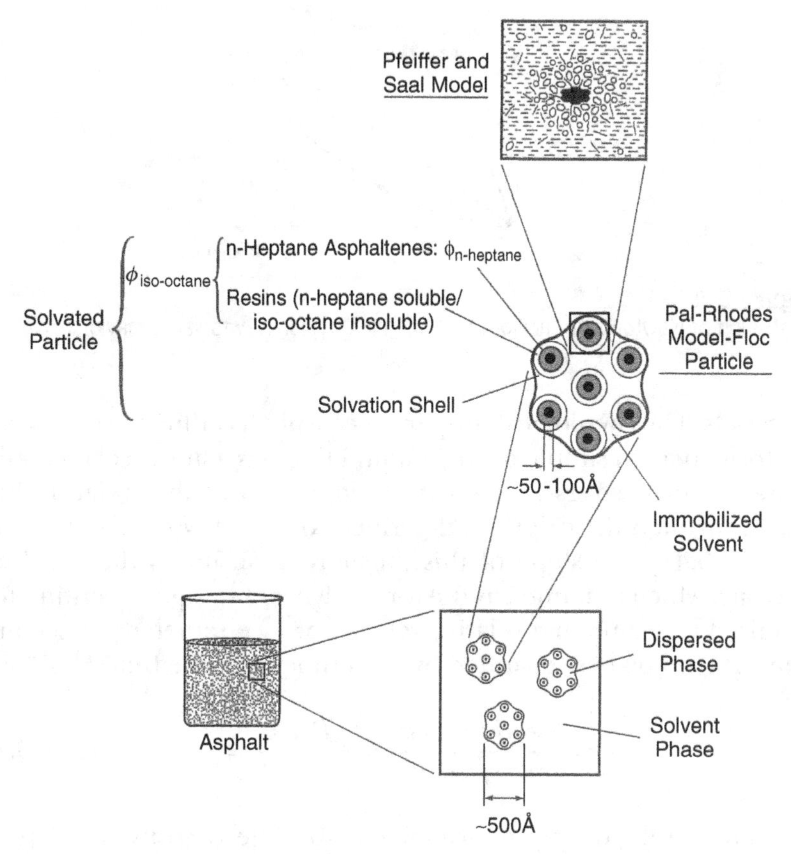


**Figure 5.3**
Schematic of Asphalt Chemical Structure (Ref. 20)

to the Pal-Rhodes model (Figure 5.3). The result is a reduction in the amount of resin in the free solvent and an increase in the amount of resin associated with the asphaltene. As temperature increases, or shear forces are applied, more resins are dissolved and more solvent joins the surrounding free solvent.

## 5.3 Preliminaries on Rheology and Viscoelasticity

### 5.3.1 Newtonian versus Non-Newtonian Behavior

It is useful to discuss the principles of rheology and linear viscoelasticty as they form the basis for understanding the behavior of asphalt binders and mixtures. Rheology is the study of the flow properties of

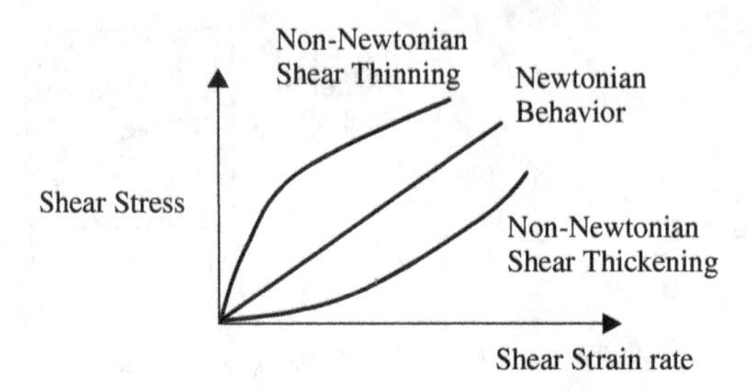


**Figure 5.4**
Illustration of Newtonian Behavior and Examples of Non-Newtonian Behavior

materials. The rheological properties of asphalts influence pavement performance. In simple terms, a liquid that has a linear relationship between shear stress and shear strain rate, and that relationship passes through the origin, is described to have a *Newtonian behavior* (Figure 5.4). The slope of this linear relationship is the absolute viscosity, which is defined as the force (dynes) per square centimeter required to maintain a relative velocity of 1 centimeter per second between two planes separated by 1 centimeter of the fluid:[14]

$$\eta = \frac{\textit{shearing stress}}{\textit{rate of shear}} = \frac{(F/A)}{(dv/dx)} = \frac{\tau}{\dot{\gamma}} \tag{5.1}$$

The linear relationship indicates that absolute viscosity is independent of shearing rate. Absolute viscosity ($\eta$) is measured by the units of Pascal.second (Pa.s), or poise, where 1 Pa.s is equal to 10 poises. In a Newtonian liquid, all the applied energy is dissipated in heat, and the liquid does not go back to its original condition when the load is removed.

The term *non-Newtonian behavior* is used to describe viscosity that is a function of shear strain rate, or a relationship between stress and shear rate that does not go through the origin. Examples of non-Newtonian behaviors are shown in Figure 5.4.

## 5.3.2 Linear Viscoelasticity

In asphalt testing and specifications, the non-Newtonian behavior is studied using linear viscoelasticity theory. A linear viscoelastic behavior indicates that the material properties change as a function of shearing rate and temperature. A viscoelastic material combines the behavior of an elastic solid and a viscous liquid. The elastic solid is represented by Hooke's law, $\tau = G\gamma$, where $G$ is the shear modulus; the viscous liquid is represented by the Newtonian viscous relationship $\tau = \eta\dot{\gamma}$, where $\eta$ is the viscostiy. The proportions of the elastic and viscous components depend on the rate of shear and temperature. The elastic behavior increases and the viscous behavior decreases as the temperature decreases and shearing rate increases.

The discussion here is limited to linear viscoelastic behavior, which means that the material properties are assumed independent of the applied stress or strain levels. The linear viscoelastic behavior under one-dimensional loading is described by its response to relaxation, creep, or steady state dynamic loading. Under a creep loading, a specimen is subjected to a constant stress, and strain is measured as a function of time. The ratio of strain $\varepsilon(t)$ to the constant stress $\sigma_o$ is the creep compliance.

$$D(t) = \frac{\varepsilon(t)}{\sigma_o} \tag{5.2}$$

Of course, creep compliance can be measured under shear loading or axial loading. It is customary to refer to the shear creep compliance as $J(t)$, and to use $D(t)$ to refer to the axial creep compliance. The response of a linear viscoelastic material under creep loading is depicted in Figure 5.5. It should be noted that a viscoelastic material can recover all the strain after the elapse of certain time, a behavior that solidlike viscoelastic materials exhibit; or it can maintain some residual strain, a behavior that liquidlike materials exhibit.

In a relaxation experiment, the material is subjected to a constant strain, $\varepsilon_o$, and stress is measured as a function to time, $\sigma(t)$. The ratio of stress to strain is referred to as the relaxation modulus.

$$E(t) = \frac{\sigma(t)}{\varepsilon_o} \tag{5.3}$$

It is also customary to denote the shear relaxation modulus as $G(t)$, and denote the axial modulus as $E(t)$. The relation between these two moduli is analogous to Hooke's law, as follows:

$$G(t) = \frac{E(t)}{2(1 + \mu(t))} \tag{5.4}$$

where $\mu(t)$ is the Poisson's ratio, which is a function of time for viscoelastic materials. The response of viscoelastic materials under relaxation loading is illustrated in Figure 5.6. It is noted that the

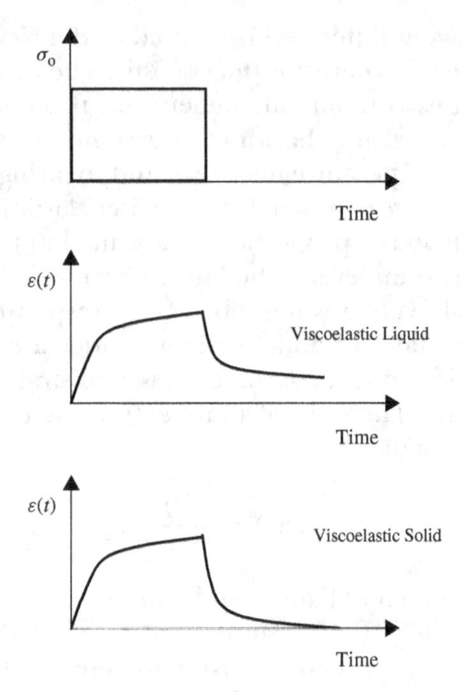


**Figure 5.5**
Creep Behavior of Viscoelastic Materials

stress required to maintain a constant strain decreases with time; it reaches zero after the lapse of certain time for liquidlike materials, and it reaches a constant value for solidlike viscoelastic materials.

Under a sinusoidal dynamic stress loading, the response has also a sinusoidal shape, but the strain response will lag behind the stress. The ratio between the stress amplitude ($\sigma_o$) and strain amplitude ($\varepsilon_o$) is known as the *dynamic modulus,* which is denoted as $|E^*|$ for axial loading or $|G^*|$ for shear loading. The inverse of the modulus is called the *dynamic creep compliance,* which is referred to as $|D^*|$ for axial loading or $|J^*|$ for shear loading. The angle that describes the lag between the stress and strain is referred to as the *phase angle* ($\delta$). An increase in phase angle indicates an increase in the viscous behavior of the material and a decrease in the elastic behavior. A Newtonian viscous liquid has $\delta$ value of $90^\circ$, while an elastic solid has a $\delta$ value of $0^\circ$. Asphalt binders have a phase angle between these two extremes. The dynamic modulus and phase angle are the two fundamental properties needed to describe the viscoelastic response under dynamic harmonic loading. These two properties

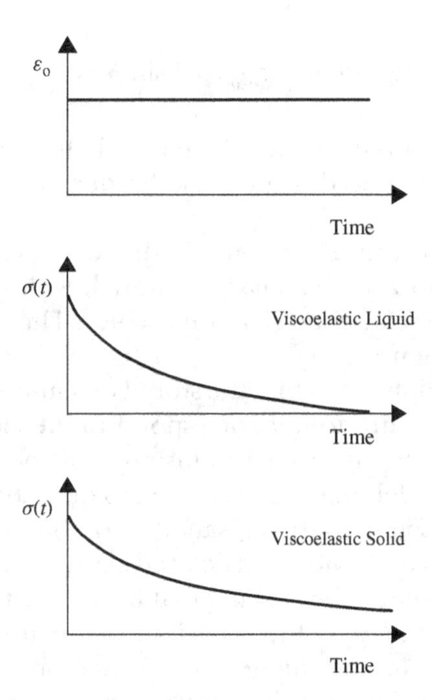


**Figure 5.6**
Relaxation Behavior of Viscoelastic Materials

are functions of the frequency of loading, which can be described by frequency ($f$) in Hertz (Hz) or angular frequency ($\omega$) in rad/s.

The dynamic properties can also be described by the storage modulus $G'$ and loss modulus $G''$ which are defined in Equations 5.5 and 5.6, respectively.

$$G' = |G^*| \cos \delta \tag{5.5}$$

$$G'' = |G^*| \sin \delta \tag{5.6}$$

There are exact and approximate methods to relate the dynamic properties to the relaxation modulus and the creep compliance. The presentation of these relationships is, however, beyond the scope of this text. The ratio of the shear dynamic modulus to the radial frequency, $\omega$, is equivalent to the apparent viscosity or shear dependent viscosity ($|\eta^*| = |G^*|/\omega$).

The supplied energy during loading is divided into a stored energy and dissipated or loss energy. The loss energy in loading cycle $i$ is given in Equation 5.7.

$$\Delta W_i = \pi \tau_{oi} \gamma_{oi} \sin \delta = \pi \gamma_{oi}^2 |G^*| \sin \delta = \frac{\pi \tau_{oi}^2}{|G^*|/\sin \delta} \tag{5.7}$$

The significance of the dissipated energy will become apparent with some of the indices used to describe binder behavior later in this chapter.

As can be seen from Equation 5.7, the loss energy is the area in the loop equal to zero for elastic material, with $\delta = 0°$; and it is maximum for a viscous material with $\delta = 90°$. The dissipated energy is illustrated in Figure 5.7.

It should be emphasized that the stored dynamic modulus and the loss dynamic modulus do not correspond to the elastic and viscous energy, respectively. In order to illustrate this point, consider the simple Burger model that has been used to describe the behavior of asphalt binders (Figure 5.8). This model combines elastic behavior, Newtonian viscous behavior, and delayed elastic behavior. The elastic behavior is represented by a spring that has an elastic modulus $E$ (or a creep compliance $D = 1/E$), which is not a function of time—or, in other words, gives an instantaneous deformation and recovery. The Newtonian viscous behavior is represented by a dashpot connected in a series to the spring. The material property that describes the behavior of the dashpot is absolute viscosity $\eta$ or fluidity $(\phi = 1/\eta)$. The spring has a $\delta$ value of 0°; the dashpot has a $\delta$ value of 90°. The delayed elastic is represented by a parallel configuration of a spring and a dashpot. This delayed elastic comes from the parallel configuration that indicates that the spring and dashpot should have the same strain. Therefore, a compromise has to be reached between the spring that has instantaneous response and the dashpot that has a time-dependent response. The compromise is a retardation of the spring deformation and a time-dependent response that is faster than that of the dashpot, but of course slower than the instantaneous response of the spring.

Under dynamic loading, the stored and loss compliances for a Burger model are given in Equations 5.8 and 5.9.

$$J' = \frac{J}{1 + \tau_v{}^2 \omega^2} + J_g \tag{5.8}$$

$$J'' = \frac{J \tau_v \omega}{1 + \tau_v{}^2 \omega^2} + \frac{\phi_f}{\omega} \tag{5.9}$$

where $\tau_v = \frac{\phi}{J}$.

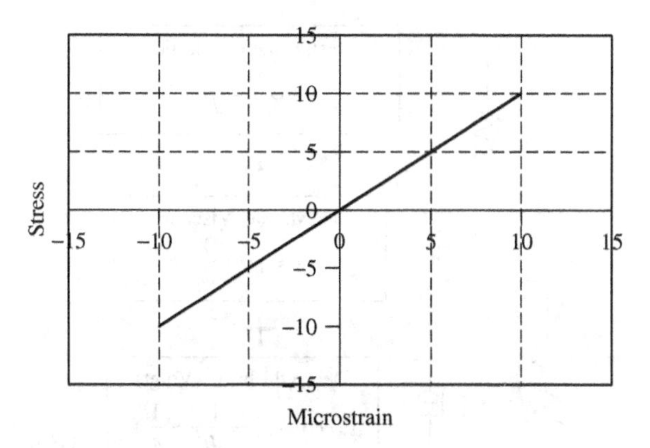


(a) Elastic Response ($\delta = 0°$)—No Dissipated Energy

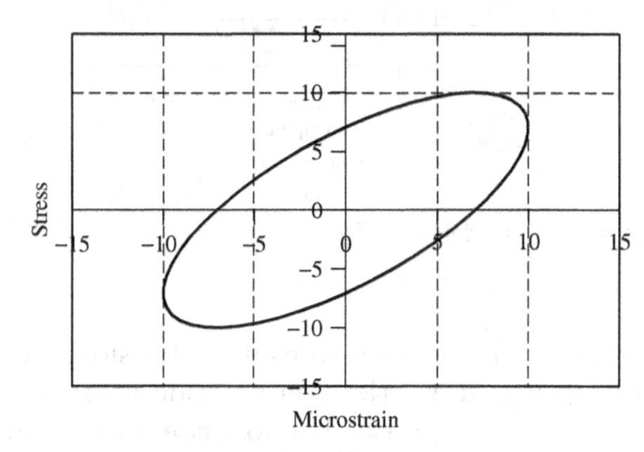


(b) Viscoelastic Response ($\delta = 45°$)

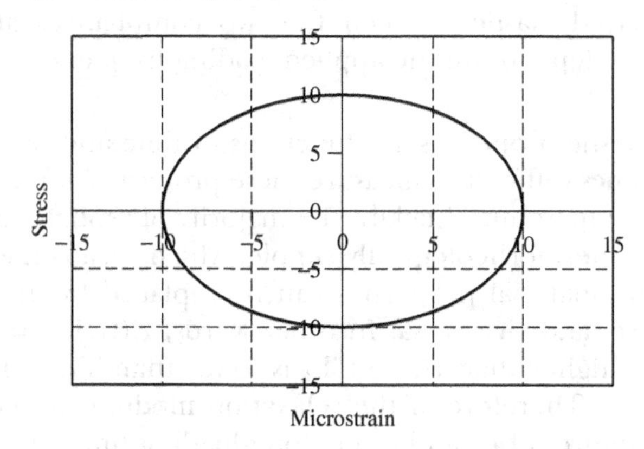


(c) Viscoelastic Response ($\delta = 90°$)

**Figure 5.7**
Dissipated Energy for Elastic, Viscoelastic, and Viscous Materials

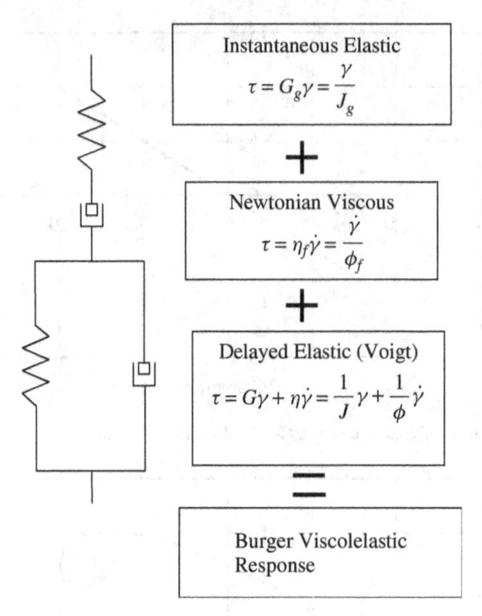


**Figure 5.8**
Description of the Burger Model

As can be seen in these relationships, the stored compliance combines the spring in a series (pure elastic response) and part of the delayed elastic response. The loss compliance includes the dashpot in a series and part of the delayed elastic. The proportion of the delayed elastic between the two compliances at a given temperature depends on the applied loading frequency.

### 5.3.3 Time-Temperature Superposition

The viscoelastic properties are functions of time and temperature. It is sometimes difficult to measure these properties for a long time at a given temperature. Luckily, the majority of asphalt binders are considered thermorheologically simple, which means the effect of time on the material properties can be replaced by the effect of temperature, and vice versa. In other words, $E(t_1, T_1) = E(t_2, T_2)$, where $t_2$ is higher than $t_1$, and $T_2$ is lower than $T_1$, as illustrated in Figure 5.9. Therefore, if the relaxation modulus is needed at a low temperature, $(T_2)$, and a very long loading time, $(t_2)$, one can increase the temperature to $T_1$ and decrease the loading time to $t_1$ to obtain the same modulus. This feature is very important to obtain

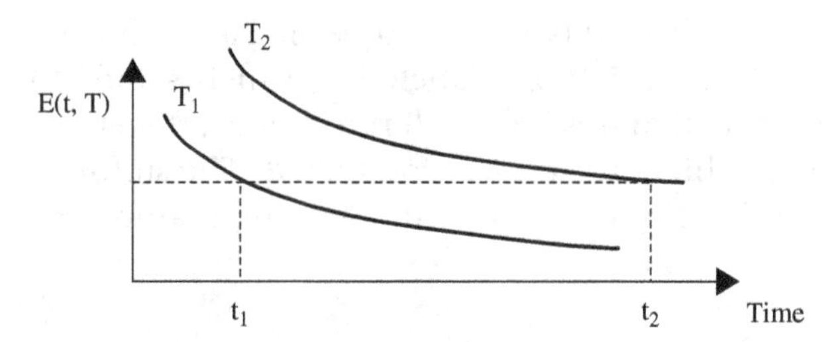


**Figure 5.9**
Illustration of the Time-Temperature Concept

the viscoelastic material properties at a wide range of temperatures and loading times.

The time-temperature superposition is applied to experimental measurements of viscoleastic properties (relaxation modulus, creep compliance, dynamic modulus, phase angle) at a wide range of temperatures and loading times (or frequencies) to obtain two relationships:

- A *master curve* that describes the viscoelastic property at a reference temperature and over a range of time or frequency
- A *shift function-temperature curve* that describes the ratio between the actual time at which the test was conducted and the reference time to which the data is shifted versus the temperature at the actual time

The master curve allows the estimation of mechanical properties over a wide range of temperatures and times (or frequencies), which could be realized in the field but are not practical to simulate in the laboratory. The concept of establishing the master curve and shift functions is illustrated next by assuming that the relaxation modulus follows a simple power law:

$$E(t) = E_1 t^{-m} \tag{5.10}$$

Taking the log of both sides of Equation 5.10 converts it to a linear function:

$$\log(E(t)) = \log E_1 - m \log t \tag{5.11}$$

If the relaxation modulus is conducted at different temperatures, the curves shown in Figure 5.10a will be produced. The curves in Figure 5.10a can be shifted horizontally to a reference temperature such as $T_2$, which results in the curve in Figure 5.10b. Note that the $x$-axis in Figure 5.10b is labeled $t_r$, which stands for reduced time, to indicate that this is the reference time after shifting, not the actual time at which the test was conducted. The shifting function is $a_T = t/t_r$, and is plotted as a function of temperature (Figure 5.11).

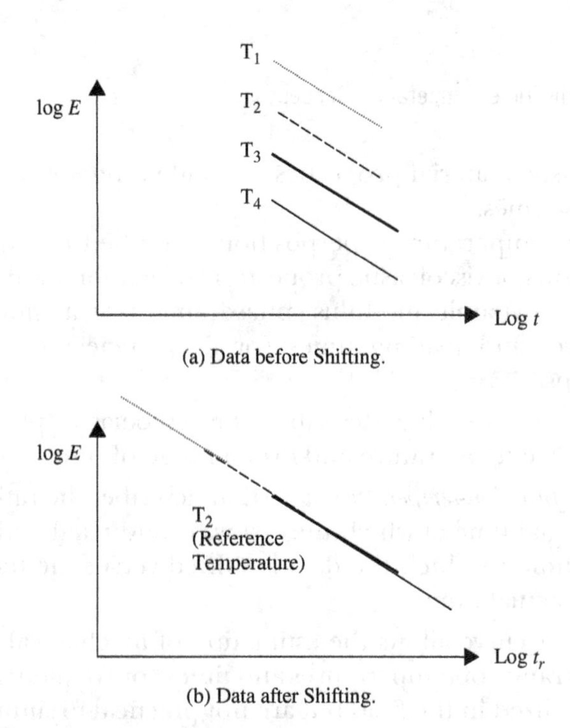


**Figure 5.10**
Data Shifting to Construct Master Curve

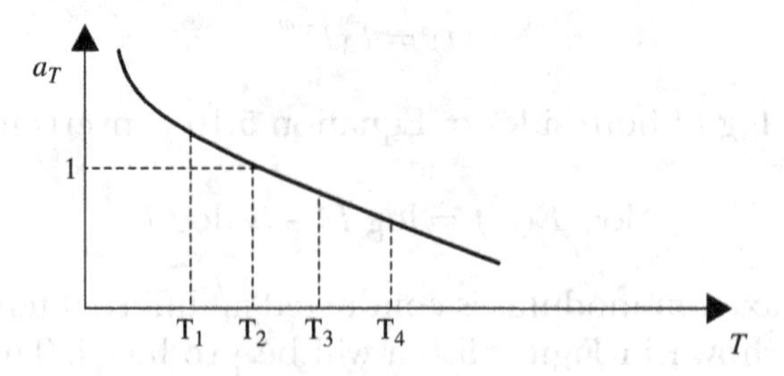


**Figure 5.11**
Shift Factor versus Temperature

## 5.4 Asphalt Binder Properties

### 5.4.1 Viscosity

Viscosity is an important rheological property for measuring the consistency of asphalt. For a Newtonian liquid, viscosity is independent of shearing rate. Asphalt binders are typically considered to exhibit Newtonian behavior at mixing temperature, which is around 160°C. However, as the temperature decreases, asphalts start to exhibit non-Newtonian behavior, where by the viscosity becomes dependent on the shear strain rate. The temperature at which this occurs depends on binder type.

There are different methods for measuring asphalt binder viscosity. Absolute viscosity can be measured at a given temperature using tube viscometers by timing the binder flow between two marks (ASTM D 2171 or AASHTO T 202). This time is multiplied by a viscometer calibration factor to obtain absolute viscosity. As discussed earlier, absolute viscosity is expressed in Pa.s (1 Pa.s = 10 poise). The viscosity is measured at 140°F (60°C). At this temperature, a binder is too viscous to flow; therefore, a vacuum pressure is applied to cause asphalt to flow.

Viscosity is also measured at 275°F (135°C), which is a high enough temperature for asphalt to flow under gravitational forces without the need to apply a vacuum pressure (ASTM D 2170 or AASHTO T 201). This measured quantity is called *kinematic viscosity*, which quantifies the resistance to flow under gravity. Kinematic viscosity is equal to the viscosity in Pa.s divided by the density in $\text{kg/m}^3$. Therefore, the unit of kinematic viscosity is:

$$\begin{aligned} \textit{Unit of Kinematic Viscosity} &= \text{Pa s}/\frac{\text{kg}}{\text{m}^3} = \frac{\text{N s}}{\text{m}^2}/\frac{\text{kg}}{\text{m}^3} \\ &= \frac{\text{kg m s}}{\text{m}^2\,\text{s}^2}/\frac{\text{kg}}{\text{m}^3} = \text{m}^2/\text{s} \end{aligned}$$

The unit Stoke = 1 $\text{cm}^2/\text{s}$ or centiStoke (cSt) = 1 $\text{mm}^2/\text{s}$ is typically used to express kinematic viscosity.

The rotational viscometer test (ASTM D 4402 or AASHTO T 316) is another method for measuring viscosity (Figure 5.12). The test is performed by rotating a spindle with a radius equal to $R_i$ and length equal to $L$ in a binder placed in a chamber with an internal radius of $R_o$. The test measures the torque required to maintain a constant rotational speed ($\omega$) at a constant temperature. The shear stress at

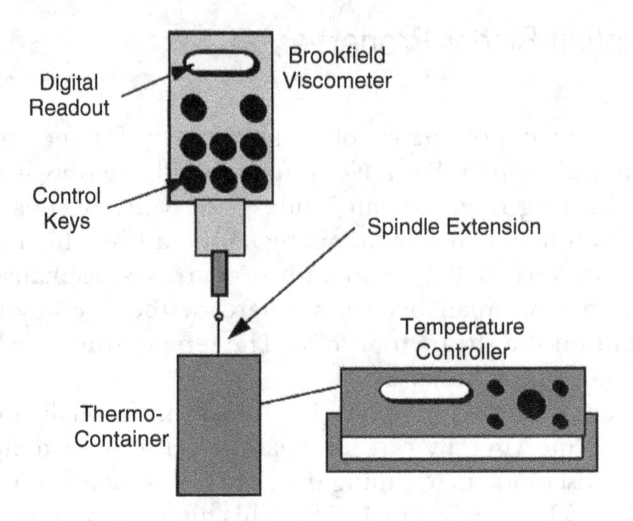


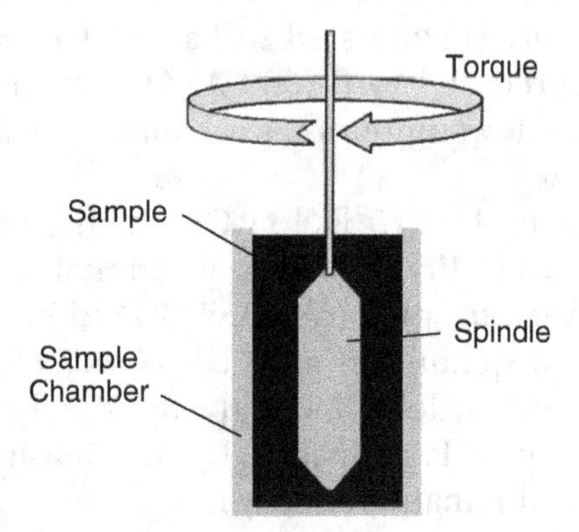


**Figure 5.12**
A Schematic of the Rotational Viscometer (Ref. 12)

the surface of the spindle is given in Equation 5.12[6].

$$\tau_b = \frac{T}{2\pi R_i^2 L} \tag{5.12}$$

The viscosity is measured in the units of Pa.s or poise and is calculated using Equation 5.13.

$$\eta = \frac{T(1/R_i^2 - 1/R_o^2)}{4\pi\omega} \tag{5.13}$$

### 5.4.2 Penetration

This is an empirical method for measuring asphalt consistency (ASTM D 5 or AASHTO T 49). It does not measure a fundamental property and it does not control rate of loading. It is expressed as the distance, in tenths of a millimeter, that a standard needle penetrates an asphalt binder under standard conditions of loading, time, and temperature.

### 5.4.3 Dynamic Shear

The dynamic shear rheometer (DSR) is used to measure the dynamic viscoelastic properties of asphalt binders ($|G^*|$, $\delta$). It is a parallel plate rheometer that is used to apply shear strain or shear stress under a controlled temperature and frequency (Figure 5.13). The test is performed according to the AASHTO T 315 procedure. The maximum shear stress and shear strain in the DSR are calculated using Equations 5.14 and 5.15, respectively:

$$\tau_{\max} = 2T/\pi r^3 \tag{5.14}$$

$$\gamma_{\max} = \theta r/h \tag{5.15}$$

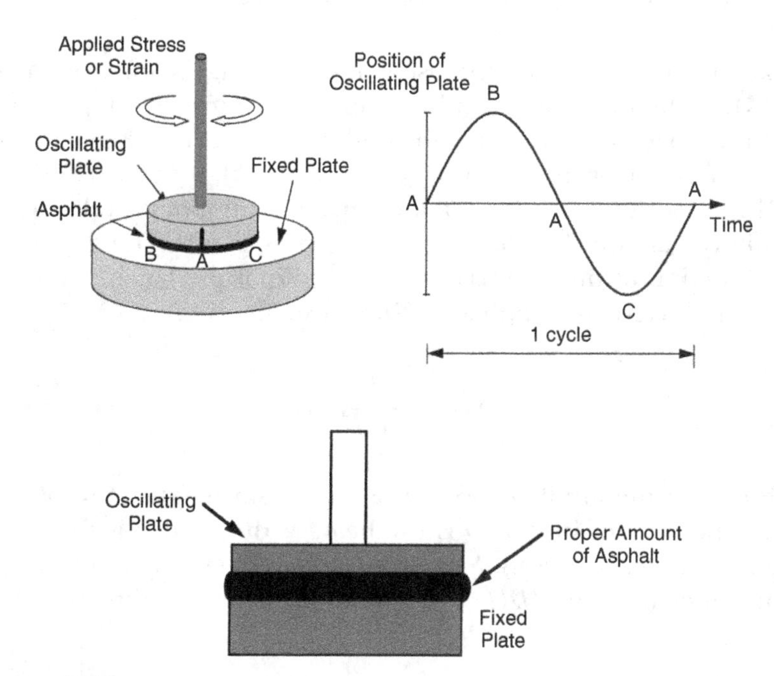


**Figure 5.13**
Schematic of the Dynamic Shear Rheometer (Ref. 12)

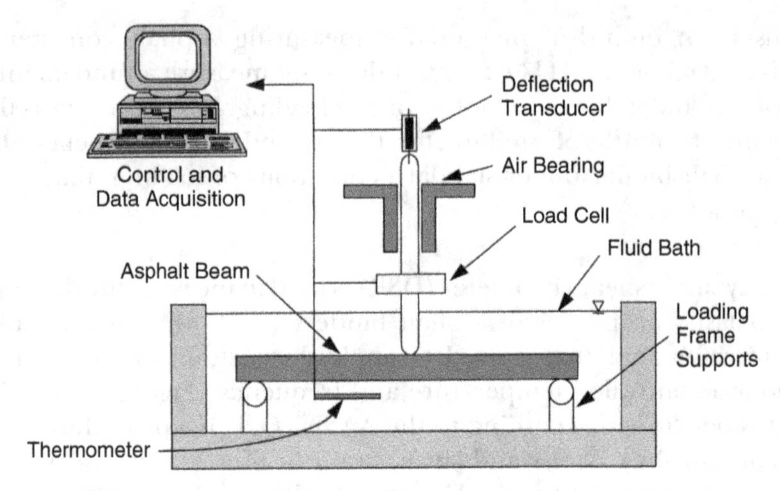


**Figure 5.14**
Schematic of the Bending Beam Rheometer (Ref. 12)

where $T$ is the maximum applied torque, $r$ is the radius of plate, $\theta$ is the deflection angle, and $h$ is specimen height.

### 5.4.4 Flexural Creep

Flexural creep is measured using the bending beam rheometer (BBR). In this test, a beam is placed under a controlled temperature, and a creep load in applied in the middle of the beam (Figure 5.14). The test is performed according to the ASTM D 6648 (AASHTO T 313) procedure. As the beam creeps, the midpoint deflection is monitored after 8, 15, 30, 60, 120, and 240 seconds. The deflection as a function of time ($\Delta(t)$) in the middle of the beam is used to compute the creep compliance ($D(t)$) using:

$$\Delta(t) = \frac{PL^3}{48I} D(t) \tag{5.16}$$

where, $P$ is the applied force, $L$ is the beam length, $I = bh^3/12$ is the moment of inertia, $b$ is the beam width, and $h$ is the beam height. The term *stiffness*, $S(t)$, is introduced here as the inverse of compliance ($S(t) = 1/D(t)$), which can be computed directly using:

$$S(t) = \frac{PL^3}{4bh^3 \Delta(t)} \tag{5.17}$$

The maximum flexural stress and flexural strain are calculated as follows:

$$\sigma_{max} = \frac{3PL}{2bh^2} \tag{5.18}$$

$$\varepsilon_{max} = \frac{6\Delta(t)h}{L^2} \tag{5.19}$$

The BBR is used for measuring the binder properties at low temperatures. It was found that at these low temperatures, the relationship between $\log(S(t))$ and $\log(t)$ can be described by a second-order polynomial, as in Equation 5.20.

$$\log(S(t)) = A + B\log(t) + C(\log(t))^2 \tag{5.20}$$

The derivative of $\log(S(t))$ with respect to $\log(t)$ gives the slope of the relationship as a function of time:

$$m(t) = \frac{d\log(S(t))}{d\log(t)} = B + 2C\log(t) \tag{5.21}$$

**Example 5.1**

The following deflection data (Table 5.2) was obtained from testing an asphalt binder using the BBR at the following conditions:

- Test temperature = −18.0°C
- Constant load applied to the beam = 0.98 N
- Beam width = 12.70 mm.
- Beam thickness = 6.35 mm
- Distance between beam supports = 102 mm

Assuming the load and length of the beam are constant throughout the test, calculate the stiffness of the material ($S(t)$) at every reported time. Plot the results.

**ANSWER**

Equation 5.17 can be used to determine the stiffness of the material with the following parameters:

- Constant load applied to the beam, $P = 100$ g (980 mN)
- Distance between beam supports, $L = 102$ mm
- Beam width, $b = 12.7$ mm
- Beam thickness, $h = 6.35$ mm

The results are presented in Table 5.3 and plotted in Figure 5.15.

**Table 5.2**
Data for Example 5.1

| Time (sec) | Deflection (mm) |
| --- | --- |
| 8 | 0.1877 |
| 15 | 0.2231 |
| 30 | 0.2730 |
| 60 | 0.3364 |
| 120 | 0.4152 |
| 240 | 0.5327 |

**Table 5.3**
Deflection and Stiffness Data for Example 5.1

| Time (*s*) | Deflection (mm) | *S*(*t*) (Mpa) |
| --- | --- | --- |
| 8 | 0.1877 | 425.97 |
| 15 | 0.2231 | 358.38 |
| 30 | 0.2730 | 292.87 |
| 60 | 0.3364 | 237.68 |
| 120 | 0.4152 | 192.57 |
| 240 | 0.5327 | 150.09 |

### 5.4.5 Tensile Strength

The asphalt tensile strength is measured by the direct tension test (DTT) using AASHTO T 314. In this system, a direct tension load is applied to maintain a constant displacement rate at a standard low temperature (Figure 5.16). The maximum load developed during the test is monitored. The tensile strain and stress in the specimen when the load reaches a maximum are reported as the failure strain and failure stress, respectively. The strain at failure is used for specifying low-temperature properties of the binder.

### 5.4.6 Surface Energy

As defined in Chapter 4, surface energy ($\gamma$) is the amount of external work done on a material to create a new unit surface area in vacuum. This property is important to determine the binder fracture property and the bond of the binder with aggregates[5,11]. The binder surface energy components have been calculated by measuring contact angles with various probe liquids using the Wilhelmy plate method.[5] Using this method, the contact angle is calculated as shown in Equation 5.22 from equilibrium considerations of a glass slide thinly

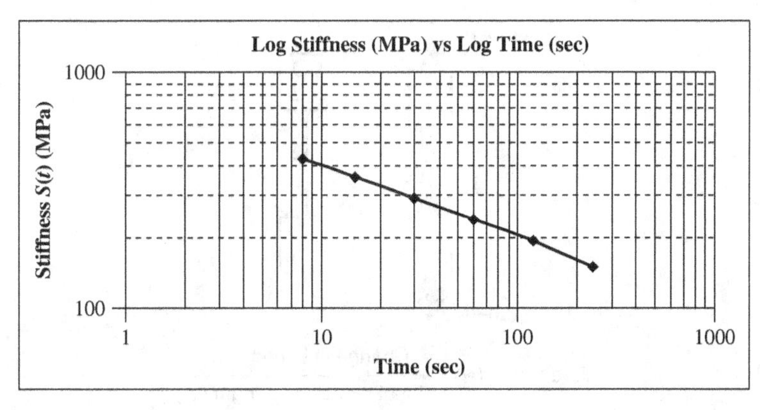


**Figure 5.15**
Stiffness of the Asphalt Binder as a Function of time

coated with the asphalt film and immersed in a probe liquid,

$$\cos\theta = \frac{\Delta F + V_{im}(\rho_L - \rho_{air})g}{P_t \gamma_P^{Tot}} \tag{5.22}$$

where, $P_t$ is the perimeter of the asphalt slide, $\gamma_P^{Total}$ is the total surface energy of the probe liquid, $\theta$ is the contact angle between the bitumen and the liquid, $V_{im}$ is the volume of the slide immersed in the liquid, $\rho_L$ is the liquid density, $\rho_{air}$ is the air density, and $g$ is the gravity acceleration. The measured contact angles and the known surface energies of the probe liquids are then used to calculate the surface energy components of asphalt.

### 5.4.7 Aging

Aging refers to the change in the binder structure and/or composition due the influence of temperature and oxygen. This change makes the binder harder and more brittle. Aging is caused by different mechanisms, such as volatilization of hydrocarbon elements at high temperatures, and breaking of intermolecular bonds to form new molecular structures, and oxidation. Oxidation combined with high temperatures can disrupt and change the strong covalent bonds, cause an increase in polarity of the molecules, and consequently, increase association among these molecules.

During construction, aging is attributed primarily to loss of volatiles and oxidation. Aging during construction is referred to as *short-term aging*, and it is simulated in the laboratory using the rolling thin-film oven (RTFO), according to ASTM D 2872

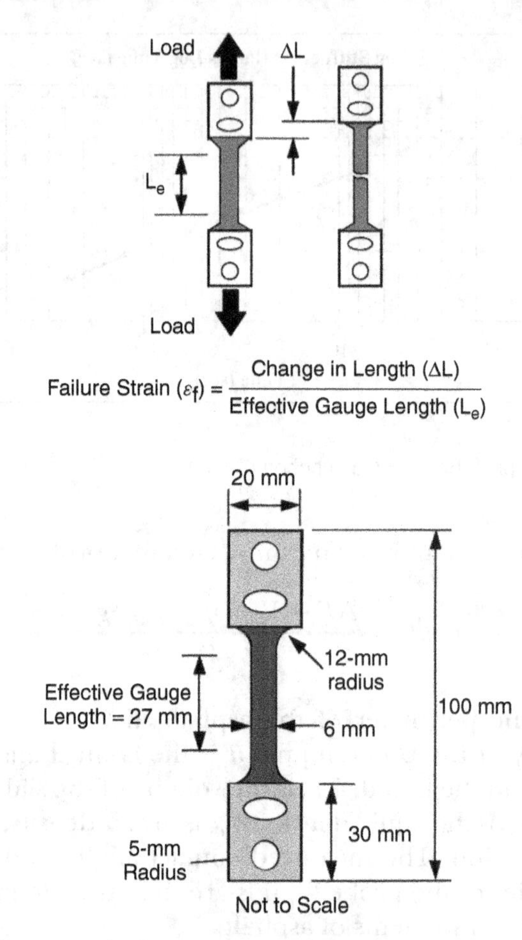


**Figure 5.16**
Schematic of the Geometry and Change in Length in the Direct Tension Specimen (Ref. 12)

(AASHTO T 240) procedure. In this test, a bottle with a small amount of asphalt is placed in a rack in an oven at a temperature similar to that used in asphalt mixing in the field. The rack rotates, and the asphalt forms a thin film covering the bottle (Figure 5.17). This thin film is subjected to an air jet during the rotation of the bottle. This process causes oxidation and volatilization of hydrocarbons. Another, less common, method for simulating short-term aging utilizes the thin film oven (TFO). This concept is similar to that of the RTFO, but the asphalt is placed in a pan that rotates horizontally. Aging in the RTFO is faster than in the TFO.

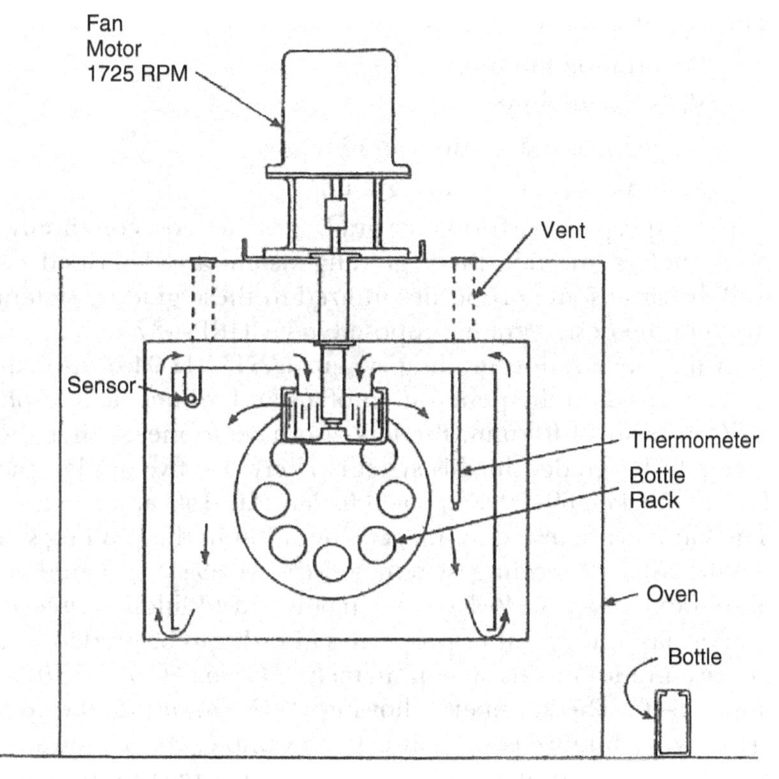


**Figure 5.17**
Schematic of the Rolling Thin Film Oven Test (Ref. 17)

Aging also occurs during the pavement service due to oxidation. This process is simulated in the laboratory by the pressure aging vessel (PAV). The PAV procedure is described in ASTM D6521. An asphalt binder is subjected to oxygen at high pressure and relatively high temperature. Aging in the field is more severe for thin asphalt films and asphalt mixtures with connected voids.

## 5.5 Asphalt Grades

Different methods have been developed over the years for grading asphalt binders. The main objective of these grading systems is to classify binders based on their rheological and mechanical properties, assuming that these properties relate to the field performance.

The asphalt grading systems are:

- Penetration grading
- Viscosity grading
- Viscosity of aged residue grading
- Superpave performance grading

The Superpave performance grading is the most commonly used. Nevertheless, the first three grading system are described here as well, because some properties utilized in these grading systems are input in the design guide proposed in NCHRP 1-37A.

In the penetration grading system (ASTM D 946), asphalts are graded based on the penetration of a standard needle in asphalt at 25°C in units of 0.1 mm. Binders also have to meet other requirements to be graded in this system. There are five grades, namely: Pen 40–50, Pen 60–70, Pen 85–100, Pen 120–150, and Pen 200–300. The binders are tested in unaged condition in this grading system.

The viscosity grading system (ASTM D 3381) is based on the absolute viscosity at 140°F (60°C) in poise. In addition to viscosity, the binder has to meet other properties in order to be graded.[17] There are five grades in this system, namely: AC-2.5, AC-5, AC-10, AC-20, and AC-40. The number following "AC" indicates the absolute viscosity in hundreds of poise. For example, AC-5 indicates that the absolute viscosity is 500 poise. Viscosity is measured on unaged binders.

The viscosity of aged residue grading system (ASTM D 3381) is based on measuring the binder viscosity after aging in the TFO. Absolute viscosity is measured in poise at a temperature of 140°F (60°C). There are five grades; AR-1000, AR-2000, AR-4000, AR-8000, AR-16000. The number in the grade refers to absolute viscosity in poise.

The Superpave performance grading system has several advantages over the other three systems. It was developed considering the influence of rate of loading on the binder properties, and the tests are conducted at temperatures that represent the geographic location in which a binder will be used. A number of tests are involved in the Superpave system to obtain the binder rheological properties at various temperatures. Recall that the penetration and viscosity grading systems are based on tests at fixed temperatures that may not represent the field temperature.

A binder grade indicates the pavement temperatures at which this binder can be used. For example, PG 64-22 indicates that this

**Table 5.4**
Superpave Binder Grades

| Low Temperature Grade (°C) | High Temperature Grade (°C) | | | | | | |
|---|---|---|---|---|---|---|---|
| | 46 | 52 | 58 | 64 | 70 | 76 | 82 |
| −10 | | ■ | | ■ | ■ | ■ | ■ |
| −16 | | ■ | ■ | ■ | ■ | ■ | ■ |
| −22 | | ■ | ■ | ■ | ■ | ■ | ■ |
| −28 | | ■ | ■ | ■ | ■ | ■ | ■ |
| −34 | ■ | ■ | ■ | ■ | ■ | ■ | ■ |
| −40 | ■ | ■ | ■ | ■ | ■ | | |
| −46 | ■ | ■ | | | | | |

binder can be used where the average seven-day maximum pavement temperature is lower than or equal to 64, and the minimum pavement temperature is higher than or equal to −22. The asphalt grades according to this system are shown in Table 5.4.

The testing conditions and the specifications the binder needs to pass in the superpave system are discussed next.

The rotational viscometer is conducted to determine the binder viscosity at a temperature of 135°C, which represents construction conditions. Viscosity should be less than 3 Pa.s for the binder to have the proper workability for use in asphalt mixtures.

The DSR is used to measure the binder viscoelastic properties ($|G^*|,\delta$) at an angular speed of 10 rad/s. The test is conducted at the maximum average seven-day pavement temperature to assess the binder resistance to permanent deformation. This temperature is obtained from climatic conditions collected from weather stations and an equation that calculates the pavement temperature from the climatic conditions. The diameter of the DSR plate used for measuring the resistance to rutting test is 25 mm. The DSR test is conducted on an unaged binder and RTFO-aged binders. For the binder to pass the rutting test at a given temperature, the value of the index $G^*/\sin\,\delta$ should be higher than 1.00 kPa for the unaged binder and more than 2.2 kPa for the aged binder. This means that is it is desirable to increase the $|G^*|$ value (higher stiffness) and decrease the value of $\delta$ decrease in viscous dissipation). According to the original Superpave developments[1], the index ($|G^*|/\sin\delta$), is inversely proportional to the energy dissipated in viscoleastic deformation. Therefore, it is postulated that permanent deformation can be decreased by decreasing the dissipated energy or increasing the $|G^*|/\sin\delta$ value.

The DSR is also used to measure the binder resistance to fatigue cracking. The test temperature is taken as 0.5 × (seven-day average maximum pavement temperature + minimum pavement temperature) + 4. For example, the fatigue test temperature for PG 64-22 binder is 25°C. The test is conducted after aging the binder in both the RTFO and PAV. The diameter of the DSR plate is 8 mm for fatigue testing. The calculated index is ($|G^*| \times \sin\delta$) and should be lower than or equal to 5000 kPa for the binder to pass the fatigue test at a given temperature.

Selection of the $|G^*| \times \sin\delta$ index suggests that it is desirable to decrease the modulus (decrease stiffness) and decrease phase angle (decrease viscous dissipation) to reduce fatigue cracking. As suggested in the development of the Superpave binder specifications,[1] the $|G^*| \times \sin\delta$ index is directly proportional to dissipated energy assuming that the analysis is conducted at a constant strain (Equation 5.7). This type of loading, however, is considered applicable to thin asphalt concrete layers only (i.e., this specification was developed considering fatigue cracking in thin pavements). It is worth mentioning, that in thick asphalt concrete layers (i.e., approximately thicker than 125 mm), it is usually found that fatigue cracking can be reduced by increasing the stiffness rather than decreasing it. The Superpave specifications for fatigue and permanent deformation are currently being revised to better relate performance indices to energy dissipation and mode of loading.

The resistance to low-temperature cracking is assessed using the BBR and the DTT. Both of these tests are conducted in a binder that is aged using both the RTFO and PAV. Also, both tests are conducted at a temperature 10°C higher than the pavement's lowest temperature. For example, a PG 64-22 is tested at −12 °C. The increase in temperature is used to reduce the testing time. Using the time-temperature superposition, it was found that the creep compliance value at 60 seconds and a temperature 10°C higher than the lowest temperature is equivalent to the creep compliance needed for low-temperature cracking, which is measured after two hours of loading at the lowest temperature. Therefore, the creep compliance can be measured within a short loading time by utilizing the time-temperature superposition.[1,3] In the BBR test, the binder stiffness $S(t)$ should be lower than 300 MPa, and the slope of the log $S(t) - \log(t)$ relationship at 60 seconds ($m(t = 60\,\text{sec})$) should be higher than 0.3 for the binder to pass the low-temperature cracking test at a given temperature. The decrease in stiffness indicates a decrease in the developed thermal stress value; an increase in $m(t)$ indicates an increase in the binder capability to release thermal stresses.

It was noticed that although some binders are stiff, they can have very good resistance to low-temperature cracking because they can exhibit high failure strain. Therefore, the DTT was developed to measure the strain at failure in the binder. The binder is pulled at a constant rate of deformation of 1 mm/min. The tensile strain and stress in the specimen, when the load reaches a maximum value, is reported as the failure strain and failure stress, respectively. The strain at failure $\varepsilon_f$ should be greater than 1% for the binder to pass. This test is conducted only if the binder stiffness in the BBR is between 300 MPa and 600 MPa. A binder that passes this BBR requirement is considered to pass the low-temperature test without running the DTT. On the other hand, a binder with a stiffness higher than 600 MPa does not pass the binder grade, regardless of the failure strain in the direct tension test. A summary of the Superave tests and requirements is shown in Table 5.5.

**Example 5.2**

Using the information obtained in Example 5.1 from a BBR test, determine whether this asphalt meets the Superpave requirements for low-temperature cracking.

**ANSWER**

To prevent low-temperature cracking, Superpave requires that the binder stiffness, $S(t)$, be lower than 300 MPa, and the slope of the log $S(t) - \log(t)$ relationship at the same time (i.e., $m(t = 60\,\text{sec})$) be higher than 0.3.

The data obtained in Example 5.2 show that at 60 seconds the stiffness of the binder, $S(t)$, is 237.68 MPa. By calculating the slope of the plot log ($S(t)$ versus $\log(t)$ at 60 seconds, an approximate value of 0.311 for $m(t)$ is obtained.

By comparing the requirements with the results from the BBR test, it can be concluded that the asphalt binder does satisfy the Superpave conditions for low-temperature cracking.

**Table 5.5**
Summary of the Superpave Test and Requirements

| | Construction | Permanent Deformation (Rutting) | | Fatigue Cracking | Low-Temperature Cracking | |
|---|---|---|---|---|---|---|
| **Test** | **RV** | **DSR** | **DSR** | **DSR** | **BBR** | **DT** |
| **Aging Condition** | **None** | **None** | **RTFO** | **RTFO + PAV** | **RTFO + PAV** | |
| Test Temperature | 135°C | Seven-day average maximum pavement temperature | Seven-day average maximum pavement temperature | 0.5 × (seven-day average maximum pavement temperature + minimum pavement temperature) + 4 | Minimum Pavement Temperature + 10°C | Minimum Pavement Temperature + 10°C |
| (Example: For PG 64–22) | | (64°C) | (64°C) | (25 °C) | (−12 °C) | (−12 °C) |
| Parameter | Viscosity | $\lvert G^*\rvert/\sin\delta$ | $\lvert G^*\rvert/\sin\delta$ | $\lvert G^*\rvert \times \sin\delta$ | $S$ ($t = 60$ sec) $m$ ($t = 60$ sec) | $\varepsilon_f$ |
| Requirement | ≤ 3 Pa s | (≥ 1.0 *kPa*) | (≥ 2.2 *kPa*) | (≤ 5000 *kPa*) | ≤ 300 *MPa* ≥ 0.3 | ≥ 1.0% |

**Example 5.3**

The following data on the dynamic shear modulus $G^*$ (Table 5.6) was obtained by performing DSR tests on an asphalt binder. The DSR test was conducted at three temperatures (12°C intervals) at frequencies ranging from 1 to 100 rad/s using the 25-mm DSR plates and at frequencies from 0.1 to 100 rad/s using the 8-mm DSR plates. Construct the master curve for the dynamic shear modulus for a reference temperature of 25°C and present the corresponding shifting factors versus temperature plot. It is noted that the symbol "| |" is omitted around the modulus $G^*$ to simplify the notation.

**ANSWER**

The Master curve should include relationships of both dynamic modulus and phase angles as functions of frequency. However, this example demonstrates only the dynamic modulus relationship. There are several methodologies for constructing a master curve. For simplicity, in this example, shifting is conducted through the following steps:

1. For each curve, a value of $a_T(\mathrm{T}_i)$ (shift factor for the i-th temperature) is assumed.
2. Using that shift factor, the reduced frequency for the whole curve is calculated (reduced frequency = $a_T(\mathrm{T}_i) \times$ frequency).
3. The plot of dynamic shear modulus versus reduced frequency is compared with the original plot of dynamic shear modulus versus frequency at the reference temperature (i.e., 25°C).
4. If the plot does not follow the same trend with the curve at the reference temperature, a new value of $a_T(\mathrm{T}_i)$ is selected and the process is repeated.
5. Once the shift factor values corresponding to each temperature curve are found, they are fitted into a William-Landel-Ferry (WLF) equation. The WLF is a common equation used in the construction of master curves:[7]

$$\log(a_T) = -\frac{C_1(T - T_R)}{C_2 + T - T_R} \quad (5.23)$$

where, $C_1$ and $C_2$ are the parameters of the curve, $T$ is the temperature at which the shift factor is needed, and $T_R$ is the reference temperature (25°C, in this case). The fitting process consists of obtaining the $C_1$ and $C_2$ parameters that minimize the sum of the

**Table 5.6**
Complex Shear Modulus (G*) Obtained from DSR Tests at Different Temperatures

| DSR with 8-mm Plate | | | | DSR with 25-mm Plate | | | |
|---|---|---|---|---|---|---|---|
| Frequency | Dynamic Shear Modulus (Pa) | | | Frequency | Dynamic Shear Modulus (Pa) | | |
| rad/sec | T4=37°C | T5=25°C | T6=13°C | rad/sec | T1=76°C | T2=64°C | T3=52°C |
| 0.10 | 6.31E+02 | 6.31E+03 | 1.00E+05 | 1.00 | 31.62 | 100.00 | 630.96 |
| 0.13 | 7.76E+02 | 7.59E+03 | 1.17E+05 | 1.26 | 39.81 | 125.89 | 794.33 |
| 0.16 | 9.55E+02 | 9.12E+03 | 1.38E+05 | 1.58 | 50.12 | 158.49 | 1000.00 |
| 0.20 | 1.17E+03 | 1.10E+04 | 1.62E+05 | 2.00 | 63.10 | 199.53 | 1258.93 |
| 0.25 | 1.45E+03 | 1.32E+04 | 1.91E+05 | 2.51 | 79.43 | 251.19 | 1584.89 |
| 0.32 | 1.78E+03 | 1.58E+04 | 2.24E+05 | 3.16 | 100.00 | 316.23 | 1995.26 |
| 0.40 | 2.19E+03 | 1.91E+04 | 2.63E+05 | 3.98 | 125.89 | 398.11 | 2511.89 |
| 0.50 | 2.69E+03 | 2.29E+04 | 3.09E+05 | 5.01 | 158.49 | 501.19 | 3162.28 |
| 0.63 | 3.31E+03 | 2.75E+04 | 3.63E+05 | 6.31 | 199.53 | 630.96 | 3981.07 |
| 0.79 | 4.07E+03 | 3.31E+04 | 4.27E+05 | 7.94 | 251.19 | 794.33 | 5011.87 |
| 1.00 | 5.01E+03 | 3.98E+04 | 5.01E+05 | 10.00 | 316.23 | 1000.00 | 6309.57 |
| 1.26 | 6.17E+03 | 4.79E+04 | 5.89E+05 | 12.59 | 398.11 | 1258.93 | 7943.28 |
| 1.58 | 7.59E+03 | 5.75E+04 | 6.92E+05 | 15.85 | 501.19 | 1584.89 | 10000.00 |
| 2.00 | 9.33E+03 | 6.92E+04 | 8.13E+05 | 19.95 | 630.96 | 1995.26 | 12589.25 |
| 2.51 | 1.15E+04 | 8.32E+04 | 9.55E+05 | 25.12 | 794.33 | 2511.89 | 15848.93 |
| 3.16 | 1.41E+04 | 1.00E+05 | 1.12E+06 | 31.62 | 1000.00 | 3162.28 | 19952.62 |
| 3.98 | 1.74E+04 | 1.20E+05 | 1.32E+06 | 39.81 | 1258.93 | 3981.07 | 25118.86 |
| 5.01 | 2.14E+04 | 1.45E+05 | 1.55E+06 | 50.12 | 1584.89 | 5011.87 | 31622.78 |
| 6.31 | 2.63E+04 | 1.74E+05 | 1.82E+06 | 63.10 | 1995.26 | 6309.57 | 39810.72 |
| 7.94 | 3.24E+04 | 2.09E+05 | 2.14E+06 | 79.43 | 2511.89 | 7943.28 | 50118.72 |
| 10.00 | 3.98E+04 | 2.51E+05 | 2.51E+06 | 100.00 | 3162.28 | 10000.00 | 63095.73 |
| 12.59 | 4.90E+04 | 3.02E+05 | 2.95E+06 | | | | |
| 15.85 | 6.03E+04 | 3.63E+05 | 3.47E+06 | | | | |
| 19.95 | 7.41E+04 | 4.37E+05 | 4.07E+06 | | | | |
| 25.12 | 9.12E+04 | 5.25E+05 | 4.79E+06 | | | | |
| 31.62 | 1.12E+05 | 6.31E+05 | 5.62E+06 | | | | |
| 39.81 | 1.38E+05 | 7.59E+05 | 6.61E+06 | | | | |
| 50.12 | 1.70E+05 | 9.12E+05 | 7.76E+06 | | | | |
| 63.10 | 2.09E+05 | 1.10E+06 | 9.12E+06 | | | | |
| 79.43 | 2.57E+05 | 1.32E+06 | 1.07E+07 | | | | |
| 100.00 | 3.16E+05 | 1.58E+06 | 1.26E+07 | | | | |

squares error between the shift factors obtained in the first part of the exercise and the shift factors estimated with the WLF equation.

The plot in Figure 5.18 presents the information provided and shows the direction in which each curve should be shifted to construct the master curve. After following the iterative process described, the shift factors shown in Table 5.7 were found for each temperature curve.

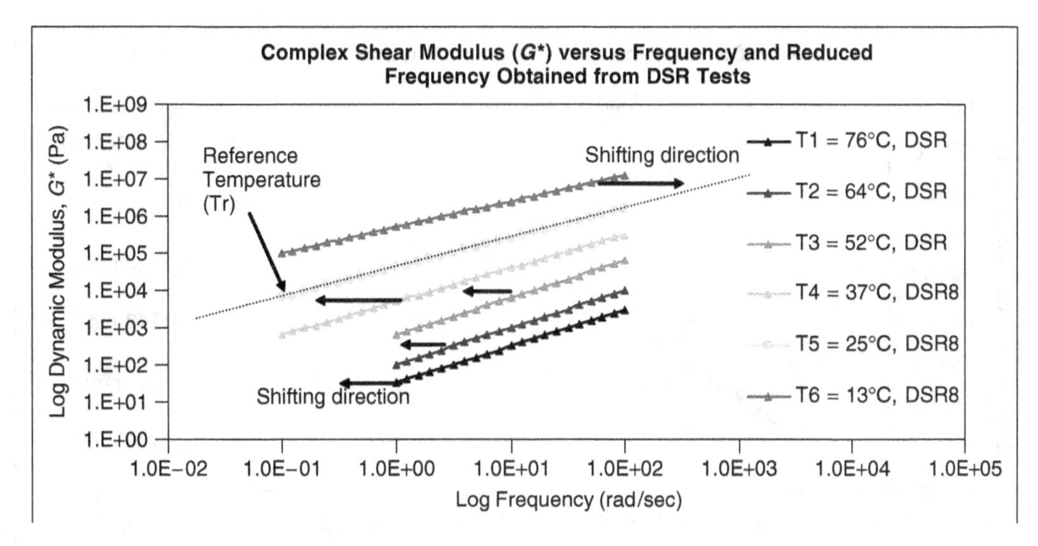


**Figure 5.18**
Dynamic Shear Modulus versus Frequency Obtained from DSR tests

**Table 5.7**
Shift Factor Calculations for Example 5.3

| Temperature | Shift Factor ($a_T$) |
|---|---|
| 13 | $2.44 \times 10^{1}$ |
| 25 | $1.00 \times 10^{0}$ |
| 37 | $9.39 \times 10^{-2}$ |
| 52 | $1.04 \times 10^{-2}$ |
| 64 | $1.67 \times 10^{-3}$ |
| 76 | $5.00 \times 10^{-4}$ |

Based on these shift factors, the master curve for the dynamic shear modulus can be obtained by plotting the dynamic shear modulus versus the reduced frequency (Figure 5.19). Note that the values of shear complex modulus do not change during the process of constructing a master curve. What changes is the horizontal position of the curve (horizontal shift) in order to follow the trend of the 25°C curve.

Table 5.8 shows the estimated shift factors obtained by fitting the data to the WLF equation. The $C_1$ and $C_2$ parameters of the WLF

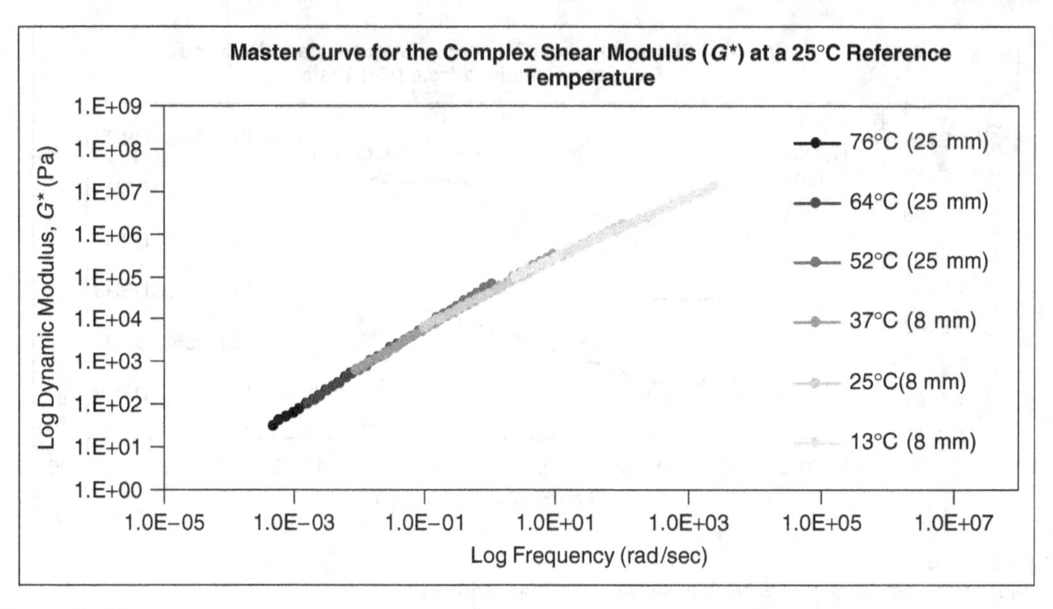


**Figure 5.19**
Master Curve for the Complex Shear Modulus ($G^*$)

**Table 5.8**
Shift Factors Obtained During Construction of the Master Curve and Estimations Using a WLF Equation (in base 10 logarithm).

| Temperature | Log (Shift Factor) [$\log(a_T)$] Obtained | Estimated with WLF Equation |
|---|---|---|
| 13 | 1.39 | 1.46 |
| 25 | 0.00 | 0.00 |
| 37 | −1.03 | −1.08 |
| 52 | −1.98 | −2.08 |
| 64 | −2.78 | −2.70 |
| 76 | −3.30 | −3.21 |

equation that minimizes the square differences between the real and the estimated values were found to be 8.20 and 79.3, respectively. The relationship between $\log(a_T)$ versus temperature is given in Figure 5.20.

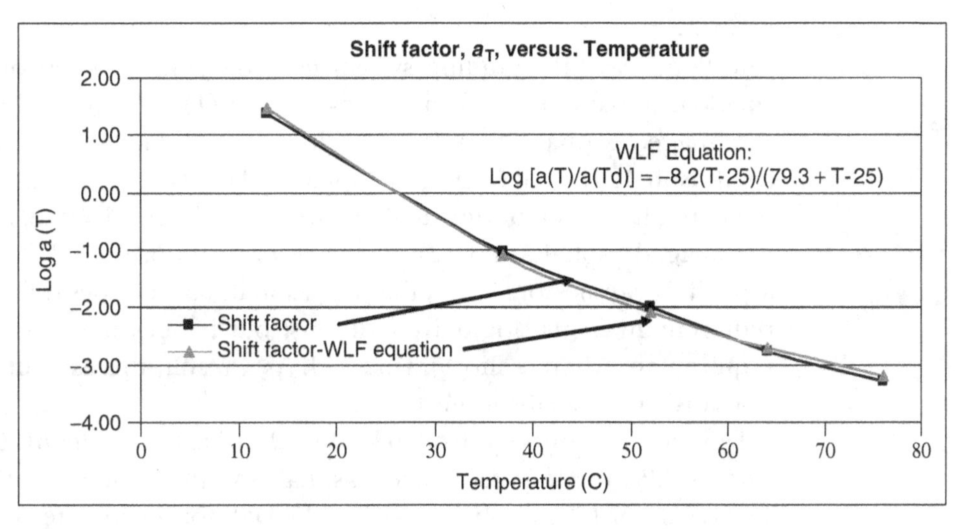


**Figure 5.20**
Log ($a_T$) versus Temperature for the Master Curve of the Dynamic Shear Modulus ($G^*$)

**Table 5.9**
Data for Example 5.4

| | Original Asphalt | RTFO-Aged Asphalt | PAV-Aged Asphalt |
|---|---|---|---|
| | | **DSR** | |
| G*(kPa) | 0.98 | 1.4 | 6000 |
| $\delta$(°) | 75 | 70 | 60 |
| | | **BBR** | |
| S (MPa) | 300 | 450 | 650 |
| m | 0.6 | 0.5 | 0.4 |
| | | **DTT** | |
| Failure Strain (%) | 2.5% | 2% | 1.2% |

**Example 5.4**

The following results were obtained from testing an asphalt binder (Table 5.9). Determine if the asphalt meets the specification requirements to resist the following distresses:

- ❑ Rutting
- ❑ Fatigue cracking
- ❑ Low-temperature cracking
- ❑ Rutting

**ANSWER**

The Superpave PG grading system controls rutting based on the rheological parameters $G^*$ and $\delta$, as follows: (1) $G^*/\sin\delta > 1.0$ kPa for unaged asphalt, and (2) $G^*/\sin\delta \geq 2.2$ kPa for short-term aged asphalt (i.e., RTFO-aged asphalt). The $G^*/\sin\delta$ values for the asphalt binder under analysis are: (1) $G^*/\sin\delta = 1.014$ kPa for unaged asphalt, and (2) $G^*/\sin\delta = 1.49$ kPa for RTFO-aged asphalt. It can be concluded that this asphalt binder does satisfy the requirement in its original conditions; but it does not satisfy the requirement when is short-term aged. As a result, this asphalt does not satisfy the specifications for rutting.

Fatigue cracking is controlled by the following requirement: $G^*\sin\delta \leq 5000$ kPa (in long-term aged asphalt samples using the PAV). The result for $G^*\sin\delta$ obtained from the DSR for the long-term aged asphalt under analysis is 5196 kPa. Therefore, it can be concluded that the asphalt binder does not meet the requirement for fatigue cracking.

Low-temperature cracking is controlled by limiting the stiffness ($S$) and the $m$-value obtained from the BBR test in long-term aged samples (PAV), as follows: (1) stiffness ($S$) $\leq 300$ MPa (in long-term aged asphalt samples), and (2) $m$-value $\geq 0.3$ (in the same conditions). The BBR results obtained for the long-term aged samples are: (1) stiffness ($S$) = 650 MPa, and (2) $m$-value = 0.4.

Overall, it can be concluded that the asphalt binder does not satisfy the requirement for stiffness, although it does satisfy the requirements for the $m$-value. Although the asphalt binder did not satisfy the performance requirements at the conditions at which the tests were conducted, it might satisfy those conditions at other temperatures.

### 5.5.1 Temperature Susceptibility

Asphalt binder properties are dependent on temperature, a phenomenon referred to as *temperature susceptibility*. Conventional tests of penetration and viscosity are conducted at one temperature and, consequently, they cannot reflect the temperature susceptibility. To overcome these limitations, some indices have been developed to determine the change in binder properties as a function of time. An index that has been used for measuring temperature susceptibility based on penetration is the penetration index ($PI$):

$$PI = \frac{20 - 500\,A}{1 + 50\,A}$$

$$A = \frac{\log Pen\ at\ T_1 - \log Pen\ at\ T_2}{T_1 - T_2} \tag{5.24}$$

where $T_1$ and $T_2$ are in degrees Centigrade. Other indices have also been used based on changes of viscosity as a function of temperature.

One of the main advantages of the Superpave system is that temperature susceptibility is accounted for by measuring the binder properties at different temperatures that represent construction, rutting, fatigue, and low-temperature cracking.

## 5.6 Binder Modification

Different processes and materials are used to enhance the binder properties. Binder modification has been driven by the increase in traffic loads, new refining technologies, enhancement in polymer technology, the increasing need to recycle waste such as rubber, and the need to meet the performance grades in the Superpave system.[10] This enhancement is realized in increasing the maximum temperature and/or reducing the minimum temperature at which a binder can be used. King et al.[10] indicate that the results of modifying asphalts depend on a number of factors, including concentration of the modifiers; molecular weight; chemical composition; particle size; and molecular orientation of the additive, crude source, refining process, and the grade of the original unmodified binder. A list of binder modifiers is given in Table 5.10.

Several studies and experience from performance of asphalt pavements have shown that the current Superpave system needs improvements to better characterize the performance of modified binders. A number of recent studies have shown progress toward the goal of enhancing the Superpave system to grade binders, including modified binders based on their performance. A brief summary of these efforts is given in this section with emphasis on the results from NCHRP.[4]

**Table 5.10**
Types of Asphalt Binder Modifiers (Ref. 9)

| Categories of Modifier | Examples of Generic Types |
|---|---|
| Thermosetting polymers | Epoxy resin<br>Polyurethane resin<br>Acrylic resin |
| Elastomeric polymers | Natural rubber<br>Volcanized (tyre) rubber<br>Styrene—butadienc—styrene (SBS) block copolymer<br>Styrene—butadiene—rubber (SBR)<br>Ethylene—propylene—diene terpolymer (EPDM)<br>Isobutene—isoprene copolymer (IIR) |
| Thermoplastic polymers | Ethylene vinyl acetate (EVA)<br>Ethylene methyl acrylate (EMA)<br>Ethylene butyl acylate (EBA)<br>Polythylene (PE)<br>Polyvinyl chloride (PVC)<br>Polystyrene (PS) |
| Chemical modifiers and extenders | Organo-managanese/cobalt compound (Chemrete)<br>Sulphur<br>Lignin |
| Fibers | Cellulose<br>Alumino-magnesium silicate<br>Glass fiber<br>Asbestos<br>Polyester<br>Polypropylene |
| Antistripping | Organic<br>amines<br>Amides |
| Natural binders | Trinidad lake asphalt (TLA)<br>Gilsonite<br>Rock asphalt |
| Fillers | Carbon black<br>Fly ash<br>Lime<br>Hydrated lime |

### 5.6.1 Resistance to Permanent Deformation

The repeated creep and recovery test is recommended to estimate the rate of accumulation of permanent strain. This test involves applying a constant stress for a period of time and then removing the load stress to allow the binder to recover the elastic and delayed elastic responses. This process is repeated for a specified number of cycles. This test can be conducted using the DSR. The Burger model is fitted to the creep compliance as a function of loading and unloading time. Then, the part of the compliance that represents the dashpot in series ($\eta_f$ in Figure 5.8) is calculated after a specified number of cycles. Some researchers recommend using the inverse of the compliance (termed "stiffness") associated with the dashpot as a parameter for quantifying resistance to rutting. A higher value of stiffness is taken as a measure of better resistance to permanent deformation.

### 5.6.2 Fatigue Cracking

The recommended fatigue parameter in NCHRP9-10 is based on plotting the ratio of the accumulated dissipated energy divided by the energy dissipated in the current cycle as a function of loading cycles.[4] The energy ratio is shown in Equation 5.25.

$$\textit{Dissipated Energy Ratio} = \frac{\sum_{i=1}^{n} \Delta W_i}{\Delta W_n} \tag{5.25}$$

where $\Delta W_i$ is the dissipated energy in cycle $i$ ($\Delta W_i = \pi \tau_{oi} \gamma_{oi} \sin \delta$), with $\tau_{oi}$ the amplitude of the stress function, $\gamma_{oi}$ the amplitude of the strain function, and $\Delta W_n$ the dissipated energy in the current cycle denoted by $n$.

Bahia et al.[4] suggested that the fatigue life consists of three stages. In the first stage, the relationship is linear, indicating there is no damage and energy is dissipated in viscoleastic deformation. In the second stage, the relationship deviates from linearity, which is attributed to crack initiation. In the third stage, crack propagation occurs and is manifested as a rapid change in the slope of the relationship. The number of cycles to failure ($N_p$) is defined by the intersections of asymptotes, as shown in Figure 5.21.

#### LOW-TEMPERATURE CRACKING

The method requires conducting the DTT at two loading rates and two temperatures. These measurements are used to establish curves

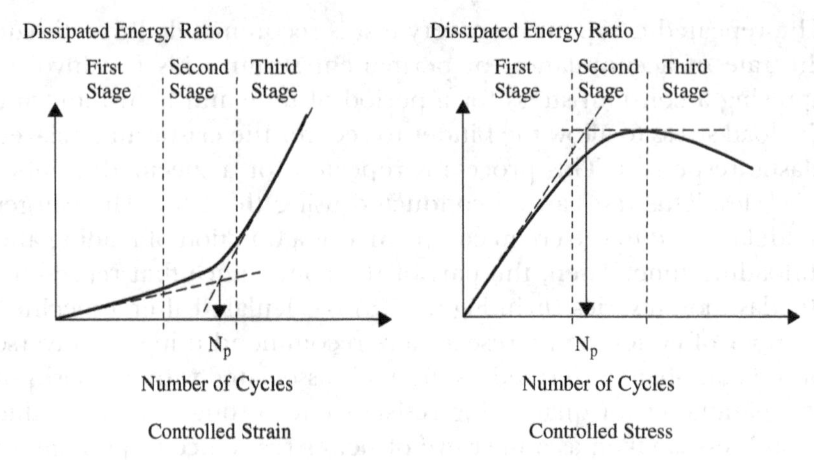


**Figure 5.21**
Definition of Fatigue Life Based on the Dissipated Energy Concept

that define the failure stress and failure strain. The BBR is used to determine the thermal stresses and strain of the binder at different temperatures. The intersection of the curves of thermal stresses and strains from BBR with the failure stress and failure strain curves, respectively, from the DTT are used to determine two cracking critical temperatures based on stress and strain. The higher of these two temperatures is used to determine the grade.

Recently, an AASHTO standard practice (AASHTO PP 42) has been developed for determining the low-temperature performance grade. This method relies on the BBR to obtain the binder relaxation modulus of the binder. This relaxation modulus, along with the changes in temperatures, is used to calculate the thermal stresses induced at low-temperatures. The thermal stresses are multiplied by a constant to determine the mixture thermal stresses and compare to the failure stress measured using the DTT to determine the critical cracking temperature.

## 5.7 Asphalt Mixture Volumetric Analysis

The volumetric composition of asphalt mixtures affects significantly their properties and performance. This section provides a brief description of these volumetrics and their calculations. The notations and equations used here are according to Roberts et al.[17].

The bulk-specific gravity of a compacted asphalt mixture ($G_{mb}$) is determined according to ASTM D 1189 and ASTM D 2726 (AASHTO T 269) procedures. Experimentally, it is calculated using Equation 5.26.

$$G_{mb} = \frac{W_D}{W_{SSD} - W_{sub}} \tag{5.26}$$

where $W_D$ is the dry weight of the compacted sample in air, $W_{SSD}$ is the weight of the asphalt mixture in saturated surface-dry condition, and $W_{sub}$ is the weight of the sample submerged in water.

The theoretical maximum specific gravity ($G_{mm}$) is the specific gravity of the mixture without voids. This specific gravity is measured using the ASTM 2041 (AASHTO T 209) procedure. One of the most important properties of the asphalt mixture is air voids in the total mix (VTM), which is calculated as follows:

$$VTM = \left(1 - \frac{G_{mb}}{G_{mm}}\right) \times 100 \tag{5.27}$$

*Voids in mineral aggregate* (*VMA*) is a term that refers to the total volume of air voids and effective asphalt. Effective asphalt is equal to the total volume of asphalt minus the asphalt absorbed by aggregates. By definition, *VMA* can be calculated as in Equation 5.28.

$$VMA = \frac{V_T - V_{Agg}(bulk)}{V_T} \tag{5.28}$$

where $V_T$ is the total volume of the compacted mixture, and $V_{Agg}(bulk)$ is the bulk volume of aggregates. The term *bulk* is used here to indicate that the aggregate volume includes the surface voids filled with asphalt. After some mathematical manipulations, *VMA* can be calculated as shown in Equation 5.29.

$$VMA = 100\left(1 - \frac{G_{mb}(1 - P_b)}{G_{sb}}\right) \tag{5.29}$$

where $G_{sb}$ is the bulk-specific gravity of aggregate in dry conditions, and $P_b$ is the asphalt content with respect to the total mixture weight.

The voids filled with asphalt (*VFA*) is the percentage of *VMA* filled with asphalt; it is calculated using Equation 5.30.

$$VFA = \frac{VMA - VTM}{VMA} \times 100 \tag{5.30}$$

An important quantity in the analysis of mixture volumetrics is the effective specific gravity of aggregate. It is equal to the weight of aggregate divided by the effective volume of aggregate. This effective volume is the volume of aggregate plus the external voids not filled with asphalt. The aggregate-effective specific gravity is related to the mixture maximum specific gravity using Equation 5.31.

$$G_{se} = \frac{1 - P_b}{\dfrac{1}{G_{mm}} - \dfrac{P_b}{G_b}} \quad (5.31)$$

where $G_b$ is the binder-specific gravity. Finally, asphalt absorption quantifies the ratio of weight of asphalt absorbed to weight of aggregates. Therefore, it is calculated as in Equation (5.32).

$$P_{ba} = \frac{W_{AAC}}{W_{Agg}} \times 100 \quad (5.32)$$

where $W_{AAC}$ is the weight of absorbed asphalt. It can be shown that the absorbed asphalt is calculated as follows:

$$P_{ba} = 100\frac{G_{se} - G_{sb}}{G_{sb}G_{se}}G_b \quad (5.33)$$

**Example 5.5**

An asphalt mixture has a bulk-specific gravity ($G_{mb}$) of 2.329. The phase diagram in Figure 5.22 shows five properties (four specific gravities and the asphalt content) of a compacted specimen of HMA that have been measured at 25°C. Using only these values, find all the volumetric properties and mass quantities as indicated on the component diagram.

**ANSWER**

The information provided is summarized in Table 5.11. Based on this information, the values of the component diagram can be obtained as follows:

- ❑ $V_{mb} = 1\ \text{cm}^3$ and $G_{mb} = 2.329$, then, $\gamma_{mb} = G_{mb}\gamma_w = 2.329\ \text{g/cm}^3 = W_{mb}/V_{mb}$, then $W_{mb} = 2.39\ \text{g/cm}^3 \times 1\ \text{cm}^3 = 2.329\ \text{g}$
- ❑ $P_b = 5\%$ by mix, then $P_b = 0.05 = W_b/W_{mb}$, then $W_b = 0.05 \times 2.329\ \text{g} = 0.1164\ \text{g}$
- ❑ $G_b = 1.015$, then $\gamma_b = G_b\gamma_w = 1.015\ \text{g/cm}^3 = W_b/V_b$; therefore, $V_b = W_b/\gamma_b = 0.1164\ \text{g}/1.015\ \text{g/cm}^3 = 0.1147\ \text{cm}^3$

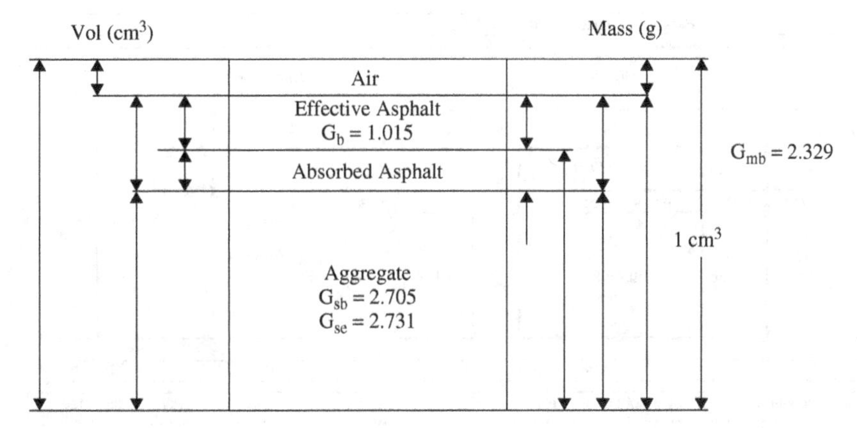


**Figure 5.22**
Asphalt Mix Phase Diagram

**Table 5.11**
Data for Example 5.7

| Material | Property |
|---|---|
| Mixture | $G_{mb} = 2.329$<br>$V_{mb} = 1\ cm^3$ |
| Effecive asphalt | $G_b = 1.015$<br>$P_b = 5\%$ by mix |
| Aggregate | $G_{agg\text{-}sb} = 2.705$<br>$G_{agg-se} = 2.731$ |

- $W_{agg} = W_{mb} - W_b = 2.329\,g - 0.1164\,g = 2.2126\,g$
- $G_{agg-se} = 2.731$, then $\gamma_{agg-se} = G_{agg-se}\gamma_w = 2.731\,g/cm^3 = W_{agg}/V_{agg-se}$, then $V_{agg-se} = W_{agg}/\gamma_{agg-se} = 2.2126\,g/2.731\,g/cm^3 = 0.8102\,cm^3$
- $V_{air} = V_{mb} - V_{agg-se} - V_b = (1 - 0.8102 - 0.11473)\,cm^3 = 0.0751\,cm^3$
- $G_{agg-sb} = 2.705$, then $\gamma_{agg-sb} = G_{agg-se}\gamma_w = 2.705\,g/cm^3 = W_{agg}/V_{agg-sb}$, then $V_{agg-sb} = W_{agg}/\gamma_{agg-sb} = 2.2126\,g/2.705\,g/cm^3 = 0.8180\,cm^3$
- Volume of $VMA = V_{mb} + V_{agg-sb} = (1 - 0.8180)\,cm^3 = 0.18203\,cm^3$. Therefore, $\%\ VMA = 18.203\%$

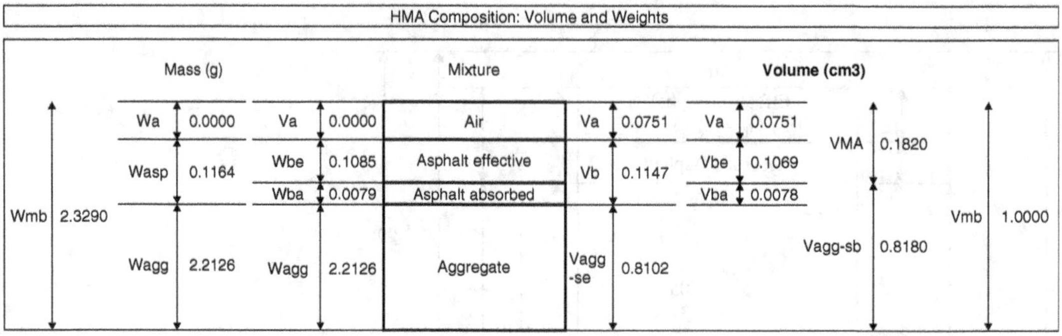

HMA Composition: Volume and Weights

| Mass (g) | | | | Mixture | | | Volume (cm3) | | |
|---|---|---|---|---|---|---|---|---|---|
| Wmb 2.3290 | Wa 0.0000 | Va 0.0000 | | Air | Va 0.0751 | Va 0.0751 | VMA 0.1820 | | Vmb 1.0000 |
| | Wasp 0.1164 | Wbe 0.1085 | | Asphalt effective | Vb 0.1147 | Vbe 0.1069 | | | |
| | | Wba 0.0079 | | Asphalt absorbed | | Vba 0.0078 | | | |
| | Wagg 2.2126 | Wagg 2.2126 | | Aggregate | Vagg-se 0.8102 | | Vagg-sb 0.8180 | | |

Volumetrics

| | | | | | |
|---|---|---|---|---|---|
| Pb | 5.00% | Percentage of binder by weight of mixture | Gagg | 2.731 | Specific gravity of aggregates |
| P'b | 5.26% | Percentage of binder by weight of aggregate | %VMA | 18.20% | Voids in mineral aggregates |
| Ps | 95.00% | Percentage of aggregate by weight of mixture | %VFA | 58.75% | Voids filled with asphalt |
| Gmb | 2.329% | Bulk-specific gravity | %Pa | 0.357% | Percentage of absorbed binder |
| Gmm | 2.51808% | Maximum specific gravity | %Pe | 4.66% | Percentage of effective binder |
| Gb | 1.01456% | Specific gravity of asphalt | Cv | 88.44% | Concentration of aggregates |

**Figure 5.23**
Mixture Volumetrics and Weight in Phase Diagram for the Given HMA

- $V_{ba} = V_{air} + V_b - VMA = (0.0751 + 0.1147 – 0.18203)\ \text{cm}^3 = 0.007787\ \text{cm}^3$
- $V_{be} = V_b - V_{ba} = (0.1147 - 0.007787)\ \text{cm}^3 = 0.10694\ \text{cm}^3$
- $W_{be} = \gamma_{be} - V_{be} = 1.015\ \text{g/cm}^3 \times 0.10694\ \text{cm}^3 = 0.1085\ \text{g}$
- $W_{ba} = W_b - W_{be} = 0.1164\ \text{g} - 0.1085\ \text{g} = 0.0079\ \text{g}$

At this point, it is possible to complete the initial component diagram and to calculate other volumetrics of the mixture. Figure 5.23 presents the summary of such results.

## 5.8 Asphalt Mixture Properties

### 5.8.1 Dynamic Modulus Test

This test is used to measure the dynamic modulus $|E^*|$ of an HMA mix at different temperatures and loading frequencies. The test was originally developed by Coffman and Pagen at Ohio State University in the 1960s. It can be conducted in a uniaxial or triaxial condition in either compression or tension. However, the majority of tests during the past years were in compression. The procedure for this test is now available in the AASHTO provisional standard TP 62. When the test is conducted in compression, the specimen experiences creep, in addition to the dynamic response. The decomposition of loading to creep and dynamic responses is shown in Figure 5.24. The strain response is shown in Figure 5.25, where the response in Figure 5.25a is the summation of the responses in Figures 5.25b and 5.25c. In the analysis, the creep response is typically ignored and the dynamic modulus is taken as the ratio of the amplitude of the dynamic stress function to the amplitude of the dynamic strain function. In the AASHTO procedure the peak stress level for measuring the dynamic modulus was chosen to maintain the total measured strain per cycle within 50 to 150 microstrain. This strain range is selected to conform to the linear viscoelastic behavior of the mixture. Three axial linear variable differential transducers (LVDTs) are used to record deformation and calculate strain. A schematic of the LVDT mounted on the specimen is shown in Figure 5.26. If the modulus is needed at a range of temperatures and a range of frequencies, the order of conducting each the test is from lowest to highest temperature and highest to lowest frequency of loading at each temperature.

The data obtained from this test can be used to construct the master curve for the mix, as shown in Example 5.3. There are several methods for constructing and mathematically representing the master curve.[21,22] Some studies found that $|E^*|$ correlated to the rutting resistance of accelerated loading tests of pavements when the test was conducted at 10 Hz and 130°F.[21,22] The 10 Hz is considered to represent highway speeds of about 60 miles per hour based on equivalent pulse-time conversion for sinusoidal loading.

Witczak et al.[21] recommended the use of the sigmoidal function in Equation 5.34 to describe the master curve.

$$\log|E^*| = \lambda + \frac{\alpha}{1 + e^{\beta + \gamma \log t_r}} \tag{5.34}$$

where $t_r$ is the reduced time at the reference temperature, $\lambda$ is minimum value of $|E^*|$, $\lambda + \alpha$ is the maximum value of $|E^*|$, and $\beta$ and $\gamma$ are parameters describing the shape of the sigmoidal function. Witczak[22] have expressed the sigmoidal function parameters as functions of mixture volumetrics and gradations.

### 5.8.2 Creep Compliance

In this test, an asphalt mix specimen is subjected to a constant stress, and strain is measured as a function of time. Creep compliance is calculated as the ratio of the measured strain to the applied

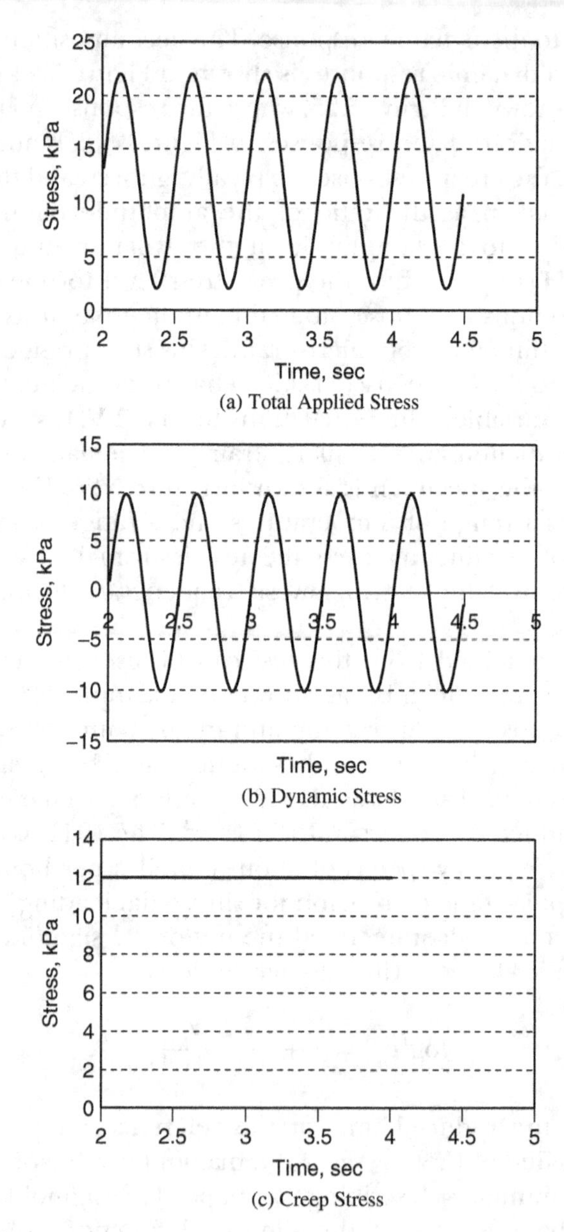


**Figure 5.24**
Decomposition of Stress Used in Dynamic Modulus into Dynamic and Creep Components

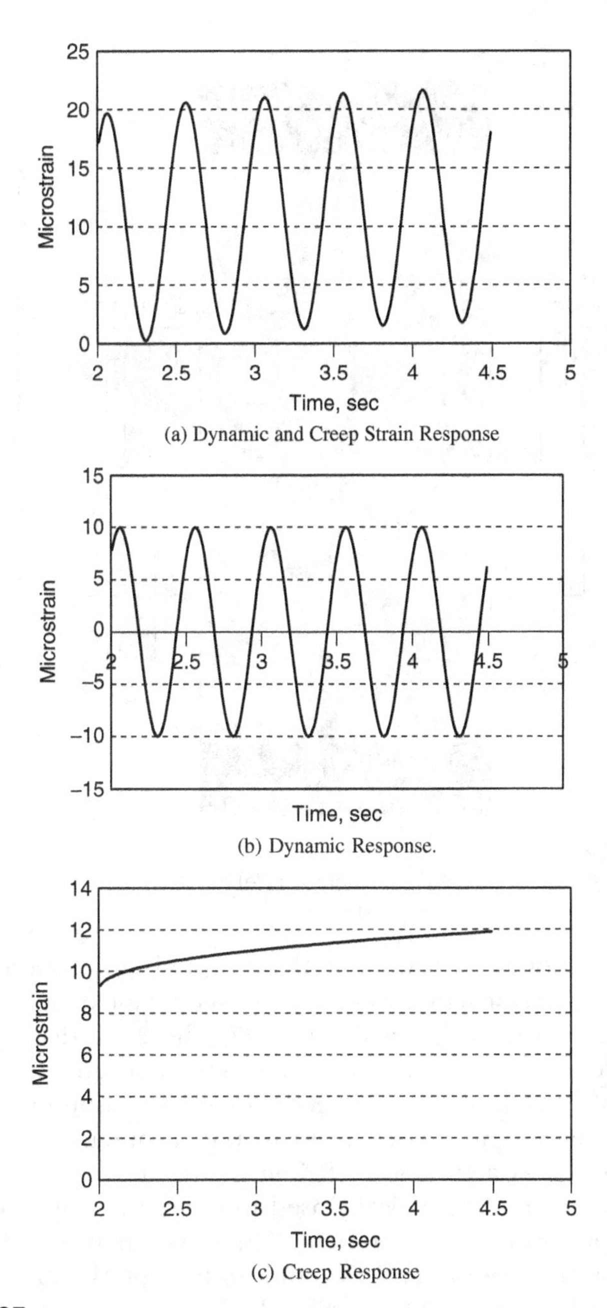


(a) Dynamic and Creep Strain Response

(b) Dynamic Response.

(c) Creep Response

**Figure 5.25**
Decomposition of Strain into Dynamic and Creep Components

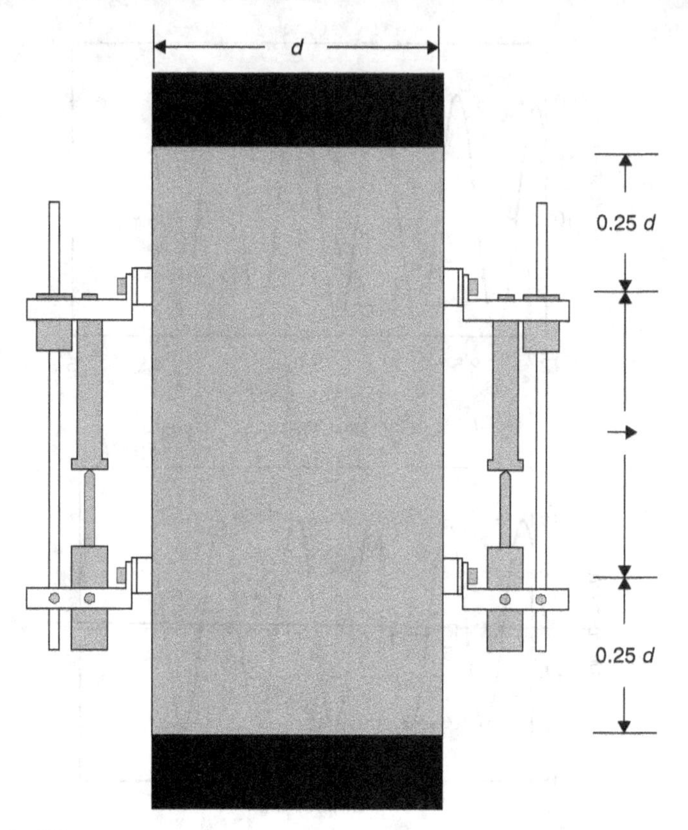


**Figure 5.26**
General Schematic of Gauge Points (not to scale) (AASHTO TP 62)

stress. The flow-time test showed very good correlation with the field rutting performance of mixes.[21] A schematic of an asphalt mix creep function is shown in Figure 5.27. There are distinct regions for the creep response, namely primary creep, secondary creep, and tertiary creep. The strain rate decreases as a function of time in the primary creep region, is constant in the secondary creep region, and increases as a function of time in the tertiary creep region. Three parameters are typically used to relate the results to asphalt mixture performance in the field. These parameters are flow-time value, flow-time slope, and flow-time intercept (Figures 5.27 and 5.28). The flow-time value is marked by the time at which the rate of change in compliance is minimum. The flow time and flow-time intercept can be obtained graphically from the log compliance versus log time plot shown in Figure 5.28.

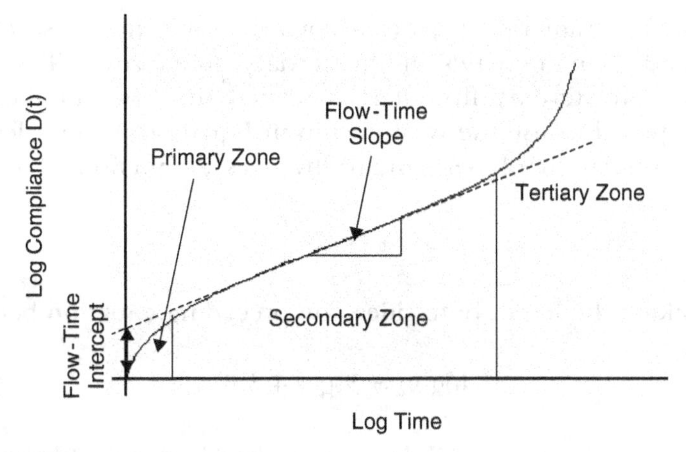


**Figure 5.27**
Compliance versus Time Curve on Log Scale

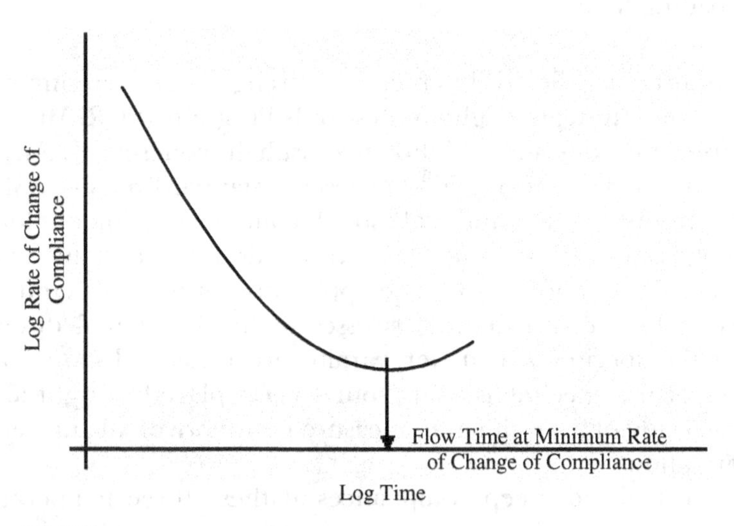


**Figure 5.28**
Rate of Change of Compliance versus Time on Log Scale

### 5.8.3 Repeated Dynamic

This test consists of applying load repetitions using a haversine pulse load of 0.1 second loading and 0.9 second of rest time. The cumulative permanent strain is measured and plotted versus the number of cycles. Similar to the creep test, the response can be defined by the primary, secondary, and tertiary zones. Permanent strain increases rapidly but at a decreasing rate in the primary zone.

Permanent strain rate reaches a constant value in the secondary zone and then increases in the tertiary creep zone. The point at which the tertiary flow starts is called the flow number.[21] It was proposed to use the relationship in Equation 5.35 to describe the accumulation of permanent strain as a function of loading cycles:

$$\varepsilon_p = aN^b \tag{5.35}$$

By taking the log of both sides, the preceding equation becomes linear:

$$\log \varepsilon_p = \log a + bN \tag{5.36}$$

The flow number value and slope are derived in exactly the same way as the flow-time value and the flow-time slope, with the exception that the number of load cycles is plotted on the $x$-axis in place of the loading time.

### 5.8.4 Indirect Tension

The indirect tensile (IDT) creep and strength tests were improved during the Strategic Highway Research Program (SHRP) to characterize the resistance of hot-mix asphalt concrete (HMA) to low-temperature cracking.[8,19] The test was standardized in AASHTO T 322. In this test, a cylindrical asphalt concrete specimen is loaded in compression through its diametrical axis at three temperatures, of 0°C, −10°C, and −20°C. The application of the diametral compression load creates tensile stresses in the horizontal direction along the specimen diameter. Strains are measured close to the center of the specimens using four LVDTs placed at right angles on each side of a specimen to measure both horizontal and vertical deformations.

The calculated creep compliances at these three temperatures are used to determine the master curve for the creep compliance. Then, mathematical conversions are used to obtain the relaxation modulus. The master curve of the relaxation modulus, along with the temperature changes in the pavement, are used to calculate thermal stresses, as explained in Chapter 11. The indirect tensile strength of the specimen for thermal cracking analysis is determined by applying a load at a rate of 12.5 mm per minute until the load starts to decrease because of specimen failure. The calculated thermal stresses are compared to the tensile strength to determine whether the mixture will experience low-temperature cracking.

### 5.8.5 Beam Fatigue

Load-associated fatigue cracking is one of the most critical distress types that occurs in flexible pavement systems. Fatigue testing of asphalt mixtures has been the focus of many studies that have utilized different sample shapes, sizes, and testing equipment. The beam fatigue test has been used to provide a measure of the laboratory fatigue life (number of cycles to failure) at different stress or strain values.[2]

The test consists of applying a repeated constant vertical strain or stress to a beam specimen in flexural tension mode until a certain criterion is met, such as complete failure or 50% reduction in initial modulus. The components of the test are shown in Figure 5.29. Under a harmonic sinusoidal loading, the energy dissipated in each cycle can be represented as in Equation 5.7. The difference is, however, that the modulus, stress, and strain correspond to the flexural condition. Flexural deflections are recorded via a single LVDT attached to the center of the specimen. Under controlled strain loading, damage is detected by the reduction in stress; under stress-controlled loading, damage is detected by the increase in strain. The measurements can be used to plot various relationships, as shown in Equations 5.37 to 5.40.

$$N_f = k_1 \left( \frac{1}{\varepsilon_o} \right)^{n_1} \tag{5.37}$$

$$N_f = k_1 \left( \frac{1}{\sigma_o} \right)^{n_1} \tag{5.38}$$

$$N_f = k_1 \left( \frac{1}{\varepsilon_o} \right)^{n_1} \left( \frac{1}{|E_o^*|} \right)^{n_2} \tag{5.39}$$

$$N_f = k_1 \left( \frac{1}{\Delta W_o} \right)^{n_1} \tag{5.40}$$

where $\varepsilon_o$ is the initial strain, $\sigma_o$ is the initial stress, $|E_o^*|$ is initial modulus, and $\Delta W_o$ is the initial dissipated energy.

**Example 5.7**

A series of flexural fatigue tests were conducted on beam specimens in strain-controlled mode. A beam specimen has the reactions 12 inches apart, and the cross-sectional dimensions are 3 inches by 3 inches. The initial flexural stiffness of the asphalt mix is 450,000 lb/in$^2$. Based on the results given in Table 5.12, determine the following:

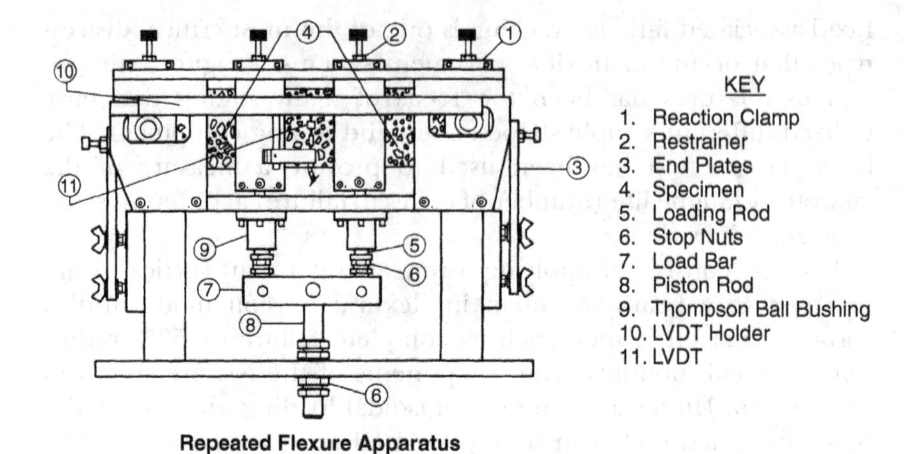


**Figure 5.29**
Components of the Beam Fatigue Test (Ref. 17)

- The initial applied load and beam deflection at each strain level
- The coefficients ($k_1$, $n_1$) of the equation that describes the relationship between strain ($\varepsilon$), and number of load applications to failure $N_f$ (i.e., $N_f = k_1\left(\frac{1}{\varepsilon}\right)^{n_1}$).

**ANSWER**

The initial applied load and beam deflection at each strain level can be obtained as follows. At each strain level ($\varepsilon$) the initial stress can be calculated using the elastic relation: $\sigma = E_o{}^*\varepsilon$, where $E_o{}^*$ is the initial flexural stiffness of the beam (450,000 lbs/in$^2$). Once the stress is known, the initial total load applied to the beam ($P$) can be obtained from:

$$P = \frac{\sigma\, b\, h^2}{3a} \tag{5.41}$$

where, $a$ is the distance between the support and the first applied load (4 in), $b$ is the specimen width (3 in), and $h$ is the specimen height (3 in). Using the values of the initial applied load ($P$), the initial dynamic deflection of the beam at the center can be determined using the expression:

$$d = \frac{P\, a\,(3l^2 - 4a^2)}{48\, IE_s} \tag{5.42}$$

**Table 5.12**
Data Obtained from a Flexural Test in Controlled Strain ($\epsilon$) Conditions

| Controlled Strain | $N_f$ |
|---|---|
| 2.00E-04 | 1.00E + 05 |
| 3.50E-04 | 7.00E + 04 |
| 5.00E-04 | 2.00E + 04 |
| 5.50E-04 | 6.50E + 04 |
| 6.00E-04 | 6.00E + 04 |
| 6.50E-04 | 8.30E + 03 |
| 7.00E-04 | 7.50E + 03 |
| 9.00E-04 | 2.50E + 03 |
| 9.50E-04 | 6.00E + 03 |
| 1.00E-03 | 6.50E + 03 |
| 1.05E-03 | 3.00E + 03 |
| 2.00E-03 | 8.00E + 02 |
| 2.50E-03 | 2.00E + 03 |
| 3.00E-03 | 3.00E + 02 |

where $I$ is the moment of inertia of the beam, $I = \frac{bh^3}{12} = 6.75\,\text{in}^4$, and $l$ is the reaction span length (12 in).

Table 5.13 shows the results for the initial stress, the initial applied load, and the initial deflection at the center of the beam for each of the tests conducted.

Taking logarithms of equation $N_f = k_1\left(\frac{1}{\varepsilon}\right)^{n_1}$, gives:

$$\log(N_f) = \log(k_1) + n_1 \log\left(\frac{1}{\varepsilon}\right) \tag{5.43}$$

The plot of $\log(1/\varepsilon)$ versus $\log(N_f)$ corresponding to the data provided is presented in Figure 5.30.

Linear regression gives the following expression:

$$\log(N_f) = -2.8315 + 2.1858 \log\left(\frac{1}{\varepsilon}\right) \tag{5.44}$$

By comparing this equation with Equation 5.43, it is clear that the two parameters, $k_1$ and $n_1$ are $10^{-2.8315} = 1.47 \times 10^{-3}$ and 2.1858, respectively.

**Table 5.13**
Initial Flexural Stress, Initial Total load, and Initial Maximum Deflection in Flexural Fatigue Tests Conducted at Different Constant Strain level

| Controlled Strain | Initial Flexural Stress (psi) | Initial $P$ (lb) | $d$ (inches) |
|---|---|---|---|
| 2.00E-04 | 90.00 | 202.50 | 2.044E-03 |
| 3.50E-04 | 158.00 | 354.38 | 3.578E-03 |
| 5.00E-04 | 225.00 | 506.25 | 5.111E-03 |
| 5.50E-04 | 248.00 | 556.88 | 5.622E-03 |
| 6.00E-04 | 270.00 | 607.50 | 6.133E-03 |
| 6.50E-04 | 293.00 | 658.13 | 6.644E-03 |
| 7.00E-04 | 315.00 | 708.75 | 7.156E-03 |
| 9.00E-04 | 405.00 | 911.25 | 9.200E-03 |
| 9.50E-04 | 428.00 | 961.89 | 9.711E-03 |
| 1.00E-03 | 450.00 | 1012.50 | 1.022E-02 |
| 1.05E-03 | 473.00 | 1063.13 | 1.073E-02 |
| 2.00E-03 | 900.00 | 2025.00 | 2.044E-02 |
| 2.50E-03 | 1130.00 | 2531.25 | 2.556E-02 |
| 3.00E-03 | 1350.00 | 3037.50 | 3.067E-02 |

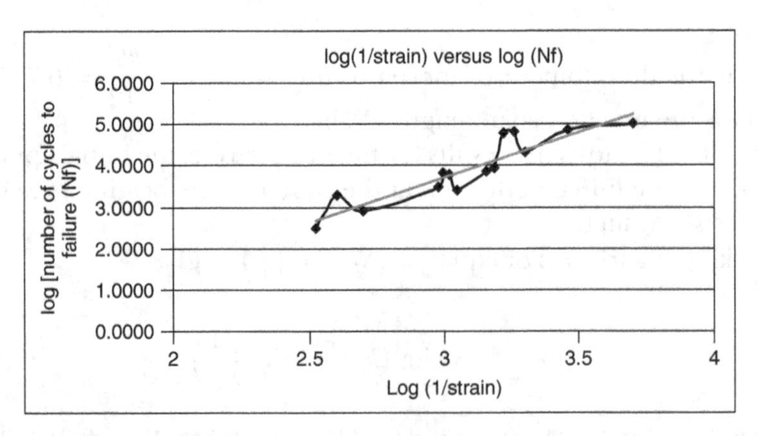


**Figure 5.30**
Log ($N_f$) versus Log($1/\varepsilon$)

## References

1 Anderson, D. A., D. W. Christensen, H. U. Bahia, R. Dongre, M. G. Sharma, C. E. Antle, and J. Button (1994). *Binder Characterization and Evaluation. Volume 3: Physical Characterization*, SHRP-A-369. National Research Council, Washington, DC.

[2] Asphalt Research Program (ARP) (1994). *Fatigue Response of Asphalt-Aggregate Mixes*, Institute of Transportation Studies, University of California, Berkeley, Strategic Highway Research Program SHRP-A-404, National Research Council, Washington, DC.

[3] Bahia, H. U., D. A. Anderson, and D. W. Christensen (1992). The Bending Beam Rheometer: A Simple Device for Measuring Low-Temperature Rheology of Asphalt Binders, *Journal of the Association of Asphalt Paving Technologists*, Vol. 61, pp. 117–153.

[4] Bahia, H., Hanson D., Zeng M., Zhai H., Khatri M., Anderson R. (2001). *Characterization of Modified Asphalt Binders in Superpave Mix Design*, National Cooperative Highway Research Program Report 459, Washington, DC.

[5] Bhasin A., Masad E., Little D., and Lytton R. (2006). Limits on Adhesive Bond Energy for Improved Resistance of Hot-Mix Asphalt to Moisture Damage, in Transportation Research Record 1970, *Journal of the Transportation Research Board*, pp. 3–13.

[6] Cheremisinoff, N. P. (1993). *An Introduction to Polymer Rheology and Processing*, CRC Press, tinden, Newjersey.

[7] Findley, W. N., Lai, J. S., and Onaran K. (1990). *Creep and Relaxation of Nonlinear Viscoelastic Materials*, Dover Publications, Inc., New York.

[8] Hiltunen, D. R., and Roque R. (1994). "A Mechanics-Based Prediction Model for Thermal Cracking of Asphaltic Concrete Pavements" *Journal of the Association of Asphalt Paving Technologists*, Vol. 63, pp. 81–117.

[9] Hunter, R. N. (2000). *Asphalts in Road Construction*, Thomas Telford Publishing, London.

[10] King G., King H., Pavlovich R., Epps A., and Kandhal P. (2001). Additives in Asphalt, *Journal of the Association of Asphalt Paving Technologists*, pp. 32–69.

[11] Little, D. N., and Bhasin A. (2006). *Using Surface Energy Measurements to Select Materials for Asphalt Pavement*, Final Report for Project 9-37, sponsored by the National Cooperative Highway Research Program, Washington, DC.

[12] Mamlouk, M. S., and Zaniewski, J. P. (1999). *Materials for Civil and Construction Engineers*, Addison-Wesley, Menlo Park, CA.

[13] McGennis, R. B., Shuler S., Bahia H. (1994). *Background of Superpave Asphalt Binder Test Methods*, Federal Highway Administration, Report No. FHWA-SA-94-069, Washington, DC.

[14] Mysels, K. J. (1959). *Introduction to Colloid Chemistry*, Interscience Publishers, Inc., New York.

[15] Peterson, J. C. (1984). "Chemical Composition of Asphalt as Related to Asphalt Durability, State-of-the-Art", in Transportation Research Record No. 999, Transportation Research Board, Washington, DC.

[16] Pfeiffer, J. P., and Saal, R. N. (1940). Asphaltic Bitumen as Colloidal System, *Journal of Physical Chemistry*, 44, pp. 139–149.

[17] Roberts, F. L., Kandhal, P. S., Brown, E. R., Lee, D. Y., and Kennedy, T. W. (1996). *Hot-Mix Asphalt Materials, Mixture, Design, and Construction*, National Asphalt Pavement Association Research and Education Foundation, Lanham, MD.

[18] Robertson, R. E. (2000). "Chemical Properties of Asphalts and Their Effects on Pavement Performance", National Cooperative Highway Research Program Report 499, Transportation Research Board, National Research Council, Washington, DC.

[19] Roque R., Hiltunen, D. R., and Buttlar, W. G. (1995). "Thermal Cracking Performance and Design of Mixtures Using Superpave". *Journal of the Association of Asphalt Paving Technologists*, Vol. 64, pp. 718–729.

[20] Western Research Institute (WRI) (2001). "Fundamental Properties of Asphalts and Modified Asphalts, Volume I: Interpretive Report", FHWA-RD-99-212, Western Research Institute, Laramie, WY.

[21] Witczak M., Kaloush K., Pellinen, T, El-Basyouny, M., and Von Quintun, H. (2002). *Simple Performance Test for Superpave Mix Design*, National Cooperative Highway Research Program Report 465, Washington, DC.

[22] Witczak M. (2002). *Simple Performance Tests: Summary of Recommended Methods and Database*, National Cooperative Highway Research Program Report 547, Washington, DC.

## Problems

5.1 Discuss the influence of oxidation on mixture molecular structure and viscosity.

5.2 Use a book on rheology to describe the behavior of a Bingham fluid and a thixotropic fluid.

5.3 Use the LTPPBind software to determine the required span of pavement temperatures and the appropriate PG binder grades in each of the following locations:

- Houston, Texas (Bush Intercontinental Airport)
- Anchorage, Alaska (Anchorage International Airport)
- New York, New York (New York JF Kennedy Airport)

The LTPPbind software can be accessed through the book Web site.

5.4 An engineer wants to determine if a certain asphalt would be graded as PG 58-28. At what temperatures should he or she run the following tests?

**Table 5.14**
Asphalt Cement Test Results—Problem 5.5

| Test | Results |
|---|---|
| **Original Properties** | |
| Flash point temperature, °C | 278 |
| Viscosity at 135°C | 0.490 Pa.s |
| Dynamic shear rheometer | |
| at 82°C | $G^* = 0.82$ kPa, $\delta = 68°$ |
| at 76°C | $G^* = 1.00$ kPa, $\delta = 64°$ |
| at 70°C | $G^* = 1.80$ kPa, $\delta = 60°$ |
| **Rolling Thin Film Oven-Aged Binder** | |
| Dynamic shear rheometer | |
| at 82°C | $G^* = 1.60$ kPa, $\delta = 65°$ |
| at 76°C | $G^* = 2.20$ kPa, $\delta = 62°$ |
| at 70°C | $G^* = 3.50$ kPa, $\delta = 58°$ |
| **Rolling Thin Film Oven-and PAV-Aged Binder** | |
| Dynamic shear rheometer | |
| 34°C | $G^* = 2500$ kPa, $\delta = 60°$ |
| 31°C | $G^* = 3700$ kPa, $\delta = 58°$ |
| 28°C | $G^* = 4850$ kPa, $\delta = 56°$ |
| Bending beam rheometer | |
| −6°C | $S = 255$ MPa, $m = 0.329$ |
| −12°C | $S = 290$ MPa, $m = 0.305$ |
| −18°C | $S = 318$ MPa, $m = 0.277$ |

- DSR for rutting analysis
- DSR for fatigue cracking analysis
- BBR

5.5 What is the PG grade of the asphalt whose results are shown in Table 5.14? Show all calculations and comparisons with Superpave requirements.

5.6 An asphalt mixture has been compacted with the Marshall hammer, using 50 blows. The following data was obtained in the laboratory:

Aggregate Blend

| | | |
|---|---|---|
| Aggregate saturated surface dry (SSD) weight | = | 459.34 gm |
| Weight of flask and water | = | 2345.67 gm |
| Weight of flask, water, and sample | = | 2640.35 gm |
| Aggregate weight after being dried in oven | = | 454.12 mg |

Asphalt Mixture

| | | |
|---|---|---|
| Weight of dry-compacted asphalt mixture in air | = | 3600.0 gm |
| Weight of SSD-compacted mixture in air | = | 3724.2 gm |
| Weight of compacted mixture in water | = | 2200.86 gm |
| Theoretical maximum specific gravity | = | 2.50 |
| Asphalt binder percent per weight of mix | = | 5.0% |
| Specific gravity of asphalt binder | = | 1.00 |

Calculate the bulk dry-specific gravity of aggregate, bulk-specific gravity of the asphalt mixture, the void, in mineral aggregate of the mix, and the percent of air voids in the compacted mix.

5.7 A uniaxial creep test was conducted on an axial mix at 40°C. The data for this test is available on the book's Web site under the name "mix creep data.xls." Use this data to calculate flow-time value, flow-time slope, and flow time intercept.

# 6 Concrete Materials

## 6.1 Introduction

Concrete can be described as "a mixture of glue (cement, water, and air) binding together fillers (aggregate)"[5]. Typically, other supplementary cementitious and chemical admixtures are added to the mixture. The combination of water and cement is referred to as *paste,* while the combination of paste and fine aggregates is referred as *mortar*. Aggregate properties were discussed in Chapter 4. This chapter discusses the properties of the other components of concrete and the properties that are relevant to the design and performance of concrete pavements. It provides a brief description of the tests used for measuring these properties, as the details of these tests are readily available in ASTM and AASHTO standards.

## 6.2 Cementitious Materials

Cementitious materials include hydraulic cements and supplementary cementitious materials. Hydraulic cements include portland cements and blended cements. Portland cement is specified in ASTM and AASHTO on the basis of either chemical/physical properties (ASTM C 150 or AASHTO M 85) or performance (ASTM

C 1157). Blended cements are classified in terms of their major constituents (ASTM C 595 and AASHTO M 240).

In terms of chemical physical properties, portland cements are classified into the following types:

- Type I: Normal resistance
- Type II: Moderate sulfate resistance
- Type III: High early strength
- Type IV: Low heat of hydration
- Type V: High sulfate resistance

In term of performance, portland cements are classified into:

- Type GU: General use
- Type MS: Moderate sulfate resistance
- Type HE: High early strength
- Type: MH: Moderate heat of hydration
- Type LH: Low heat of hydration

In terms of constituents, blended cements are classified into:

- Type IS: Portland blast-furnace slag cement
- Type IP and P: Portland-pozzolan cement
- Type I(PM) Pozzolan-modified portland cement
- Type S: Slag cement
- Type I(SM): Slag-modified portland cement

The major compounds of portland cement are shown in Table 6.1. Supplementary cementitious materials such as fly ash, silica fumes, and blast-furnace slag are used to enhance some of the concrete properties. The majority of these supplementary materials are by-products of industrial processes. Their effects on concrete properties are summarized in Table 6.2.

## 6.3 Hydration

Hydration refers to the reaction of cement and water leading to the hardening of the paste. Hydration is an exothermic process that leads to heat generation. The hydration process is divided into five stages, as shown in Figure 6.1.

In the mixing stage, significant heat is generated due primarily to the immediate reaction between tricalcium aluminate ($C_3A$) and

**Table 6.1**
Major Compounds of Portland Cement

| Group | Compound | Amount (%) | Primary Influence on Concrete Properties |
|---|---|---|---|
| Aluminates | Tricalcium Aluminate ($C_3A$) | 5–10 | Can cause premature stiffening |
| | Tetracalcium aluminoferrite ($C_4AF$) | 5–15 | |
| Silicates | Tricalcium silicate ($C_3S$) | 50–70 | Contributes to early strength |
| | Dicalcium silicate ($C_2S$) | 15–30 | Contributes to long-term strength |
| Sulfates | Calcium sulfate dihydrate or Gypsum ($CSH_2$)<br>Calcium sulfate hemihydrate or Bassanite ($CSH_{1/2}$)<br>Anhydrous calcium hydrate ($CS$) | 3–5 | Reduces the chance of premature stiffening |

water, which produces calcium aluminate hydrate (CAH). The heat of hydration is tested in accordance to ASTM C 186. The generated heat could cause flash set, defined as early stiffening of the concrete. However, gypsum (a source of sulfate) also dissolves and reacts with the dissolved aluminate to produce ettringite (C–A–S–H). This reaction reduces the amount of heat and decreases the possibility of the flash set. Therefore, hydration in the first 15 minutes is a delicate balance between the aluminate and the sulfate in solution. Once the gypsum is depleted, the ettringite layer reacts with $C_3A$ to form monosulfate, which has minimal effect on concrete physical properties.

In the second stage (dormancy), the concrete remains plastic without heat generation. This is the stage during which concrete should be placed and finished. This stage could last from two to four hours. The formation of ettringite is responsible for this dormant period, as it slows the hydration and the heat generation, but it contributes to the early concrete strength.

The third stage, hardening, is when concrete starts to stiffen, and hydration products continue to increase. The hardening stage is dominated by the silicate reactions that produce calcium silicate hydrate (C–S–H). The $C_3S$ reaction is responsible for the

**Table 6.2**
Effects of Supplementary Cementitious Materials on Concrete Properties (Adapted from Refs. 1, 5 and 7)

| | Supplementary Cementitious Materials | | | | | | |
|---|---|---|---|---|---|---|---|
| **Property** | **Class F Fly Ash** | **Class C Fly Ash** | **GGBF Slag** | **Silica Fume** | **Calcinated Shale** | **Calcinated Clay** | **Metakaolin** |
| Water Requirements | −2 | −2 | −1 | 2 | 0 | 0 | 1 |
| Workability | 1 | 1 | 1 | −2 | 1 | 1 | −1 |
| Bleeding and Segregation | −1 | −1 | | −2 | 0 | 0 | −1 |
| Air Content | −2 | −1 | −1 | −2 | 0 | 0 | −1 |
| Heat of Hydration | −1 | | −1 | 0 | −1 | −1 | −1 |
| Setting Time | 1 | | 1 | 0 | 1 | 1 | 0 |
| Finishability | 1 | 1 | 1 | | 1 | 1 | 1 |
| Pumpability | 1 | 1 | 1 | 1 | 1 | 1 | 1 |
| Plastic Shrinkage Cracking | 0 | 0 | 0 | 1 | 0 | 0 | 0 |
| Early Strength | −1 | 0 | −1 | 2 | −1 | −1 | 2 |
| Long-Term Strength | 1 | 1 | 1 | 2 | 1 | 1 | 2 |
| Permeability | −1 | −1 | −1 | −2 | −1 | −1 | −2 |
| Chloride Ingress | −1 | −1 | −1 | −2 | −1 | −1 | −2 |
| ASR | −2 | | −2 | −1 | −1 | −1 | −1 |
| Sulfate Resistance | 2 | | 2 | 1 | 1 | 1 | 1 |
| Freezing and Thawing | 0 | 0 | 0 | 0 | 0 | 0 | 0 |
| Abrasion Resistance | 0 | 0 | 0 | 0 | 0 | 0 | 0 |
| Drying Shrinkage | 0 | 0 | 0 | 0 | 0 | 0 | 0 |

Index 2 = significantly increased, 1 = increased, 0 = no significant change, −1 = reduced, −2 = significantly reduced, ☐ = varies.

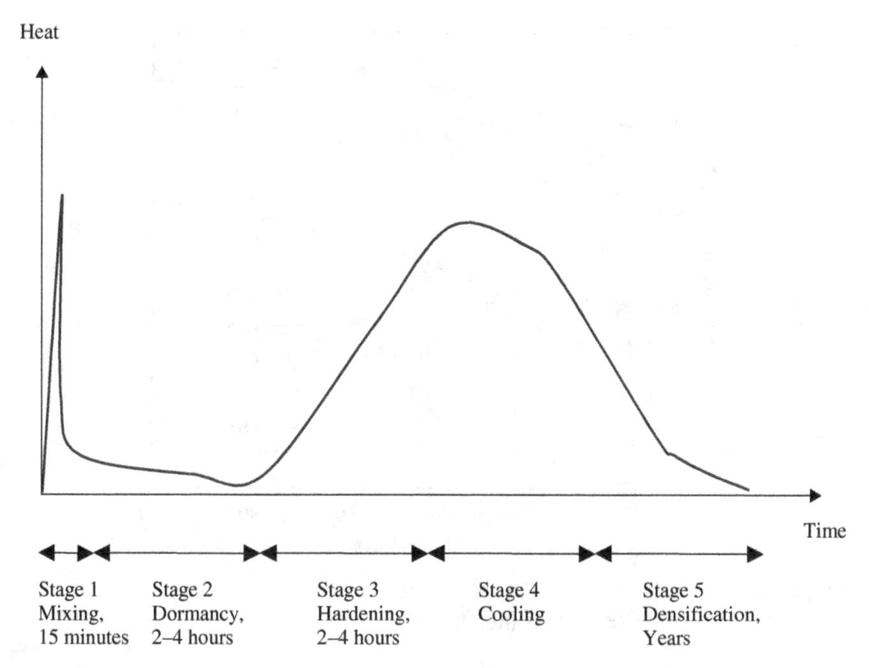


**Figure 6.1**
Illustration of the Different Stages of Hydration (Ref. 5)

early strength gain, while the $C_2S$ reaction contributes to long-term strength gain and low permeability. Calcium silicate hydrate (C–S–H) is a major contributor to concrete strength and low permeability. Calcium hydroxide (CH) is another product of the reactions of $C_3S$ and $C_2S$ with water. This product causes a weak plane in the concrete structure, and provides a high pH value that allows C–S–H to be stable. Figure 6.2 shows the rate of reaction of the main components of cement.

In the fourth stage, cooling, concrete shrinks. This shrinkage is restrained by friction from the underlying layers, which results in the buildup of tensile stresses in concrete slabs. Cracking can be avoided by sawing the concrete to relieve the stresses, as described in Chapter 8.

The last stage is densification, during which hydration continues as long as cement and water are present in the mix. Curing compounds can be used early in this stage to keep the concrete moist, and assists in the continuation of the hydration process.

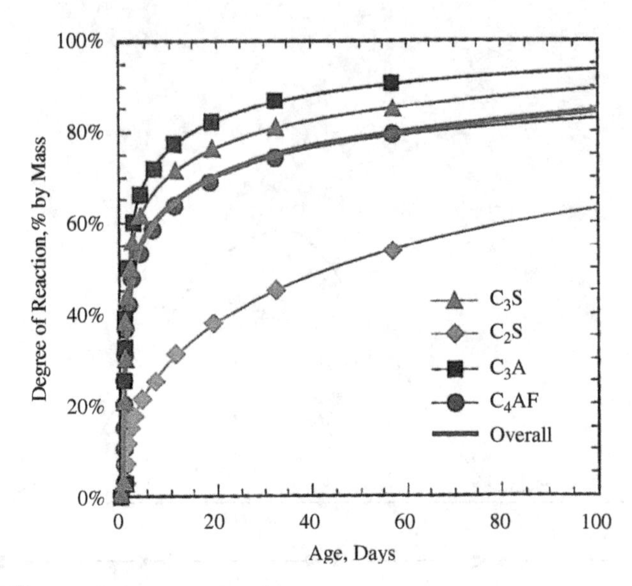


**Figure 6.2**
Reactivity of Cement Compounds (Ref. 6)

## 6.4 Chemical Admixtures

Admixtures are materials added to concrete mixtures to modify concrete properties such as air content, water requirement, and setting time[5]. Air-entraining admixtures are specified by ASTM C 260. These admixtures are used to create small air bubbles in the paste and improve the concrete resistance to freezing/thawing cycles and scaling. Water-reducing admixtures are used to reduce the water content without reducing slump. These admixtures are specified in the ASTM C 494 and AASHTO M 194 procedures. Admixtures are also specified in these procedures for decreasing the rate of hydration (retarders) or increasing the rate of hydration (accelerators).

## 6.5 Properties of Cement, Paste, and Mortar

The rate of hydration and, consequently, the gain in strength increase with an increase in cement particle size fineness. However, an increase in fineness can lead to an increase in permeability in the long term. In general, 95% of cement particles are smaller than

45 micrometers, while the average particle is around 15 micrometers[1]. The Blaine test (ASTM C 204 or AASHTO T 153) is typically used for measuring fineness of cement particles. The underlying principle of this test is that the permeability of a powder decreases as the size of particles decreases. The test measures the time required for a standard volume of air to flow through a standard volume of portland cement, and compares it to the time required for the air to flow through a reference material. Typical values for Blaine fineness of portland cement range from 300 to 450 $m^2/kg$ (3,000 to 4,500 $cm^2/g$)[5]. Higher values of Blaine fineness indicate a finer-graded cement.

The density of cement particles without air between them is measured using the ASTM C 188 or AASHTO T 133 procedure. This density ranges from 3100 to 3250 $kg/m^3$. The bulk density includes the volume of air between particles, and depends on the level of compaction. Uncompacted portland cement has a density of about 830 $kg/m^3$; it has a density of about 1650 $kg/m^3$ when consolidated with vibration[8].

Another important property is the consistency of paste or mortar. The Vicat plunger shown in Figure 6.3 (ASTM C 187 or AASHTO T 129) is used to test the consistency of paste. The water content is varied in the paste until a penetration of 10 mm is achieved. The mortar consistency is measured using the flow table test described in ASTM C 230 or AASHTO M 152 and in ASTM C 1437. Both the Vicat test and flow table test are used to prepare paste and mortar, respectively, at a given consistency for further testing.

Cement specifications include limits on the initial set, when the paste loses its fluidity, and the final set, when the paste attains some hardness. These two time limits are determined using the Vicat test according to the ASTM C 191 or AASHTO T 131 procedure. As discussed earlier in this chapter, the setting time is influenced by the balance between the sulfate content (primarily from gypsum) and the aluminate compounds. Cements are tested for early paste stiffening using ASTM C 451 or AASHTO T 186, and early mortar stiffening using ASTM C 359 or AASHTO T 185.

A powerful tool that has been used to identify the cement composition and its interaction with other paste and concrete compounds is thermal analysis. The main concept in this analysis is heating a small sample to high temperatures. The increase in temperature

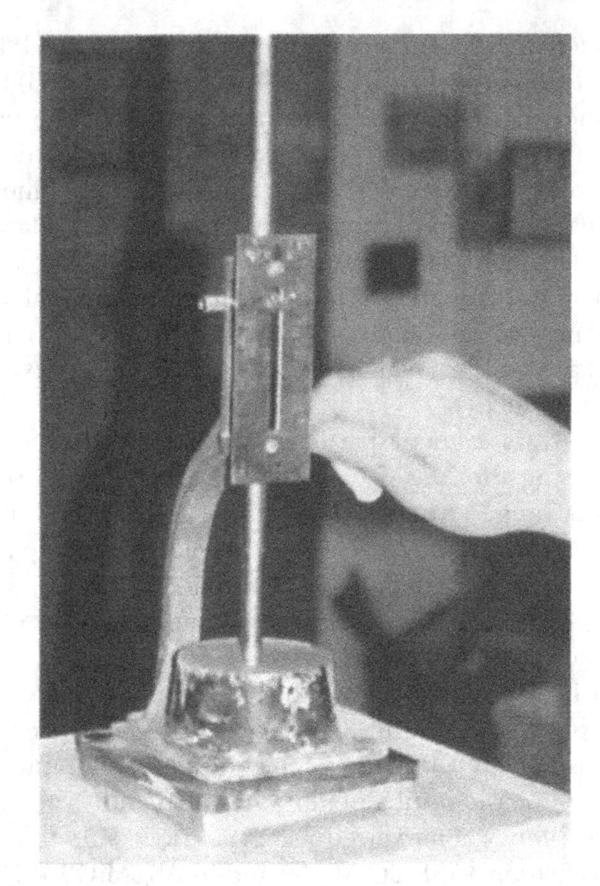

**Figure 6.3**
Vicat Plunger Used for Measuring Paste Consistency (Ref. 1)

causes some compounds to react or decompose. The analysis identifies these processes by recording the time and temperature at which these changes take place.

Thermogravimetric analysis (TGA) measures the change in mass of a sample under a change in temperature. Each compound has a specific temperature at which it decomposes. Consequently, a change in mass within a specific temperature range identifies the presence of a particular chemical compound. For example, calcium hydroxide decomposes to water vapor and calcium oxide at a temperature between 400°C and 500°C. Therefore, the amount of weight loss at this temperature can be used to determine how much calcium hydroxide was originally present in the sample, and gives an indication of the degree of hydration that has taken place in a sample.

Another method of thermal analysis is the differential scanning calorimetry (DSC). This method measures the heat released or absorbed to identify the presence of a compound. It does not rely on the loss of mass, therefore it can still identify a compound if it melts without vaporizing. DSC can be used to determine which compounds are present at different stages of hydration[1].

The compressive strength of mortar is measured on cubes using the ASTM C 109 or AASHTO T 106 procedure. However, mortar strength does not have strong correlation with concrete strength, because the latter is influenced by many other factors, including aggregate properties and environmental conditions.

## 6.6 Properties of Concrete

Workability refers to the consistency, mobility, and compactibility of fresh concrete. Good workability leads to easier finishing and more uniform properties of the pavement. It also affects the properties of concrete after it hardens. Workability depends on the physical characteristics of aggregates and cement, proportioning of the concrete components, water content, the equipment used, and the construction conditions such as pavement thickness and reinforcement. The slump test (ASTM C 143 or AASHTO T 119) is still the most popular method for measuring consistency. Other methods were also developed that rely on measuring the applied torque required to rotate an impeller in concrete as a function of rotational speed. An increase in torque is an indication of a decrease in concrete workability.

An increase in cement fineness at a given water content causes a decrease in workability. An increase in water-to-cement ratio ($w/c$) causes an increase in workability. A deficiency of fine aggregate can lead to a harsh mix that is difficult to work, while an increase in water content will increase the flow and compactability of the mix. An increase in the aggregate-to-cement ratio results in a decrease in workability for a fixed $w/c$ ratio[2]. The increase in aggregate texture, angularity and elongation causes a decrease in workability. Entrained air helps improving the workability of concrete[1]. However, excessive amounts of entrained air can make a mixture difficult to finish and may reduce concrete strength. Water-reducing admixtures are used to improve workability without increasing the $w/c$ ratio.

Bleeding is a characteristic related to the properties of fresh concrete, and it refers to the presence of water at the pavement surface when the concrete is fresh. The main cause of bleeding is the settlement of cement and aggregates and the migration of water to the top[1]. Bleeding causes the presence of fine cement particles at the surface, which form a weak layer susceptible to scaling[3].

### 6.6.1 Strength

Strength is an important property of concrete, which is more influenced by the paste strength than the aggregate strength, for the majority of concrete mixtures. Theoretically, maximum strength is expected when the mix contains just sufficient water for hydration. However, low $w/c$ ratios have adverse effects on fresh concrete properties, (e.g., on workability and compactability).

Air-entrained concrete is typically used to protect pavements from the effect of freeze/thaw cycles. At a given $w/c$ ratio, an increase in entrained air causes a decrease in strength. However, typically $w/c$ ratio is reduced for air-entrained concretes, hence comparable strengths with non air-entrained concretes are achieved.

Compressive strength is measured according to ASTM C 39 on cylindrical specimens of 150 mm diameter and 300 mm height. The failure load is divided by the cross-sectional area to give the compressive strength at a given curing period.

$$f_c' = \frac{P}{(\pi/4)D^2} \tag{6.1}$$

where $P$ is the failure load and $D$ is the diameter of the cylinder.

Tensile strength is an important property in determining the concrete resistance to cracking under shrinkage and temperature changes and loads in plain concrete pavements. The two common methods for measuring tensile strength are the flexural test and the split cylinder test. In the split cylinder test (ASTM C 496), a cylindrical specimen of minimum 50 mm diameter is placed on its axis in a horizontal plane and is subjected to a uniform line load along the length of the specimen. The tensile strength is calculated as follows:

$$f_t = \frac{2P}{\pi LD} \tag{6.2}$$

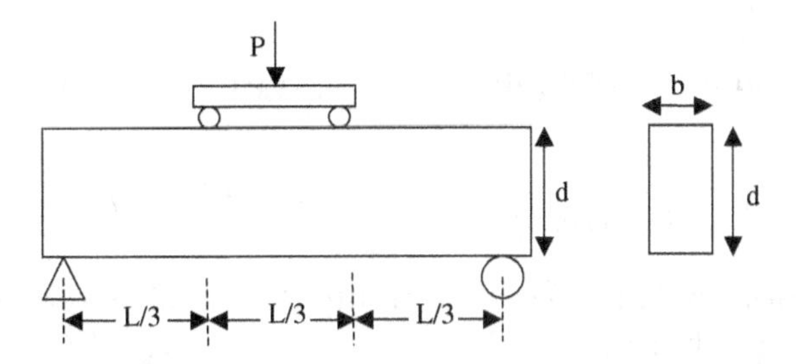


**Figure 6.4**
Schematic of the Flexural Test

where $P$ is the failure load, $L$ is the length, and $D$ is the diameter of the cylinder. The tensile strength of concrete can be estimated from its compressive strength using empirical equations. For normal weight concrete, it can be computed as:

$$f_t = 0.556\sqrt{f_c'} \tag{6.3}$$

where the strength is expressed in MPa.

In the flexure test illustrated in Figure 6.4, a concrete beam is tested using third-point loading (ASTM C 78). From the failure load, the modulus of rupture ($f_r$), which describes the tensile strength, is calculated as follows:

$$f_r = \frac{PL}{bd^2} \tag{6.4}$$

where $L$ is the span length, $P$ is the failure load, $b$ is the beam width, and $d$ is the beam depth.

The following empirical relationship is commonly used to predict the modulus of rupture from the compressive strength:

$$f_r = 0.75\sqrt{f_c'} \tag{6.5}$$

where $f_r$ and $f_c'$ are in MPa. The variable $Sc'$ is used for the modulus of rupture.

**Example 6.1**

A 150-mm diameter portland concrete cylinder failed under a 487-kN compressive load. Compute its compressive strength and estimate its tensile and flexural strengths.

**ANSWER**

Use Equation 6.1 to compute the compressive strength:

$$f'_c = \frac{487}{(\pi/4)0.15^2} = 27{,}580 \text{ kPa} (4000 \text{ lb/in}^2)$$

Use Equations 6.3 and 6.5 to compute the tensile strength and the flexural strength, respectively:

$$f_t = 0.556\sqrt{27.58} = 2.92 \text{ MPa} \quad (426 \text{ lb/in}^2)$$

$$f_r = 0.75\sqrt{27.58} = 3.94 \text{ MPa} \quad (571 \text{ lb/in}^2)$$

### 6.6.2 Modulus of Elasticity and the Poisson's Ratio

The stress-strain relationship for concrete is used to determine the modulus of elasticity and the Poisson's ratio. These properties are important in the structural analysis and design of rigid pavements (Chapters 8 and 12). As shown in Figure 6.5, there are different definitions for the modulus of elasticity. The tangent modulus is the slope of a tangent at a point in the stress-strain curve. If the tangent is taken at the origin, then the slope is called the initial tangent modulus. The most common method to measure the modulus is the secant modulus. The secant modulus is measured according to

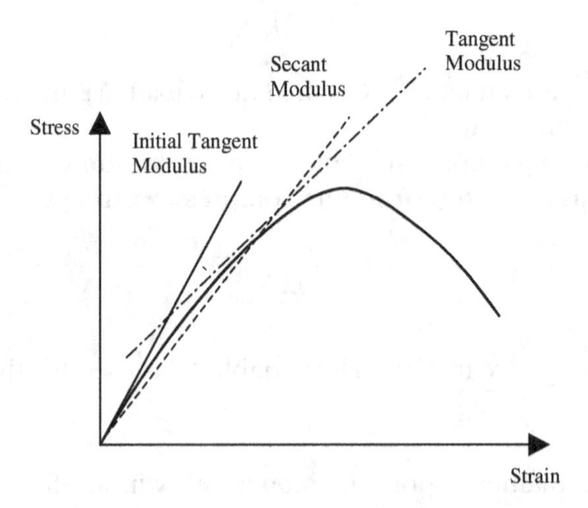


**Figure 6.5**
Illustrations of the Different Types of Elastic Moduli

ASTM C 469 procedure, and is defined as follows:

$$E = \frac{S_2 - S_1}{\varepsilon_2 - \varepsilon_1} \tag{6.6}$$

where $S_2$ is the stress corresponding to 40% of the ultimate load, $S_1$ is the stress at a strain, $\varepsilon_1$, which is taken at 0.0005, and $\varepsilon_2$ is the strain produced at stress $S_2$.

According to American Concrete Institute (ACI) 318, the modulus of elasticity of concrete can be calculated using the following empirical equation:

$$E = 0.043\ \rho^{1.5}\sqrt{f_c'} \tag{6.7}$$

where $E$ is the modulus in MPa, $\rho$ is the density of concrete in kg/m$^3$, and $f_c'$ is the compressive strength of the concrete in MPa.

Special meters are available to measure temperature over time during the hydration process. This information is used to relate concrete properties (strength and modulus of elasticity) to time and temperature. This relationship is known as the *maturity curve.* The Poisson's ratio is defined as the ratio of the lateral strain to the axial strain measured in a unixial test of concrete. It is used in predicting the behavior of early-age concretes and in the structural design of rigid pavements.

### 6.6.3 Shrinkage and Creep

The loss of water from concrete causes a reduction in volume or shrinkage. However, this reduction is restrained by friction between the concrete pavement and the supporting layer. As a result, stresses develop in the pavement, leading to shrinkage cracks at the surface, as will be more fully described in Chapter 8. The loss of water can occur early in the life of the pavement, when it is plastic, or during subsequent drying, referred to as *plastic shrinkage* and *drying shrinkage*, respectively. Plastic shrinkage is about 5 to 10 times larger than drying shrinkage. The decrease in the amount of coarse aggregate and the increase in sand content increases shrinkage. Conversely, too much coarse aggregate restrains the shrinkage deformation, resulting in excessive tensile stresses[4]. Plain concrete pavements experience more shrinkage than reinforced concrete. This reinforcement does not prevent cracks from occurring, but it does serve the purpose of controlling the location and width of cracks resulting from both shrinkage and temperature effects, as described in Chapter 8.

Another type of shrinkage is associated with the fact that some of the hydration products of cement occupy less space than the original materials. This is referred to as *autogenous shrinkage*, and it becomes significant when the $w/c$ ratio is below 0.40. Shrinkage can also be caused by concrete contraction under cooling temperatures.

Creep, which is defined as the time-dependent deformation under load, occurs mostly in the paste of hardened concrete. Creep strain is different from the elastic strain, which occurs instantaneously when loads are applied. Creep strain is several times larger than the initial or elastic strain[4]. When load is removed, the elastic strain recovers instantaneously. Some of the creep strain recovers with time, and is referred to as *creep recovery*, while some of the strain remains permanently. ASTM C 157 and AASHTO T 160 are commonly used to determine the length change in unrestrained concrete due to drying shrinkage.

### 6.6.4 Durability

Durability refers to the concrete resistance to environmental and chemical exposure. The main factors that influence concrete pavement durability are permeability, freezing and thawing, temperature variation, influence of chemicals such as deicers, and chemical reactions such as alkali-silica reaction and sulfate attack.

Permeability is a property that quantifies the transport of fluids in concrete. This is an important property because most of the durability-related problems in concrete pavements are associated with the transport of harmful materials through the concrete. These transport mechanisms include water penetration, causing freezing-related cracking; salt movement, causing scaling at the surface; penetration of sulfates, causing sulfate attacks; and flow of oxygen, moisture, and chlorides, causing steel corrosion[5]. There are several methods for measuring chloride ion penetration (ASTM C 1202 or AASHTO T 277, AASHTO TP 64), chloride resistance of concrete (ASTM C 1543 and ASTM C 1556), and capillary absorption of concrete (ASTM C 1585). The flow of water carrying some of the harmful materials can be quantified by the permeability coefficient determined from Darcy's law, which is described in detail in Chapter 10.

Harmful materials can also penetrate concrete through the diffusion process (diffusion of ions of chloride, carbon dioxide, and oxygen). Diffusion can be described by Fick's second law:

$$\frac{dc}{dt} = D\,\frac{d^2c}{dx^2} \tag{6.8}$$

where $c$ is the concentration of the diffusing material, $dc/dt$ is the rate of diffusion, $dc/dx$ is the gradient of the concentration of the material, and $D$ is the diffusion coefficient, which depends on the diffusing material, concrete properties, and environmental conditions. An application of this expression in estimating the time to corrosion initiation is given in Example 6.2.

**Example 6.2**

A reinforced concrete pavement is subjected to chloride concentration of 2.2 kg/m$^3$ at the surface. The critical chloride threshold at which steel corrosion is initiated is 1.1 kg/m$^3$. The steel cover is 70 mm, and the diffusion coefficient is $1.1 \times 10^{-11}$ m$^2$/s. Determine the time (years) to steel corrosion initiation.

**ANSWER**

The chloride diffusion in concrete is governed by Fick's second law, shown in Equation 6.8. The solution for this equation is:

$$\left(\frac{C(x,t) - C_{initial}}{C_{surface} - C_{initial}}\right) = 1 - erf\left(\frac{x}{\sqrt{4Dt}}\right)$$

where $C(x, t)$ is the chloride concentration at time $t$ and distance $x$ from the surface, $C_{initial}$ is the initial concentration of chloride prior to the application of the surface chloride, $C_{surface}$ is the chloride concentration at the surface, $D$ is the diffusion coefficient, and $erf$ is an "error function" given in Table 6.3 for $z = \frac{x}{\sqrt{4Dt}}$.

To simplify the solution, it is assumed that the concrete is exposed to a constant concentration of chloride ($C_{surface}$) that does not change with time, and the initial concentration ($C_{initial}$) is zero. Therefore, the solution to Equation 6.8 becomes:

$$\frac{C(x,t)}{C_{surface}} = 1 - erf\left(\frac{x}{\sqrt{4Dt}}\right)$$

$$\frac{1.1}{2.2} = 1 - erf\left(\frac{70 \times 10^{-3}}{\sqrt{4 \times 1.1 \times 10^{-11} \times t}}\right)$$

$$0.500 = erf\left(\frac{10553}{\sqrt{t}}\right)$$

**Table 6.3**
Values of the Error Function *erf* for Solving Equation

| z | erf(z) | z | erf(z) | z | erf(z) |
|---|---|---|---|---|---|
| 0 | 0 | 0.55 | 0.5633 | 1.30 | 0.9340 |
| 0.025 | 0.0282 | 0.60 | 0.6039 | 1.40 | 0.9523 |
| 0.05 | 0.0564 | 0.65 | 0.6420 | 1.50 | 0.9961 |
| 0.10 | 0.1125 | 0.70 | 0.6778 | 1.60 | 0.9763 |
| 0.15 | 0.1680 | 0.75 | 0.7112 | 1.70 | 0.9838 |
| 0.20 | 0.2227 | 0.80 | 0.7421 | 1.80 | 0.9891 |
| 0.25 | 0.2763 | 0.85 | 0.7707 | 1.90 | 0.9928 |
| 0.30 | 0.3286 | 0.90 | 0.7970 | 2.00 | 0.9953 |
| 0.35 | 0.3794 | 0.95 | 0.8209 | 2.10 | 0.9981 |
| 0.40 | 0.4284 | 1.00 | 0.8427 | 2.20 | 0.9993 |
| 0.45 | 0.4755 | 1.10 | 0.8802 | 2.30 | 0.9998 |
| 0.50 | 0.5205 | 1.20 | 0.9103 | 2.40 | 0.9999 |

From Table 6.3, $z = 0.4772$ for $erf = 0.5$. Therefore, time (sec) is obtained by solving:

$$0.4772 = \left(\frac{10553}{\sqrt{t}}\right)$$

which translates to 15.5 years.

Concrete pavement durability is further compromised by the effect of frost. As temperature drops, water begins to freeze, and increase in volume. Water does not all freeze at the same temperature since a lower temperature is needed for water to freeze in smaller pores. The increase in volume of frozen water causes pressure on the concrete, causing fracture where this pressure exceeds the concrete strength. This durability problem is manifested as cracks near the joints, which are known as *D-cracks*. The concrete resistance to freezing and thawing can be measured using the ASTM C 666 procedure. The resistance to freezing and thawing is related to the air-void system. The possibility of concrete fracture decreases if there is sufficient volume of air bubbles where water can freeze without exerting high pressure on concrete. The air-void system is measured on hardened concrete using a microscope that gives the size of voids and their spacing (ASTM C 457).

Scaling or deterioration of the concrete pavement surface is caused by the chemical reaction of deicers with concrete. As discussed in Chapter 4, some aggregates such as cherts, are susceptible to deterioration from freezing and thawing. The use of entrained air, low *w/c* ratio, and low-permeability concrete assist in reducing the effect of freeze/thaw cycles. Resistance to salt-induced scaling is determined using the ASTM C 672 procedure.

One important aspect of concrete pavement durability is the reaction between alkali from cement with silica compounds in aggregates in the presence of moisture. This reaction causes swelling of the aggregates due to the formation of sodium silicate gel, leading to cracks. This problem can be controlled by avoiding the use of aggregates with soluble silica.

Corrosion of steel in reinforced concrete pavements is caused in the presence of water with high salinity or deicing salts. Corrosion is an electrochemical reaction involving four components of an electron cell, namely an anode, a cathode, an electrolyte, and a conductor. One part of the reinforcement bar acts like an anode where ions go into solution and electrons are released, while another part of the bar acts like a cathode where electrons are consumed[4]. This movement of electrons combined with moisture in the concrete causes reinforcement corrosion[4], which can progress at a fast rate. Good protection against corrosion is the use of cements with higher $C_3A$.

Sulfate attack is caused by the reaction of sulfates (such as sodium sulfate, magnesium sulfate, and calcium sulfate) from soil and seawater with the free calcium hydroxide and aluminates in the cement. The reaction causes an increase in volume, leading to cracking of the concrete. Sulfate attack can be minimized using cement Types II and V. Also, the use of fly ash and other mineral admixtures causes a reduction in the potential of sulfate attack[4]. Tests on the sulfate resistance of cements are performed on mortars using the ASTM C 452 and ASTM C 1012 procedures. There are no standard tests for the sulfate resistance of concrete.

### 6.6.5 Curling and Warping

Thermal conductivity controls the temperature distribution in concrete pavements. The uniform change in temperature causes shrinkage due to cooling or expansion due to heating. The resistance to these changes due to friction with the underlying layers causes transverse cracks. Also, the variation in temperature or moisture with depth causes curling/warping, as will be described in detail in

Chapter 8. Curling and warping cause loss of support under concrete slabs, which in turn causes an increase in stresses developed under applied traffic loads.

## References

[1] Kosmatka, S., Kerkhoff, B., and Panarese, W. (2002). *Design and Control of Concrete Mixtures*, EB001.14, Portland Cement Association, Skokie, IL.

[2] Mindess, S., and Young. J. (1981). *Concrete*, Prentice-Hall, Inc., Englewood Cliffs, NJ.

[3] Neville, A. (1996). *Properties of Concrete*, John Wiley & Sons, Inc., New York.

[4] Somyaji, S. (2001). *Civil Engineering Materials*, 2nd Ed., Prentice Hall, Englewood Cliffs, NJ.

[5] Taylor, P., Kosmatka, S., Voigt, G., et al. (2004). *Integrated Materials and Construction Practices for Concrete Pavements: A State-of-the-Art Manual*, Federal Highway Administration Publication No. HF-07-004, U.S. Department of Transportation, Washington, DC.

[6] Tennis, P. and Jennings, H. (2000). A Model for Two Types of Calcium Silicate Hydrate in the Microstructure of Portland Cement Pastes, *in Cement and Concrete Research*, Pergamon, pp. 855–863.

[7] Thomas, M. and Wilson, M. (2002). *Supplementary Cementing Materials for Use in Concrete*, CD-ROM CD038, Portland Cement Association, Skokie, IL.

[8] Toler, H. R. (1963). *Flowability of Cement*, Research Department Report MP-106, Portland Cement Association, Skokie, IL.

## Problems

6.1 Describe the influence of the concentrations of the different compounds of cement on the false set and flash set of concrete.

6.2 Discuss the factors that influence the concrete resistance to sulfate attack.

6.3 Discuss the relationship between flexural strength and the performance of unreinforced or plain concrete pavements.

6.4 Discuss the relationship between concrete shrinkage behavior and potential cracking.

6.5 The compressive strength obtained from testing cylindrical specimens of portland concrete was 31,000 kPa (4500 lbs/in$^2$). Estimate the tensile and flexural strengths of this concrete.

6.6 A reinforced concrete pavement is subjected to chloride concentration of 0.75 kg/m$^3$ at the surface. The critical chloride threshold at which steel corrosion is initiated is 1.25 kg/m$^3$. The steel cover is 100 mm, and the diffusion coefficient is 1.1 $10^{-11}$ m$^2$/s. Determine the time in years when steel corrosion is initiated.

6.7 Plot the following relationships considering ranges of compressive strength and density for normal weight concrete:

a. Flexural strength versus compressive strength
b. Tensile strength versus compressive strength
c. Modulus of elasticity versus compressive strength

# 7 Flexible Pavement Analysis

## 7.1 Introduction

Flexible pavements are modeled as layered elastic systems with infinite lateral dimensions. These layers rest on the subgrade, which is often modeled as an elastic layer of infinite depth. Elasticity implies that all the pavement layers and the subgrade can be described by their elastic Young's modulus $E$ and their Poisson's ratio $\mu$. Furthermore, the layers are assumed to be homogeneous and isotropic. Tire loads are modeled as either point loads or circular loads of uniform pressure using Equation 2.1. Under these conditions, the stress state is axisymmetric; that is, it exhibits rotational symmetry around the center axis of the load and, as a result, it is easier to describe using a radial coordinate system. Pavement responses, (i.e., stress, strains, and deflections) are calculated using relationships from the theory of elasticity. The responses from multiple loads are calculated by superimposing the stresses from the individual tires, according to D'Alembert's superposition principle. Analyzing these responses is essential for the mechanistic design of asphalt concrete pavements, as described in Chapter 11.

The following discussion describes solutions for single-layer, two-layer and multilayer flexible pavement systems. The granular layers are treated as linear elastic and nonlinear elastic, (i.e., having moduli that are stress-independent and stress-dependent, respectively), while the asphalt concrete is treated as either linear elastic or linear viscoelastic.

## 7.2 Single-Layer Elastic Solutions

### 7.2.1 Point Load

The simplest loading condition is that of a single-point load, $P$, applied on a semi-infinite elastic space illustrated in Figure 7.1, which shows the nonzero stresses at a location defined by a depth $z$ and a radial offset $r$. Obviously, due to axial symmetry, the radial location defined by the angle $\theta$ is not relevant. The stresses are defined as:

$\sigma_z$ = vertical normal stress
$\sigma_r$ = radial normal stress
$\sigma_\theta$ = tangential normal stress
$\tau_{zr}$ = horizontal shear stress in the radial direction

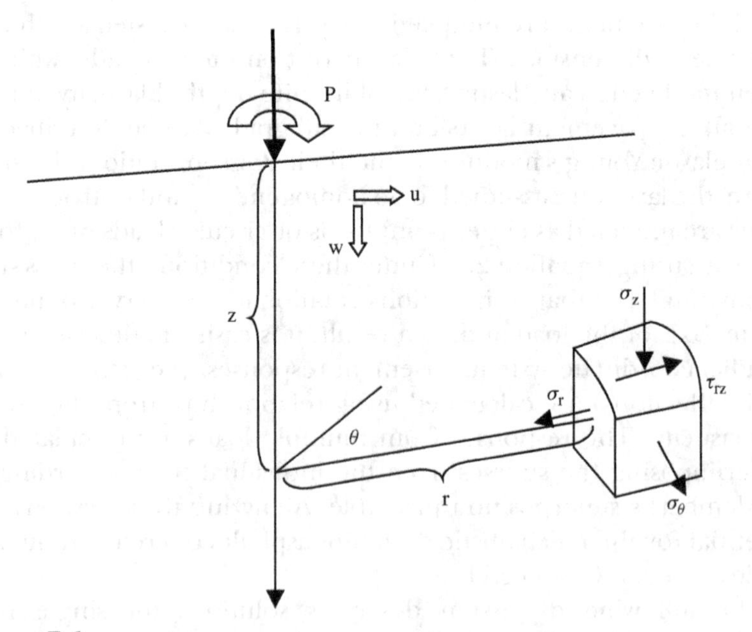


**Figure 7.1**
Axisymmetric Stresses State in an Elastic Half Space

The corresponding strains are:

$\varepsilon_z$ = vertical normal strain
$\varepsilon_r$ = radial normal strain
$\varepsilon_\theta$ = tangential normal strain
$\gamma_{zr}$ = horizontal shear strain in the radial direction.

In defining these strains, it should be noted that the displacement field is two-dimensional; that is, a point in this semielastic space can move only vertically or horizontally, denoted by $w$ and $u$, respectively, as shown in Figure 7.1. Hence:

$$\varepsilon_z = \frac{\partial w}{\partial z} \tag{7.1a}$$

$$\varepsilon_r = \frac{\partial u}{\partial r} \tag{7.1b}$$

$$\varepsilon_\theta = \frac{u}{r} \tag{7.1c}$$

$$\gamma_{zr} = \frac{\partial u}{\partial z} + \frac{\partial w}{\partial r} \tag{7.1d}$$

where Equation 7.1c suggests that the tangential normal strain is, in essence, the change in the perimeter of the circle with radius $r$ divided by the original perimeter, that is, $[2\pi(r+u) - 2\pi r]/(2\pi r)$.

The closed-form solution to this problem was originally developed by Boussinesq, circa 1880[11], and adapted by Taylor[10] in the following form:

$$\sigma_z = -\frac{P}{2\pi}\frac{3z^3}{(r^2+z^2)^{5/2}} \tag{7.2a}$$

$$\sigma_r = -\frac{P}{2\pi}\left[\frac{3r^2 z}{\left(r^2+z^2\right)^{5/2}} - \frac{1-2\mu}{r^2+z^2+z\sqrt{r^2+z^2}}\right] \tag{7.2b}$$

$$\sigma_\theta = \frac{P}{2\pi}(1-2\mu)\left[\frac{z}{\left(r^2+z^2\right)^{3/2}} - \frac{1}{r^2+z^2+z\sqrt{r^2+z^2}}\right] \tag{7.2c}$$

$$\tau_{zr} = \frac{P}{2\pi}\frac{3rz^2}{\left(r^2+z^2\right)^{5/2}} \tag{7.2d}$$

Note that for normal stresses, the sign notation is minus for compression and positive for tension. Note also that directly under the point of load application, (i.e., $r = 0$, $z = 0$), the stresses are undefined. The strain components can be calculated from the stress components through generalized Hooke's law.

$$\varepsilon_z = \frac{1}{E}\left(\sigma_z - \mu\left(\sigma_r + \sigma_\theta\right)\right) \tag{7.3a}$$

$$\varepsilon_r = \frac{1}{E}\left(\sigma_r - \mu\left(\sigma_z + \sigma_\theta\right)\right) \tag{7.3b}$$

$$\varepsilon_\theta = \frac{1}{E}\left(\sigma_\theta - \mu\left(\sigma_r + \sigma_z\right)\right) \tag{7.3c}$$

$$\gamma_{zr} = \frac{2\tau_{zr}\left(1+\mu\right)}{E} = \frac{\tau_{zr}}{G} \tag{7.3d}$$

where $G$ is the shear modulus of the elastic medium. These stress-strain relationships can be written in matrix form as:

$$\begin{Bmatrix} \sigma_z \\ \sigma_r \\ \sigma_\theta \\ \tau_{zr} \end{Bmatrix} = \frac{E}{\left(1+\mu\right)\left(1-2\mu\right)} \begin{bmatrix} (1-\mu) & \mu & \mu & 0 \\ \mu & (1-\mu) & \mu & 0 \\ \mu & \mu & (1-\mu) & 0 \\ 0 & 0 & 0 & \frac{1-2\mu}{2} \end{bmatrix} \begin{Bmatrix} \varepsilon_z \\ \varepsilon_r \\ \varepsilon_\theta \\ \gamma_{zr} \end{Bmatrix} \tag{7.4}$$

The vertical and horizontal deflections, $w$ and $u$, at any point, are computed by integrating the vertical and horizontal strains, respectively. The resulting expressions are:

$$w = \frac{P}{2\pi E}\left[\left(1+\mu\right)z^2\left(r^2+z^2\right)^{-3/2} + 2\left(1-\mu^2\right)\left(r^2+z^2\right)^{-1/2}\right] \tag{7.5a}$$

$$u = P\frac{\left(1+\mu\right)\left(1-2\mu\right)}{2\pi r E}\left[z\left(r^2+z^2\right)^{-1/2} - 1 + \frac{1}{1-2\mu}r^2 z\left(r^2+z^2\right)^{-3/2}\right] \tag{7.5b}$$

It should be noted that the surface (i.e., $z = 0$), the vertical deflection is:

$$w = \frac{P\left(1-\mu^2\right)}{\pi E r} \tag{7.6}$$

This expression defines the so-called solid or Boussinesq subgrade foundation model for rigid pavements, which is discussed further in Chapter 8. Using these closed-form stress, strain and deflection expressions are straightforward, as explained in the following example.

**Example 7.1**

Compute the stresses and strains from a point load of 40 kN resting on a semi-infinite elastic space. The location of interest is at a depth of 0.1 m and a radial offset of 0.2 m. Given, $E = 140$ MPa and $\mu = 0.4$.

**ANSWER**

Substituting the specified values into Equation 7.2 gives the stresses:

$$\sigma_z = -\frac{40}{2\pi}\frac{3\,0.1^3}{\left(0.2^2+0.1^2\right)^{5/2}} = -34.16\,\text{kPa}$$

$$\sigma_r = -\frac{40}{2\pi}\left[\frac{3\,0.2^2\,0.1}{\left(0.2^2+0.1^2\right)^{5/2}} - \frac{1-20.4}{0.2^2+0.1^2+0.1\sqrt{0.2^2+0.1^2}}\right] = -119.06\,\text{kPa}$$

$$\sigma_\theta = \frac{40}{2\pi}(1-20.4)\left[\frac{0.1}{\left(0.2^2+0.1^2\right)^{3/2}} - \frac{1}{0.2^2+0.1^2+0.1\sqrt{0.2^2+0.1^2}}\right] = -6.21\,\text{kPa}$$

$$\tau_{zr} = \frac{40}{2\pi}\frac{3\,0.2\,0.1^2}{\left(0.2^2+0.1^2\right)^{5/2}} = 68.36\,\text{kPa}$$

Equation 7.3 gives the strains:

$$\varepsilon_z = \frac{1}{140000}(-34.16-0.4(-119.06-6.21)) = 113.9\;10^{-6}$$

$$\varepsilon_r = \frac{1}{140000}(-119.06-0.4(-34.16-6.21)) = -735\;10^{-6}$$

$$\varepsilon_\theta = \frac{1}{140000}(-6.21-0.4(-119.06-34.16)) = 393\;10^{-6}$$

$$\gamma_{zr} = \frac{268.36\,(1+0.4)}{140000} = 1367.2\;10^{-6}$$

## 7.2.2 Circular Load with Uniform Vertical Stress

The response under a uniformly distributed stress $p$ on a perfectly flexible circular area of radius $a$ (e.g., an idealized tire imprint, as defined in Chapter 2) is obtained by integrating the stress components given by Equation 7.2. For points on the centerline of the load (i.e., $r = 0$), these stress expressions are[11]:

$$\sigma_z = p\left[-1 + \frac{z^3}{\left(a^2 + z^2\right)^{3/2}}\right] \tag{7.7a}$$

$$\sigma_r = \sigma_\theta = \frac{p}{2}\left[-(1 + 2\mu) + \frac{2(1+\mu)z}{\sqrt{a^2 + z^2}} - \frac{z^3}{\left(a^2 + z^2\right)^{3/2}}\right] \tag{7.7b}$$

$$\tau_{zr} = 0 \tag{7.7c}$$

and the vertical deflection under the centerline of the load is given by:

$$w = \frac{2(1 - \mu^2)}{E}\, pa \tag{7.8}$$

**Example 7.2**

Compute the stresses from a tire inflated to 600 kPa, carrying 30 kN resting on a semi-infinite elastic space. The location of interest is at a depth of 0.1 m and a radial offset of 0.0 m. Also, compute the surface deflection, (i.e., $z = 0.00$) under the same tire. Given, $E =$ 140 MPa and $\mu = 0.4$.

**ANSWER**

The radius of the tire imprint is computed from Equation 2.1.

$$a = \sqrt{\frac{30}{600\,\pi}} = 0.126 \text{ m}$$

The stresses are computed by substituting the given data into Equation 7.7.

$$\sigma_z = 600\left[-1 + \frac{0.1^3}{\left(0.126^2 + 0.1^2\right)^{3/2}}\right] = -455.9\,\text{kPa}$$

$$\sigma_r = \sigma_\theta = \frac{600}{2}\left[-(1 + 2\;0.4) + \frac{2(1+0.4)0.1}{\sqrt{0.126^2 + 0.1^2}} - \frac{0.1^3}{\left(0.126^2 + 0.1^2\right)^{3/2}}\right]$$

$$= -89.9\,\text{kPa}$$

The surface vertical deflection is computed from Equation 7.8:

$$w = \frac{2(1 - 0.4^2)}{140000} 600\, 0.126 = 0.907\ 10^{-3}\mathrm{m}$$

## 7.3 Two-Layer Elastic Solutions

This system consists of a finite thickness layer placed on top of another layer of infinite thickness. These two layers have different elastic properties, as shown in Figure 7.2. This is an idealized representation of a simple pavement consisting of a stiffer layer (e.g., asphalt concrete) resting on a weaker foundation (i.e., subgrade). Burmister[1] developed the solution for the surface deflection of this system under the centerline of a uniform vertical stress $p$ distributed over a circular area of radius $a$, assuming a Poisson's ratio of 0.5. In condensed form, this is expressed as:

$$w = \frac{1.5\, pa}{E_2} F_w \left[ \frac{a}{h}, \frac{E_2}{E_1} \right] \tag{7.9}$$

where $F_w$ is a function that depends on the ratios $a/h$ and $E_2/E_1$, where $h$ is the thickness of the finite layer. Burmister produced a chart for $F_w$ for selected ratios of $a/h$ and $E_2/E_1$ based on the theory of elasticity (Figure 7.3).

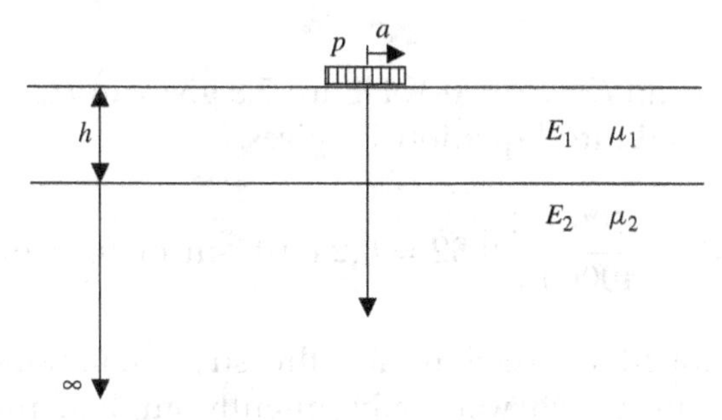


**Figure 7.2**
Schematic of Two-Layer Elastic System

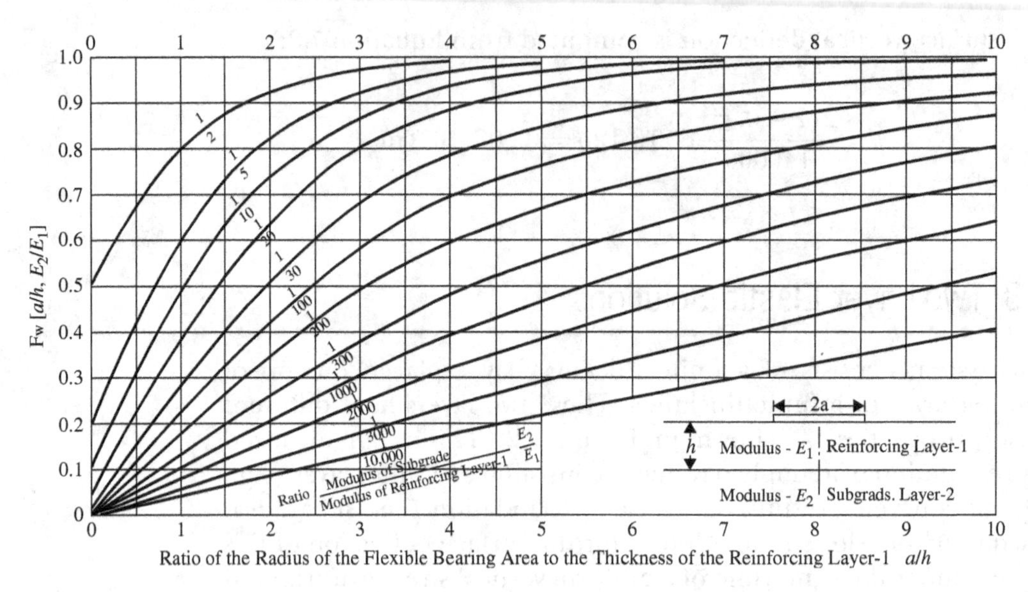


**Figure 7.3**
$F_w$ Factors for Computing Surface Deflection at the Centerline of a Circular Imprint Carrying Uniform Stress (Ref. 1 Reproduced by Permission)

It is worthwhile to note the form similarities between Equation 7.9 and 7.8 when substituting a $\mu$ value of 0.5.

**Example 7.3** Compute the surface deflection from a circular tire imprint with 0.1 m radius carrying a pressure of 700 kPa resting on a 0.2 m thick asphalt concrete layer placed on top of a subgrade of infinite depth. The layer moduli are 1400 MPa and 140 MPa, respectively, and $\mu$ is 0.5 for both layers.

**ANSWER**

For $a/h = 0.5$ and $E_2/E_1 = 0.1$, Figure 7.3 gives an $F_w$ value of 0.32, which substituted into Equation 7.9 gives:

$$w = \frac{1.5\ 700\ 0.1}{140000} 0.32 = 0.24\ 10^{-3} \text{m or } 0.24 \text{ mm}.$$

Fox[4] developed expressions for the stress components in the two-layer system, which were subsequently implemented through nomographs (e.g., reference 2). Advent of modern computers, however, enabled solving this problem for multiple layers, as described next, rendering such nomographs obsolete. It is, nevertheless, worthwhile mentioning an approximate method for solving the elastic layered system problem attributed to Odemark[8,9]. It consists of translating multiple layers of different moduli into an equivalent single layer, hence known as the method of equivalent thicknesses. For a system of two layers, such as the one shown in Figure 7.2, the top layer with thickness $h$ can be translated into an equivalent thickness $h_e$, with a modulus $E_2$. For $\mu_1 = \mu_2$, the equivalent layer thickness of the top layer is given by:

$$h_e = 0.9\left(\frac{E_1}{E_2}\right)^{\frac{1}{3}} h \tag{7.10}$$

where 0.9 is an approximation factor. This allows utilizing the single-layer solutions in computing pavement responses in the lower layer.

**Example 7.4**

Compute the stresses at the bottom of a flexible pavement surface layer 0.3 m thick resting on a semi-infinite subgrade layer. The load consists of a circular tire with a 0.1 m radius carrying a uniform pressure of 700 kPa. The stresses are to be computed under the centerline of the load. The layer moduli are 1400 MPa and 140 MPa, respectively, and $\mu$ is 0.5 for both layers.

**ANSWER**

Equation 7.10 gives the equivalent thickness of the top layer in terms of the modulus of the bottom layer as:

$$h_e = 0.9\left(\frac{1400}{140}\right)^{\frac{1}{3}} 0.3 = 0.582 \text{ m}$$

which allows using Equation 7.7 to compute the stresses in the subgrade. At the bottom of the top layer, they are:

$$\sigma_z = 700\left[-1 + \frac{0.582^3}{\left(0.1^2 + 0.582^2\right)^{3/2}}\right] = -29.9 \text{ kPa}$$

$$\sigma_r = \sigma_\theta = \frac{700}{2}\left[-(1 + 2\,0.5) + \frac{2(1+0.5)0.582}{\sqrt{0.1^2 + 0.582^2}} - \frac{0.582^3}{\left(0.1^2 + 0.582^2\right)^{3/2}}\right] = -0.218 \text{ kPa}$$

This approach can be extended to translate the thickness of multiple layers $i$ to an equivalent thickness with a modulus equal to the modulus of the bottom layer $E_n$.

$$h_e = f\sum_{i=1}^{n-1}\left\{h_i\left(\frac{E_i}{E_n}\right)^{\frac{1}{3}}\right\} \tag{7.11}$$

where $f$ is the value of an approximation factor that depends on the ratio of the layer moduli and the relative magnitude of the layer thicknesses, with respect to the radius of the load.[9]

## 7.4 Multilayer Linear Elastic Solutions

A multilayer elastic system consists of multiple finite-thickness layers resting on a subgrade of infinite thickness (Figure 7.4). It is an idealized representation of multiple pavement layers, such as asphalt concrete friction and leveling layers, base layers and sub-base layers, each having different elastic properties. Elastic response solutions to this system were developed by extending the Burmister analytical approach for the two layer system (1) to multiple layers. A variety of software implements such solutions, such as ELSYM5[7], DAMA[5], KENLAYER[6], and EVERSTRESS[3]. These programs have, to a great

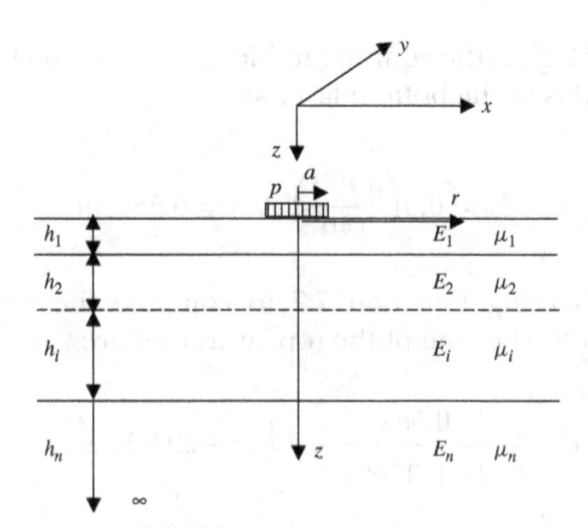


**Figure 7.4**
Schematic of Multilayer Elastic System

extent, similar features and input requirements and can handle up to five layers, including an infinite depth subgrade. They accept multiple circular tires and compute stresses at any location in the layered system. The calculations are made in the radial coordinate system, and then translated to a Cartesian coordinate system, with its origin in the middle of a tire imprint (Figure 7.4). In the following examples, the computer program EVERSTRESS 5.0[3] is used. Suggestions on available layered analysis software and their sources are given on the Web site for this book, at www.wiley.com/go/pavement.

**Example 7.5**

Compute the stresses, strains, and deflections in a layered system under a set of four tires as shown in Figure 7.5. The locations of interest are the bottom of the asphalt concrete layer and the top of the subgrade, at coordinates (x = 0, y = 0). Assume full friction between the layers.

**ANSWER**

The output of EVERSTRESS is shown in Table 7.1, which lists stresses, strains, and deflections at the locations requested. They

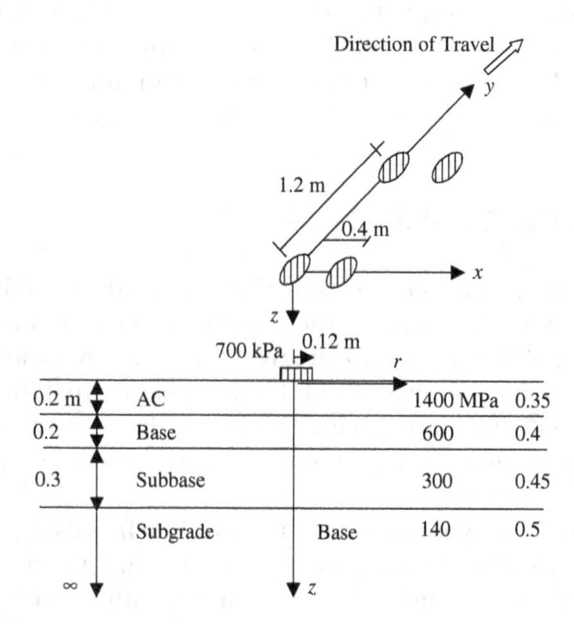


**Figure 7.5**
Layered Elastic System for Example 7.7

**Table 7.1**
Results of Problem 7.5: Responses under (x = 0, y = 0) Computed with EVERSTRESS (Ref. 3)

| Depth (cm) | $\sigma_{xx}$(kPa) | $\sigma_{yy}$(kPa) | $\sigma_{zz}$(kPa) | $\tau_{xy}$(kPa) | $\tau_{yz}$ (kPa) | $\tau_{zx}$(kPa) |
|---|---|---|---|---|---|---|
| 19.999 | 134.1 | 176.4 | −197.4 | 2.7 | 23.3 | −0.7 |
| 70.001 | −5.9 | −7.4 | −29.9 | 2.9 | 4.5 | −0.5 |
| **Depth (cm)** | **$\varepsilon_{xx}$ ($10^{-6}$)** | **$\varepsilon_{yy}$ ($10^{-6}$)** | **$\varepsilon_{zz}$ ($10^{-6}$)** | **$w_x$ ($\mu$m)** | **$w_y$ ($\mu$m)** | **$w_z$ ($\mu$m)** |
| 19.99 | 101.4 | 141.8 | −218.6 | −8.1 | −0.8 | 399.1 |
| 70.001 | 91.2 | 74.9 | −166.1 | −19.8 | −26.1 | 304.4 |

were obtained by setting the coefficient of friction between the layers to 1.0. As discussed later in Chapter 11, traditional flexible pavement mechanistic design is based on controlling two of these response parameters, namely the tensile strain at the bottom of the asphalt concrete layer and the compressive strain at the top of the subgrade. They are associated with bottom-up fatigue cracking and subgrade plastic deformation (i.e., rutting), respectively. As shown in Table 7.1, the magnitude of the asphalt concrete bottom tensile strains is 101.4 $10^{-6}$ and 141.8 $10^{-6}$ in the transverse and longitudinal directions, respectively, while the subgrade top compressive strain is 166.1 $10^{-6}$. It should be noted that the maximum strain under this set of tires may not be under the (x = 0, y = 0) location.

## 7.5 Multilayer Nonlinear Elastic Solutions

As discussed in Chapter 3, granular materials exhibit a stress-dependent behavior; that is, their resilient (i.e., elastic) modulus is a function of the stress state at any given location. Coarse-grained layers (e.g., base, subbase) exhibit an exponential dependence to bulk stress, while fine-grained/cohesive layers (e.g., subgrade) exhibit an exponential dependence to deviatoric stress (Equations 3.4 and 3.8, respectively).

Layer elastic analysis can handle this nonlinearity using a piecewise linear iterative algorithm. Seed moduli are initially assigned; the layered analysis is conducted to compute initial stresses; and in subsequent iterations, the moduli are updated on the basis of the calculated stresses, as schematically shown in Figure 7.6. The process

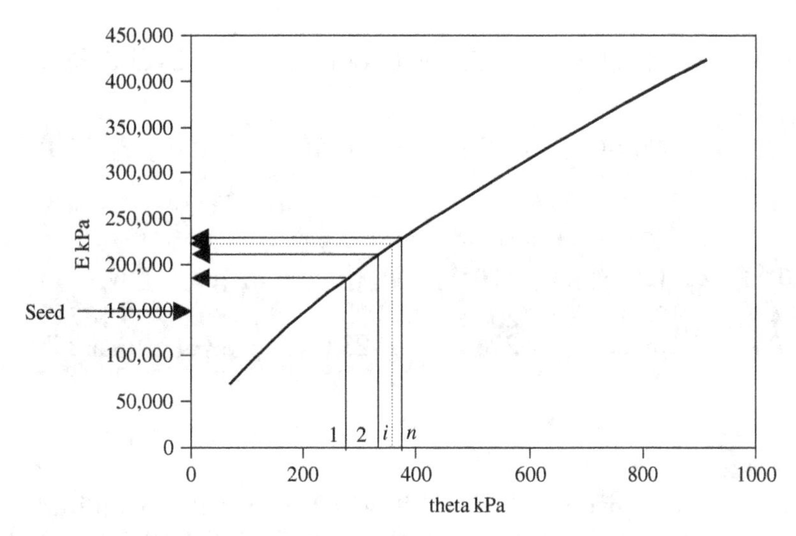


**Figure 7.6**
Schematic Representation of the Iterative Procedure Used to Handle Granular Layer Nonlinearity

is repeated until the computed moduli in two successive iterations are within a prescribed tolerance. This process can be refined by subdividing each granular layer into sublayers and considering the stresses in the middle of each sublayer. In doing so, the weight of the layers should also be considered. EVERSTRESS[3] handles nonlinear granular material moduli.

**Example 7.6**

Repeat the solution of the previous example by considering the stress dependence of the base/subbase and subgrade layers given by:

$$E = 400\,\theta^{0.6} \tag{7.12}$$

$$E = 120\,(\sigma_1 - \sigma_3)^{0.2} \tag{7.13}$$

respectively, where $E$ is the elastic modulus in MPa and $\theta$ is the bulk stress in atmospheres (i.e., $\sigma_1 + \sigma_2 + \sigma_3$), and $\sigma_1$, $\sigma_2$, and $\sigma_3$ are the principal stresses in atmospheres. The unit weight of the pavement layers are 22.8, 20.5, 19.7, and 18.9 kN/m$^3$ (i.e., these are used for computing the stress from the overburden to be added to the load-induced stresses to obtain the bulk and deviator stresses).

**Table 7.2**
Results of Example 7.6: Responses under (x = 0, y = 0) Computed with EVERSTRESS (Ref. 3)

| Depth (cm) | $\sigma_{xx}$(kPa) | $\sigma_{yy}$(kPa) | $\sigma_{zz}$(kPa) | $\tau_{xy}$(kPa) | $\tau_{yz}$ (kPa) | $\tau_{zx}$(kPa) |
|---|---|---|---|---|---|---|
| 19.999 | 238.6 | 296.4 | −168.2 | 3.7 | 21.9 | −1.5 |
| 70.001 | −5.4 | −6.9 | −27.2 | 2.8 | 4 | −0.5 |
| **Depth (cm)** | **$\varepsilon_{xx}$ ($10^{-6}$)** | **$\varepsilon_{yy}$ ($10^{-6}$)** | **$\varepsilon_{zz}$ ($10^{-6}$)** | **$w_x$ ($\mu$m)** | **$w_y$ ($\mu$m)** | **$w_z$ ($\mu$m)** |
| 19.99 | 138.4 | 194.1 | −253.9 | −13.4 | −2.9 | 587.2 |
| 70.001 | 131.9 | 106.3 | −238.2 | −28.5 | −39.6 | 460.7 |

**ANSWER**

The answer is given in Table 7.2, which shows the magnitude of the asphalt concrete bottom tensile strains is 138.4 $10^{-6}$ and 194.1 $10^{-6}$ in the transverse and longitudinal directions, respectively, while the subgrade bottom compressive strain is 238.1 $10^{-6}$. The solution was achieved in seven iterations, satisfying a tolerance in $E$ lower than 1%. After convergence, the moduli of the base, subbase and subgrade layers were 374.3 MPa, 238.61 MPa and 88.31 MPa, respectively.

## 7.6 Viscoelastic Solutions

The previous discussion assumed elastic material behavior. As described in Chapter 5, however, asphalt concretes exhibit viscoelastic behavior, hence their response is time-dependent. Their response to a time-dependent (e.g., moving) load is computed through Boltzmann's superposition principle[12], assuming linear viscoelastic behavior. In the time domain, this is expressed by the following convolution integral:

$$\varepsilon(t) = \int_{0}^{t} D\left(t - \xi'\right) \frac{\partial \sigma\left(\xi'\right)}{\partial \xi'} d\xi' \tag{7.14}$$

where, $\varepsilon(t)$ is the strain at time $t$, $D(t - \xi')$ is the creep compliance or retardation modulus of the asphalt concrete layer after a lapsed time of $(t - \xi')$, and $\sigma(\xi')$ is the stress history as a function of time

$\xi'$ ranging from 0 to $t$. Plainly stated, Equation 7.14 means that the strain at time $t$, under an arbitrary stress history $\sigma(\xi')$, is the linear sum of all the strain increments experienced from the beginning of the stress imposition to time $t$. The creep compliance is the inverse of modulus and has units of 1/stress (i.e., 1/kPa or 1/lbs/in.$^2$).

Given a mechanistic model for the creep compliance, the integral in Equation 7.14 can be computed numerically, as demonstrated in the following example. It should be noted that this convolution integral can be written similarly in terms of stress rather than strain, expressed as:

$$\sigma(t) = \int_0^t E\left(t-\xi'\right)\frac{\partial\varepsilon\left(\xi'\right)}{\partial\xi'}d\xi' \tag{7.15}$$

where, $\sigma(t)$ is the stress at time $t$, $E(t-\xi')$ is the stiffness or relaxation modulus of the asphalt concrete layer, and $\varepsilon(\xi')$ is the applied strain history. Mechanistic models can be fitted to describe the creep compliance or the relaxation modulus based on data obtained from creep testing or relaxation testing, respectively. Hence, the viscoelastic behavior of asphalt binders and asphalt concretes can be effectively modeled through state-of-the art laboratory testing.

### 7.6.1 Single Semi-Infinite Layers

Implementing Equation 7.14 in a semi-infinite elastic space is straightforward, because of the availability of closed-form solutions for the stress (Equation 7.2). An example of computing the response of a semi-infinite viscoelastic layer to a moving load is given in the following example.

**Example 7.7**

Consider a 40 kN point load traveling at a speed of 22.22 m/sec (80 km/h) on a semi-infinite viscoelastic layer with a creep compliance given by:

$$D(t) = \frac{1}{E_0 + E_1 e^{-\frac{t}{T_1}} + E_2 e^{-\frac{t}{T_2}}} \tag{7.16}$$

with $E_0 = 5{,}000{,}000$ kPa, $E_1 = 2{,}000{,}000$ kPa, $E_2 = 4{,}000{,}000$ kPa, $T_1 = 0.05$ sec, and $T_2 = 0.005$ sec. Compute and plot the vertical strain $\varepsilon_z$ as a function of time at a depth of 0.1 m, assuming a constant Poisson's ratio of 0.40.

**ANSWER**

A load influences layer response at a particular location before and after it reaches this location. The total distance of load influence depends on the magnitude of the load and the stiffness of the layer. For this example, assume that the total length of load influence is 1.0 m (i.e., the response is substantial for locations of the load 0.5 m before and 0.5 m after a point of interest). At 22.22 m/sec, the load takes 0.045 sec to traverse this 1.0 m length of load influence (i.e., 1.0/22.22). Obviously, the resulting loading frequency is 22.22 Hz. According to Equation 7.3a, the vertical strain $\varepsilon_z$ can be computed from the stress history of the three stress components $\sigma_z$, $\sigma_r$, and $\sigma_\theta$. These can be computed from Equations 7.2 for the given depth $z$ of 0.10 m and any desired offset $r$ ranging from −0.5 to 0.5 m (note that only stresses for the positive or the negative offsets need to be computed due to axial symmetry). Hence, according to Equation 7.14, and given a constant Poisson's ratio, the vertical strain can be computed using:

$$\varepsilon_z(t) = \int_0^t D\left(t - \xi'\right) \frac{\partial}{\partial \xi'} \left(\sigma_z\left(\xi'\right) - \mu\left(\sigma_r\left(\xi'\right) + \sigma_\theta\left(\xi'\right)\right)\right) d\xi' \quad (7.17)$$

This equation was solved numerically using a spreadsheet and time/distance increments of 0.00045 sec/0.01 m, as shown in Table 7.3. Note that the value of the creep compliance is computed from Equation 7.16 for values of time, increasing from 0 to 0.045 sec, while the stress history is calculated by positioning the 40 kN point load at successive locations 0.01 m apart and computing the change in the stress components. The resulting vertical strain versus time is plotted in Figure 7.7 and has a maximum compressive value of about 330 $10^{-6}$.

### 7.6.2 Multilayer Systems

Modeling the response of layered systems under moving loads is more complex because there are no closed-form solutions for the stress components, hence layered analysis software has to be used to compute them (e.g., EVERSTRESS). It is not practical, however, doing so for every distance/time increment (it would involve 100 sets of analysis for the example just presented, with different surface layer moduli for each set). Alternatively, pavement response can be

**Table 7.3**
Numerical Integration for Example 7.7

| $r$ | $\xi'$ | $\sigma_z$ Equation 7.2a | $\sigma_r$ Equation 7.2b | $\sigma_\theta$ Equation 7.2c | Stress $\sigma_z - \mu(\sigma_r + \sigma_\theta)$ | Stress Increase $\Delta\sigma(\xi)$ | $D(t-\xi')$ Equation 7.16 | Strain increase $D(t-\xi')\Delta\sigma(\xi')\varepsilon_z(t)$ | Equation 7.17 |
|---|---|---|---|---|---|---|---|---|---|
| (m) | (sec) | (kPa) | (kPa) | (kPa) | (kPa) | (kPa) | (1/kPa.$10^{-6}$) | (microns) | (microns) |
| −0.5 | 0 | −0.55 | −9.76 | −3.13 | 4.60 | 0 | 0.09 | 0.42 | 0.42 |
| −0.49 | 0.00045 | −0.61 | −10.42 | −3.22 | 4.85 | 0.24 | 0.09 | 0.02 | 0.44 |
| −0.48 | 0.0009 | −0.67 | −11.13 | −3.32 | 5.11 | 0.26 | 0.10 | 0.03 | 0.47 |
| −0.47 | 0.00135 | −0.75 | −11.90 | −3.42 | 5.38 | 0.28 | 0.10 | 0.03 | 0.49 |
| −0.46 | 0.0018 | −0.83 | −12.74 | −3.52 | 5.68 | 0.30 | 0.10 | 0.03 | 0.52 |
| −0.45 | 0.00225 | −0.92 | −13.66 | −3.62 | 5.99 | 0.32 | 0.11 | 0.03 | 0.56 |
| −0.44 | 0.0027 | −1.02 | −14.65 | −3.73 | 6.33 | 0.34 | 0.11 | 0.04 | 0.59 |
| −0.43 | 0.00315 | −1.14 | −15.73 | −3.85 | 6.69 | 0.36 | 0.11 | 0.04 | 0.63 |
| −0.42 | 0.0036 | −1.27 | −16.91 | −3.96 | 7.08 | 0.39 | 0.11 | 0.04 | 0.68 |
| −0.41 | 0.00405 | −1.43 | −18.20 | −4.09 | 7.49 | 0.41 | 0.12 | 0.05 | 0.73 |
| −0.4 | 0.0045 | −1.60 | −19.62 | −4.21 | 7.93 | 0.44 | 0.12 | 0.05 | 0.78 |
| And so on. | | | | | | | | | |

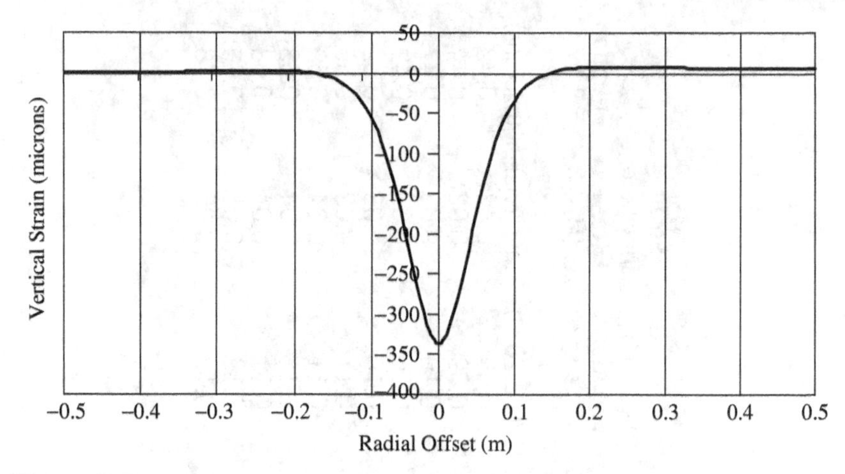


**Figure 7.7**
Vertical Strain $\varepsilon_1$ versus Time for Example 7.7

computed at selected locations and values in between approximated through interpolation, thus making it possible to use the same approach followed in the previous example.

**Example 7.8** Consider a pair of identical tires moving at a speed of 16.667 m/sec (i.e., 60 km/h) on a pavement structure, as shown in Figure 7.8. Assume that these loads influence the response of the pavement within a radius of 1.00 m around their instantaneous location, (i.e., load influence length of 2.0 m). Compute and plot the transverse and longitudinal strains (i.e., $\varepsilon_{xx}$ and $\varepsilon_{yy}$) at the bottom of the asphalt concrete layer as a function of time.

**ANSWER**

At 16.667 m/sec, the load takes 2.0/16.667 = 0.12 sec to traverse its influence length (i.e., a loading frequency of 8.33 Hz). The stresses were computed at the bottom of the asphalt concrete layer (i.e., 0.152 m) in Cartesian coordinates using EVERSTRESS at $y$ intervals of 0.10 m for the pavement layer layout shown in Figure 7.8. Stress values between these points were obtained through linear interpolation to yield stress estimates at increments of 0.01 m. Table 7.4 shows excerpts of the spreadsheet used to compute the stress history and the creep compliance values necessary for integrating numerically

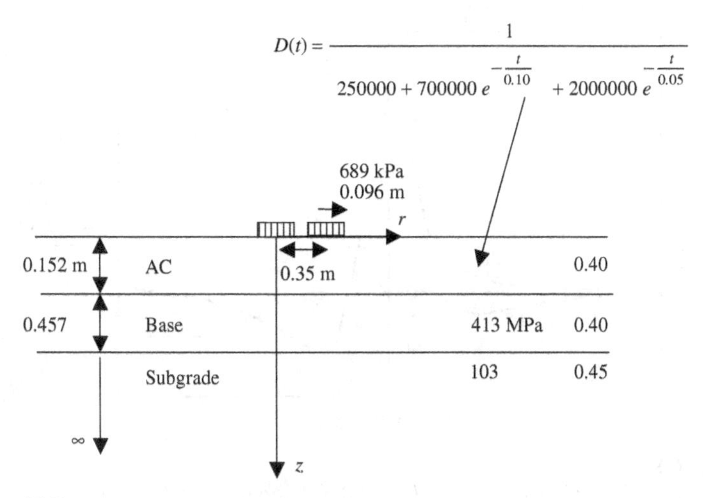


**Figure 7.8**
Layer Viscoelastic System for Example 7.8

**Table 7.4**
Numerical Integration for Example 7.8

| $y$ | $\xi'$ | $\sigma_{xx} - \mu(\sigma_{yy} + \sigma_{zz})$ | $\Delta\sigma_{xx}(\xi')$ | $D(t - \xi')$ | $D(t - \xi')\,\Delta\sigma_{xx}(\xi')$ | $\varepsilon_{xx}(t)$ |
|---|---|---|---|---|---|---|
| (m) | (sec) | (kPa) | (kPa) | (1/kPa $10^6$) | (microns) | (microns) |
| −1 | 0 | 2.71 | — | 0.34 | 0.92 | 0.92 |
| −0.99 | 0.0006 | 2.83 | 0.12 | 0.34 | 0.04 | 0.96 |
| −0.98 | 0.0012 | 2.95 | 0.12 | 0.35 | 0.04 | 1.00 |
| −0.97 | 0.0018 | 3.07 | 0.12 | 0.35 | 0.04 | 1.04 |
| −0.96 | 0.0024 | 3.19 | 0.12 | 0.35 | 0.04 | 1.09 |
| −0.95 | 0.003 | 3.31 | 0.12 | 0.36 | 0.04 | 1.13 |
| −0.94 | 0.0036 | 3.43 | 0.12 | 0.36 | 0.04 | 1.17 |
| −0.93 | 0.0042 | 3.55 | 0.12 | 0.36 | 0.04 | 1.22 |
| −0.92 | 0.0048 | 3.67 | 0.12 | 0.37 | 0.04 | 1.26 |
| −0.91 | 0.0054 | 3.79 | 0.12 | 0.37 | 0.04 | 1.30 |
| −0.9 | 0.006 | 3.91 | 0.12 | 0.37 | 0.04 | 1.35 |
| And so on. | | | | | | |

Equation 7.17. The resulting transverse and longitudinal strains are plotted in Figure 7.9.

An approximate method was proposed for computing pavement response under moving loads. It involves direct superposition of

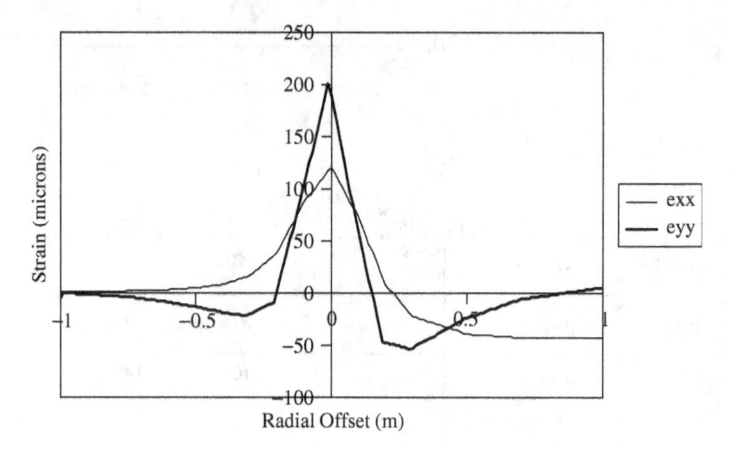


**Figure 7.9**
Transverse and Longitudinal Strains ($\varepsilon_{xx}$ and $\varepsilon_{yy}$) versus Time for Example 7.8

strains weighed by a loading function[13], expressed as:

$$R(t) = \int_{0}^{t} I(t - \xi') \frac{\partial L(\xi')}{\partial \xi'} d\xi' \tag{7.18}$$

where $R(t)$ is the pavement response parameter at time $t$ (e.g., the transverse strain $\varepsilon_{xx}$ at the bottom of the asphalt concrete layer), $I(t)$ is the influence function for that pavement response parameter, and $L(t)$ is the loading function. For a moving load of constant magnitude $A$, a common form for the loading function $L(t)$ is one pulse-shaped

$$L(t) = A\sin^2\left(\frac{\pi}{2} + \frac{\pi t}{D}\right) \qquad \text{with} -\frac{D}{2} \leq t \leq \frac{D}{2} \tag{7.19}$$

where $D$ is the time interval required to traverse the length of the road being affected by the load in any particular location. Differentiating Equation 7.19, substituting it into Equation 7.18, and setting the integration limits from $-D/2$ to time $t$ gives:

$$R(t) = -\int_{-D/2}^{t} I(t - \xi') \frac{A\pi}{D} \sin\left(\frac{2\pi\xi'}{D}\right) d\xi' \tag{7.20}$$

which can be computed numerically.

## References

[1] Burmister, D.M. (1943). *The Theory of Stresses and Displacements in Layered Systems and Applications to the Design of Airport Runways,* Highway Research Board Proceedings, Washington, DC.

[2] Burmister, D.M. (1958). "Evaluation of Pavement Systems of the WASHO Road Test by Layered Systems Methods," Bulletin 177, Highway Research Board, Washington, DC.

[3] Everseries Pavement Analysis Programs (1999). Washington State Department of Transportation, Olympia, WA.

[4] Fox, L. (1948). "Computation of Traffic Stresses in a Simple Road Structure," (1948). Department of Scientific and Industrial Research, Road Research Technical Paper 9, Oxford, UK.

[5] Hawng, D., and M.W. Witzack (1979). Program DAMA (Chevron), User's Manual, Department of Civil Engineering, University of Maryland, College Park, MD.

[6] Huang, Y.H. (2004). *Pavement Analysis and Design,* 2nd ed. Pearson-Prentice Hall, Upper Saddle River, NJ.

[7] Kooperman, S., G. Tiller and M. Tseng (1986). ELSYM5, "Interactive Microcomputer Version, User's Manual," Report FHWA-TS-87-206, Federal Highway Administration, Washington, DC.

[8] Odemark N. (1949). *Undersökning av Elasticitetegenskaperna hos Olika Jordarter samt Teori för Beräkning av Belägningar Enligt Elasicitetsteorin,* Statens Väginstitut, Meddelande 77 (in Swedish).

[9] Ullidtz, P. (1987). *Pavement Analysis,* Elsevier, Amsterdam, New York.

[10] Taylor, D.W. (1963). *Fundamentals of Soil Mechanics,* John Wiley & Sons, Inc., New York.

[11] Timoshenko, S. P. and J. N. Goodier (1987). *Theory of Elasticity,* 3rd ed., McGraw-Hill Inc., New York.

[12] Tschoegl, N.W. (1989). *The Phenomenological Theory of Linear Viscoelastic Behavior: An Introduction,* Springer-Verlag, Berlin Heidelberg.

[13] FHWA (1978). *VESYS User's Manual: Predictive Design Procedures,* FHWA Report FHWA-RD-77-154, Federal Highway Administration, Washington, DC.

## Problems

7.1 Compute the stresses and strains from a point load of 20 kN resting on a semi-infinite elastic space. The location of interest is at a depth of 0.10 m and a radial offset of 0.25 m. Given, $E =$ 150 MPa and $\mu = 0.35$. What is the deflection at the surface?

7.2 Compute the stresses from a tire inflated to 750 kPa, carrying 40 kN resting on a semi-infinite elastic space. The location of interest is at a depth of 0.25 m and a radial offset of 0.0 m. Given, $E = 150$ MPa and $\mu = 0.35$.

7.3 Compute the surface deflections of a semi-infinite elastic space under a point load of 50 kN at radial offsets of 0.10, 0.25, 0.50, 0.75, and 1.00 m. Given, $E = 135$ MPa and $\mu = 0.45$.

7.4 Use the approximate Odemark approach to determine the surface deflection under the centerline of a tire with a 0.15 m radius, carrying a load of 40 kN resting on a 0.4 m thick asphalt concrete layer placed on top of a subgrade of infinite depth. The layer moduli are 1300 MPa and 150 MPa, respectively, and $\mu$ is 0.5 for both layers.

7.5 Repeat the computations for the previous question using the Burmister approach and layered-elastic analysis software. Compare the results.

7.6 For the layered system shown in Figure 7.10, determine and plot the strains at the bottom of the asphalt concrete layer versus radial offsets of 0.0, 0.05, 0.10, 0.15, 0.20, 0.25, 0.3, 0.35, and 0.4.

7.7 For the layered system of the previous question, determine and plot the surface deflections versus the same radial offsets.

7.8 For the layered system shown in Figure 7.8, compute the strains at bottom of the asphalt concrete layer for a radial offset of 0.0, considering that the base and subgrade moduli are stress-dependent, as defined by:

$$E = 380\ \theta^{0.5}$$

$$E = 100\ (\sigma_1 - \sigma_3)^{0.3}$$

respectively, where the moduli are in MPa and the stresses in atmospheres.

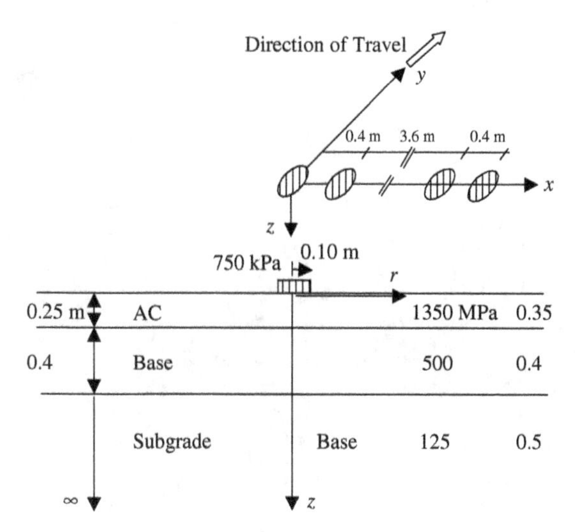


**Figure 7.10**
Layer Viscoelastic System for Problem 7.6

7.9 Consider a point load of 30 kN traveling at a speed of 16.667 m/sec (60 km/h) on a semi-infinite viscoelastic layer with a creep compliance given by:

$$D(t) = \frac{1}{E_0 + E_1 e^{-\frac{t}{T_1}} + E_2 e^{-\frac{t}{T_2}}}$$

with $E_0 = 80000$ kPa, $E_1 = 100000$ kPa, $E_2 = 300000$ kPa, $T_1 = 0.25$ sec, and $T_2 = 0.05$ sec. Compute and plot the radial strain $\varepsilon_r$ as a function of time at a depth of 0.15 m. Assume that the load affects pavement response within a radius of 0.6 m (i.e., from $-0.6$ m to $+0.6$ m) and that the Poisson's ratio is constant and equal to 0.45.

# 8 Rigid Pavement Analysis

## 8.1 Introduction

Rigid pavements consist of portland concrete slabs resting on a base course or directly on the subgrade. The modulus of the portland concrete, which is in the order of 28,000 MPa, is much higher than the moduli of the underlying layers, which typically range from 80 to 600 MPa. As a result, rigid pavements derive much of their load-carrying capacity through plate action (i.e., two-directional slab bending on the $x$-$y$ plane, as shown in Figure 8.1), while being supported by the reaction of the lower layers.

As discussed in Chapter 1, unreinforced slabs tend to crack transversely where thermally induced tensile stresses exceed the tensile strength of the concrete. Hence, they require either transverse joints or tensile reinforcement. Jointed pavements provide vertical load transfer between adjacent slabs through either aggregate interlock or dowel bars. They are referred to as *jointed plain concrete pavements* (JPCPs) and jointed *dowel-reinforced concrete pavements* (JDRCPs), respectively. Continuous reinforcement keeps the tension cracks in the portland concrete closed and hence maintains the integrity of the slab. These pavements are called *continuously reinforced concrete pavements* (CRCP). The various design of joints

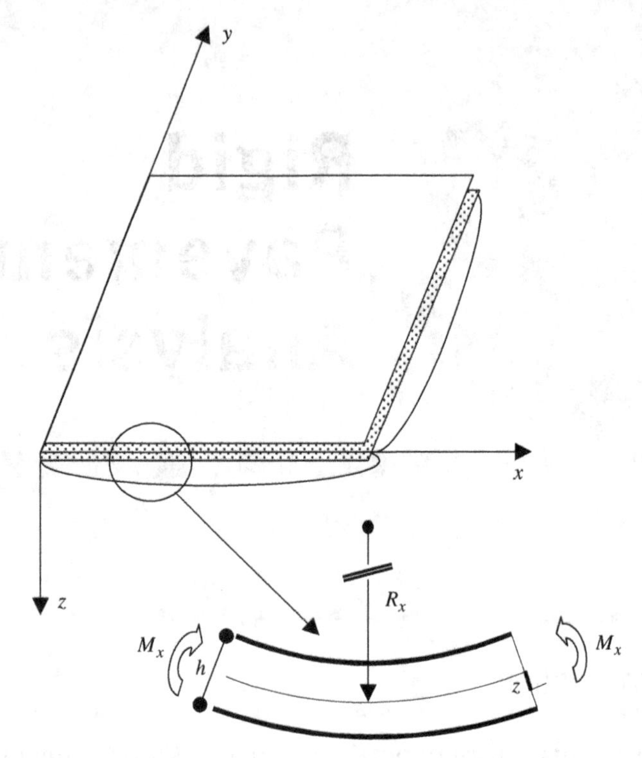


**Figure 8.1**
Pure Bending of Slabs

were shown in Figures 1.3 and 1.4. Details on their actual design can be found in reference 6.

Stresses in concrete pavements are the result of the interaction of a number of factors, which can be grouped into three main categories:

1. Environmental (i.e., effect of temperature and moisture changes in the slab)
2. Traffic loading
3. Base/subgrade support of the slab (i.e., slab curling and volume changes or erosion of the subgrade)

One of the most common simplifications used in analyzing concrete pavements concerns the subgrade and the way it supports the slab. The subgrade is modeled either as a series of non interacting linear springs (Figure 8.2) or as an homogeneous and isotropic continuum of infinite depth (Figure 8.3).

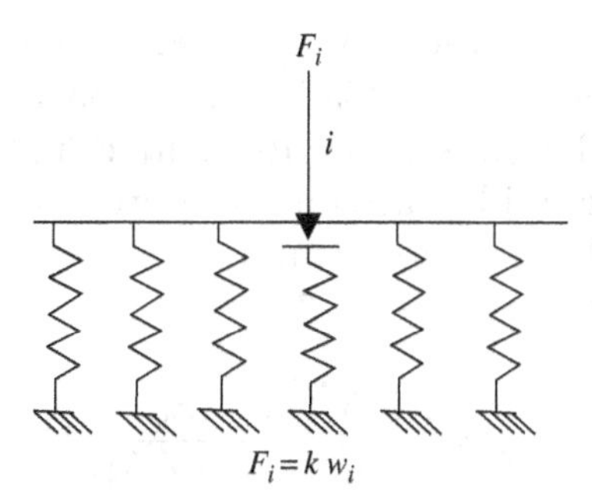


**Figure 8.2**
Liquid Concrete Pavement Foundation Model

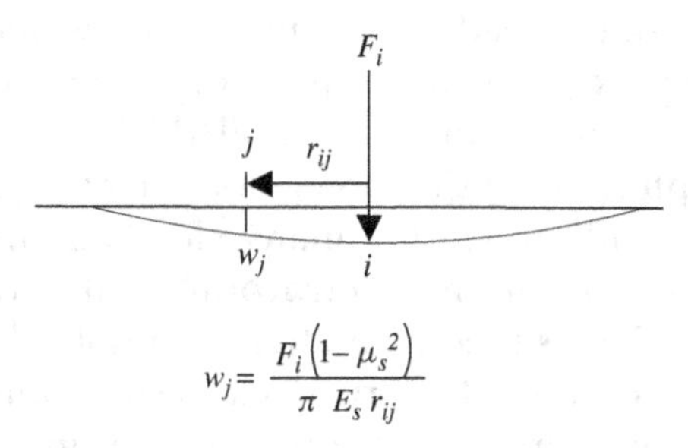


**Figure 8.3**
Solid Concrete Pavement Foundation Model

The first foundation model is characterized by the elastic constant of the springs, referred to as the *modulus of subgrade reaction*, denoted by $k$. It is defined as the ratio of the stress divided by the deflection, and measured through plate-loading tests in units of MPa/m (Equation 3.19). This foundation, referred to as Winkler, or "liquid," implies that load at a particular point generates subgrade deflection only directly underneath that point. The second foundation model is characterized by the elastic modulus and the Poisson's ratio of the subgrade, denoted by $E_s$ and $\mu_s$, respectively (this is to differentiate them from the elastic constants of the slab itself, denoted by $E$ and $\mu$). These subgrade properties are determined through either triaxial laboratory testing (see Chapter 3) or through back-calculation based on in-situ surface deflection measurements, (see Chapter 9). This foundation, referred to as Boussinesq, or "solid," implies that load at a particular point generates subgrade deflections at and around that point. The relationship between load and deflection is shown in Figure 8.3, where $r_{ij}$ is the distance between the location of the load application $i$ and the deflection location $j$. Note that this is identical to Equation 7.6 presented earlier. Clearly, the solid foundation is more realistic than the liquid foundation model. The most commonly used expression for the relationship between $k$ and $E_s$ is[18]:

$$k = \left(\frac{E_s}{E}\right)^{1/3} \frac{E_s}{(1 - \mu_s^2)h} \tag{8.1}$$

where $h$ is the thickness of the portland concrete slab.

It should stated from the outset that closed-form solutions for the structural analysis of concrete pavements are available only for a limited number of simple loading circumstances, which are described next. More complex problems require the use of numerical methods, such as the Finite Element Method (FEM), that is introduced later in this chapter. The FEM allows modeling the slab and the subgrade as a whole and can readily analyze load and moment transfer across joints under a variety of environmental and traffic-loading input. It is particularly suited to analyzing complex boundary conditions, such as those caused by varying base/subgrade support under a slab. The structural analysis of concrete slabs in response to each of the three groups of factors just outlined is presented in detail next. This discussion begins with a review of the theory of elasticity on the pure bending of plates.[15]

## 8.2 Overview of the Elastic Theory on Plates

Consider a Cartesian coordinate system with its origin on the neutral axis of a slab with infinite $x$-$y$ dimensions and finite thickness $h$, as shown in Figure 8.1. One-dimensional pure bending of this slab, say on the $x$-$z$ plane, yields a deformed slab shape resembling a cylindrical section and represents a plane-strain state. The resulting strains at a distance $z$ from the neutral axis are:

$$\varepsilon_x = \frac{z}{R_x} \tag{8.2a}$$

$$\varepsilon_y = \varepsilon_z = 0 \tag{8.2b}$$

where $R_x$ is the radius of curvature of the slab on the $x$-$z$ plane, and $E$, $\mu$ are its elastic modulus and Poisson's ratio, respectively.

The corresponding stresses are:

$$\sigma_x = E\frac{z}{R_x} \tag{8.3a}$$

$$\sigma_y = \mu\sigma_x = \mu E\frac{z}{R_x} \tag{8.3b}$$

$$\sigma_z = 0 \tag{8.3c}$$

Note that the maximum values of stress/strain occur at the upper and lower boundaries of the slab (i.e., for $z = h/2$ and $z = -h/2$). Note also that Equation 8.3b is a consequence of the plane-strain condition and the generalized Hooke's law; that is:

$$\varepsilon_y = 0 = \frac{1}{E}(\sigma_y - \mu\sigma_x) \tag{8.4}$$

Taking into account Equation 8.3b, allows expressing the strain $\varepsilon_x$ in terms of stress:

$$\varepsilon_x = \frac{1}{E}(\sigma_x - \mu\,\sigma_y) = \frac{1}{E}(\sigma_x - \mu^2\,\sigma_x) = \frac{\sigma_x}{E}(1 - \mu^2) \tag{8.5}$$

Solving Equation 8.5 for stress, and considering Equation 8.3b, gives the following stress-strain expressions for bending in the $x$ direction only:

$$\sigma_x = \frac{E\varepsilon_x}{(1 - \mu^2)} \tag{8.6a}$$

$$\sigma_y = \mu\frac{E\varepsilon_x}{(1 - \mu^2)} \tag{8.6b}$$

Similar expressions can be written for bending in the $y$ direction only.

Consider now combined bending in both the $x$ and $y$ directions. The combined stresses are obtained by superimposing the stresses from the two stress states just described.

$$\sigma_x = \frac{E\varepsilon_x}{(1 - \mu^2)} + \mu\frac{E\varepsilon_y}{(1 - \mu^2)} = \frac{E\,z}{(1 - \mu^2)}\left(\frac{1}{R_x} + \mu\frac{1}{R_y}\right) \tag{8.7a}$$

$$\sigma_y = \frac{E\varepsilon_y}{(1 - \mu^2)} + \mu\frac{E\varepsilon_x}{(1 - \mu^2)} = \frac{E\,z}{(1 - \mu^2)}\left(\frac{1}{R_y} + \mu\frac{1}{R_x}\right) \tag{8.7b}$$

The corresponding strains are:

$$\varepsilon_x = \frac{z}{R_x} = \frac{\sigma_x}{E}\left(1-\mu^2\right) \tag{8.8a}$$

$$\varepsilon_y = \frac{z}{R_y} = \frac{\sigma_y}{E}\left(1-\mu^2\right) \tag{8.8b}$$

Consider now the relationship between the radius of curvature $R$ and the bending moment $M$:

$$M = \int_{-h/2}^{h/2} \sigma\ z\ dA = \int_{-h/2}^{h/2} \frac{E}{R} z^2\ dA = \frac{E\ I}{R} \tag{8.9}$$

where $A$ is the area of the cross section of the slab per unit width, and $I$ is the moment of inertia of the slab per unit width, being equal to $h^3/12$. Equation 8.9 can be written in the more familiar form:

$$\frac{1}{R} = \frac{M}{EI} = \frac{12\ M}{E\ h^3} \tag{8.10}$$

Hence, taking the area integral of both sides of Equation 8.7 gives:

$$M_x = \frac{E\ h^3}{12\left(1-\mu^2\right)}\left(\frac{1}{R_x} + \mu\frac{1}{R_y}\right) \tag{8.11a}$$

$$M_y = \frac{E\ h^3}{12\left(1-\mu^2\right)}\left(\frac{1}{R_y} + \mu\frac{1}{R_x}\right) \tag{8.11b}$$

where $M_x$ and $M_y$ are the bending moments in the $x$ and $y$ directions, respectively. Note that the common factor in these expressions is referred to as the flexural rigidity of the plate, and it is denoted by $D$.

$$D = \frac{E\ h^3}{12(1-\mu^2)} \tag{8.12}$$

Equation 8.11 allows solving for the radii of curvature:

$$\frac{1}{R_x} = \frac{12}{Eh^3}\left(M_x - \mu M_y\right) \tag{8.13a}$$

$$\frac{1}{R_y} = \frac{12}{Eh^3}\left(M_y - \mu M_x\right) \tag{8.13b}$$

which, in turn, are related to vertical deflections $w$ by:[15]

$$\frac{1}{R_x} = -\frac{\partial^2 w}{\partial x^2} \tag{8.14a}$$

$$\frac{1}{R_y} = -\frac{\partial^2 w}{\partial y^2} \tag{8.14b}$$

As discussed next, slabs exposed to temperatures of equal magnitudes and opposite signs at their top and bottom are under pure bending, if their interaction with the subgrade is ignored.

## 8.3 Environment-Induced Stresses

### 8.3.1 Stresses Due to Temperature Gradients

Temperature variation within the slab thickness generates variation in the length of slab fibers, which results in curling. Consider a slab of length $L$. Changes in temperature $\Delta T$ at any particular depth in this slab result in changes of length $\Delta L$ at this depth, assuming that expansion and contraction are unimpeded, according to:

$$\Delta L = La_t\, \Delta T \tag{8.15}$$

where, $L$ is the length of the slab in the direction of the expansion or contraction (i.e., axis $x$ or $y$) and $a_t$ is the coefficient of linear thermal expansion of the portland concrete, typically about $9 \times 10^{-6}/^\circ\mathrm{C}$. This implies a strain $\varepsilon$ given by:

$$\frac{\Delta L}{L} = \varepsilon = a_t\, \Delta T \tag{8.16}$$

When the temperatures at the upper half or the slab are lower than the temperatures at the lower half, as is the case in the evening, slabs have a concave upper shape (Figure 8.4a). The opposite is true in early morning, resulting in a convex upper shape (Figure 8.4b). The weight of the slab acting on these deformed shapes, and the varying reaction from the subgrade, generates stresses in the slab. When the slab has a concave upper shape, there is tension at the top and compression at the bottom. The opposite is true when the slab has a convex upper shape.

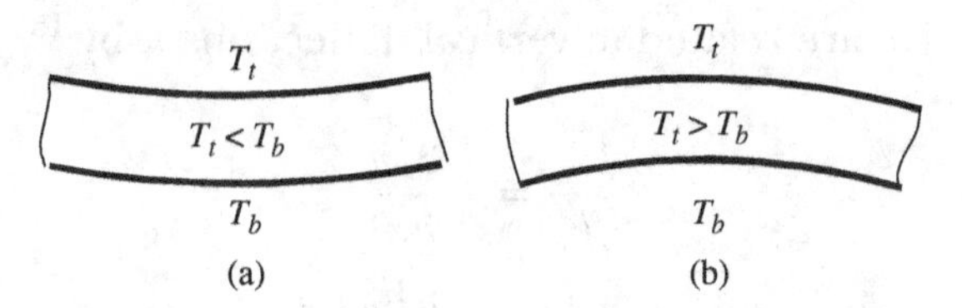


**Figure 8.4**
Slab Warping under Temperature Gradients

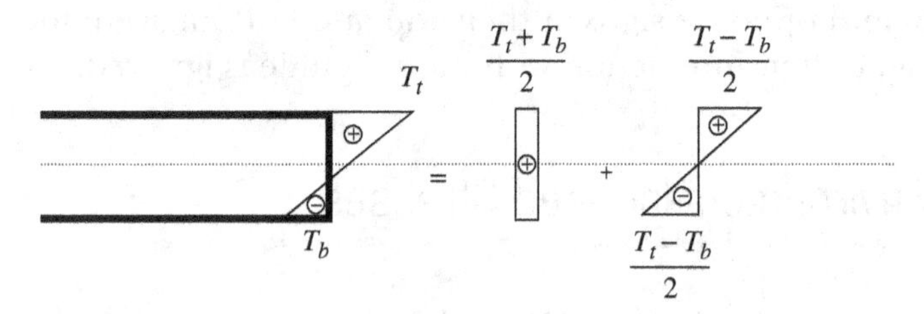


**Figure 8.5**
Disaggregating Temperature Gradients

The distribution of temperatures within the slab can be computed from the boundary temperatures (i.e., the temperatures at the top and the bottom of the slab, denoted by $T_t$ and $T_b$, respectively), using heat diffusion principles (Chapter 10). However, it can be assumed that temperature varies linearly with depth. Consider a slab of infinite *x-y* dimensions with the linear distribution of temperatures shown in Figure 8.5.

It can be disaggregated into a uniform temperature change, and one where temperatures at the top and bottom have the same magnitude but opposite signs. The first temperature component causes no bending stresses (i.e., if unimpeded by adjacent slabs, it is simply resisted by subgrade friction, as discussed later). The second temperature component has an outer fiber temperature of magnitude $T_a$:

$$T_a = \frac{T_t - T_b}{2} \tag{8.17}$$

and results in pure bending. The magnitude of the corresponding outer fiber strains is given by Equation 8.16.

$$\varepsilon_x = \varepsilon_y = a_t \, T_a \tag{8.18}$$

Substituting Equation 8.18 into Equations 8.7 gives the outer fiber slab stresses as:

$$\sigma_y = \sigma_x = \frac{Ea_t\, T_a}{(1-\mu^2)} + \mu\frac{Ea_t\, T_a}{(1-\mu^2)} = \frac{Ea_t\, T_a}{(1-\mu^2)}(1+\mu) = \frac{Ea_t\, T_a}{(1-\mu)} \quad (8.19)$$

Bradbury[3] extended this pure bending formulation to slabs of finite dimensions by weighing the contribution of stresses from bending in the two axes through the variables $C_x$ and $C_y$, resulting in the following stress expressions:

$$\sigma_x = \frac{Ea_t T_a}{(1-\mu^2)}(C_x + \mu C_y) \quad (8.20a)$$

$$\sigma_y = \frac{Ea_t T_a}{(1-\mu^2)}(C_y + \mu C_x) \quad (8.20b)$$

where $C_x$ and $C_y$ are obtained from Figure 8.6.

To obtain these variables, the normalized dimensions of the slab need to be computed by dividing its dimensions in the $x$ and $y$

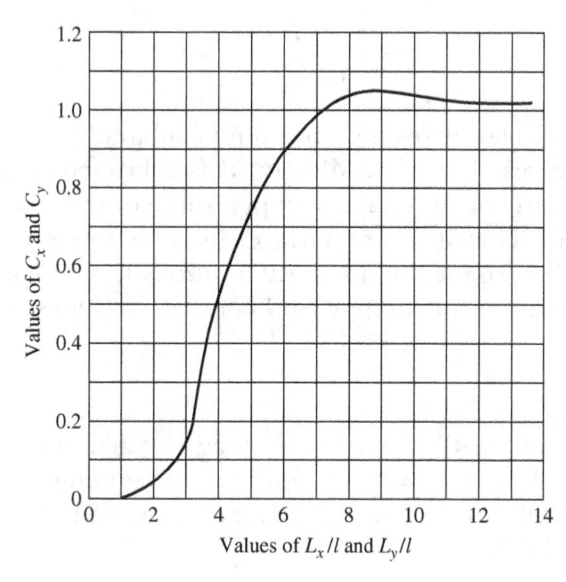


**Figure 8.6**
Variables $C_x$ and $C_y$ for Computing Pure Bending Stresses of Finite-Dimension Slabs (Ref. 3)

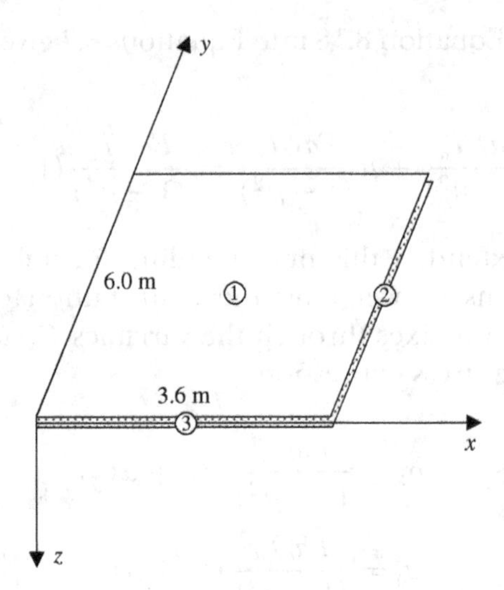


**Figure 8.7**
Slab Layout for Example 8.1

direction by the radius of relative stiffness $\ell$, which for a liquid foundation, is defined by:

$$\ell = \left( \frac{E\, h^3}{12(1 - \mu^2)k} \right)^{1/4} \tag{8.21}$$

**Example 8.1** A 20 cm thick slab is resting on a liquid subgrade with a modulus of subgrade reaction of 80 MPa/m. It is subjected to an increase in temperature of 10°C at its upper surface and a temperature decrease of 5°C at its lower surface. Its dimensions are shown in Figure 8.7. Determine the stresses at locations 1, 2, and 3. Additional information given for the portland concrete includes: $E = 28$ GPa, $\mu = 0.15$, and $a_t = 9 \times 10^{-6}/^{\circ}\mathrm{C}$.

**ANSWER**

The temperature distribution is disaggregated into a uniform increase of $(10 - 5)/2 = 2.5^{\circ}\mathrm{C}$ and a pure bending inducing temperature $T_a$ of $(10 + 5)/2 = 7.5^{\circ}\mathrm{C}$. Equation 8.21 gives the radius of relative stiffness.

$$\ell = \left( \frac{28000\ 0.2^3}{12(1 - 0.15^2)80} \right)^{1/4} = 0.70 \text{ m}$$

This gives normalized slab dimensions of $3.6/0.7 = 5.14$ and $6.0/0.7 = 8.57$ in the $x$ and $y$ directions, respectively. Accordingly, Figure 8.6 yields $C_x$ and $C_y$ values of 0.77 and 1.07, respectively. The stresses are computed using Equations 8.20 At location 1:

$$\sigma_y = \frac{28\,10^6\,9\,10^{-6}\,7.5}{(1-0.15^2)}(1.07 + 0.15\,0.77) = 2292.2\,\text{kPa}$$

$$\sigma_x = \frac{28\,10^6\,9\,10^{-6}\,7.5}{(1-0.15^2)}(0.77 + 0.15\,1.07) = 1799.1\,\text{kPa}$$

These stresses are outer fiber stresses, compressive at the top and tensile at the bottom.

At location 2, there is obviously no stress in the $x$ direction, hence no contribution from it to the stress in the $y$ direction. Accordingly:

$$\sigma_y = \frac{28\,10^6\,9\,10^{-6}\,7.5}{(1-0.15^2)}(1.07) = 2068.5\,\text{kPa}$$

$$\sigma_x = 0$$

At location 3, the reverse is true for location 3, hence:

$$\sigma_y = 0$$

$$\sigma_x = \frac{28\,10^6\,9\,10^{-6}\,7.5}{(1-0.15^2)}(0.77) = 1488.8\,\text{kPa}$$

### 8.3.2 Stresses Due to Subgrade Friction

Consider now uniform expansions/contractions of slabs along their depth. These can be the result of uniform temperature changes, as already described, or due to post-construction concrete shrinkage. This expansion/contraction is impeded only by the shear stresses generated by subgrade friction, assuming that there is no contact between adjacent slabs. Subgrade friction is a function of the amount of slippage between the slab and the subgrade, which increases from the centerline of the slab toward the edges. The resulting free-body diagram of a unit width of slab is shown in Figure 8.8. Only half the slab of length $L$ is shown (i.e., the free-body diagram is symmetric about the centerline axis of the slab). The distribution of the shear subgrade friction stresses can be approximated by a uniform

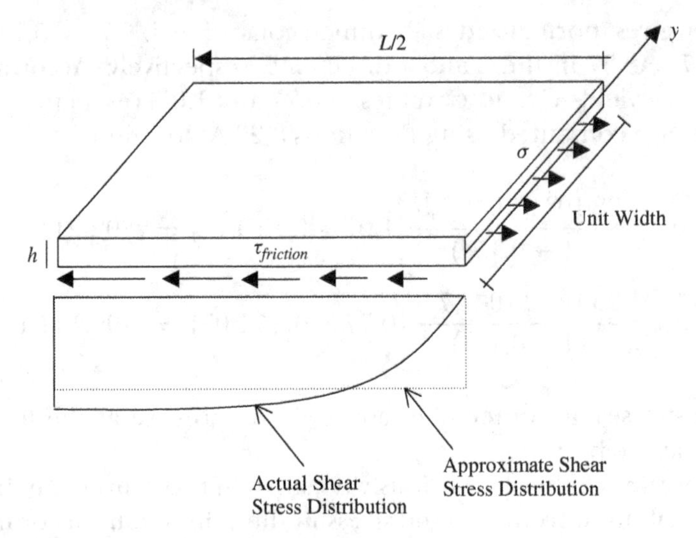


**Figure 8.8**
Free-Body Diagram of Half a Slab Subjected to Subgrade Friction

distribution. Equilibrium of horizontal forces gives:

$$h\frac{L}{2}\gamma\, f = \sigma\ h \tag{8.22}$$

where $\gamma$ is the unit weight of concret, (about $22.5\,\mathrm{kN/m^3}$) and $f$ is the coefficient of friction between the slab and the subgrade (typically around 1.5). The resulting tensile concrete stresses are:

$$\sigma = \frac{L}{2}\gamma f \tag{8.23}$$

**Example 8.2** Compute the tensile stresses generated by subgrade friction in a 6.0 m long concrete slab. How high is this stress compared to the tensile strength of concrete, which is given equal to 2.44 MPa. The unit weight of the concrete is equal to $22.5\,\mathrm{kN/m^3}$.

**ANSWER**
Equation 8.23 gives:

$$\sigma = 3.0\ 22.5\ 1.5 = 101.25\ \mathrm{kPa}$$

This value is considerably lower than the concrete tensile strength of 2.44 MPa specified. This is the case for the typical slab lengths encountered in jointed concrete pavements (i.e., JPCP and JDRCP). Hence, subgrade-friction-induced stresses do not control slab length in jointed pavements. As discussed next, joint opening is the controlling factor in selecting JPCP slab length.

Consider now situations where two slabs need to be tied together, as is the case along longitudinal construction joints (e.g., between driving lane and shoulder or between two adjacent driving lanes). Since the joint carries no subgrade friction-induced tensile stresses, they need to be carried by deformed steel bars (i.e., tiebars). Consult again the free-body diagram shown in Figure 8.8 and rewrite the equilibrium of horizontal forces by considering only the steel reinforcement of cross-sectional area $A_r$ per unit width of slab carry tension:

$$h\frac{L}{2}\gamma f = A_r f_r \tag{8.24}$$

where $f_r$ is the allowable stress of the steel reinforcement (it ranges between 186 and 320 MPa, depending on steel quality). Equation 8.24 allows calculation of the required area of tiebar steel per unit width of slab. It can be generalized to reflect the required area of steel reinforcement, where tiebars used to tie adjacent lanes:

$$A_r = \frac{L' h \gamma f}{f_r} \tag{8.25}$$

where $L'$ is the effective length of slab, defined as the distance between the tiebar location and the farthest free concrete edge. The corresponding average bond stress between a tiebar and the concrete, $u$, is given by:

$$u = \frac{L' h \gamma f}{n \Sigma_o (t/2)} \tag{8.26}$$

where $n$ is the number of tiebars per unit width of slab, $t$ is their length, and $\Sigma_o$ is their circumference. Typically, the allowable maximum bonding stress between tiebars and concrete is 2.4 MPa.

**Example 8.3**

Consider the rigid pavement layout shown in Figure 8.9. The slabs are 0.25 m thick and were poured in two halves by two separate passes of a slip-form paver (the construction joint indicated by a dotted line). Compute the necessary area of tiebar steel across the construction joint and the average bond stress between the tiebars

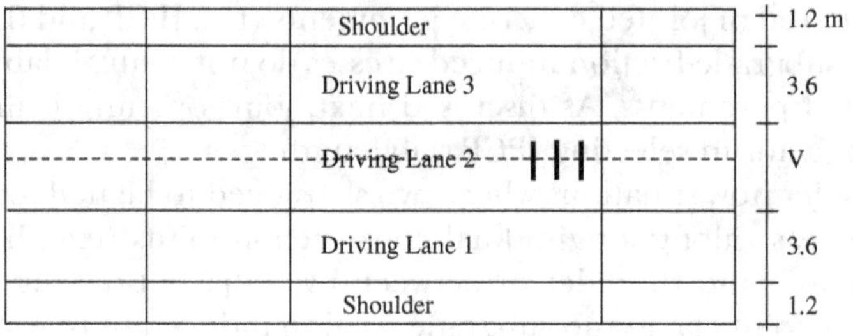

**Figure 8.9**
Slab Layout for Example 8.3

and the concrete. The allowable stress of the steel $f_r$ is given as 186 MPa, and the length of the tiebars is 0.6 m.

**ANSWER**

The distance $L'$ between the construction joint and the free pavement edge is 6.6 m. Substituting the values for the given variables into Equation 8.25 yields:

$$A_r = \frac{6.6 \ 0.25 \ 22.5 \ 1.5}{186000} = 0.0003 \text{ m}^2/\text{m width} = 3.00 \text{ cm}^2/\text{m width}$$

Selecting three tiebars per meter of width yields a bar diameter of $(3.00/3 \times 4/\pi)^{0.5} = 1.1$ cm. The corresponding bond stress between a tiebar and the concrete is given by Equation 8.26:

$$u = \frac{6.6 \ 0.25 \ 22.5 \ 1.5}{3 \ 0.0346 \ (0.6/2)} = 1788 \text{ kPa.}$$

### 8.3.3 Joint Opening Due to Uniform Temperature Changes or Shrinkage

Joint opening is particularly important to JPCPs because they derive their vertical load transfer only through aggregate interlock. Hence, as slabs contract from uniform temperature reductions or from postconstruction shrinkage, the joints open and their vertical load transfer efficiency declines. Practically, plain joints with openings larger than 0.1 cm are not effective in transferring vertical loads across adjacent slabs. Consider the change in slab length $\Delta L$ due to a combined reduction in temperature $\Delta T$ and post-construction shrinkage. Expanding Equation 8.15 to incorporate concrete curing shrinkage gives:

$$\Delta L = c\ L(a_t\ \Delta T + \varepsilon) \quad (8.27)$$

where $c$ is a scaling factor accounting for subgrade resistance (typically around 0.65) and $\varepsilon$ is additional strain due to shrinkage, ranging from $0.5\ 10^{-4}$ to $2.510^{-4}$, depending on curing practices. $\Delta T$ is approximated as the difference between the construction temperature and the coldest mean monthly temperature.[4] Note that slab shrinkage takes place symmetrically about its center axis (see Figure 8.8). The reason for not including half the length of the slab in Equation 8.27 is that the joint opening is made up of two displacement components from the identical half slabs on either side of the joint opening.

**Example 8.4**

Compute the maximum feasible slab length for a JPCP pavement experiencing a temperature difference between construction and the coldest winter month of 35°C. Given, $a_t$ equal to $9 \times 10^{-6}/°\mathrm{C}$ and $\varepsilon$ equal to $0.5\ 10^{-4}$.

**ANSWER**

Substitute the maximum joint opening of 0.001 m into Equation 8.27, along with the given variables, and solve for $L$:

$$0.001 = 0.6\ L(9\ 10^{-6}\ 35 + 0.5\ 10^{-4})$$

which gives a maximum slab length of 4.5 m. Hence, indeed, slab length in JPCPs is controlled by the need to limit joint opening.

## 8.4 Load-Induced Stresses

As mentioned earlier, closed-form solutions of load-induced stresses in portland concrete pavements are available for a limited number of simple loading situations, assuming a liquid subgrade. These are described next, to be followed by a demonstration of a FEM computer program capable of analyzing more complex situations.

### 8.4.1 Stresses under Concentrated Point Loads

Consider a point load applied to a jointed portland concrete pavement. The location of the load likely to cause the highest stresses is near the corner of the slab (i.e., at the edge of a joint). Stress solutions for this problem were developed[12] by analyzing the free-body

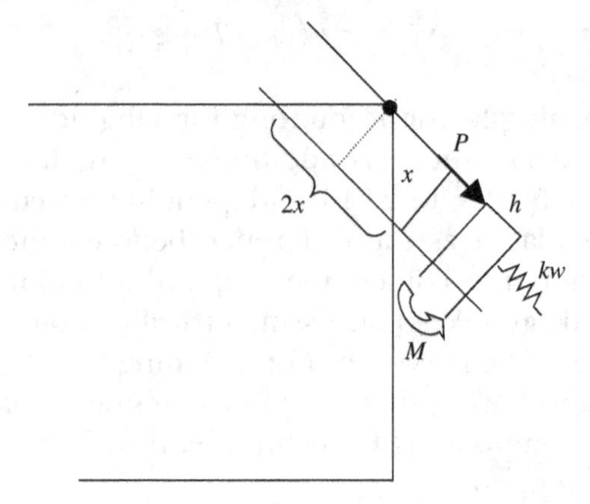


**Figure 8.10**
Stresses under a Concentrated Point Load at the Corner of a Slab

diagram of a slab section oriented 45° to the direction of travel (Figure 8.10). Ignoring the reaction from the subgrade, this section can be analyzed as a cantilevered beam, whereby the bending moment $M$ computed from the triangular stress distribution balances the moment from the point load $P$:

$$\frac{1}{2}\sigma_c\frac{h}{2}2x\frac{2}{3}h = Px \tag{8.28}$$

which gives an outer-fiber concrete corner stress $\sigma_c$ as:

$$\sigma_c = \frac{3\ P}{h^2} \tag{8.29}$$

Note that the subgrade reaction would reduce this stress, hence ignoring it is conservative. Also note that the computed bending stress is not a function of the location of the slab section considered.

**Example 8.5** Compute the maximum tensile stress on a 0.2 m thick slab of a JPCP under a corner point load of 20 kN.

ANSWER

Substituting the given values into Equation 8.29 gives:

$$\sigma_c = \frac{3\ 20}{0.2^2} = 1500\text{ kPa}$$

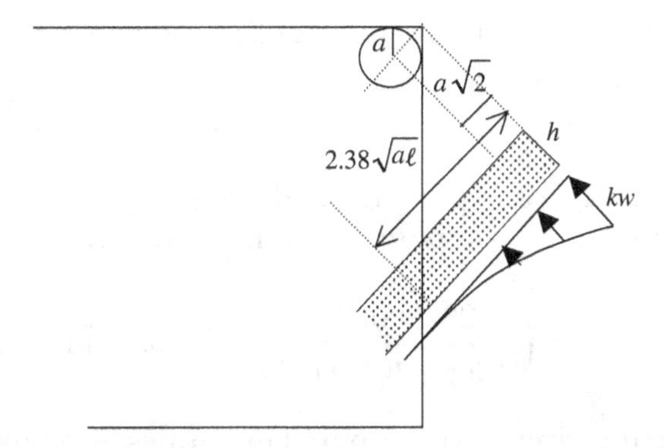


**Figure 8.11**
Stresses under a Circular Imprint Load at the Corner of a Slab

Comparing this stress to the 2440 kPa tensile strength of concrete (Example 8.2), suggests that this load-induced stress is substantial, hence fatigue damage is an issue. As discussed in Chapter 12, fatigue is one of the considerations in the mechanistic design of rigid pavements.

### 8.4.2 Stresses under Uniform Circular Stresses

For loads uniformly distributed over a circular area of radius $a$, the cornermost location of the load is shown in Figure 8.11 (i.e., moving the center of the load closer to the corner would transfer part of it to the adjacent slabs). Work by Westergard[20] determined that, assuming a Poisson's ratio of 0.15, the maximum stress occurs at a distance of $2.38\sqrt{a\ell}$ from the corner, and amounts to:

$$\sigma_c = \frac{3\,P}{h^2}\left[1 - \left(\frac{a\sqrt{2}}{\ell}\right)^{0.6}\right] \quad (8.30)$$

where $\ell$ is the *radius of relative stiffness* of the slab (Equation 8.21). It has also provided expressions for the corner deflection $\Delta_c$.

$$\Delta_c = \frac{P}{k\ell^2}\left[1.1 - 0.88\left(\frac{a\sqrt{2}}{\ell}\right)\right] \quad (8.31)$$

**Example 8.6**

Determine the maximum tensile stress and the corner deflection under a circular load of 0.12 m radius carrying a 600 kPa pressure, given a slab thickness of 0.22 m, a modulus of subgrade reaction of 40 MPa/m, a concrete modulus of 28 GPa, and a Poisson's ratio of 0.15.

**ANSWER**

Use Equation 2.1 to compute the load $P$ as $\pi \times 0.12^2 \times 600 = 27$ kN. Use Equation 8.21 to compute the radius of relative stiffness.

$$\ell = \left(\frac{280000.22^3}{12(1 - 0.15^2)40}\right)^{1/4} = 0.89 \text{ m}$$

Substitute the given and computed quantities into Equations 8.30 and 8.31.

$$\sigma_c = \frac{3\ 27}{0.22^2}\left[1 - \left(\frac{0.12\sqrt{2}}{0.89}\right)^{0.6}\right] = 1054.4\,\text{kPa}$$

$$\Delta_c = \frac{27}{40000\ 0.89^2}\left[1.1 - 0.88\left(\frac{0.12\sqrt{2}}{0.89}\right)\right] = 0.00079\,\text{m or } 0.8\,\text{mm}$$

Westergard[20] also developed expressions for the stresses under uniform circular stresses applied to interior points, as well as at the edge of slabs, denoted by $\sigma_i$ and $\sigma_e$, respectively. For a Poisson's ratio of 0.15, these expressions are:

$$\sigma_i = \frac{0.316\,P}{h^2}\left[4\log\left(\frac{l}{b}\right) + 1.069\right] \tag{8.32}$$

$$\sigma_e = \frac{0.572\,P}{h^2}\left[4\log\left(\frac{l}{b}\right) + 0.359\right] \tag{8.33}$$

where:

$$b = a \qquad \text{for } a \geq 1.724\,h \tag{8.34a}$$

$$b = \sqrt{1.6\,a^2 + h^2} - 0.675\,h \qquad \text{for } a < 1.724\,h \tag{8.34b}$$

**Example 8.7**

Determine the stresses in the interior and edge point of a 0.20 m thick slab under a uniform stress of 600 kPa distributed over a circular area with a radius of 0.15 m. Given, a modulus of subgrade reaction of 60 MPa/m, an elastic modulus of concrete of 28 GPa, and a Poisson's ratio of concrete of 0.15.

**ANSWER**

Use Equation 8.21 to compute the radius of relative stiffness of the slab.

$$\ell = \left( \frac{28000\ 0.2^3}{12(1 - 0.15^2)60} \right)^{1/4} = 0.75 \text{ m}$$

Compute $1.724h = 0.34$ m, which is larger than $a = 0.15$ m; hence compute $b$ using Equation 8.34b.

$$b = \sqrt{1.6\ 0.15^2 + 0.2^2} - 0.675\ 0.2 = 0.141 \text{ m}$$

The interior and edge stresses are computed using Equations 8.32 and 8.33, where

$$P = 600\pi 0.15^2 = 42.41 \text{ kN}$$

$$\sigma_i = \frac{0.316\ 42.41}{0.2^2} \left[ 4 \log \left( \frac{0.75}{0.141} \right) + 1.069 \right] = 1330.9 \text{ kPa}$$

$$\sigma_e = \frac{0.572\ 42.41}{0.2^2} \left[ 4 \log \left( \frac{0.75}{0.141} \right) + 0.359 \right] = 1978.5 \text{ kPa}$$

Pickett and Ray[13] and Pickett and Badaruddin[14] produced solutions of rigid pavement response to load in the form of influence charts for liquid and solid foundations, respectively, for both corner and interior slab points. An example of these charts is shown in Figure 8.12 for the deflection $\Delta$ under a load in the interior point of a slab, with a Poisson's ratio of 0.15, assuming a liquid foundation. This influence chart is used by drawing the tire imprint(s) to the scale dictated by the length of the radius of the relative stiffness of the slab shown on the chart, counting the number of blocks $N$ and using:

$$\Delta = \frac{0.0005\ p\ l^4\ N}{D} \tag{8.35}$$

where $p$ is the uniform contact pressure, $\ell$ is the radius of relative stiffness (Equation 8.21), and $D$ is the flexural rigidity of the plate (Equation 8.12). Note that imperial units need to be used in this influence chart. Additional influence charts for bending moments and stresses can be found in the original papers or in earlier textbooks (e.g., reference 16). As will be pointed out later, the advent of numerical methods (e.g., FEM) has rendered such influence-function-based solutions somewhat impractical for routine use.

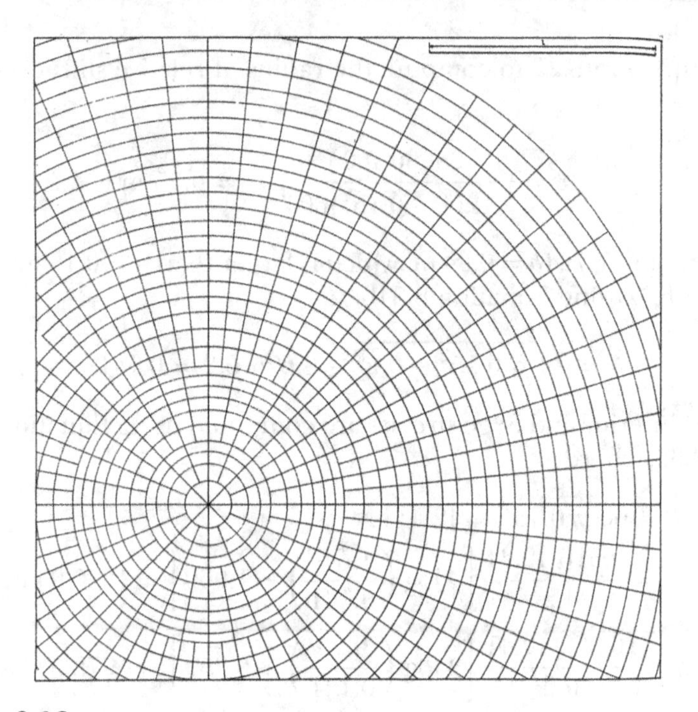

**Figure 8.12**
Influence Chart for Computing Surface Slab Deflections (Ref. 13)

### 8.4.3 Dowel-Bar-Induced Stresses in JDRCPs

Dowel bars in JDRCPs provide vertical load transfer between adjacent slabs. The design challenge in transferring these vertical loads is to prevent fracture of the concrete supporting the dowel bars. The allowable bearing stress of concrete $f_b$ under a dowel bar is adapted from an expression given in reference 2 as:

$$f_b = \left(\frac{4 - 0.3937\, d}{3}\right) f_c' \tag{8.36}$$

where $f_b$ and $f_c'$ are in kPa, and the diameter of the dowel bar $d$ is in cm. Friberg[7] developed a solution for the vertical deflection of a dowel bar at the outer edge of a joint, denoted by $y_0$, (Figure 8.13). This was done by assuming that the concrete functions as a liquid foundation, with a spring constant $K_c$ in supporting the dowel bar (typically, between 80,000 and 400,000 MPa/m). This deflection is calculated as:

$$y_0 = \frac{P_t}{4\beta^3\, E_r\, I_r}(2 + \beta z) \tag{8.37}$$

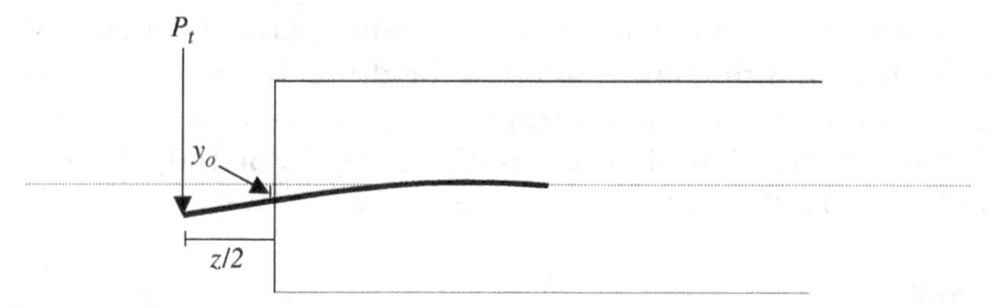


**Figure 8.13**
Dowel Bar and Concrete Deflection Under Load $P_t$ (not to scale)

where $E_r$ and $I_r$ are the modulus and moment of inertia of the dowel bar reinforcement, respectively; $P_t$ is the load transferred by the dowel bar, $z$ is the opening of the joint, and $\beta$ is given by:

$$\beta = \left(\frac{K_c d}{4\,E_r\,I_r}\right)^{0.25} \tag{8.38}$$

The corresponding bearing concrete stress at the joint face is computed as:

$$\sigma = K_c\,y_0 \tag{8.39}$$

In practice, the length of the dowel bars is decided from the thickness of the concrete slab and the diameter of the dowel, as shown in Table 8.1.

**Table 8.1**
Dowel Bar Lengths for JDRCPs (Ref. 6)

| Slab Thickness, *h* (cm) | Dowel Diameter (cm) | Dowel Bar Length (cm)* |
|---|---|---|
| 12.5 | 1.6 | 30 |
| 15 | 1.9 | 36 |
| 18 | 2.2 | 36 |
| 20 | 2.5 | 36 |
| 23 | 2.9 | 40 |
| 25 | 3.2 | 45 |
| 28 | 3.5 | 45 |
| 30 | 3.8 | 50 |

*Dowels are assumed to be spaced 0.30 m center to center.

**Example 8.8** A 2.5 cm diameter dowel bar is transferring a vertical load of 3500 N across a 0.5 cm wide joint. Compute the dowel bar deflection at the edge of the joint and the corresponding concrete bearing stresses. Can the concrete handle this stress? Given, $K_c$ of 100,000 MPa/m, $E_r$ of 200,000 MPa, and $f_c'$ of 28 MPa.

**ANSWER**

Compute the moment of inertia of the circular cross section of the dowel bar.

$$I_r = \frac{\pi d^4}{64} = \frac{\pi 0.025^4}{64} = 1.92\ 10^{-8}\ \mathrm{m}^4$$

Compute the quantity $\beta$:

$$\beta = \left(\frac{100000\ 0.025}{4\ 200000\ 1.92\ 10^{-8}}\right)^{0.25} = 20.11\ \mathrm{m}^{-1}$$

Substituting into Equation 8.37 gives:

$$y_0 = \frac{3.5}{4\ 20.11^3\ 20010^6\ 1.91\ 10^{-8}}(2 + 20.11\ 0.005) = 0.000059\ \mathrm{m} = 0.059\ \mathrm{mm}$$

The concrete bearing stress at the edge of the joint is computed from Equation 8.39.

$$\sigma = 100\ 10^6\ 0.000059 = 5916.2\ \mathrm{kPa}$$

Comparing this concrete stress to the allowed bearing stress of concrete (Equation 8.36):

$$f_b = \left(\frac{4 - 0.3937\ 2.5}{3}\right)28000 = 28147\ \mathrm{kPa}$$

suggests that the concrete can indeed handle the applied stress.

In practice, load is transferred across the joints of JDRCPs by the combined action of a group of dowels. Work by Friberg[7] established that load affects dowels up to a distance of $1.8\ell$ away from its location, where $\ell$ is the radius of relative stiffness of the slab (Equation 8.21). Assuming that the load is distributed across a group of dowels in a linear fashion allows calculation of the load carried by each dowel, as demonstrated in the following example.

**Example 8.9**

Consider a JDRCP consisting of slabs 20 cm thick and 3.6 m wide, resting on a subgrade with a modulus of subgrade reaction of 60 MPa/m. An axle load consisting of two identical tires 1.8 meters apart, each carrying 40 kN, is located at the edge of the joint, 0.30 m from the edge of the slab. Load across the joint is carried by 25 mm diameter dowel bars placed at 0.3 m center-to-center distances, as shown in the Figure 8.14. Compute the load carried by each dowel bar. Given, $E$ for the portland concrete of 28,000 MPa and a Poisson's ratio of 0.15. Assume that the tires apply point loads, that the load transfer across the slabs is 50/50, and that the distribution of load varies linearly with the distance from each tire load location.

**ANSWER**

Compute first the radius of relative stiffness of the slabs.

$$\ell = \left(\frac{28000\ 0.2^3}{12(1 - 0.15^2)60}\right)^{1/4} = 0.75 \text{ m}$$

Hence, the loads influence dowels within a distance of $1.8 \times 0.75 = 1.35$ m away from their location. Given a 50/50 load transfer, half of the 40 kN load applied by each tire is transferred by the dowels to the adjacent slab, (i.e., the 50/50 load transfer is ideal; in practice, less than half of the load is transferred to the adjacent slab). Hence, a 20 kN load is carried by the dowel bars within the 1.35 m range computed earlier around the location of each tire. The linear distribution of load allows plotting the distribution of load under each tire (Figure 8.15).

Using simple geometry allows computing the fraction of the loads $x$ and $y$ corresponding to each dowel bar location. Equilibrium of

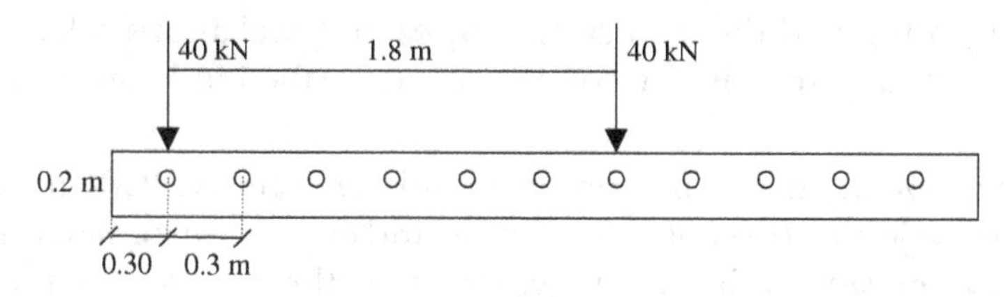


**Figure 8.14**
Slab and Dowel Bar Layout for Example 8.9

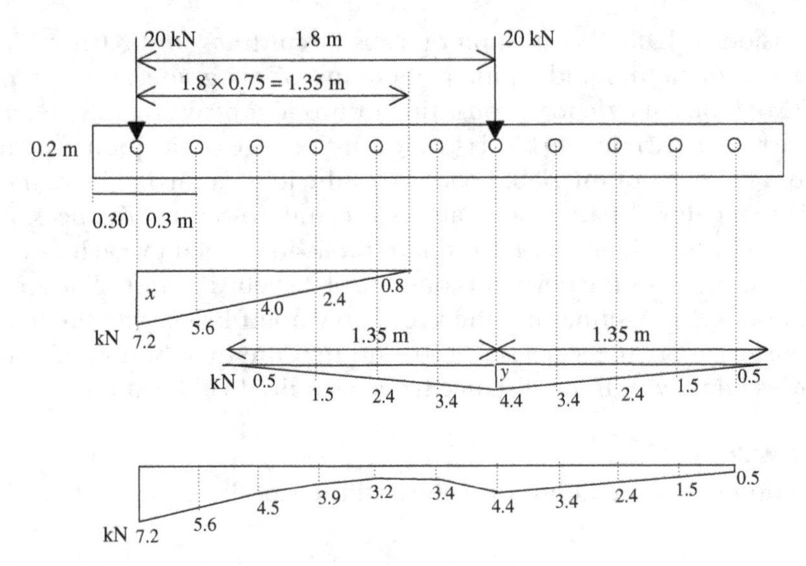


**Figure 8.15**
Load Diagrams for Dowel Bars for Example 8.9

forces for the left-hand-side load gives:

$$x\left(1 + \frac{3.5}{4.5} + \frac{2.5}{4.5} + \frac{1.5}{4.5} + \frac{0.5}{4.5}\right) = 20$$

which results in a value for $x$ of 7.2 kN. Similarly, equilibrium of forces for the right-hand-side load gives:

$$2y\left(\frac{3.5}{4.5} + \frac{2.5}{4.5} + \frac{1.5}{4.5} + \frac{0.5}{4.5}\right) + y = 20$$

which results in a value for $y$ of 4.4 kN. The combined diagram of dowel bar loads from the two loads is included in Figure 8.15. Having computed the load carried by each dowel allows calculation of the bearing stress in the concrete, as described in Example 8.8.

### 8.4.4 Reinforcement Stresses in CRCPs

As mentioned in the beginning of this chapter, CRCPs have no transverse joints, (except construction and expansion joints), relying instead on steel reinforcement to carry the stresses that result from subgrade friction in response to post-construction concrete shrinkage and in-service temperature changes. This reinforcement consists of substantial amounts of deformed steel (i.e., rebar) located at the neutral axis of the slabs. The steel bars are spliced by overlap to function as continuous reinforcement. Obviously, the portland concrete will develop transverse cracks where its tensile strength is reached, which is at predictable intervals. The function of the reinforcement is to hold these cracks closely together, hence provide functional continuity for the slab, (transverse crack intervals ranging between 1 and 3 meters have exhibited excellent performance).

Theoretical work by Vetter[19] provided expressions for the required percentage of rebar reinforcement steel, $p$, and the corresponding distance interval of concrete cracks $L$. For post-construction concrete shrinkage, these expressions are:

$$p = \left( \frac{f_t}{S_r + \varepsilon \, E_r - r \, S'_c} \right) 100 \tag{8.40}$$

where $f_t$ is the tensile strength of the concrete measured through indirect tension tests, $S_r$ is the elastic limit of steel, $\varepsilon$ is the concrete shrinkage strain (typically ranging from 0.5 $10^{-4}$ to 2.5 $10^{-4}$), and $r$ is the ratio of the elastic modulus of the steel reinforcement $E_r$ divided by the elastic modulus of the concrete $E$:

$$r = \frac{E_r}{E} \tag{8.41}$$

$$L = \frac{(f_t)^2}{r \, p^2 \frac{\Sigma_o}{A_r} u \left( \varepsilon \, E - S'_c \right)} \tag{8.42}$$

where $u$ is the average bond stress between the steel reinforcement and the concrete, $A_r$ is the area of steel reinforcement per unit width of slab, and $\Sigma_0$ is the perimeter of the rebar selected.

For an in-service uniform temperature drop $t$, these expressions are:

$$p = \left( \frac{f_t}{2 \, (S_r - t \, a_t \, E_r)} \right) 100 \tag{8.43}$$

$$L = \frac{(f_t)^2}{r \, p^2 \dfrac{\Sigma_o}{A_r} u \left( t \, a_t E - S'_c \right)} \tag{8.44}$$

where $a_t$ is the coefficient of linear thermal expansion of the portland concrete (approximately 9.0 $10^{-6}$/°C).

**Example 8.10** Design the amount of rebar reinforcement required for a CRCP 0.20 m thick slab subjected to a temperature difference between pouring and the coldest winter day of 30°C. Given, a coefficient of linear thermal expansion for the concrete of $9.0\ 10^{-6}/°C$, a concrete tensile strength of 3 MPa, a steel elastic modulus of 200000 MPa, a concrete elastic modulus of 28000 MPa, a steel elastic limit of 320 MPa, and an allowable bonding stress between steel and concrete of 2.4 MPa.

**ANSWER**

Using Equation 8.43 gives:

$$p = \left(\frac{3}{2(320 - 30\ 9 10^{-6} 200000)}\right) 100 = 0.56\%$$

Considering the area of a unit width of slab of $0.2 \times 1 = 0.2\,m^2$ gives a required area of steel of $0.2 \times 0.0056 = 1.12\ 10^{-3}\,m^2 = 11.2\,cm^2$, which, divided between three reinforcing bars per meter width of slab, gives a diameter of $(11.2/3 \times 4/\pi)^{0.5} = 2.2\,cm$ and a rebar circumference of 6.91 cm.

In practice, a minimum of 0.6% of steel reinforcement is used in CRCPs. Obviously, the free ends of CRCP slabs exhibit large horizontal displacements, which must be accommodated, where the pavement transitions to another fixed structure, (e.g., a bridge abutment). This is done through expansion joints that provide no resistance to the horizontal slab displacement, hence transmit no horizontal forces to the fixed structure (see Figure 1.4b).

## 8.5 Finite Element Method Solutions

The Finite Element Method (FEM) is the recommended approach for analyzing the complex behavior of slabs under load/environmental input and subgrade support. The method is based on discretizing the problem (i.e. slabs, subgrade, and reinforcement) into basic elements, defining the stresses and the strains inside each element as a function of its nodal displacements, solving for the nodal displacements of all the elements simultaneously, and, finally, computing element stresses and strains from the known displacements. Slabs can be modeled as two-dimensional thin plates or three-dimensional solids, depending on the desired level of detail.

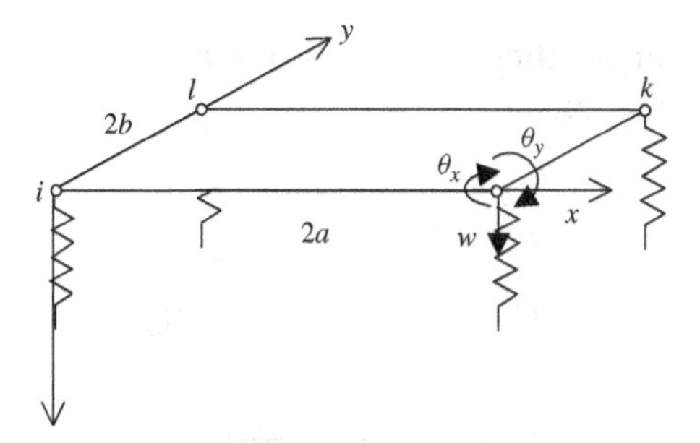


**Figure 8.16**
Degrees of Freedom of Plate Element

### 8.5.1 Element Stiffness

Let us consider the simplest slab representation, which is a two-dimen sional thin plate. An element of this thin plate is shown in Figure 8.16, having dimensions $2a$ by $2b$. Its thickness, $h$, is considered negligible, compared to its other two dimensions.

The four nodes of this element are denoted by $i$, $j$, $k$, and $l$. Each node is supported by a spring and has three degrees of freedom, as shown for node $j$; that is, it can displace vertically, rotate around axis $x$, and rotate around axis $y$, which are denoted by $w_j$, $\theta_{jx}$, and $\theta_{jy}$, respectively. Hence, this basic thin plate element has 12 degrees of freedom overall. The forces/moments corresponding to these degrees of freedom are denoted by $F$, $M_x$, and $M_y$, with additional subscripts denoting the node in question. The relationship between the nodal displacements/rotations and the forces/moments is given by:

$$\begin{Bmatrix} F_i \\ M_{ix} \\ M_{iy} \\ F_j \\ M_{jx} \\ M_{jy} \\ F_k \\ M_{kx} \\ M_{ky} \\ F_l \\ M_{lx} \\ M_{ly} \end{Bmatrix} = [s] \begin{Bmatrix} w_i \\ \theta_{ix} \\ \theta_{iy} \\ w_j \\ \theta_{jx} \\ \theta_{jy} \\ w_k \\ \theta_{kx} \\ \theta_{ky} \\ w_l \\ \theta_{lx} \\ \theta_{ly} \end{Bmatrix} \tag{8.45}$$

where $s$ is a $12 \times 12$ matrix that depends on the geometry of the element and the elastic properties of the concrete. For a rect- angular element and an orthotropic material (i.e., material that exhibits three perpendicular planes of elastic symmetry), $s$ is given in Table 8.2.[17] where:

$$p = \frac{a}{b} \tag{8.46}$$

$$D_x = D_y = \frac{E\,h^3}{12\left(1-\mu^2\right)} \tag{8.47}$$

$$D_1 = \frac{E\,h^3}{12\left(1-\mu^2\right)}\mu \tag{8.48}$$

$$D_{xy} = \frac{E\,h^3}{24\left(1+\mu\right)} \tag{8.49}$$

**Example 8.11** Consider a thin plate element with dimensions $2a = 2.0$ m and $2b = 1.0$ m. Compute the stiffness matrix of this element. Given, thickness $h = 0.10$ m, $E = 28000$ MPa, and $\mu = 0.15$.

**ANSWER**

Compute the rigidity-related parameters of the slab, using Equations 8.46 to 8.49.

$$p = \frac{1.0}{0.5} = 2$$

$$D_x = D_y = \frac{28000\ 0.1^3}{12\left(1-0.15^2\right)} = 2.387\,\text{MN/m}$$

$$D_1 = \frac{28000\ 0.1^3}{12\left(1-0.15^2\right)}0.15 = 0.358\,\text{MN/m}$$

$$D_{xy} = \frac{28000\ 0.1^3}{24\left(1+0.15\right)} = 1.014\,\text{MN/m}$$

Substituting these values into the stiffness matrix described by Table 8.2 gives:

$$s = \frac{1}{410.5}\begin{bmatrix}
11.46 & -2.86 & 9.55 & -4.30 & -1.43 & 0 & -7.16 & 0 & 4.77 & 0 & 0 & 0 \\
29.72 & -19.10 & 1.43 & -28.46 & -9.55 & 0 & -1.07 & 0 & 0.72 & 0 & 0 & 0 \\
-2.03 & 2.03 & -4.06 & 2.03 & 0 & 4.06 & 2.03 & -2.03 & 0 & -2.03 & 0 & 0 \\
-4.3 & 1.43 & 0 & 11.46 & 2.86 & 9.55 & 0 & 0 & 0 & -4.30 & 0 & 4.77 \\
-28.64 & 9.55 & 0 & 29.72 & 19.10 & 1.43 & 0 & 0 & 0 & -4.30 & 0 & 0.72 \\
-2.03 & 0 & -4.06 & 2.03 & 2.03 & 4.06 & 2.03 & 0 & 0 & -2.03 & -2.03 & 0 \\
-7.16 & 0 & -4.77 & 0 & 0 & 0 & 11.46 & -2.86 & -9.55 & -4.30 & -1.43 & 0 \\
-1.07 & 0 & -.72 & 0 & 0 & 0 & 29.72 & -19.10 & -1.43 & -4.30 & -9.55 & 0 \\
-2.03 & 2.03 & 0 & 2.03 & 0 & 0 & 2.03 & -2.03 & -4.06 & -4.77 & 0 & 4.06 \\
0 & 0 & 0 & -7.16 & 0 & -2.03 & -4.3 & 1.43 & 0 & 11.46 & 2.86 & -9.55 \\
0 & 0 & 0 & -1.07 & 0 & -0.72 & -28.64 & 9.55 & 0 & 29.72 & 19.10 & -9.55 \\
-2.03 & 0 & 0 & 2.03 & 4.77 & 0 & 2.03 & 0 & -4.06 & -2.03 & -4.77 & 4.06
\end{bmatrix}$$

**Table 8.2**
Slab Stiffness Matrix (Ref. 17)

$$
s = \frac{1}{4ab}\begin{bmatrix}
6p^{-1}D_x+6pD_1 & -8aD_1 & 8bD_x & -6pD_1 & -4aD_1 & 0 & -6p^{-1}D_x & 0 & 4bD_x & 0 & 0 & 0 \\
6pD_y+6p^{-1}D_1 & -8aD_y & 8bD_1 & -6pD_y & -4aD_y & 0 & -6p^{-1}D_1 & 0 & 4bD_1 & 0 & 0 & 0 \\
-2D_{xy} & 4bD_{xy} & -4aD_{xy} & 2D_{xy} & 0 & 4aD_{xy} & 2D_{xy} & -4bD_{xy} & 0 & -2D_{xy} & 0 & 0 \\
-6pD_1 & 4aD_1 & 0 & 6p^{-1}D_x+6pD_1 & 8aD_1 & 8bD_x & 0 & 0 & 0 & -6p^{-1}D_x & 0 & 4bD_x \\
-6pD_y & 4aD_y & 0 & 6pD_y+6p^{-1}D_1 & 8aD_y & 8bD_1 & 0 & 0 & 0 & -6p^{-1}D_1 & 0 & 4bD_1 \\
-2D_{xy} & 0 & -4aD_{xy} & 2D_{xy} & 4bD_{xy} & 4aD_{xy} & 2D_{xy} & 0 & 0 & -2D_{xy} & -4bD_{xy} & 0 \\
-6p^{-1}D_x & 0 & -4bD_x & 0 & 0 & 0 & 6p^{-1}D_x+6pD_1 & -8aD_1 & -8bD_x & -6pD_1 & -4aD_1 & 0 \\
-6p^{-1}D_1 & 0 & -4bD_1 & 0 & 0 & 0 & 6pD_y+6p^{-1}D_1 & -8aD_y & -8bD_1 & -6pD_y & -4aD_y & 0 \\
-2D_{xy} & 4bD_{xy} & 0 & 2D_{xy} & 0 & 0 & 2D_{xy} & -4bD_{xy} & -4aD_{xy} & -2D_y & 0 & 4aD_{xy} \\
0 & 0 & 0 & -6p^{-1}D_x & 0 & -4bD_x & -6pD_1 & 4aD_1 & 0 & 6p^{-1}D_x+6pD_1 & 8aD_1 & -8bD_x \\
0 & 0 & 0 & -6p^{-1}D_1 & 0 & -4bD_1 & -6pD_y & 4aD_y & 0 & 6pD_y+6p^{-1}D_1 & 8aD_y & -8bD_1 \\
-2D_{xy} & 0 & 0 & 2D_{xy} & 4bD_{xy} & 0 & 2D_{xy} & 0 & -4aD_{xy} & -2D_{xy} & -4bD_{xy} & 4aD_{xy}
\end{bmatrix}
$$

Although assembling the element stiffness matrix shorthand is tedious, it can be readily done through a computer algorithm.

### 8.5.2 Subgrade Support Stiffness

Consider now the subgrade support of the thin plate element just described, in terms of elastic sprigs, one under each node. The relationship between vertical nodal forces and nodal displacements defines the type of foundation supporting the thin plate element. For a liquid (i.e., Winkler) foundation, the vertical forces at a particular node affect the deflection under that node alone. This results in the following foundation stiffness matrix:

$$\begin{Bmatrix} F_i \\ F_j \\ F_k \\ F_l \end{Bmatrix} = \begin{bmatrix} k & 0 & 0 & 0 \\ 0 & k & 0 & 0 \\ 0 & 0 & k & 0 \\ 0 & 0 & 0 & k \end{bmatrix} \begin{Bmatrix} w_i \\ w_j \\ w_k \\ w_l \end{Bmatrix} \tag{8.50}$$

where $k$ is the modulus of subgrade reaction. For a solid (i.e., Boussinesq) foundation, a vertical force on the subgrade affects deflections not only directly under it but also in its vicinity, according to:

$$w_j = \frac{(1 - {\mu_s}^2)}{\pi\ E_s\ r_{ij}} F_i \tag{8.51}$$

where $E_s$ and $\mu_s$ are the elastic properties of the subgrade. This expression can be used for relating nodal displacement to nodal forces for all load-deflection combinations that do not coincide (i.e., for $i \neq j$). Obviously, where $i = j, r_{ij}$ is zero, hence the deflection directly under a point load would be infinite. This limitation is circumvented by distributing the load $F_i$ uniformly over an area in the vicinity of node $i$ (i.e., a quarter of the plate element, as shown in Figure 8.17) and computing the resulting deflections at node $i$ through numerical integration.

The resulting uniform pressure is $F_i/(a\ b)$, and the deflection at point $i$ is computed as:

$$\begin{aligned} w_j &= \frac{(1 - \mu_s^2)}{\pi E_s} \frac{F_i}{a\ b} \int\int_A \frac{1}{r_{ij}} dA \\ &= \frac{(1 - \mu_s^2)}{\pi\ E_s} \frac{F_i}{a\ b} \int_0^a \left[ \int_0^b \left(x^2 + y^2\right)^{-0.5} dy \right] dx \end{aligned} \tag{8.52}$$

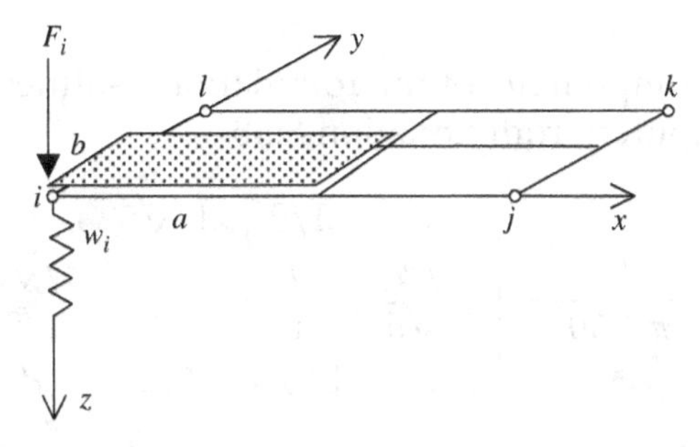


**Figure 8.17**
Distributing a Point Load over an Area for the Purpose of Computing the Deflection under It in a Solid Foundation

where $A$ is the area of the quarter of the plate element, and $x$ and $y$ are the Cartesian coordinates of the plate element (ranging from 0 to $a$ and 0 to $b$, respectively). The computation of this integral is done numerically using a fifth order Gaussian quadrature,[8] as described in the following example. Hence, the nodal displacement versus nodal force relationship for a solid foundation can be written in matrix form as:

$$\begin{Bmatrix} w_i \\ w_j \\ w_k \\ w_l \end{Bmatrix} = \frac{(1-\mu_s^2)}{\pi E_s} \begin{bmatrix} d & 1/r_{ij} & 1/r_{ik} & 1/r_{il} \\ 1/r_{ji} & d & 1/r_{jk} & 1/r_{jl} \\ 1/r_{ki} & 1/r_{kj} & d & 1/r_{kl} \\ 1/r_{li} & 1/r_{lj} & 1/r_{lk} & d \end{bmatrix} \begin{Bmatrix} F_i \\ F_j \\ F_k \\ F_l \end{Bmatrix} \tag{8.53}$$

where the value of the diagonal elements $d$ is given by:

$$d = \frac{1}{a\,b} \int_0^a \left[ \int_0^b \left(x^2 + y^2\right)^{-0.5} dy \right] dx \tag{8.54}$$

The latter is simply a function of the geometry of the plate element.

Finally, the stiffness matrix of the solid foundation is obtained by inverting the square matrix in Equation 8.53, to obtain nodal forces as a function of nodal displacements.

**Example 8.12**

Compute the solid foundation stiffness matrix for the same plate element described in Example 8.11. Given, subgrade $E_s$ of 250 MPa and $\mu_s$ of 0.35.

**ANSWER**

Nondiagonal components of the foundation "softness" matrix can be readily computed from Equation 8.53.

$$\begin{Bmatrix} w_i \\ w_j \\ w_k \\ w_l \end{Bmatrix} = \frac{(1-0.35^2)}{\pi\ 250} \begin{bmatrix} d & 1/2 & 1/\sqrt{5} & 1 \\ 1/2 & d & 1 & 1/\sqrt{5} \\ 1/\sqrt{5} & 1 & d & 1/2 \\ 1 & 1/\sqrt{5} & 1/2 & d \end{bmatrix} \begin{Bmatrix} F_i \\ F_j \\ F_k \\ F_l \end{Bmatrix}$$

The diagonal components can be computed using MathCad™ or shorthand through numerical integration via a fifth order Gaussian quadrature. This approach suggests that the integral of any function $f$ between $-1$ and 1 can be approximated by:[8]

$$\int_{-1}^{1} f(t)\,dt \approx 0.5555f(-0.7746) + 0.88889f(0) + 0.5555f(0.7746) \tag{8.55}$$

For integration limits other than $-1$ to 1 (for this problem, it is either from 0 to $a$ or from 0 to $b$), a variable transformation is necessary.

$$\int_{0}^{a} f(x)\,dx = \frac{a}{2}\int_{-1}^{1} f\left(\frac{a\,t+a}{2}\right) dt \text{ or } \int_{0}^{b} f(x)\,dx = \frac{b}{2}\int_{-1}^{1} f\left(\frac{b\,t+b}{2}\right) dt \tag{8.56}$$

Equations 8.55 and 8.56 allow computation of the double integral in the expression for the diagonal elements $d$ of the "softness" matrix in two steps. Let us demonstrate the first step by computing the inside integral of Equation 8.54, after substituting $b = 0.5$ m.

$$\left[\int_{0}^{0.5} \left(x^2+y^2\right)^{-0.5} dy\right] = \frac{0.5}{2}\int_{-1}^{1} \left(x^2 + \left(\frac{0.5t+0.5}{2}\right)^2\right)^{-0.5} dt$$

$$= 0.25\left\{0.5555\left[x^2 + (0.25\,(-0.7746) + 0.25)^2\right]^{-0.5}\right.$$
$$+0.8888\left[x^2 + (0 + 0.25)^2\right]^{-0.5}$$
$$\left.+\ 0.5555\left[x^2 + (0.25\,(0.7746) + 0.25)^2\right]^{-0.5}\right\}$$

This function of $x$ integrated in the same fashion in the other direction (i.e., for $x$ ranging from 0 to $a = 1.0$ m) gives an area integral of 1.203 m$^2$. Hence, according to Equation 8.54, the value of the diagonal components of the "softness" matrix $d$ is 1.203/(1 × 0.5) = 2.406, and the stiffness matrix is obtained from:

$$\begin{Bmatrix} F_i \\ F_j \\ F_k \\ F_l \end{Bmatrix} = \frac{\pi\ 250}{(1-0.35^2)} \begin{bmatrix} 2.406 & 1/2 & 1/\sqrt{5} & 1 \\ 1/2 & 2.406 & 1 & 1/\sqrt{5} \\ 1/\sqrt{5} & 1 & 2.406 & 1/2 \\ 1 & 1/\sqrt{5} & 1/2 & 2.406 \end{bmatrix}^{-1} \begin{Bmatrix} w_i \\ w_j \\ w_k \\ w_l \end{Bmatrix}$$

$$= 895.04 \begin{bmatrix} 0.515 & -0.058 & -0.031 & -0.197 \\ -0.058 & 0.515 & -0.197 & -0.031 \\ -0.031 & -0.197 & 0.515 & -0.058 \\ -0.197 & -0.031 & -0.058 & 0.515 \end{bmatrix} \begin{Bmatrix} w_i \\ w_j \\ w_k \\ w_l \end{Bmatrix}$$

Again, assembling the stiffness matrix for the solid foundation shorthand is tedious, but it can readily be done through a computer algorithm. Note that the units of the foundation stiffness matrix are MN/m.

### 8.5.3 Overall Element Stiffness and Slab Stiffness

The overall stiffness of a plate element is computed by adding the stiffness matrix due to bending and the stiffness matrix due to the foundation. In doing so, it should be noted that these two matrices have different dimensions (12 and 4, respectively), and adding them involves adding stiffness elements that correspond to forces and displacements only.

Assembling the element stiffness of an entire concrete slab $S$ is done by superimposing the stiffness matrices of the individual plate elements in the slab, according to standard FEM procedures.[17] The corresponding force/moment versus deflection/rotation relationship is given in Equation 8.57, where $n$ is the number of nodes in a particular slab.

$$\begin{Bmatrix} F_1 \\ M_{1x} \\ M_{1y} \\ \cdot \\ \cdot \\ \cdot \\ \cdot \\ \cdot \\ \cdot \\ F_n \\ M_{n\,x} \\ M_{n\,y} \end{Bmatrix} = [S] \begin{Bmatrix} w_1 \\ \theta_{1x} \\ \theta_{1y} \\ \cdot \\ \cdot \\ \cdot \\ \cdot \\ \cdot \\ \cdot \\ w_n \\ \theta_{n\,x} \\ \theta_{n\,y} \end{Bmatrix} \tag{8.57}$$

After modifying for boundary conditions and externally applied loads, the $3n$ linear equations shown are solved simultaneously to yield the deflections/rotations at the nodes, which allow calculation of the stresses/strains at any location within each plate element.

### 8.5.4 Joint Stiffness

Analysis of jointed rigid pavements requires handling the interaction between adjacent slabs across each joint. This involves transferring vertical loads and possibly moments across the joints. JPCP joints transfer vertical loads only, while JDRCP joints transfer both vertical loads and moments. A schematic of the load transfer mechanism across a JPCP resting on a liquid foundation is shown in Figure 8.18, where $k$ is the modulus of the subgrade reaction and $c_w$ is the constant of an idealized linear spring connecting the two slabs.

Consider a load being applied to the right of the joint and the resulting deflection measurements to the loaded and unloaded edges of a joint, denoted by $w_l$ and $w_{ul}$. Note that such measurements can be readily obtained by FWD testing, as described in Chapter 9. Clearly, the force transmitted across the joint to the unloaded slab is equal to $c_w(w_l - w_{ul})$. Ignore the slab weight, the free-body diagram of the unloaded slab on the left (Figure 8.18) gives:

$$c_w(w_l - w_{ul}) = k\ w_{ul} \tag{8.58}$$

which is a practical way of quantifying $c_w$, given the modulus of subgrade reaction.

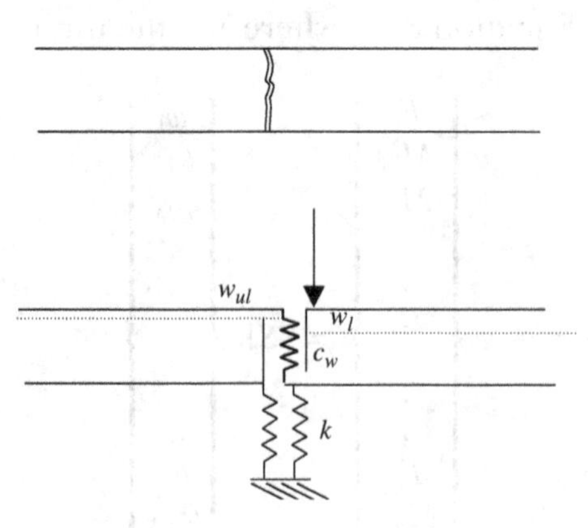


**Figure 8.18**
Vertical Load Transfer across a PJCP Joint (not to scale)

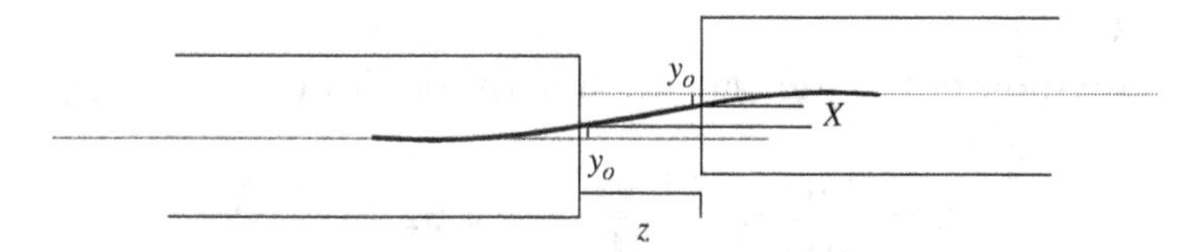


**Figure 8.19**
Vertical Load Transfer across a DRJCP Joint (not to scale)

The vertical load transfer mechanism across JDRCP joints involves forces/deflections within the concrete itself and within the dowel rebar. A schematic of these deflection components is shown in Figure 8.19 and denoted by $y_o$ and $x$, respectively. As shown, the total relative vertical deflection between adjacent slabs $w_d$ is:

$$w_d = 2\ y_0 + x \tag{8.59}$$

An expression for the deflection in the concrete $y_o$ was given earlier (Equation 8.37), while a simplified expression for $x$ is:

$$x = \frac{P_t\ z}{G_r\ A_r} \tag{8.60}$$

where $z$ is the opening of the joint, $A_r$ is the cross-sectional area of the dowel bar, and $G_r$ is the shear modulus of steel. Hence, the expression for the relative deflection between adjacent slabs becomes:

$$w_d = 2\left[\frac{P_t}{4\beta^3\ E_r\ I_r}(2+\beta\ z)\right] + \frac{P_t\ z}{G_r\ A_r} \tag{8.61}$$

which gives the following expression for the elastic constant of an idealized linear spring connecting two adjacent slabs across a JDRCP joint.

$$c_w = \frac{P_t}{w_d} = \frac{1}{\left[\frac{(2+\beta\ z)}{2\beta^3\ E_r\ I_r}\right] + \frac{z}{G_r\ A_r}} \tag{8.62}$$

**Example 8.13**

Compute the coefficients of force transmission across a dowel in a JDRPC pavement. Given, slab thickness of 0.20 m, dowel bar diameter of 0.025 m, joint opening of 0.005 m, elastic modulus of the dowel bar steel 200000 MPa, Poisson's ratio of the dowel bar steel of 0.4, and spring constant of concrete support $K_c$ 100000 MPa/m.

**ANSWER**

Compute moment of inertia and area of the dowel bar.

$$I_r = \frac{\pi\ d^4}{64} = \frac{\pi\ 0.025^4}{64} = 1.92\ 10^{-8}\ \text{m}^4$$

$$A_r = \frac{\pi\ 0.025^2}{4} = 4.91\ 10^{-4}\ \text{m}^2$$

Compute the quantity $\beta$ from Equation 8.38.

$$\beta = \left(\frac{100000\ 0.025}{4\ 200000\ 1.92\ 10^{-8}}\right)^{0.25} = 20.11\ \text{m}^{-1}$$

Compute the shear modulus of steel as:

$$G_r = \frac{E_r}{2(1+\mu)} = \frac{200000}{2(1+0.4)} = 71429\ \text{MPa}$$

Use Equation 8.62 to compute the constant of transmitting vertical load $c_w$.

$$c_w = \frac{1}{\left[\frac{(2 + 20.11\ 0.005)}{2\ 20.11^3\ 200000\ 1.92\ 10^{-8}}\right] + \frac{0.005}{71429\ 4.91\ 10^{-4}}}$$
$$= 29.61\ \text{MN/m}$$

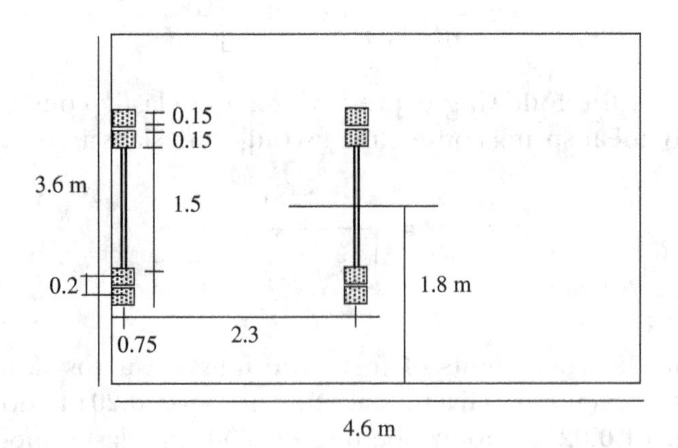


**Figure 8.20**
Layout for Example 8.14

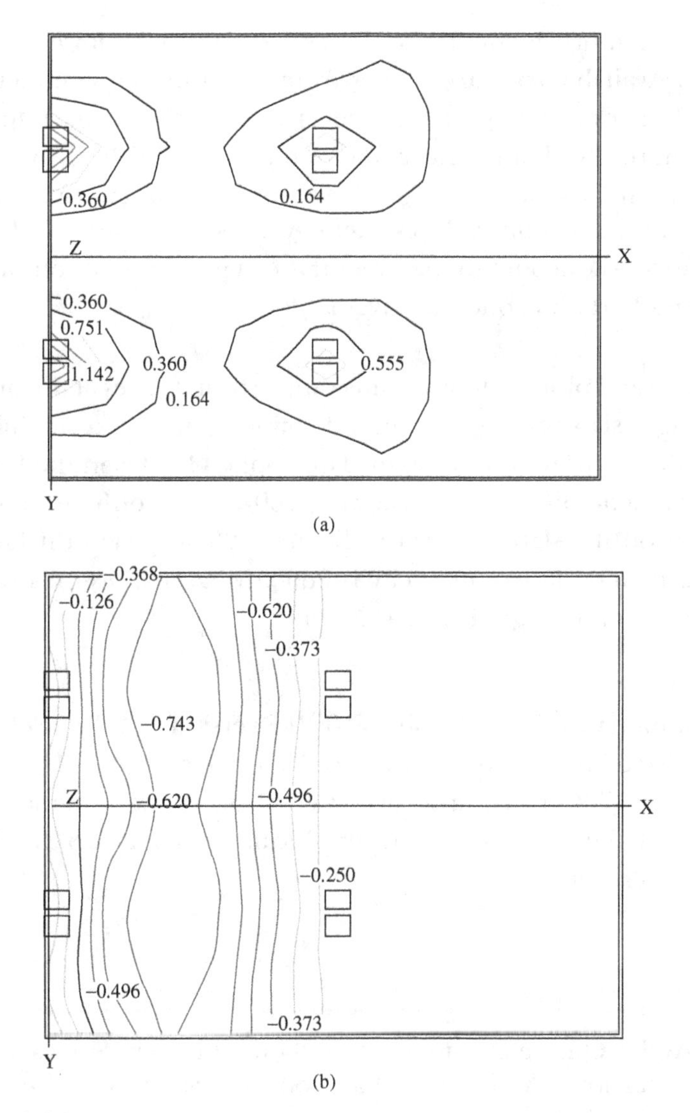


**Figure 8.21**
Stress Plots for Example 8.14, Produced by EVERFE (Ref. 5)

**USING AVAILABLE FEM SOFTWARE**

A variety of FEM codes are available to implement the plate element formulation just described. Other codes implement more complex elements, such as three-dimensional ''brick'' elements involving 8 nodes and 48 degrees of freedom. Some codes allow modeling of the steel reinforcement, hence allow stress analysis of dowel bars and tiebars. Available software ranges from sophisticated commercially available packages (e.g., reference 1) to freeware, (e.g., reference 9). Example 8.14 uses the computer model EVERFE, to handle rigid pavements with multiple base/sub base layers on a liquid foundation, and utilizes brick elements.[5] See the web site for the book, www.wiley.com/go/pavement, for suggestions on software for analyzing Portland concrete pavements.

**Example 8.14**

Compute and plot the tensile and compressive stresses at the bottom of a single slab resting on a liquid foundation with a modulus of subgrade reaction of 30 MPa/m. The applied load is in the form of a tandem axle on dual tires carrying 160 kN. Its configuration and location on the slab is shown in Figure 8.20. The slab thickness is 0.25 m, its modulus of elasticity is 28000 MPa, its Poisson's ratio is 0.2, and its unit weight is 23.54 kN/$m^3$.

**ANSWER**

The solution was obtained using EVERFE version 2.22.[5] The element grid utilized involved $10 \times 12 \times 2 = 240$ brick elements. The results for the tensile and compressive stresses on the $x$–$y$ plane at the bottom of the slab are shown graphically in Figures 8.21(a) and 8.21(b), respectively.

## References

1. ABAQUS Online Documentation: Finite Element Software Package, Version 6.2, Habbitt, Karlsson and Sorrenson, Inc., Pawtucket, RI. 2001.
2. American Concrete Institute (1954). "Structural Design Considerations for Pavement Joints," Report of ACI Committee 325, *ACI Journal*, July.
3. Bradbury, R. D. (1938). Reinforced Concrete Pavements, Wire Reinforcement Institute, Washington DC.
4. Darter, M. I., and E. J. Barenberg (1977). "Design of Zero Maintenance Plain Concrete Pavement," Report FHWA-RD-78-111, Volume 1, Federal Highway Administration, Washington, DC.

[5] Davids, W. G., Wang, Z. M., Turkiyyah, G., Mahoney, J. and Bush, D. (2003). "Finite Element Analysis of Jointed Plain Concrete Pavement with EVERFE 2.2". Transportation Research Record 1853: Journal of the Transportation Research Board, TRB, National Research Council, Washington, D.C., pp 92–99.

[6] PCA (1980). *Joint Design for Concrete Highways and Street Pavements*, Portland Cement Association, Skokie IL.

[7] Friberg, B. F. (1940). *Design of Dowels in Transverse Joints of Concrete Pavements*, Transactions of ASCE, Vol. 105, Reston VA.

[8] Gerald, F. C., and P. O. Wheatley (1985). *Applied Numerical Analysis*, 3rd ed., Addison Wesley, Reading, MA.

[9] Gobat, J. I., and D. C. Atkinson (1994). *The FELT System: User's Guide and Reference Manual*, Computer Science Technical Report CS94-376, University of California, San Diego.

[10] Goldbeck, A. T. (1917). *Thickness of Concrete Slabs*, Public Roads, U.S. Dept of Agriculture, Washington DC.

[11] Lytton, R. L., D. E. Pufhal, C. H. Michalak, H. S. Liang, and B. J. Dempsey (1993). *An Integrated Model of the Climatic Effects on Pavements*, FHWA RD-90-033, Washington, DC.

[12] Older, C. (1924). *Highway Research in Illinois*, Transactions of the ASCE, Vol. 87, Washington, DC.

[13] Pickett, G. and G. K. Ray (1951). *Influence Charts for Rigid Pavements*, Transactions of ASCE, Vol. 116, pp. 49–73.

[14] Pickett, G., and S. Badaruddin (1956). "Influence Charts for Bending of a Semi-infinite Slab," *Proceedings of the 9th International Congress on Applied Mechanics*, Vol. 6, pp. 396–402, Brussels, Belgium.

[15] Timoshenko, S. P., and J. N. Goodier (1987). Theory of Elasticity, 3rd ed., McGraw-Hill Inc., New York.

[16] Yoder, E. J., and M. W. Witczak, (1975). *Principles of Pavement Design*, 2nd ed., John Wiley & Sons Inc., New York.

[17] Zienkiewicz, O. C. (1977). *The Finite Element Method*, 3rd ed., McGraw-Hill, New York.

[18] Vesic, A. S. and K. Saxena (1974). "Analysis of Structural Behavior of AASHO Road Test Rigid Pavements," NCHRP Report N. 97, Highway Research Board, Washington, DC.

[19] Vetter, V. P. (1933). *Stresses in Reinforced Concrete Due to Volume Changes*, Transactions of ASCE, Vol. 98, Reston, VA.

[20] Westergard, H. M. (1926). *Stresses in Concrete Pavements Computed by Theoretical Analysis, Public Roads,* Vol. 7, U.S. Dept. of Agriculture, Washington, DC.

## Problems

8.1 A 28 cm thick isolated slab, with dimensions of 3.6 × 4.2 m, is resting on a liquid subgrade with a modulus of subgrade reaction of 60 MPa/m. It is subjected to a decrease in temperature of −12°C at its upper surface and an increase of +7°C on its lower surface. Determine and plot the stresses versus slab depth at midslab, as well as at midspan of the two free boundaries. Additional information given for the portland concrete includes: $E = 28$ GPa, $\mu = 0.15$, and $a_t = 9.0\ 10^{-6}/°C$.

8.2 Compute the tensile stresses generated by subgrade friction in a 8.0 m long concrete slab. How high is this stress compared to the tensile strength of concrete, given that the 28-day compressive strength of the concrete $f_c'$ is 20 MPa, and $\gamma$ for the concrete is equal to 22.5 kN/m$^3$.

8.3 Consider the rigid pavement layout shown in Figure 8.22. The slabs are 0.20 m thick and were poured in two halves by two separate passes of a slip-form paver (the construction joint is indicated by dotted line). Compute the necessary area of

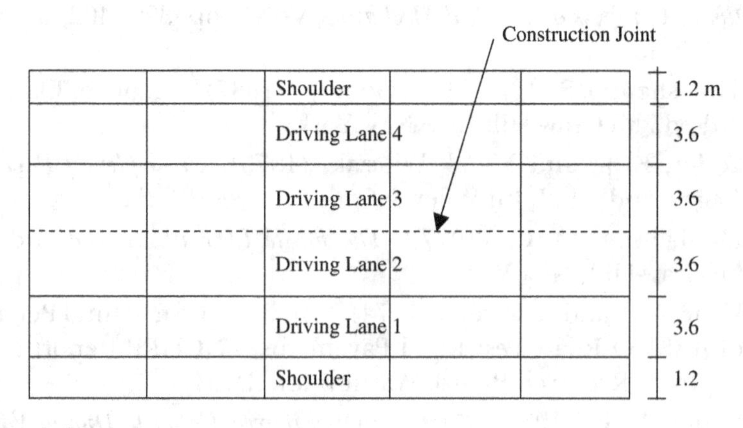


**Figure 8.22**
Layout for Problem 8.3

tiebar steel across the construction joint and the average bond stress between the tiebars and the concrete. The allowable stress of the steel $f_r$ is given as 200 MPa, the length of the tiebars is 1.00 m.

8.4 Compute the maximum tensile stress on a 0.25 m thick slab of a JPCP m under a corner point load of 40 kN.

8.5 Determine the maximum tensile stress and the corner deflection under a circular load of 0.15 m radius carrying 700 kPa pressure, given a slab thickness of 0.22 m, a modulus of subgrade reaction of 60 MPa/m, a concrete modulus of 28 GPa, and a Poisson's ratio of 0.15.

8.6 A 3.0 cm diameter dowel bar is transferring a vertical load of 4500 N across a 0.35 cm wide joint. Compute the dowel bar deflection at the edge of the joint and the corresponding concrete bearing stresses. Can the concrete handle this stress? Given, $K_c$ of 120,000 MPa/m, $E_r$ of 200,000 MPa, and $f_c'$ 30 MPa.

8.7 Consider a JDRCP, consisting of slabs 25 cm thick and 3.6 m wide, resting on a subgrade with a modulus of subgrade reaction of 50 MPa/m. An axle load consisting of two identical tires 1.8 meters apart, each carrying 44 kN, is located at the edge of the joint, 0.30 m from the edge of the slab. The load across the joint is carried by 30 mm diameter dowel bars placed at 0.3 m center-to-center distances, as shown in Figure 8.15. Compute the load carried by each dowel bar. Given, $E$ for the portland concrete of 28,000 MPa and a Poisson's ratio of 0.15. Assume that the tires apply point loads, the load transfer across the slabs is 50/50, and the distribution of load varies linearly with the distance from each tire load location.

8.8 Design the amount of rebar reinforcement required for a CRCP 0.25 m thick slab subjected to a temperature difference between pouring and the coldest winter day of $-40°$C, and estimate the anticipated average spacing of the transverse concrete cracks. Given, a coefficient of thermal expansion for the concrete of 9.0 $10^{-6}/°$C, a tensile concrete strength of 3.5 MPa, a steel elastic modulus of 200000 MPa, a concrete elastic modulus of 28000 MPa, a steel elastic limit of 340 MPa, and an allowable bonding stress between steel and concrete of 2.6 MPa.

8.9 Given the stiffness matrix of the plate element described in Example 8.11, compute the slab stiffness for the slab below made up of four identical plate elements, laid out as shown in Figure 8.23.

8.10 Expand the stiffness matrix of the slab in problem 8.9, to account for a liquid foundation, given a modulus of subgrade reaction of 40 MPa/m.

8.11 Compute the solid foundation stiffness matrix for a plate element with dimensions $2a = 4.0$ m and $2b = 1.5$ m. Given, subgrade $E_s$ of 300 MPa and $\mu_s$ of 0.40.

8.12 A FWD load of 44 kN is applied to the right-hand-side slab in the middle of a JPCP joint. The deflection measurements obtained in the right and the left of slabs are 2.08 and 1.82 mm, respectively. Compute the elastic constant of normal load transmission $c_w$ across this joint, given that the modulus of subgrade reaction $k$ is 60 MPa/m.

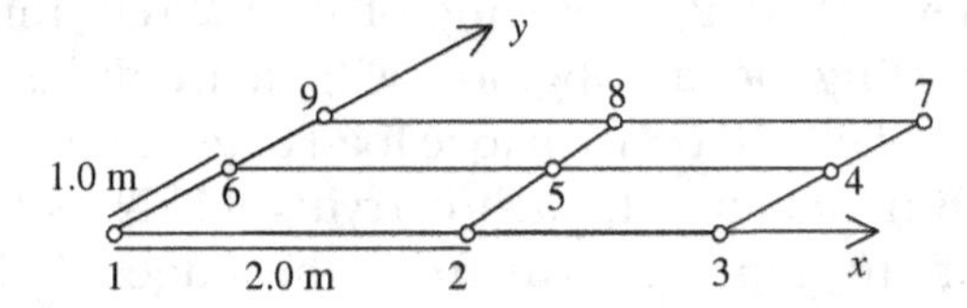


**Figure 8.23**
Layout for Problem 8.9

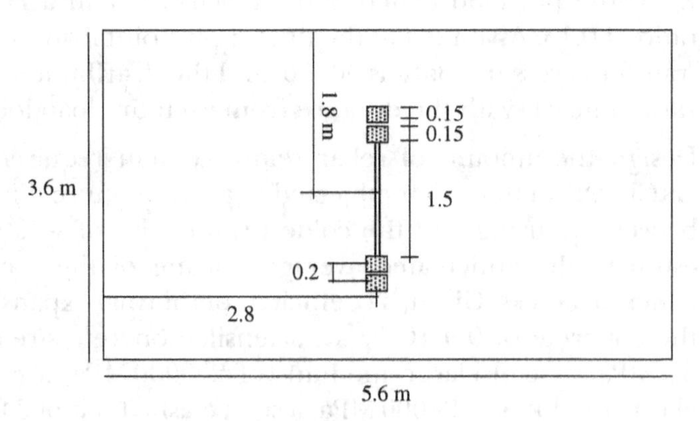


**Figure 8.24**
Layout for Problem 8.14

8.13 Compute the coefficient of force transmission across a dowel in a JDRPC pavement. Given, slab thickness of 0.25 m, dowel bar diameter of 0.030 m, joint opening of 0.01 m, elastic modulus of the dowel bar steel 250,000 MPa, a Poisson's ratio of the dowel bar steel of 0.4, and a spring constant of concrete support $K_c$ 100,000 MPa/m.

8.14 Utilizing a FEM software package, compute the stresses at the bottom of a 0.20 m thick portland concrete slab subjected to a combined load of 100 kN and a thermal gradient consisting of a reduction at the surface of 5°C and an increase at the bottom of 5°C. The layout of the slab and the load is shown in Figure 8.24. Given, modulus of subgrade reaction of 80 MPa/m, tensionless subgrade, slab modulus of elasticity of 28000 MPa, Poisson's ratio of 0.15, unit weight of 23.54 kN/m$^3$, and coefficient of linear thermal expansion $a_t$ of 9.0 $10^{-6}$/°C.

# 9 Pavement Evaluation

## 9.1 Introduction

Pavement evaluation encompasses a range of qualitative and quantitative measurements intended to capture the structural and functional condition of pavements. The information collected provides a ''report card'' of pavement condition at a particular point in time, while changes in pavement condition define pavement performance. The data associated with pavement performance is primarily technical in nature, but it is often summarized in a format meaningful to the layperson. The latter is essential in conveying funding needs to legislative bodies that appropriate funds for pavement 4-R activities (Chapter 13).

Traditionally, the pavement evaluation information collected is grouped into four broad categories, namely:

- Serviceability
- Structural capacity
- Surface distress and
- Safety

This chapter describes each of these four pavement evaluation components in detail, presents the state-of-the art devices utilized for obtaining physical measurements, and describes how these physical measurements are processed and summarized.

## 9.2 Serviceability

### 9.2.1 Definitions—Introduction

The concept of serviceability arises from the well-accepted principle that pavements are built for serving the traveling public and, therefore, the quality of service they provide is best judged by them.[7] This principle has motivated the use of a rating scale for pavement serviceability, ranging from 0 to 5, whereby 0 signifies very poor and 5 signifies very good. Representative samples of public ratings of pavement serviceability can be obtained through the use of pavement serviceability evaluation panels, which consist of a group of evaluators, not necessarily technical, riding a popular model passenger car while rating the serviceability of a pavement section using a standard evaluation form (e.g., Figure 9.1).

Such forms usually solicit additional input on whether the pavement section provides service that is acceptable. The information collected by such panel ratings is averaged and reported as the Present Serviceability Rating (*PSR*). A standard method for conducting panel serviceability ratings is given in reference 37. The *PSR*, although widely understood, is largely qualitative and subjective in nature. It is clearly a function of the pavement condition that the panel evaluators are accustomed to, hence would vary widely

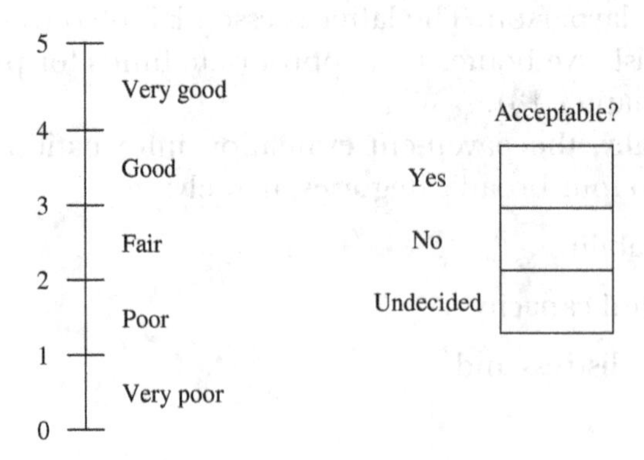


**Figure 9.1**
Pavement Serviceability Evaluation Form

between jurisdictions (e.g., industrialized versus developing countries). Nevertheless, panel ratings have established that *PSR* values between 2.0 and 2.50 reflect the lower limit of acceptable pavement serviceability.

### 9.2.2 Early Efforts in Predicting Serviceability

Considerable efforts were made in the 1950s to devise an objective means of estimating pavement serviceability through measurements of pavement condition. These early efforts correlated panel serviceability ratings to pavement condition attributes such as roughness, cracking, and rutting. One of the earliest forms of such relationships for flexible and rigid pavements, respectively,[2] is:

$$PSI = 5.03 - 1.91\log(1 + \overline{SV}) - 0.01\sqrt{C+P} - 1.38(\overline{RD})^2 \quad (9.1a)$$

$$PSI = 5.41 - 1.80\log(1 + \overline{SV}) - 0.09\sqrt{C+P} \quad (9.1b)$$

where *PSI* is the Present Serviceability Index (i.e., an estimate of the panel-obtained *PSR* index ranging from 0 to 5), $\overline{SV}$ is a pavement roughness summary statistic referred to as the slope variance (rad$^2$), $C+P$ is the relative extent of cracking and patching in the wheel-path (ft$^2$/1000 ft$^2$), and $\overline{RD}$ is the average rut depth in the left and right wheel-paths (inches). The *SV* was measured in the left and right wheel-paths through a device shown in Figure 9.2 and averaged. This devise consisted of a long metal frame carried by a set of large wheels behind a towing vehicle. A set of smaller wheels, closely spaced, traced the pavement surface in the wheel path. The variance of the slope of the line defined by the centers of the two small tracing wheels defined the *SV*. As will be described in Chapter 11, this device was used for measuring pavement roughness

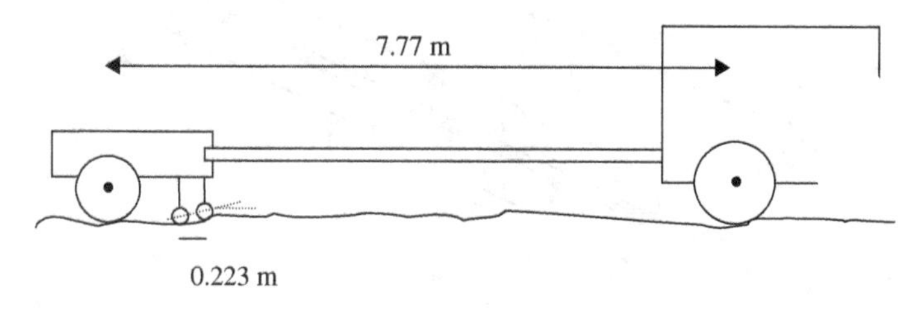


**Figure 9.2**
Schematic of the Pavement Roughness Measuring Devise Used for Obtaining *SV*

during the American Association of State and Highway Officials (AASHO) Road Test, which is one of the most significant past pavement performance studies.

Equation 9.1 provided early evidence that pavement roughness is by far the most significant contributor to serviceability loss (e.g., wheel-paths totally covered with cracks contribute only 0.01 and 0.09 to serviceability loss, for flexible and rigid pavements, respectively). Hence, early in the development of pavement serviceability prediction models, it was surmised that roughness was the main contributor to the public perception of pavement serviceability.

In this context, pavement roughness can be defined as the variation in the longitudinal (i.e., in the direction of travel) pavement elevation in the wheel-paths that excites traversing vehicles.[32] This definition associates the longitudinal elevation in the wheel paths, referred to as the *pavement profile*, to the riding response of a vehicle traversing the pavement. A schematic representation of the pavement profile in the wheel-paths is shown in Figure 9.3. Figure 9.4 shows a pavement profile with the vertical dimension exaggerated.

Figure 9.4 shows that some of the variations in elevation have periodic features (i.e., repeat themselves at regular intervals, referred as wavelengths). It demonstrates that of pavement profile consists of a wide range of wavelengths ranging from several centimeters to tens of meters, with varying amplitudes. As described later in this chapter, these wavelengths affect the excitation of the various vehicles traversing the road in different ways, depending on their traveling speed

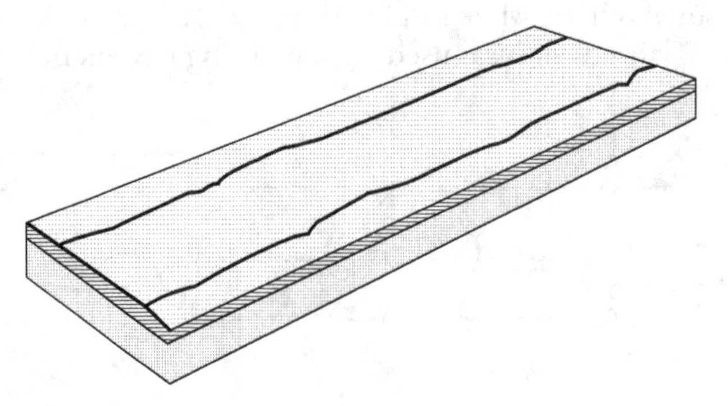

**Figure 9.3**
Schematic of the Pavement Roughness Profile in the Wheel-paths

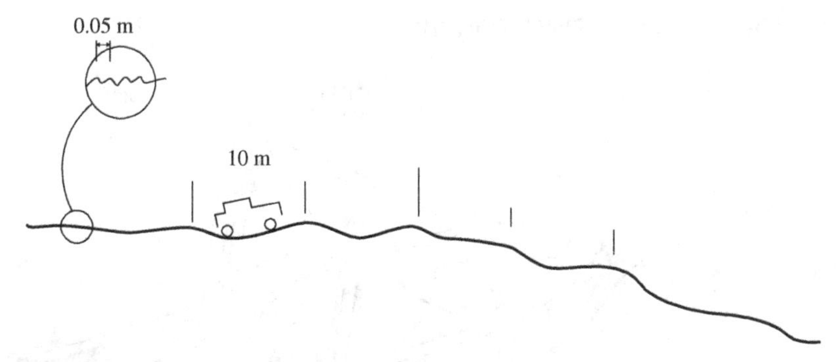


**Figure 9.4**
Wavelength Content of Pavement Roughness Profile

and dynamic characteristics (e.g., suspension configuration, wheel and frame inertial properties, and so on).

The device used for measuring the *SV* in the AASHO Road Test (Figure 9.2) was one of the predecessors of mechanical devices that measured roughness in terms of their response to the pavement roughness profile excitation. These devices are to be distinguished from another group of devices that collect actual profile elevation data in the wheel-paths by sampling the pavement profile at regular intervals. These two groups of devices are referred to as *response-type* and *profilometer-type*, respectively, and are described next, in detail.

### 9.2.3 Response-Type Pavement Roughness Measuring Devices

The idea of measuring roughness through the response of "reference" mechanical systems driven or towed over the pavement predates the roughness measuring device used in the AASHO Road Test for measuring *SV*. One of the earliest such devices is the roughness meter developed in the 1940s by the U.S. Bureau of Public Roads, referred as the BPR Roughness Meter (Figure 9.5). This was a single-wheel trailer, referred to as a quarter-car, towed behind a vehicle through a hitch that controlled roll. The wheel was supported by a pair of struts equipped with springs and shock absorbers. The device was equipped with a mechanical integrator of the relative displacement of the axle with respect to the frame of the trailer driven at 32 km/h in a wheel-path. Roughness was reported in terms of this accumulated relative displacement per unit length of pavement traveled (in/mi or m/km).

Another type of response roughness measuring device consists of pairs of such quarter-cars combined to create half-car trailers.

METHODS OF MEASURING CONDITION

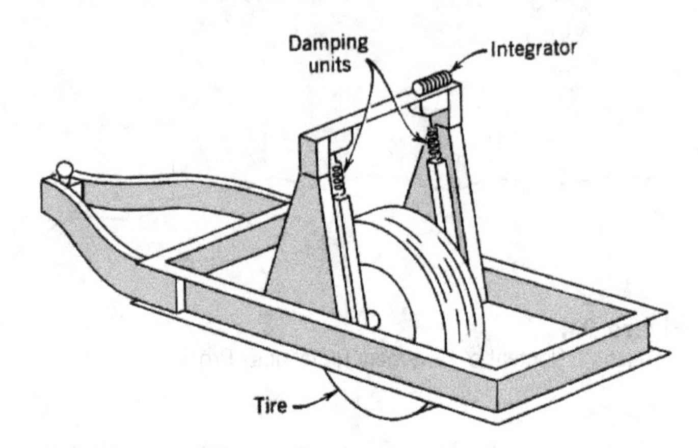


**Figure 9.5**
Components of the Roughometer. Yoder, 1975, Courtesy of John Wiley & Sons, Inc.

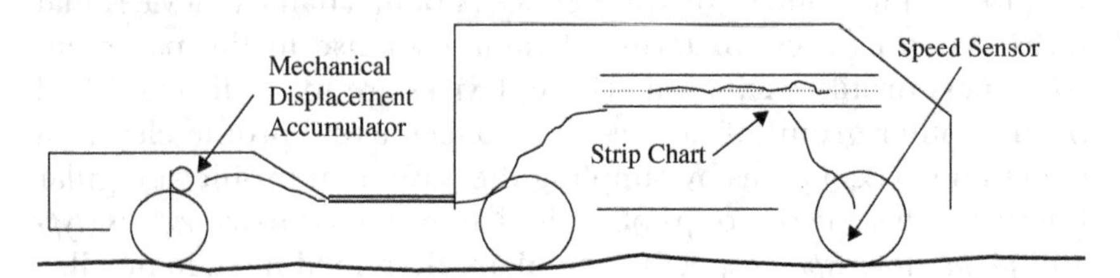


**Figure 9.6**
Schematic of a Mays Ride Meter (MRM)

The most common example of such devices is the Mays Ride Meter (MRM), still in production today (Figure 9.6). The device is driven over the pavement at speeds of either 56 km/h or 80 km/h. It yields a continuous trace of the relative displacement of the middle of the axle with respect to the frame of the trailer, which is not to be confused with the road profile. This signal is integrated either mechanically or electronically, divided by the distance traveled, and reported in units of m/km or in./mi.

Variations of this type of device have subsequently emerged, incorporating the relative displacement integrating hardware into the rear axle of a vehicle, and thus doing away with the need to tow a separate vehicle. These devices, shown in Figure 9.7, are known

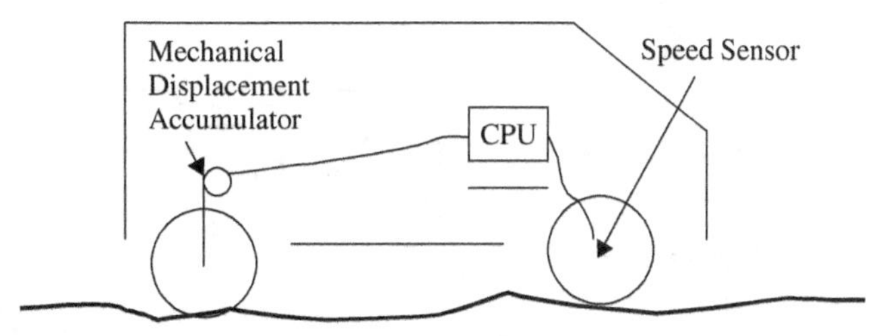


**Figure 9.7**
Schematic of a Car Road Meter (CRM)

as Car Road Meters (CRM), and often incorporate a variety of instruments that collect other pavement evaluation data in addition to pavement roughness (e.g., surface distress data).

The response-type roughness measuring systems just described have a number of limitations. The main one is that the roughness statistic reported depends on the properties of the mechanical system used. This means that the unavoidable variations in tire inflation pressure, spring elastic properties, and shock absorber damping properties affect device response. As a result, the output of these type of devices is neither stable (i.e., repeatable for a particular device) nor universal (i.e., comparable between devices of the same manufacturer and model).

An entirely different generation of response-type roughness measuring devices are based on the principle of the rolling straightedge (RSE), shown in Figure 9.8. In its simplest form, this device consists of a beam carried by two wheels at either end and a middle wheel that traces the road surface. The device is hand-pushed in the wheel-path and yields a continuous trace of the distance between the center of the beam and the tracing wheel. Clearly, this continuous trace is not the pavement profile, because the measurements are affected by the length of the RSE beam. This is explained through the following example.

**Example 9.1**

A road profile consists of one sinusoidal wavelength of 10.00 m and an amplitude of 0.10 m. Calculate and plot the trace of a RSE with a length of 5.00 meters, and repeat it for an RSE with a length of 8.00 meters. Assume that the transport wheels and the tracing wheels are small enough to neglect their dimensions.

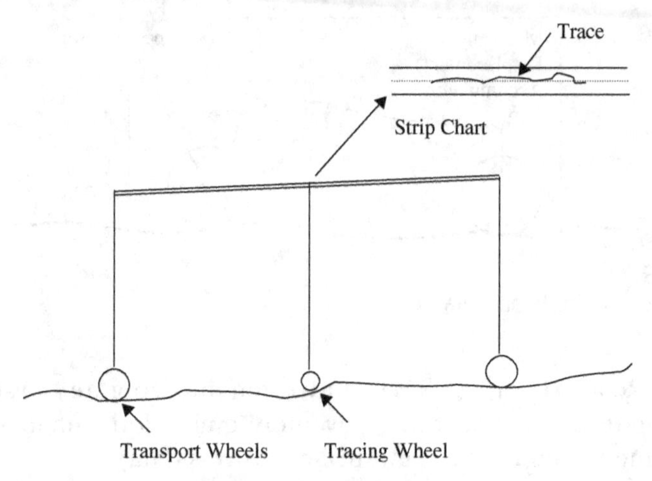


**Figure 9.8**
Principle of the Rolling Straightedge (RSE)

**ANSWER**

The road profile elevation $p(x)$ in the wheel-path can be expressed as a function of the distance traveled $x$ as:

$$p(x) = 0.1 \sin\left(x \frac{\pi}{5}\right)$$

and given a RSE beam of 5.00 m. The elevation of the pavement under the middle of the rolling straightedge is given by:

$$p(x + 2.5) = 0.1 \sin\left((x + 2.5)\frac{\pi}{5}\right)$$

while the elevation under the downstream transport wheel of the RSE is given by:

$$p(x + 5) = 0.1 \sin\left((x + 5)\frac{\pi}{5}\right)$$

Hence, the trace being recorded can be calculated as the difference in elevations between the middle of the beam and $p(x + 2.5)$, expressed as:

$$trace = \frac{p(x) + p(x + 5)}{2} - p(x + 2.5)$$

The calculations for the RSE of 8.00 m are similar. The pavement profile and the traces for the two RSEs are plotted next (Figure 9.9). It is evident that the output of the RSE depends on the combination of beam length and the wavelength of the pavement profile being measured.

To reduce this type of bias, commercially available RSEs are equipped with multiple transport wheel assemblies for carrying the beam. The most common such device is the California Profilograh, which consists of a metal truss 7.62 m (25 ft) long, being supported by a set of six staggered wheels on each end (Figure 9.10).

The trace of the device is obtained on either a strip chart or electronic format and is typically plotted in vertical scale 300 times larger than the horizontal. A set of lines is overlaid to this trace,

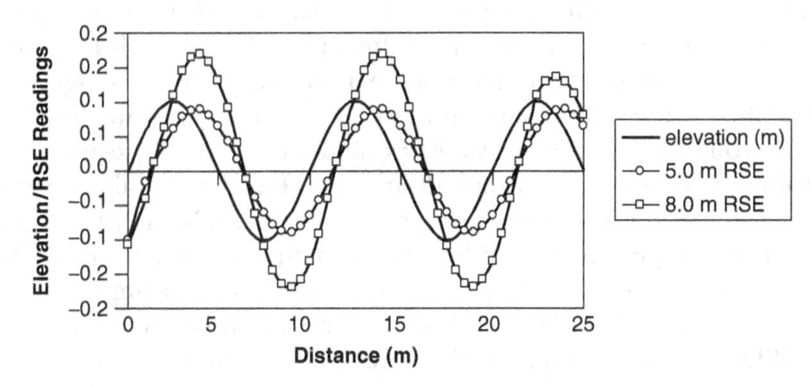


**Figure 9.9**
RSE Measurements of Sinusoidal Road Profile with Different-Length Beams

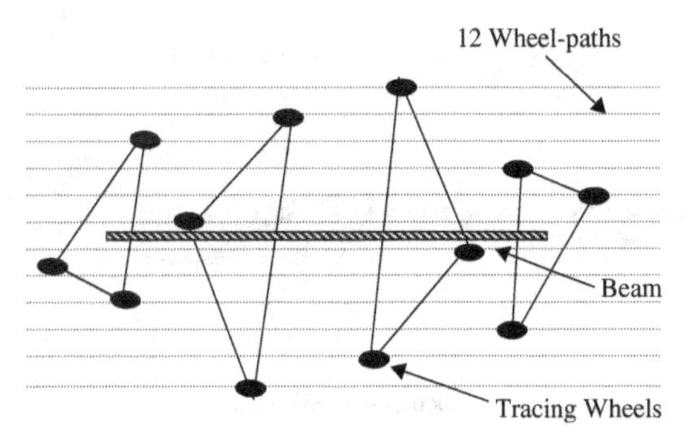


**Figure 9.10**
Schematic Plane View of the California Profilograph

forming a ''blanking'' band; excursions of the trace above or below the blanking band are measured and their magnitude summed to calculate a Profile Index (PI). Excursions larger than a specified magnitude indicate locations where the contractor must grind/fill the pavement to reduce roughness. The methodology used for obtaining the PI is shown in Figure 9.11.

Several variations of this type of profilograph are currently being manufactured by companies such as James Cox and Sons Inc., McCracken Concrete Pipe Machinery Co., Soiltest, and Ames Engineering. They are used for post-construction pavement quality control (QC) rather than routine pavement network evaluation. The PI obtained with the method just described is used for determining post-construction pay factors to contractors. Their advantage is that they are relatively inexpensive, lightweight (this is particularly desirable for obtaining early roughness measurements in rigid pavements), and provide a trace that allows identifying rough locations following the procedure described earlier. Their disadvantages are that they are cumbersome and slow to operate and do not produce roughness summary statistics comparable to those used for evaluating pavements through their lives. The advent of lightweight profilometers, as described later, has allowed simulating the output of profilographs form profile measurements using software. This makes it possible to identify pavement rough spots using a procedure similar to the one described earlier without running an actual profilograph at the construction site. The additional advantage of lightweight profilometers is that they can produce conventional

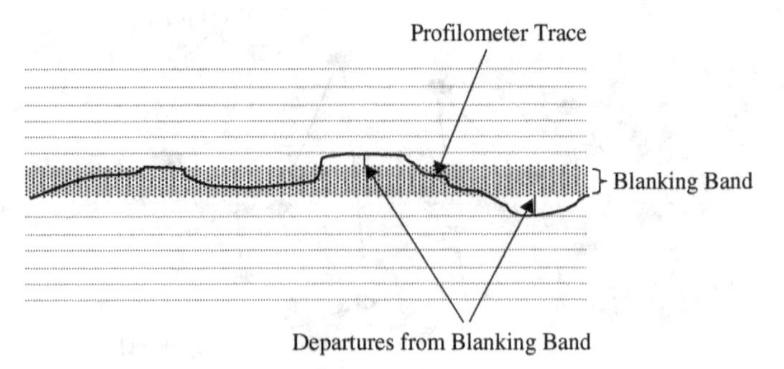


**Figure 9.11**
Example Processing the California Profilograph Trace for Calculating the PI

profile-based roughness statistics similar to those used for routine pavement network evaluation, as described later.

### 9.2.4 Profilometer-Type Pavement Roughness Measuring Devices

Response-type pavement roughness measuring devices lack the stability and the universality desired. Furthermore, establishing relationships between the output of any pair of such devices is possible only through regression, the outcome of which depends on the characteristics of the pavement sections used for obtaining the measurements. Moreover, pavement profile data can be used to simulate the output of any response-type pavement roughness measuring device, if so desired. For these reasons, measuring the actual profile of the pavement is preferable. This is effectively done by sampling the pavement elevation in the wheel-paths at regular intervals (e.g., every 0.25 or 0.50 m).

Obtaining such measurements can be accomplished with off-the-shelf surveying equipment (i.e., rod-and-level). This approach, however, is rather labor-intensive, requires lane closures, and is limited to a productivity of several kilometers per day, hence it is not practical for routine road network evaluation. Nevertheless, using the rod-and-level approach is a viable alternative for obtaining pavement profile data for calibration purposes.[42] The following sections describe the variety of profilometer-type pavement roughness devices available.

### 9.2.5 High-Speed Profilometers

The development of devices capable of measuring the actual road profile at high speeds date back to the 1960s, led by General Motors, intended to provide input to the dynamic simulation of cars and trucks. The device developed was the GMR Profilometer.[36] Its principle of operation is shown in Figure 9.12.

The GMR profilometer utilized a tracing wheel to measure the distance between the pavement surface in the wheel-paths and the vehicle frame; an accelerometer was used for measuring the vertical acceleration of the frame. Double integration of the signal from the latter yielded the vertical displacement of the vehicle frame. The device was equipped with a set of sensors, one over each wheel-path. For each wheel-path, pavement profile elevation was calculated by subtracting the distance between the tracing wheels and the frame from the vertical displacement of the frame. Obviously, this type of output is a digitized sample of the continuous (i.e., analogue) pavement profile, obtained at selected space intervals. Later versions of

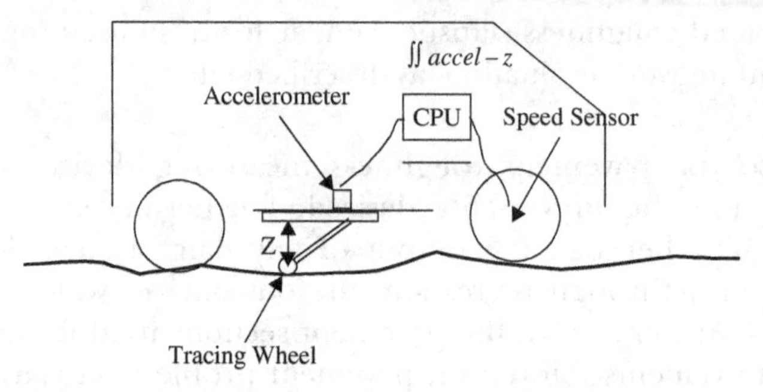


**Figure 9.12**
Operating Principle of the GMR Profilometer

this device employed non-contact optical sensors for measuring the distance between the pavement surface and the vehicle frame and included vehicle speed sensors and computerized data reduction onboard the vehicle. In a similar configuration they are marketed by Dynatest Inc.

In 1984, the South Dakota DOT developed a road profilometer in-house, utilizing similar principles to the ones just described.[16] The major difference was the use of laser sensors for measuring the distance between the pavement surface and the vehicle frame. More recently, other companies have manufactured similar profilometric devices, among them, International Cybernetics Corporation, Infrastructure Management Services, Pathway and Roadware. An example of one of these devices is shown in Figure 9.13.

These devices are equipped with a variety of noncontact sensors, including laser, ultrasonic, or infrared, which sample different widths of the wheel-path at any location in measuring elevation. A number of them includes additional capabilities, such as, multiple sensors transversely mounted, to measure rut depth. A number of these manufacturers produce self-standing assemblies of accelerometer and elevation sensors that can be mounted on the bumper of a vehicle. An overview of the technical characteristics of these profilometers can be found in reference 29. Although all these high-speed profilometers share roughly the same operating principle, they vary considerably in data collection method and processing. The latter involves digital filtering of the profile measurements, which affects significantly the pavement profile being recorded, as described later in this chapter.

**Figure 9.13**
Modern High-Speed Profilometer. (Courtesy of Dynatest Inc.)

### 9.2.6 Low-Speed Profilometers

The rod-and-level approach mentioned earlier qualifies as the earliest low-speed profilometric method. A number of purpose-built devices has been developed in recent years for obtaining pavement profile measurements at low speeds. One of the earlier examples of such a device is the Dipstick, which is marketed by Face Industries (Figure 9.14).

The Dipstick consists of a beam 0.30 m (1 ft) long supported by two legs. The device is equipped with a high-resolution inclinometer that measures the difference in elevation between the two supporting legs. Once the inclinometer has stabilized, the difference in elevation is recorded in the attached computer and the device audibly signals the operator to advance the device to the next measuring location by rotating the handle by 180°. Each measurement takes several seconds, yielding a productivity of several wheel-path kilometers per day, depending on the operator. Another inclinometer-based low-speed profilometer-type device is manufactured by the Australian Road Research Board and marketed in the United States by Trigg Industries International. It consists of a small pushable cart, resembling a lawn mower, that records the difference in elevation between its front and rear wheels, spaced 0.24 m, apart and records the data in an onboard computer.

Another generation of low-speed profilometer-type pavement roughness devices is emerging in the form of small self-propelled

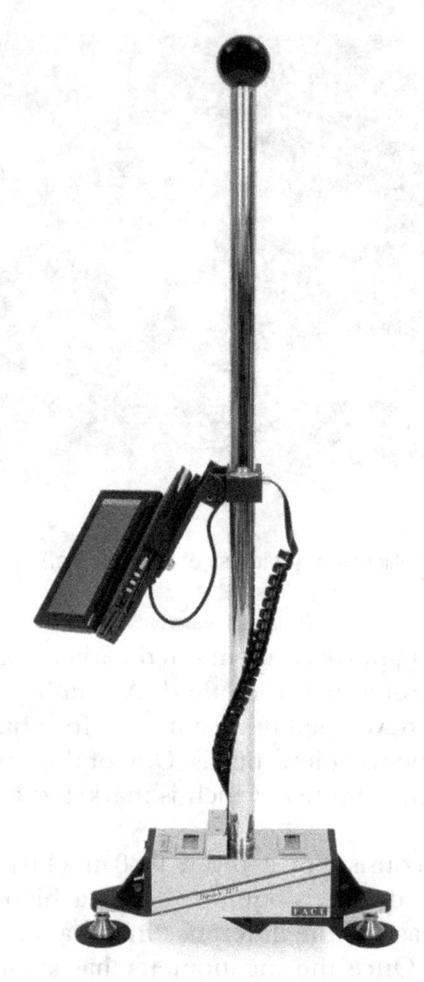

**Figure 9.14**
The Dipstick (Courtesy of Face Industries)

buggies that resemble golf carts. These are equipped with sensors similar to the ones installed in high-speed profilometers, namely accelerometers, for measuring the displacement of the frame; non-contact sensors, for measuring the distance between the pavement surface and the frame; and distance log sensors, for providing spatial reference. The signal processing for these systems is adapted to account for the slower speed of operation.

Although none of these slow-speed profilometers can be used without controlling the traffic away from the lane being measured, they involve lower labor costs, and result in increased productivity compared to the rod-and-level method. Their characteristics make them ideally suited for calibrating high-speed profilometers and for post-construction pavement roughness QC. For the latter, the pavement profile obtained can be used to analytically simulate the response of the RSE type profilographs described earlier. The advantage of using the same type of pavement roughness measuring system and data throughout the life cycle of pavements is obvious.

### 9.2.7 Processing Profilometer Measurements

All the profilometer-type devices described earlier record digital electronic signals, process them into elevation, and filter them to extract the information that is relevant to the ride quality of vehicles traversing the pavement. Although all these calculations are performed electronically by different proprietary software onboard each of these devices, it is desirable to have an understanding of how these calculations are carried out. The following discussion is a generic treatment of profilometer signal processing and profile data filtering.

**PROFILE FILTERING**

The raw pavement elevation data calculated from the electronic signals need to be filtered to eliminate wavelengths that do not affect the ride quality of traversing vehicles, such as wavelengths shorter than the dimensions of pavement macrotexture and longer than roadway geometric features perceived as longitudinal slope or curvature. Additional filtering of the elevation data can emphasize or eliminate certain pavement profile features (e.g., faulting or warping in rigid pavement slabs). Although there is no universally accepted practice for pavement profile filtering, most data manipulation is carried out by the moving average (MA) technique.[31] It consists of replacing the elevation of a number of consecutive points with the average of their elevation. This is mathematically expressed as:

$$p_f(i) = \frac{1}{N} \sum_{j=i-\frac{B}{2\Delta x}}^{j=i+\frac{B}{2\Delta x}} p(j) \tag{9.2}$$

where, $p(i)$ is the raw profile elevation at point $i$, $p_f(i)$ is the filtered profile elevation at point $i$, $B$ is the base length used for calculating the MA, and $\Delta x$ is the distance increment used in sampling the profile. If, for example, the base length used for calculating the MA is twice as long as the sampling interval, the elevation of the filtered profile is obtained by replacing the elevation of each point in the raw profile by the average of the elevation of the three points, including itself and the two adjacent points. Needless to say, if $B/2\Delta x$ is not an integer, some simplifying assumption needs to be made in applying Equation 9.2. This type of filter removes short wavelengths from the profile (i.e., smoothens the profile), hence is refered to as a *low-pass filter* (Figure 9.15).

The reverse effect is achieved (i.e., the profile can be filtered to appear rougher) if for each elevation point, the MA-filtered value calculated earlier is subtracted from the raw profile value, which is expressed as:

$$p_f(i) = p(i) - \frac{1}{N} \sum_{j=i-\frac{B}{2\Delta x}}^{j=i+\frac{B}{2\Delta x}} p(j) \tag{9.3}$$

and is referred to as a *high-pass filter* (Figure 9.16). The difference in the elevation scale between Figures 9.15 and 9.16 should be noted.

**Example 9.2** Consider an artificial pavement elevation road profile synthesized of random elevations between −0.005 and +0.005 m superimposed on three in-phase sinusoidal waveforms characteristics given in Table 9.1. Using a 0.25 m interval, plot the raw pavement profile

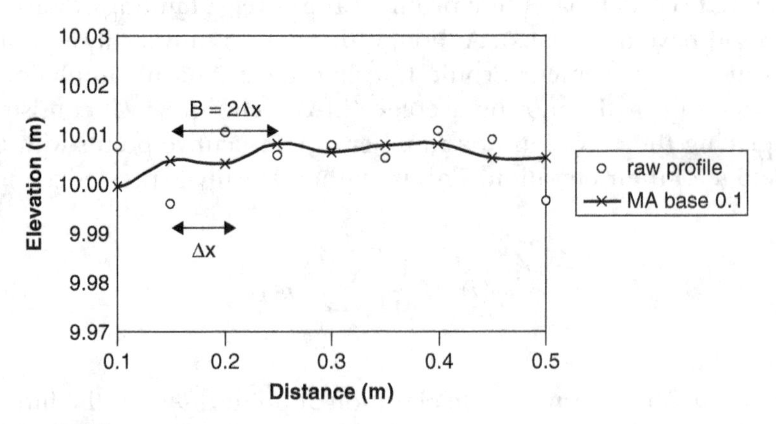


**Figure 9.15**
Example of a Low-Pass MA Filter with B = 2 δx

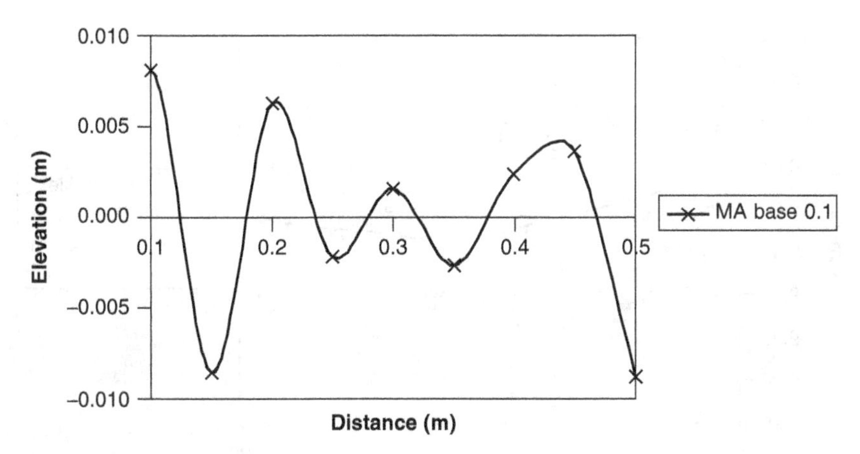


**Figure 9.16**
Example of a High-Pass MA Filter (continues from Figure 9.15)

**Table 9.1**
Waveform Characteristics for Example 9.2

| Amplitude (m) | Wavelength (m) |
|---|---|
| 0.02 | 5 |
| 0.015 | 10 |
| 0.01 | 20 |

and the filtered profiles using low-pass and high-pass MA filters with base lengths of 1, 5, and 10 m.

ANSWER

The mathematical expression for the specified profile is:

$$p(x) = 0.02 \sin\left(x\frac{\pi}{2.5}\right) + 0.015 \sin\left(x\frac{\pi}{5}\right) + 0.01 \sin\left(x\frac{\pi}{10}\right) + r(x)$$

where $r(x)$ is the specified random component. For plotting purposes, the latter was simulated using a random number generator (computer spreadsheets have built-in functions for generating random numbers; for example, the function RAND in Excel). Given $\Delta x$ of 0.25 m, low-pass filtering for base lengths of 1, 5, and 10 m

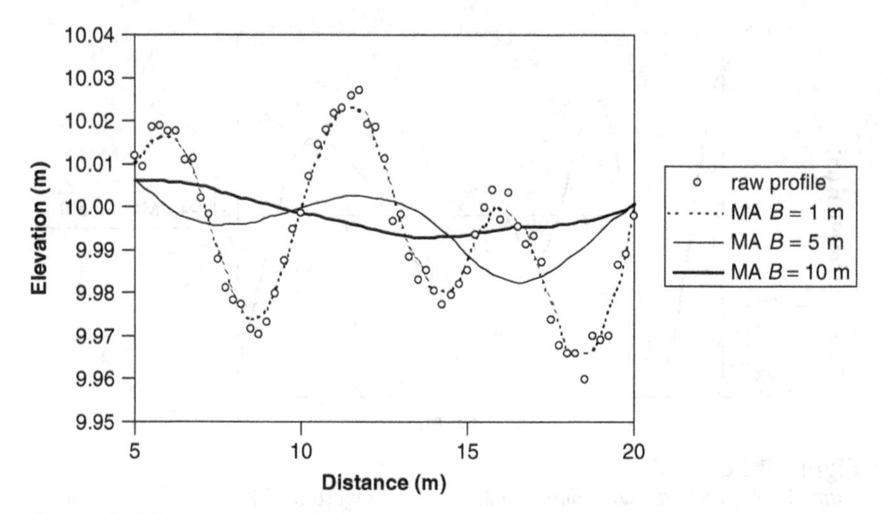


**Figure 9.17**
Effect of Low-Pass MA Filtering of Various Base Lengths

is achieved using Equation 9.2, with $B/(2Dx)$ values equal to 2, 10, and 20, respectively. This means that for the $B = 1$ m filter, for example, the elevation of each point is replaced with the average of five elevations, corresponding to the point itself, its two immediately preceding points, and its two immediately subsequent points. The low-pass filtering results are shown in Figure 9.17. The high-pass filtering results are calculated as per Equation 9.3 and plotted in Figure 9.18.

These figures demonstrate the effects of filtering. A low-pass filter with a base length considerably smaller than the shortest profile wavelength (i.e., 1.0 m) provides a smoothing of the profile without significantly compromising profile amplitude. This is not the case for the other two low-pass filters, which involve base lengths similar to the other two basic wavelengths of this profile. This problem is called *aliasing*, and it can avoided by using filters that have base lengths smaller than half the smallest wavelength of interest, $\lambda$. The high-pass filters, on the other hand, amplify the differences in variation between points. Figure 9.18 shows that the high-pass filter with a base length of 1.00 m removes most high-amplitude variation, while the one with a base length of 10.0 m removes the small amplitude/high frequency variation. Clearly, the type of filtering used affects significantly the filtered profile. As mentioned earlier, there are no established standards for pavement profile

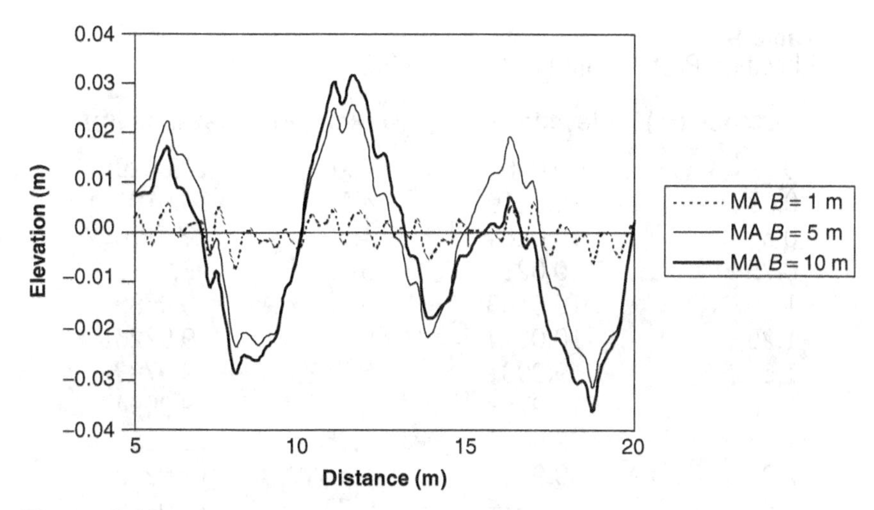


**Figure 9.18**
Effect of High-Pass MA Filtering of Various Base Lengths

filtering; rather, the selection of filters depends on the application at hand. More information on analyzing random data can be found in reference 4.

#### PROFILE SPECTRAL CONTENT

An artificial pavement profile was studied, consisting of three sinusoidal components. Theoretically, an actual pavement profile can be simulated by an infinite number of sinusoidals of various wavelengths and amplitudes. Often, the opposite problem is of interest, that is, transforming an actual pavement profile into its constituent sinusoidals. This transformation from the distance domain into the frequency domain can be done through Fourier analysis.[23] It should be noted that the pavement profile can be a expressed as a function of time, rather than distance, considering a vehicle traversing it at a constant speed. The resulting profile representation, referred as a power spectral density (PSD), is in terms of elevation amplitude squared versus frequency. The latter is defined in terms of either distance (i.e., cycles/m), referred to as the wave number, or time (i.e., cycles/sec or Hz).

**Example 9.3**

Consider the pavement profile data shown in Table 9.2, which is plotted in Figure 9.19. Compute and plot the PSD for this profile

**Table 9.2**
Elevation Profile Data For Example 9.3

| Distance (m) | Elevation (m) | Distance (m) | Elevation (m) |
|---|---|---|---|
| 0 | 10.0000 | 4.25 | 10.0229 |
| 0.25 | 10.0219 | 4.5 | 10.0101 |
| 0.5 | 10.0264 | 4.75 | 9.9918 |
| 0.75 | 10.0216 | 5 | 9.9835 |
| 1 | 10.0163 | 5.25 | 9.9788 |
| 1.25 | 10.0074 | 5.5 | 9.9726 |
| 1.5 | 9.9901 | 5.75 | 9.9783 |
| 1.75 | 9.9783 | 6 | 9.9994 |
| 2 | 9.9818 | 6.25 | 10.0219 |
| 2.25 | 9.9917 | 6.5 | 10.0272 |
| 2.5 | 9.9936 | 6.75 | 10.0218 |
| 2.75 | 9.9913 | 7 | 10.0169 |
| 3 | 9.9994 | 7.25 | 10.0076 |
| 3.25 | 10.0081 | 7.5 | 9.9895 |
| 3.5 | 10.0077 | 7.75 | 9.9788 |
| 3.75 | 10.0071 | 8 | 9.9825 |
| 4 | 10.0165 | — | — |

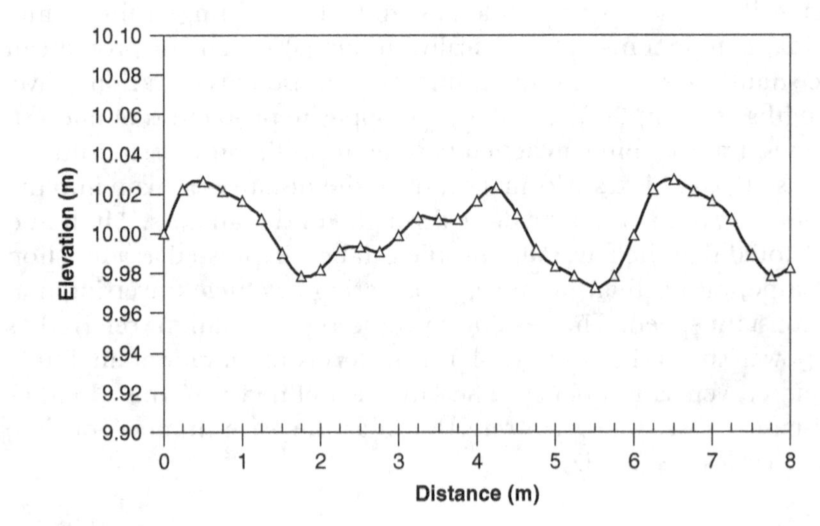


**Figure 9.19**
Pavement Elevation Profile for Example 9.3

and determine the frequencies and amplitudes of its constitutive waveforms.

**ANSWER**

A Fast-Fourier Transform (FFT) of the 32 pavement profile elevation measurements given can be carried out by a variety of means, but the most accessible is by spreadsheet (e.g., in Excel, it can be found under Tools/Data Analysis, after selecting the Analysis ToolPak from the Add-in list). The results are given in Table 9.3, whereby the wavelength is calculated as the ratio of the total length analyzed divided by the number of distance increments (i.e., 8/1, 8/2, 8/3, and so on); the wave number is simply its inverse. For each frequency, the FFT transformation is an imaginary number, with an absolute value proportional to the amplitude of the corresponding sinusoidal wave. The actual sinusoidal wave amplitude is calculated by dividing the absolute value by the number of distance increments (16 in this example).

**Table 9.3**
Spectral Analysis for the Profile in Example 9.3

| Distance Increment | Wavelength m/cycle | Wave Number cycles/m | FFT Transformation | Profile Amplitude \|Imaginary\| | Normalized Amplitude (Amplitude/16) |
|---|---|---|---|---|---|
| 0 | — | — | 320.0429 | 320.0429 | — |
| 1 | 8.0000 | 0.1250 | 0.0493+0.0276i | 0.0565 | 0.0035 |
| 2 | 4.0000 | 0.2500 | 0.0900+0.0977i | 0.1329 | 0.0083 |
| 3 | 2.6667 | 0.3750 | −0.1093−0.2367i | 0.2607 | 0.0163 |
| 4 | 2.0000 | 0.5000 | 0.0999−0.1784i | 0.2044 | 0.0128 |
| 5 | 1.6000 | 0.6250 | 0.0026−0.0377i | 0.0378 | 0.0024 |
| 6 | 1.3333 | 0.7500 | 0.0063−0.0268i | 0.0275 | 0.0017 |
| 7 | 1.1429 | 0.8750 | 0.0077−0.0207i | 0.0221 | 0.0014 |
| 8 | 1.0000 | 1.0000 | 0.0913−0.0209i | 0.0937 | 0.0059 |
| 9 | 0.8889 | 1.1250 | 0.0040−0.0132i | 0.0138 | 0.0009 |
| 10 | 0.8000 | 1.2500 | 0.0141−0.0087i | 0.0166 | 0.0010 |
| 11 | 0.7273 | 1.3750 | 0.0109−0.0087i | 0.0139 | 0.0009 |
| 12 | 0.6667 | 1.5000 | 0.0131−0.0103i | 0.0166 | 0.0010 |
| 13 | 0.6154 | 1.6250 | 0.0123−0.0044i | 0.0131 | 0.0008 |
| 14 | 0.5714 | 1.7500 | 0.0144−0.0021i | 0.0146 | 0.0009 |
| 15 | 0.5333 | 1.8750 | 0.0146−0.0036i | 0.0151 | 0.0009 |
| 16 | 0.5000 | 2.0000 | 0.0157+0.0000i | 0.0157 | 0.0010 |

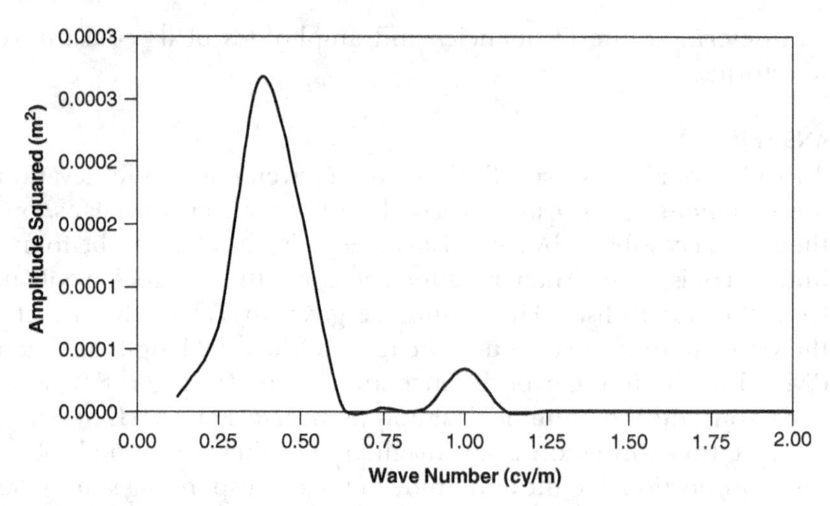


**Figure 9.20**
PSD of Pavement Profile Elevation Data for Example 9.3

The resulting pavement profile PSD is shown in Figure 9.20 in terms of wave number. It reveals two main constitutive sine waves with wave numbers of 0.4 and 1.0 cycles/m, which suggests wavelengths of 2 and 1.0 m/cycle, with corresponding amplitudes of 0.016 and 0.006 m. Clearly, the pavement profile described in the previous example was artificially created. The profile of an actual pavement consists of a multitude of waveforms consisting of a variety of wavelengths and amplitudes, as shown in Figure 9.21. The particular PSD was obtained by analyzing inertial profilometer measurements obtained at intervals of 0.1524 m (6 in) over a distance of 156.06 m. This resulted in 1024 data points that were analyzed using the FFT approach just described. This procedure can be applied to any $2^n$ observations of any time series, where $n$ is an integer.

### 9.2.8 Indices Summarizing Pavement Roughness

Regardless of the technology used for measuring pavement roughness, there is a need to summarize the results into a pavement roughness index, which in turn could be related to public perception of serviceability (i.e., *PSI*). Efforts to establish such an index date back to the 1950s, (i.e., the *SV* used in the AASHO Road Test[2] was perhaps one of the earliest pavement roughness indices). More recently, efforts to establish a "universal" and "transportable" index for quantifying pavement roughness were led by the World Bank.[32,33] This work produced the so-called International Roughness Index

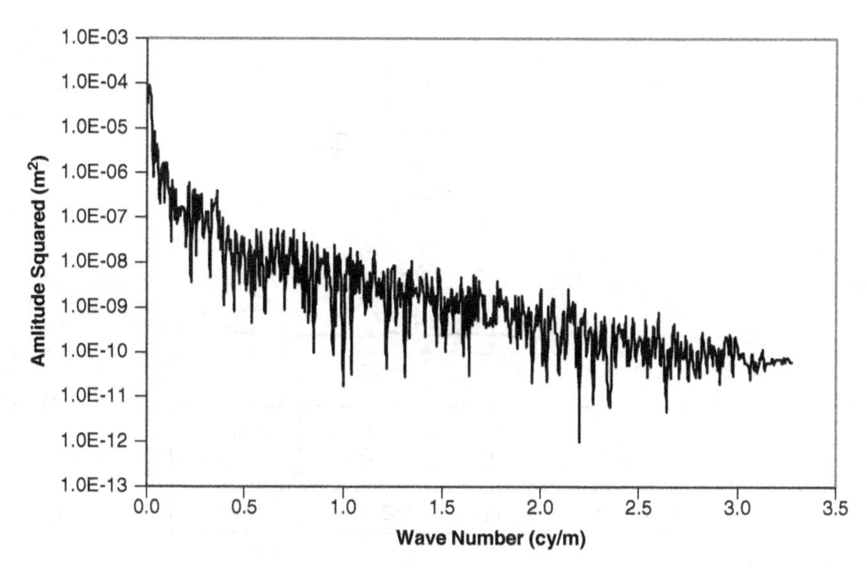


**Figure 9.21**
Example of a PSD of an Actual Pavement Elevation Profile

(*IRI*). Another index was develop to directly capture passenger car ride quality, referred to as the Ride Number (*RN*).[20,21] Although the search for the optimum pavement roughness index is still on, these two indices are most commonly used in North American practice, hence are the ones described in detail next.

**IRI**

The *IRI* was conceived as an index that simulates analytically the response of one of the earliest mechanical roughness measuring devices, namely the BPR roughness meter (Figure 9.5). The analytical model describing this quarter-car was a simple two-mass arrangement with linear springs and a linear dash pot (Figure 9.22). Its mechanical constants were derived experimentally to simulate the dynamic response of a common North American car, which at that time was a 1969 Chevrolet Impala.[11] The *IRI* was defined as the cumulative relative displacement of the axle with respect to the frame of this reference quarter-car per unit distance traveled over the pavement profile at a speed of 80 km/h. It is expressed in m/km or in/mi at selected intervals, (e.g., every 100 m). Figure 9.23 shows examples of the frequency distributions of *IRI* for selected states in the United States. It can be seen that IRI values on U.S. highways range between 1 and about 5 m/km. Higher *IRI* values may be

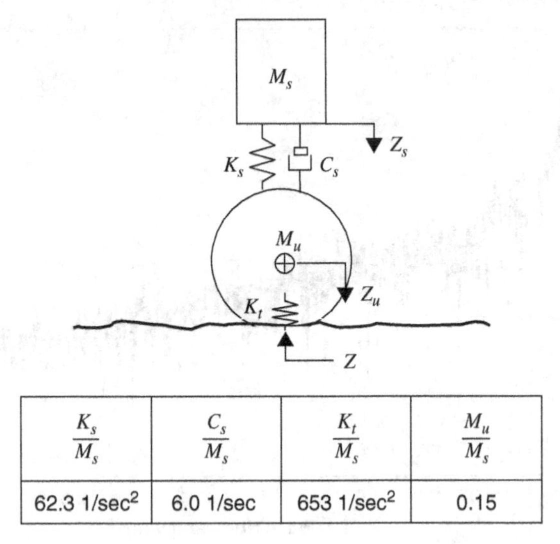


| $\frac{K_s}{M_s}$ | $\frac{C_s}{M_s}$ | $\frac{K_t}{M_s}$ | $\frac{M_u}{M_s}$ |
|---|---|---|---|
| 62.3 1/sec$^2$ | 6.0 1/sec | 653 1/sec$^2$ | 0.15 |

**Figure 9.22**
Reference Quarter-Car Used to Define the IRI

encountered on other U.S. paved roads, while in developing countries *IRI* values higher than 8 m/km may be reached, approaching the roughness of unpaved roads.

The IRI meets the universal and transportable requirements, although the actual statistic being calculated depends to some extent on the pavement profile sampling and filtering method used. The *IRI* can be used to predict the subjective pavement serviceability index described earlier (i.e., the *PSI*). A variety of regression equations relating these two indices can be found in the literature, such as Equation 9.4, which is plotted in Figure 9.24.[27]

$$PSI = 5.0\ e^{-0.18\ IRI} \tag{9.4}$$

These relationships are developed by conducting panel serviceability ratings on a number of pavement sections, measuring their profile, computing the *IRI*, and conducting regression analysis using the *PSI* and *IRI* measurements as dependent and independent variables, respectively. Equation 9.4, for example, suggests that for a minimum acceptable *PSI* the value of 2.0 corresponding *IRI* value is of 5.1 m/km, a roughness level that would dictate pavement rehabilitation.

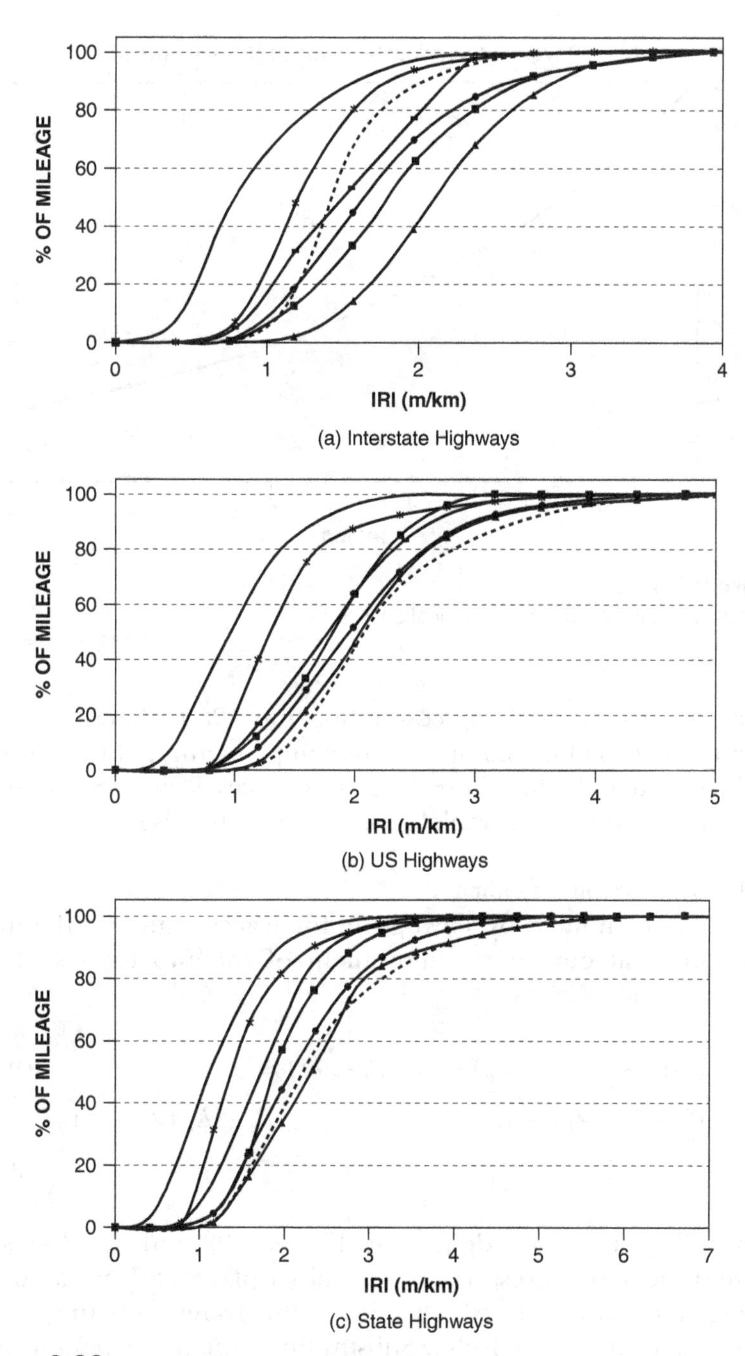


**Figure 9.23**
Frequency Distribution of IRI for Selected States by Highway Classification (1996 HPMS)

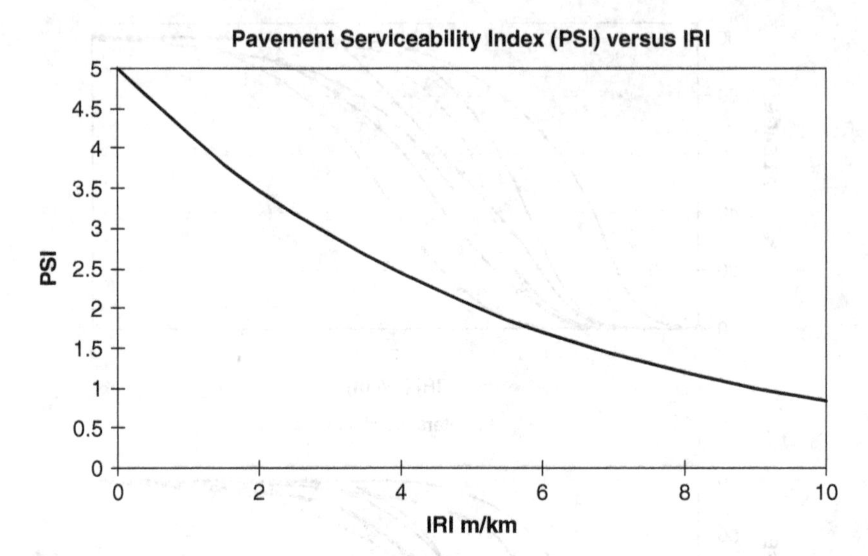


**Figure 9.24**
Relationship between IRI and PSI (Equation 9.4)

The algorithm used for computing the *IRI* is described next. Several versions of the computer code implementing this algorithm can be found in the literature, (e.g., references 33 and 38). A variety of computer software is available to incorporate this code.

The IRI Computation Algorithm

The *IRI* is computed by solving the force-acceleration differential equations that govern the movement of the two masses of the quarter-car model:[32,33]

$$\ddot{Z}_s M_s + C_s(\dot{Z}_s - \dot{Z}_u) + K_s(Z_s - Z_u) = 0 \quad (9.5a)$$

$$-C_s(\dot{Z}_s - \dot{Z}_u) - K_s(Z_s - Z_u) + M_u \ddot{Z}_u + K_t(Z_u - Z) = 0 \quad (9.5b)$$

where all variables are defined in Figure 9.22 and the dots symbolize time derivatives, (i.e., the subscripts $s$ and $u$ stand for sprung and unsprung, which denote the frame and the tire of the quarter-car, respectively). Substituting Equation 9.5a into 9.5b and dividing through by $M_s$ gives an equivalent set of differential equations:

$$\ddot{Z}_s + \frac{C_s}{M_s}(\dot{Z}_s - \dot{Z}_u) + \frac{K_s}{M_s}(Z_s - Z_u) = 0 \tag{9.6a}$$

$$\ddot{Z}_s + \frac{M_u}{M_s}\ddot{Z}_u + \frac{K_t}{M_s}(Z_u - Z) = 0 \tag{9.6b}$$

which can be translated into a set of four linear differential equations:

$$\dot{Z}_3 + \frac{C_s}{M_s}(Z_3 - Z_4) + \frac{K_s}{M_s}(Z_1 - Z_2) = 0 \tag{9.7a}$$

$$\dot{Z}_3 + \frac{M_u}{M_s}\dot{Z}_4 + \frac{K_1}{M_s}(Z_2 - Z) = 0 \tag{9.7b}$$

$$Z_3 - \dot{Z}_1 = 0 \tag{9.7c}$$

$$Z_4 - \dot{Z}_2 = 0 \tag{9.7d}$$

where:

$$\begin{Bmatrix} Z_1 \\ Z_2 \\ Z_3 \\ Z_4 \end{Bmatrix} = \begin{Bmatrix} Z_s \\ Z_u \\ \dot{Z}_s \\ \dot{Z}_u \end{Bmatrix} \tag{9.8}$$

The relative displacement of the tire with respect to the frame of the quarter-car is calculated as the difference between $Z_4$ and $Z_3$ accumulated over distance or time, which is mathematically expressed as:

$$IRI = \frac{1}{T}\int_0^T |Z_3 - Z_4|\,dt = \frac{1}{T}\int_0^T |\dot{Z}_s - \dot{Z}_u|\,dt \tag{9.9}$$

Hence, there is need to calculate only $Z_3$ and $Z_4$ and their derivatives to advance this solution in distance or time. This is done in conjunction with Equations 9.7a–d with a method referred to as the *state transition method*, whereby the value of these variables at time $t + dt$ is calculated from their values at the previous time step $t$, expressed as:

$$\begin{Bmatrix} Z_3 \\ \dot{Z}_3 \\ Z_4 \\ \dot{Z}_4 \end{Bmatrix}_{t+dt} = [S]\begin{Bmatrix} Z_3 \\ \dot{Z}_3 \\ Z_4 \\ \dot{Z}_4 \end{Bmatrix}_t + \{P\}\,\dot{Z} \tag{9.10}$$

where $[S]$ is the $4 \times 4$ state transition matrix of constants to be determined, $\{P\}$ is a $4 \times 1$ matrix of constants to be determined, and $\dot{Z}$ is the distance/time derivative of the pavement profile. Equation 9.10 can be abbreviated as:

$$\{\hat{Z}\}_{t+dt} = [S]\{\hat{Z}\}_t + \{P\}\dot{Z} \tag{9.11}$$

where:

$$\{\hat{Z}\} = \begin{Bmatrix} Z_3 \\ \dot{Z}_3 \\ Z_4 \\ \dot{Z}_4 \end{Bmatrix} \tag{9.12}$$

and $\{\hat{Z}\}$ is defined in terms of $Z_1$, $Z_2$, $Z_3$, and $Z_4$ through Equations 9.7a–d as:

$$\begin{Bmatrix} Z_3 \\ \dot{Z}_3 \\ Z_4 \\ \dot{Z}_4 \end{Bmatrix} = \begin{bmatrix} 0 & 1 & 0 & 0 \\ -\frac{K_s}{M_s} & -\frac{C_s}{M_s} & \frac{K_s}{M_s} & \frac{C_s}{M_s} \\ 0 & 0 & 0 & 1 \\ \frac{K_s}{M_u} & \frac{C_s}{M_u} & -\frac{K_s + K_t}{M_u} & -\frac{C_s}{M_u} \end{bmatrix} \begin{Bmatrix} Z_1 \\ Z_3 \\ Z_2 \\ Z_4 \end{Bmatrix} + \begin{Bmatrix} 0 \\ 0 \\ 0 \\ \frac{K_t}{M_u} \end{Bmatrix} Z \tag{9.13}$$

which can be abbreviated as:

$$\{\hat{Z}\} = [A] \begin{Bmatrix} Z_1 \\ Z_3 \\ Z_2 \\ Z_4 \end{Bmatrix} + \{B\} Z \tag{9.14}$$

where $[A]$ and $\{B\}$ are matrices of constants that depend on the mechanical constants of the quarter-car model, defined as:

$$[A] = \begin{bmatrix} 0 & 1 & 0 & 0 \\ -\frac{K_s}{M_s} & -\frac{C_s}{M_s} & \frac{K_s}{M_s} & \frac{C_s}{M_s} \\ 0 & 0 & 0 & 1 \\ \frac{K_s}{M_u} & \frac{C_s}{M_u} & -\frac{K_s + K_t}{M_u} & -\frac{C_s}{M_u} \end{bmatrix} \tag{9.15}$$

and,

$$\{B\} = \begin{Bmatrix} 0 \\ 0 \\ 0 \\ 0 \\ \frac{K_t}{M_u} \end{Bmatrix} \tag{9.16}$$

Hence, the problem reduces in expressing the matrices $[S]$ and $\{P\}$ in terms of the matrices $[A]$ and $\{B\}$. This is done expanding Equation 9.14 using a Taylor series and assuming that the derivative of the pavement profile $Z$ is constant within $dt$ (i.e., the slope of the profile does not change between successive elevation points).

$$\{\hat{Z}\}_{t+dt} = \{\hat{Z}\}_t + [A]\left[\begin{Bmatrix}\dot{Z}_1\\ \dot{Z}_2\\ \dot{Z}_3\\ \dot{Z}_4\end{Bmatrix}_t dt + \begin{Bmatrix}\ddot{Z}_1\\ \ddot{Z}_2\\ \ddot{Z}_3\\ \ddot{Z}_4\end{Bmatrix}_t \frac{dt^2}{2!} + \begin{Bmatrix}\dddot{Z}_1\\ \dddot{Z}_2\\ \dddot{Z}_3\\ \dddot{Z}_4\end{Bmatrix}_t \frac{dt^3}{3!} + \ldots..\right]$$
$$+ \{B\}\,\dot{Z}dt \quad (9.17)$$

Note that:

$$\begin{Bmatrix}\dot{Z}_1\\ \dot{Z}_2\\ \dot{Z}_3\\ \dot{Z}_4\end{Bmatrix} = \begin{Bmatrix}Z_3\\ Z_3\\ Z_4\\ \dot{Z}_4\end{Bmatrix} = \{\hat{Z}\}, \begin{Bmatrix}\ddot{Z}_1\\ \ddot{Z}_2\\ \ddot{Z}_3\\ \ddot{Z}_4\end{Bmatrix} = [A]\{\hat{Z}\} + \{B\}\,\dot{Z},$$

$$\begin{Bmatrix}\dddot{Z}_1\\ \dddot{Z}_2\\ \dddot{Z}_3\\ \dddot{Z}_4\end{Bmatrix} = [A]\left([A]\{\hat{Z}\} + \{B\}\,\dot{Z}\right), \ldots \quad (9.18)$$

allows writing Equation 9.17 as:

$$\{\hat{Z}\}_{t+dt} = \{\hat{Z}\}_t + [A]\left[\{\hat{Z}\}_t dt + [A]\{\hat{Z}\}_t \frac{dt^2}{2!} + \{B\}\,\dot{Z}\frac{dt^2}{2!}\right.$$
$$\left. + [A]\,[A]\{\hat{Z}\}_t \frac{dt^3}{3!} + [A]\{B\}\,\dot{Z}\frac{dt^3}{3!}\ldots..\right] \quad (9.19)$$

which can be written as:

$$\{\hat{Z}\}_{t+dt} = \{\hat{Z}\}_t\left[[I] + [A]\,dt + [A]\,[A]\frac{dt^2}{2!} + [A]\,[A]\,[A]\frac{dt^3}{3!} + \ldots..\right]$$
$$+\dot{Z}\,\{B\}\left[dt + [A]\frac{dt^2}{2!} + [A]\,[A]\frac{dt^3}{3!} + \ldots.\right] \quad (9.20)$$

where $[I]$ is a $4 \times 4$ unit matrix. Considering that:

$$\left[[I] + [A]dt + [A][A]\frac{dt^2}{2!} + [A][A][A]\frac{dt^3}{3!} + \ldots..\right] = e^{[A]\mathrm{dt}} \quad (9.21)$$

allows reducing Equation 9.20 into:

$$\{\hat{Z}\}_{t+dt} = \{\hat{Z}\}_t e^{[A]dt} + \dot{Z}\{B\}[A]^{-1}\left(e^{[A]dt} - [I]\right) \tag{9.22}$$

Comparing Equation 9.11 and 9.22 gives:

$$[S] = e^{[A]dt} \tag{9.23}$$

and,

$$\{P\} = \{B\}[A]^{-1}\left(e^{[A]dt} - [I]\right) = \{B\}[A]^{-1}([S] - [I]) \tag{9.24}$$

As will be shown in the following example, a finite number of terms of Equation 9.21, (typically, between 5 and 10) is sufficient for calculating the matrix $[S]$.

The initial conditions of the quarter-car simulation are specified as:

$$\begin{Bmatrix} Z_3 \\ \dot{Z}_3 \\ Z_4 \\ \dot{Z}_4 \end{Bmatrix}_0 = \begin{Bmatrix} \frac{Z(k)-Z(0)}{0.5} \\ 0 \\ \frac{Z(k)-Z(0)}{0.5} \\ 0 \end{Bmatrix} \tag{9.25}$$

where $k$ is the integer part of the ratio $0.5/dt$. The physical interpretation of these initial conditions is that the quarter-car is running for a period of 0.5 seconds on a perfectly smooth pavement with a slope equal to the average slope defined by the initial and the $k_{th}$ point of the pavement profile.

**Example 9.4**

Calculate the values of matrices $[S]$ and $\{P\}$ for a pavement profile obtained at a distance increment, $dx$, of 0.25 m, and a quarter-car speed of 80 km/h. The values of the quarter-car constants are given in Figure 9.22.

**ANSWER**

The time interval $dt$ corresponding to the vehicle speed selected is 0.01125 seconds. Equation 9.15 gives the following value for $[A]dt$, $[A]\,[A]dt^2/2$ and so on.

$$[A]\,dt = \begin{bmatrix} 0 & 1 & 0 & 0 \\ -62.3 & -6.0 & 62.3 & 6 \\ 0 & 0 & 0 & 1 \\ 415.33 & 40.0 & -4768.7 & -40.0 \end{bmatrix} 0.01125$$

$$[A]\,[A]\,\frac{dt^2}{2!}$$

$$= \begin{bmatrix} -62.3 & -6 & 62.3 & 6 \\ 2865.78 & 213.7 & -28986 & -213.7 \\ 415.33 & 40 & -4768.7 & -40 \\ -19105.2 & -1424.67 & 193240 & -2928.7 \end{bmatrix} 0.63310^{-4}$$

$$[A]\,[A]\,[A]\,\frac{dt^3}{3!}$$

$$= \begin{bmatrix} 2865.78 & 213.7 & -28986 & -213.7 \\ -102069.5 & -6964.42 & 1032384.7 & -19155.8 \\ -19105.2 & -1424.67 & 193240 & -2928.7 \\ -1127620 & -127705.2 & 13877334 & 301839.9 \end{bmatrix}$$

$$\times\ 0.00237310^{-4}$$

Terms should be added until changes in the cell values of the matrix $[S]$ are lower than a prescribed tolerance. The matrix $[S]$ here was obtained by adding six of these terms:

$$[S] = \begin{bmatrix} 0.9967 & 0.0109 & -0.0021 & 0.0003 \\ -0.5476 & 0.9439 & -0.8406 & 0.0506 \\ 0.0212 & 0.0021 & 0.7512 & 0.0082 \\ 3.2827 & 0.3374 & -39.0795 & 0.4350 \end{bmatrix}$$

Accordingly, Equation 9.24 gives matrix $\{P\}$ as:

$$\{P\} = \begin{Bmatrix} 0.0055 \\ 1.3881 \\ 0.2276 \\ 35.7962 \end{Bmatrix}$$

Matrices $[S]$ and $\{P\}$ can be used to advance $\{\hat{Z}\}$ from $t$ to $t + dt$ as per Equation 9.11, which subsequently leads to an *IRI* computation by numerically integrating Equation 9.9.

**RN**

The Ride Number (*RN*) was conceived from the outset as a predictor of pavement serviceability (i.e., *PSI*).[20,21] It is an index, in the familiar scale from 0 to 5, computed from pavement elevation profile data. The pavement profile is processed through a band-pass filter, which is a combination of a low-pass and a high-pass filter, to eliminate pavement profile wave numbers outside the range that affects passenger car excitation, hence the perception serviceability. These wave number limits were established through extensive panel rating surveys to be about 0.41 cycles/m (0.125 cycles/ft) and 2.1 cycles/m (0.63 cycles/ft), respectively. The corresponding pavement profile wavelengths are 2.44 m/cycle and 0.48 m/cycle.

The band-pass filtered pavement profile elevation is summarized in terms of its root-mean-square (RMS), referred to as the Profile Index (*PI*):

$$PI = \sqrt{\frac{\sum_{i=1}^{m} (Z(i) - \overline{Z})^2}{m}} \tag{9.26}$$

where $Z(i)$ is the elevation at point $i$, and $\overline{Z}$ is the average of the elevation of the $m$ points in the profile being analyzed. Hence, this *PI* has the same units as the profile elevation. Note that this *PI* is not related to the PI computed with a rolling straightedge type of profilograph described earlier. The *PI* calculated through Equation 9.26 is used to predict *RN* through a regression equation:

$$RN = -1.74 - 3.03 \log(PI) \tag{9.27}$$

where the *PI* is in inches. The algorithm used for computing the *PI* and, consequently, the *RN* is described next.

#### The *RN* Computation Algorithm

The *PI* could be calculated from the time-domain pavement profile elevation data according to Equation 9.26, after filtering it with a low-pass filter and a high-pass filter with wave numbers of 0.41 cycles/m and 2.1 cycles/m, respectively. However, it is preferable to perform these calculations in the frequency-domain, by considering Parseval's formula. This states that the mean square of a function is equal to the sum of the squares of the absolute values of the coefficients of its Fourier transformation.[44] The absolute value of these coefficients is the amplitude of the PSD of the pavement profile, at a given wave number. Summing the square of these values within the range of wave numbers of interest (between 0.41 and 2.1 cycles/m) gives the mean square of the profile elevation, without having to filter it. This is expressed as:

$$\frac{\sum_{i=1}^{m} (Z(i) - \overline{Z})^2}{m} = \frac{1}{2} \sum_{k} |C_k|^2 \tag{9.28}$$

where, $k$ is the number of Fourier coefficients $C_k$ within the wave number range of interest. More details on the development of the *RN* are provided in reference 20. An example of calculating the *RN* is given next.

**Example 9.5**

Calculate the *RN* for the pavement profile, for which the PSD is shown in Figure 9.21.

**ANSWER**

The sum of the squares of the amplitude of the PSD of the profile was computed between the limits of 0.41 and 2.1 cycles/m, as shown in Figure 9.25. The value obtained was 9.1326 $10^{-7}$ m$^2$, and its square root was 0.0009556 m (0.0376 in.). Inputting this value into Equation 9.27, gives:

$$RN = -1.74 - 3.03 \log(0.0376) = 2.58,$$

which suggests a rough pavement in the 0 to 5 serviceability scale.

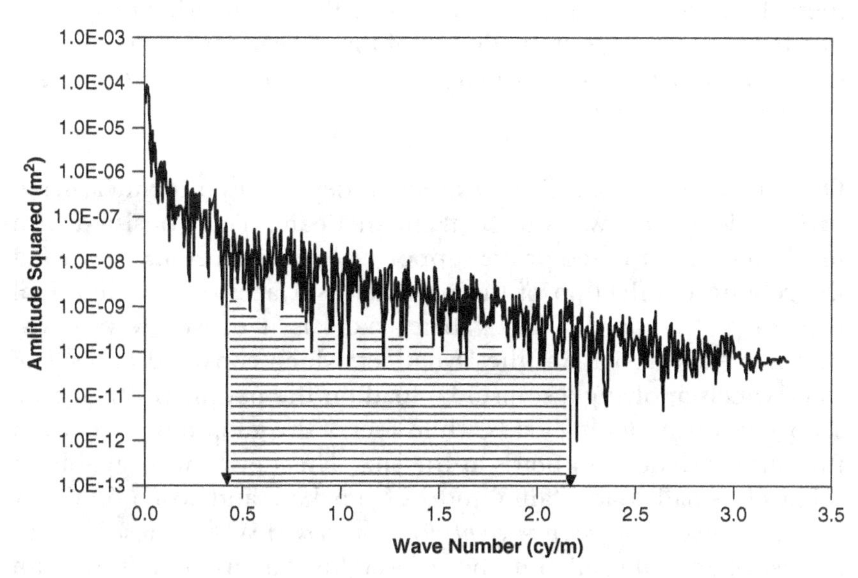


**Figure 9.25**
*PI* Calculation by Numerically Integrating the Pavement Profile PSD.

As for the *IRI*, a variety of computer software is available to incorporate the *RN* algorithm just described. For suggestions on the most recent software, see the Web site for this book, at www.wiley.com/go/pavement.

## 9.3 Structural Capacity

### 9.3.1 Definitions—Introduction

Structural capacity is defined as the capability of a pavement to physically handle the traffic loads anticipated over its life. Historically, surface deflection has been considered as the pavement attribute associated with structural capacity. There is a variety of commercially available devices for measuring in-situ pavement deflections, referred to as *deflectometers*. State-of-the-art deflectometers apply a known load to the surface and use geophones (i.e., vertical speed sensors, whose signal needs to be integrated once to give deflection), arranged to yield a ''bowl'' of deflection measurements. These devices provide information not only on the structural capacity of pavement sections but also on the structural properties of their layers and the subgrade. The latter is done through back-calculation, which is presented later in this chapter, in detail. Hence, interpreting deflection data offers guidance in selecting applicable future pavement 4-R treatments, as well as providing the engineering properties needed for carrying out the design of these treatments (e.g., overlay design). As such, it is of primary interest to the engineer/manager of the roadway system.

### 9.3.2 Devices for Measuring Surface Deflection

One of the earlier and least expensive devices used for measuring surface deflection was the Benkelman Beam. This consisted of a mechanical lever-based probe supported by three legs that recorded the rebound deflection of the pavement surface between the dual tires of a ''standard'' 80 kN truck axle as it drove away slowly. Another early device was the Dynaflect, which consisted of a set of steel wheels applying a sinusoidal load on the pavement through a set of eccentric rotating loads, while a set of five geophones recorded the surface deflection bowls at the site. This device was capable of relatively small loads (amplitude of 4.5 kN) and lost popularity with the advent of *falling-weight deflectometers* (FWD). These devices consist of an adjustable number of weights that are let fall from an adjustable height onto the pavement surface or on a ''buffer'' placed

**Figure 9.26**
A Falling Weight Deflectometer (Courtesy of Dynatest Inc.)

on the pavement surface (a set of plates supported by springs, which translates the impact load to a pulse, to more realistically simulate the passages of a tire). Deflection bowl data is obtained from multiple geophones, typically seven or nine, arranged at various distances from the point of load application. A computerized data acquisition system collects and reduces the data. The methodology for carrying out these deflection measurements was standardized.[41] There is a variety of commercially available FWDs, such as those made by Dynatest, Kuab, Phoenix, as well as the Thumper, developed by the FHWA. A picture of one of these devices is shown in Figure 9.26.

One of the limitations of all these deflection measuring systems is that they need to be stationary in order to take measurements; as a result, their productivity is low and their operation is labor-intensive (it involves lane closures through traffic control). This discourages some highway agencies from conducting such measurements networkwide. Instead, they resort to conducting project-specific structural evaluation surveys, to determine the range of pavement

4-R treatments applicable; and, if needed, they use the deflection measurements for overlay design.

There have been efforts to develop a "rolling weight" deflectometer that allows deflection measurements at highway speeds.[14] This device consists of a heavy truck equipped with a scanning laser system that is capable of measuring the elevation of a given spot on the pavement surface under unloaded conditions and under the axle load of that truck. Although a working prototype of this device has been developed, further refinements are needed to enhance the agreement between its measurements and those obtained with conventional FWD.

### 9.3.3 Processing Deflection Data

For flexible pavements, deflection measurements are taken every 50 to 100 meters. This data needs to be processed and a statistic calculated to reflect the structural capacity of that site. The most commonly used method for reducing flexible pavement deflection data was developed by the Asphalt Institute.[3] It was based on Benkleman Beam rebound deflection measurements. Deflection measurements obtained with other types of deflection measuring devices need to be converted to equivalent Benkelman Beam deflection measurements. Developing such a correlation is not trivial, given the differences in loading levels and loading rates between the Benkelman Beam and other deflection measuring devices. Jurisdictions need to perform field experimentation to establish this correlation using the actual devices and procedures used in the field. Where this is not possible, a conversion factor of 1.61 can be applied to FWD measurements to estimate equivalent Benkleman Beam measurements, provided that the FWD load applied is 40 kN (9,000 lbs) and the radius of its loading plate is 0.152 m (6 in).[3] The equivalent Benkleman Beam deflection measurements thus obtained need first to be adjusted for temperature to bring them all to the same reference temperature of 21°C. This is done by determining the temperature at the top, bottom, and middle of the pavement layer and averaging the three temperatures. The nomograph shown in Figure 9.27 is used for this purpose, whereby temperature is indexed by the sum of the ambient temperature at the time of the deflection measurements plus the average ambient temperature in the five days that preceded the deflection measurements. The corresponding deflection adjustment factors are obtained from Figure 9.28, as a function of the thickness of the granular base.

**Example 9.6**

The Benkelmam Beam equivalent deflection measurements and the corresponding temperatures listed in Table 9.4 were obtained at a pavement site, with asphalt concrete and base layer thicknesses of 100 mm and 250 mm, respectively. Adjust these deflection measurements for temperature, given that the average ambient temperature the five days prior to the deflection measurements was 16.8°C.

**ANSWER**

The calculation steps are shown in Table 9.5, where the temperature at the middle and the bottom of the slab were computed using Figure 9.27, and the adjustment factors were computed from Figure 9.28.

Once all deflection measurements have been adjusted for temperature, as shown in the preceding example, the following statistic is calculated, referred to as the *representative rebound deflection* (*RRD*):

$$RRD = \overline{x} + 2SD \tag{9.29}$$

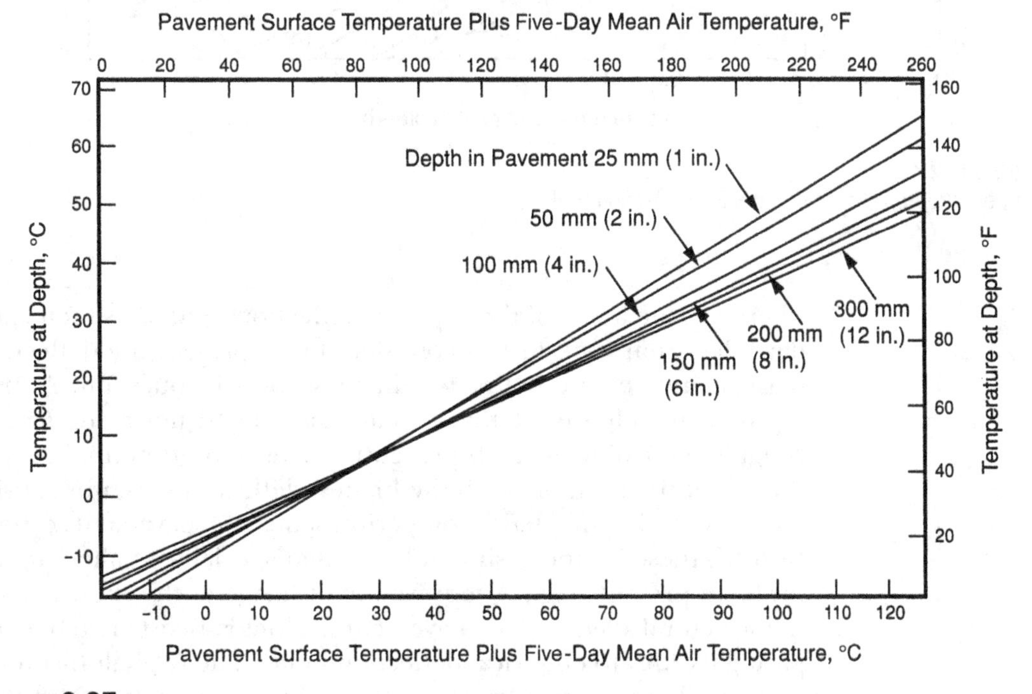


**Figure 9.27**
Temperature Distribution in Asphalt Concrete Layer (Ref. 3)

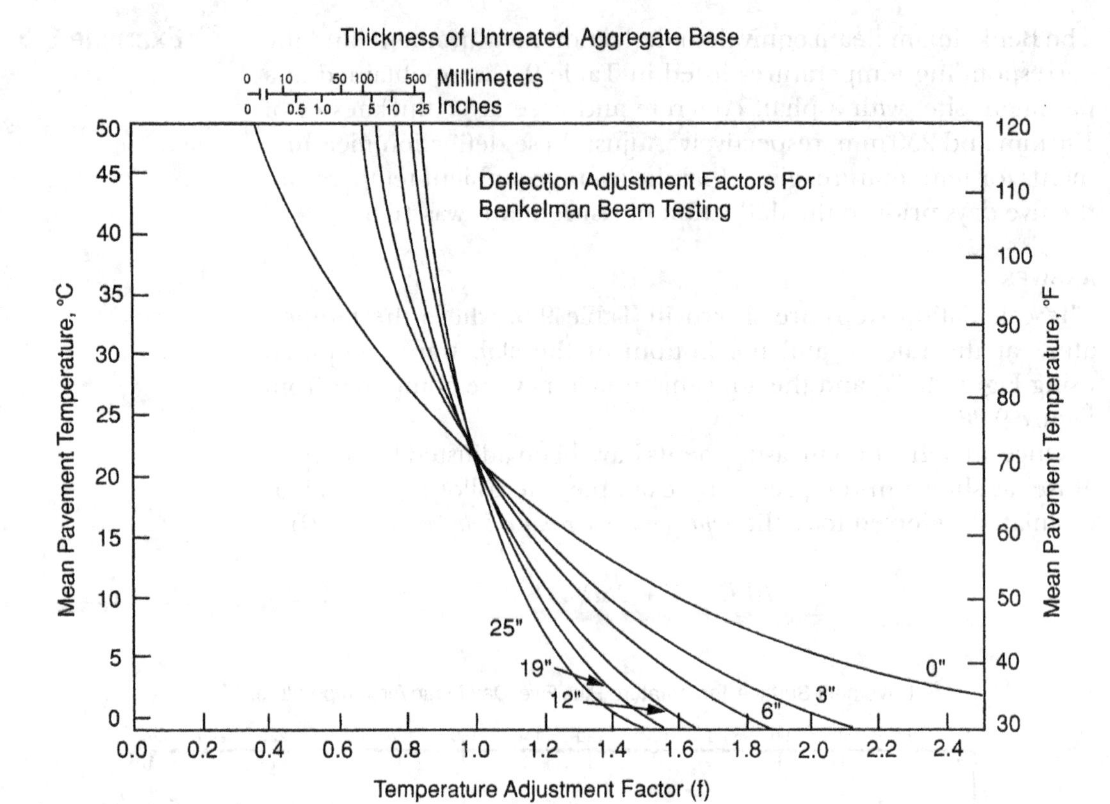


**Figure 9.28**
Temperature Adjustment Factors for Deflection (Ref. 3)

where $\overline{x}$ is the mean of the adjusted deflections and *SD* is their standard deviation. The *RRD* is considered to represent the deflection performance of the entire pavement section in question. Assuming that the adjusted deflections are distributed normally, *RRD* is roughly their ninety-eighth percentile. This definition implies that the 2% of the locations with the highest deflections are not considered in defining the deflection performance of a pavement section. Instead, these locations should be treated specially, to alleviate the problem present (e.g., a improve the drainage at the site). Finally, the structural adequacy of a pavement sections is ascertained by comparing its *RRD* to empirical deflection limits. The Asphalt Institute[3] developed such limits, referred to as the *design rebound deflection*

**Table 9.4**
Deflection Measurements and Temperatures for Example 9.6

| Deflection (mils) | Temperature (°C) |
|---|---|
| 1.440 | 17.8 |
| 1.209 | 18.3 |
| 1.405 | 18.9 |
| 1.245 | 18.9 |
| 1.209 | 18.9 |
| 1.369 | 20.0 |
| 1.636 | 20.0 |
| 1.316 | 20.6 |
| 1.476 | 21.1 |
| 0.907 | 21.1 |
| 0.676 | 21.1 |
| 0.729 | 23.9 |
| 0.836 | 25.0 |
| 0.996 | 25.6 |
| 0.853 | 25.6 |
| 1.013 | 25.6 |
| 1.031 | 26.7 |
| 0.871 | 27.2 |
| 0.907 | 27.2 |

(*DRD*), as a function of the traffic level anticipated over the future life of a pavement section, expressed in terms of *ESALs* (Figure 9.29).
This comparison can have one of three outcomes:

- $RRD < DRD$, which suggests that the section is more than structurally adequate.
- $RRD = DRD$, which suggests that the section is just structurally adequate.
- $RRD > DRD$, which suggests that the section is structurally inadequate.

The third outcome would suggest that the section requires an improvement in structural capacity (e.g., a structural overlay) if it is to sustain the anticipated traffic loads.

**Table 9.5**
Adjusting Deflections for Temperature for Example 9.6

| Deflection (mm) | Temperature +5 Days | Temperature Middle | Temperature Bottom | Average Slab | Adjustment Factor | Adjusted Deflection |
|---|---|---|---|---|---|---|
| 1.440 | 34.58 | 19.0 | 16.0 | 17.6 | 1.06 | 1.527 |
| 1.209 | 35.13 | 19.0 | 16.0 | 17.8 | 1.06 | 1.282 |
| 1.405 | 35.69 | 19.0 | 16.0 | 18.0 | 1.05 | 1.475 |
| 1.245 | 35.69 | 19.0 | 16.0 | 18.0 | 1.04 | 1.294 |
| 1.209 | 35.69 | 19.0 | 16.0 | 18.0 | 1.03 | 1.245 |
| 1.369 | 36.80 | 20.0 | 18.0 | 19.3 | 1.02 | 1.396 |
| 1.636 | 36.80 | 20.0 | 18.0 | 19.3 | 1.02 | 1.668 |
| 1.316 | 37.36 | 20.0 | 18.0 | 19.5 | 1.01 | 1.329 |
| 1.476 | 37.91 | 20.0 | 18.0 | 19.7 | 1.01 | 1.490 |
| 0.907 | 37.91 | 20.0 | 18.0 | 19.7 | 1.01 | 0.916 |
| 0.676 | 37.91 | 20.0 | 18.0 | 19.7 | 1.01 | 0.682 |
| 0.729 | 40.69 | 21.0 | 19.0 | 21.3 | 1 | 0.729 |
| 0.836 | 41.80 | 22.0 | 20.0 | 22.3 | 0.98 | 0.819 |
| 0.996 | 42.36 | 23.0 | 21.0 | 23.2 | 0.97 | 0.966 |
| 0.853 | 42.36 | 23.0 | 21.0 | 23.2 | 0.97 | 0.828 |
| 1.013 | 42.36 | 23.0 | 21.0 | 23.2 | 0.97 | 0.983 |
| 1.031 | 43.47 | 24.0 | 22.0 | 24.2 | 0.95 | 0.980 |
| 0.871 | 44.02 | 24.0 | 22.0 | 24.4 | 0.95 | 0.828 |
| 0.907 | 44.02 | 24.0 | 22.0 | 24.4 | 0.95 | 0.861 |

**Example 9.7**

Given the deflection measurements shown in Example 9.6 and that the section is required to carry 1 million *ESALs* over its remaining life, decide on its structural adequacy.

**ANSWER**

The mean and the standard deviation of the temperature-adjusted deflections computed in the previous example was 1.121 mm and 0.307 mm, respectively, which gives an *RRD* of 1.735 mm. For the traffic level of 1 million *ESALs*, Figure 9.29 gives a *DRD* of 1 mm. Hence, the *RRD* is larger than the *DRD*, therefore the section is not structurally adequate.

### 9.3.4 Assigning an Index to Structural Capacity

Often, it is desirable to express pavement structural adequacy through an index understood by nontechnical people (e.g., a 0 to 5 or 0 to 10 scale). There is no widely accepted method for doing

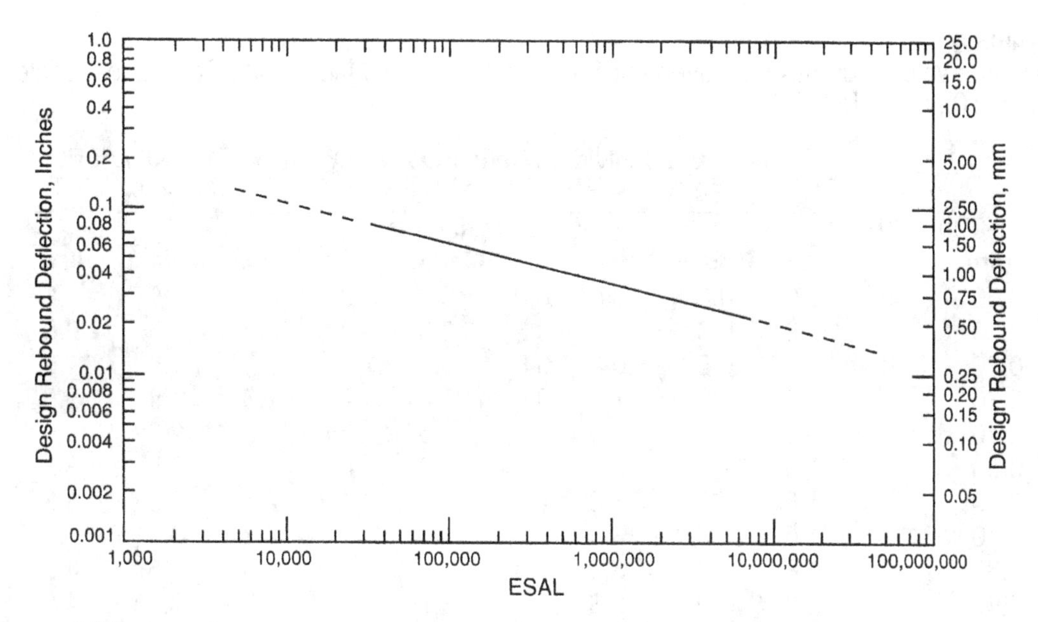


**Figure 9.29**
Design Rebound Deflection (*DRD*) as a Function of *ESALs* (Ref. 3)

so, although most methods are based on comparisons of the *RRD*, as computed earlier, and the *DRD*. One of the methods used assigns a midscale value (2.5 or 5, respectively) to the condition of *RRD* = *DRD*, a lower than the midscale value for the condition *RRD* > *DRD*, and a higher than the midrange value for the condition *RRD* < *DRD*.[12] Table 9.6 lists deduct values as a function of the difference between *RRD* and *DRD*, the traffic level (low, medium, and high for daily lane traffic volumes lower than 1000, between 1000 and 3000, and higher than 3000 vehicles per day respectively), as well as the frequency at which individual deflection measurements exceed the *DRD*. The procedure for computing such a Structural Adequacy Index (*SAI*) is explained through the following example.

**Example 9.8**

Compute the SAI for the deflection measurements given in Examples 9.6 and 9.7, given that the daily lane traffic volume at the site is 2000 vehicles/day.

**ANSWER**

The difference between *RRD* and *DRD* is $1.735 - 1.0 = 0.735$, rounded to 0.7 mm. Of the 19 adjusted deflection measurements,

**Table 9.6**
Deduct Values for Flexible Pavement Structural Adequacy Index (SAI) Calculation (scale 1 to 10), (after Ref. 12)

| —RRD-DRD— | Frequency of Individual Deflection Observations Exceeding DRD | | | | | | | | |
|---|---|---|---|---|---|---|---|---|---|
| | <30 % | | | 30%–60% | | | >60 % | | |
| | Traffic Level | | | | | | | | |
| mm | Low | Medium | High | Low | Medium | High | Low | Medium | High |
| 0.00 | 0.0 | 0.0 | 0.0 | 0.0 | 0.0 | 0.0 | 0.0 | 0.0 | 0.0 |
| 0.10 | 0.0 | 0.3 | 0.3 | 0.3 | 0.3 | 0.5 | 0.3 | 0.5 | 0.8 |
| 0.20 | 0.3 | 0.50 | 0.8 | 0.5 | 0.8 | 1.0 | 0.8 | 1.0 | 1.5 |
| 0.30 | 0.5 | 0.8 | 1.5 | 0.8 | 1.0 | 2.0 | 1.3 | 1.5 | 2.5 |
| 0.40 | 0.8 | 1.0 | 2.0 | 1.3 | 1.5 | 2.5 | 1.5 | 2.0 | 3.0 |
| 0.50 | 1.0 | 1.5 | 2.5 | 1.5 | 2.0 | 2.8 | 2.0 | 2.5 | 3.3 |
| 0.60 | 1.3 | 2.0 | 3.0 | 2.0 | 2.8 | 3.3 | 2.5 | 3.0 | 3.8 |
| 0.70 | 1.5 | 2.8 | 3.3 | 2.0 | 3.0 | 3.5 | 2.8 | 3.3 | 4.0 |
| 0.80 | 1.8 | 2.8 | 3.5 | 2.3 | 3.3 | 4.0 | 2.8 | 3.5 | 4.3 |
| 0.90 | 2.0 | 3.0 | 3.8 | 2.5 | 3.5 | 4.3 | 3.0 | 3.8 | 4.5 |
| 1.00 | 2.0 | 3.0 | 4.3 | 2.5 | 3.8 | 4.5 | 3.0 | 4.0 | 4.8 |
| 1.10 | 2.3 | 3.3 | 4.5 | 2.8 | 3.8 | 4.5 | 3.3 | 4.3 | 4.8 |
| 1.20 | 2.3 | 3.3 | 4.5 | 2.8 | 4.0 | 4.8 | 3.3 | 4.3 | 5.0 |
| 1.30 | 2.5 | 3.3 | 4.5 | 3.0 | 4.0 | 4.8 | 3.5 | 4.5 | 5.0 |
| 1.40 | 2.5 | 3.5 | 4.8 | 3.0 | 4.3 | 5.0 | 3.5 | 4.8 | 5.0 |
| 1.50 | 2.5 | 3.5 | 4.8 | 3.0 | 4.3 | 5.0 | 3.8 | 4.8 | 5.0 |
| 1.60 | 2.8 | 3.8 | 4.8 | 3.3 | 4.5 | 5.0 | 3.8 | 4.8 | 5.0 |
| 1.70 | 2.8 | 3.8 | 5.0 | 3.3 | 4.5 | 5.0 | 3.8 | 5.0 | 5.0 |
| 1.80 | 2.8 | 4.0 | 5.0 | 3.5 | 4.8 | 5.0 | 4.0 | 5.0 | 5.0 |
| 1.90 | 3.0 | 4.0 | 5.0 | 3.5 | 4.8 | 5.0 | 4.0 | 5.0 | 5.0 |
| 2.00 | 3.0 | 4.5 | 5.0 | 3.8 | 5.0 | 5.0 | 4.0 | 5.0 | 5.0 |
| 2.10 | 3.0 | 4.5 | 5.0 | 3.8 | 5.0 | 5.0 | 4.0 | 5.0 | 5.0 |
| 2.20 | 3.3 | 4.8 | 5.0 | 3.8 | 5.0 | 5.0 | 4.3 | 5.0 | 5.0 |
| 2.30 | 3.3 | 4.8 | 5.0 | 4.0 | 5.0 | 5.0 | 4.3 | 5.0 | 5.0 |
| 2.40 | 3.3 | 5.0 | 5.0 | 4.0 | 5.0 | 5.0 | 4.3 | 5.0 | 5.0 |
| 2.50 | 3.5 | 5.0 | 5.0 | 4.3 | 5.0 | 5.0 | 4.5 | 5.0 | 5.0 |
| 2.60 | 3.5 | 5.0 | 5.0 | 4.3 | 5.0 | 5.0 | 4.5 | 5.0 | 5.0 |
| 2.70 | 3.8 | 5.0 | 5.0 | 4.3 | 5.0 | 5.0 | 4.5 | 5.0 | 5.0 |

6 exceed the *DRD*; that is, 31.6%, which, through Table 9.4 gives a deduct value of 3.0. Hence, the SAI for this section is 5.0 − 3.0 = 2.0, which suggests a serious structural deficiency (i.e., need for an overlay).

In circumstances where *RRD* < *DRD* (i.e., the section is more than structurally adequate), a credit value needs to be calculated and added to the value of 5. Table 9.4 is used for calculating these credit values. This is done by reversing the meaning of the traffic level columns (i.e., utilizing high-traffic volume for low, and vice versa, while leaving the mid-traffic level unchanged) and reversing the meaning of the frequency of individual observations exceeding *DRD* to indicate the frequency of individual observations below the *DRD*.

**BACK-CALCULATING FLEXIBLE PAVEMENT LAYER ELASTIC MODULI**

Another significant application of flexible pavement deflection data is in estimating the elastic moduli of the pavement layers. This process, called *back-calculation*, is possible through software that implement layer elastic theory solutions, as described in Chapter 7. Two general approaches are used for this purpose. The first uses an iterative algorithm, which assumes the layer moduli, computes surface deflections at the same locations where the FWD data were obtained, and compares them to the measured deflections. This process is repeated until predicted and measured layer moduli are within a prescribed tolerance. Several computer programs have been purpose-designed to carry out this task, incorporating different layered elastic software and search algorithms for adjusting the layer moduli in each iteration (e.g., MODCOMP, WESDEF, and ISSEM4). The second approach utilizes a database of deflection bowl predictions covering a range of elastic layer moduli to obtain a closed-form solution for the subgrade modulus by minimizing the errors between measured deflections $\omega_i^m$ and predicted deflections $\omega_i^p$, where $i$ identifies the sensor (ranging from 1 to $s$, where $s$ is either 7 or 9, depending on the FWD system). This approach is implemented into the software package MODULUS 4[34] and is described in detail next.

The approach seeks to minimize the following error, $e$:

$$e = \sum_{i=1}^{s} \left( \frac{\omega_i^m - \omega_i^p}{\omega_i^m} \right)^2 w_i = \sum_{i=1}^{s} \left( 1 - \frac{\omega_i^p}{\omega_i^m} \right)^2 w_i \tag{9.30}$$

where $w_i$ are sensor-specific weights, which for the remaining discussion will be assumed equal to 1.0. To minimize the error $\varepsilon$, Equation 9.30 is differentiated with respect to the subgrade modulus $E_{SG}$ and set equal to 0.

$$\frac{\partial \varepsilon}{\partial E_{SG}} = \sum_{i=1}^{s} 2\left(1 - \frac{\omega_i^p}{\omega_i^m}\right)\left(-\frac{1}{\omega_i^m}\frac{\partial \omega_i^p}{\partial E_{SG}}\right) = 0 \tag{9.31}$$

The predicted deflections for sensor $i$ are functions of the layer moduli $E_k$, the layer Poison ratios $\mu_k$, the layer thicknesses $h_k$, the radius $a$ of the load applied, and the contact pressure $p$, expressed as:

$$\omega_i^p = f_i(E_k, \mu_k, h_k, a, p) \tag{9.32}$$

where $k$ refers to the number of the pavement layer, ranging from 1 for the asphalt concrete layer to $N + 1$ for the subgrade. Equation 9.32 can be written as follows, by dividing the layer moduli by the modulus of the subgrade $E_{SG}$ and taking into account that the $h_k$ are known at a location, and $\mu_k$ can be safely assumed, while $a$ and $p$ are constants for a set of measurements.

$$\omega_i^p = \frac{p\,a}{E_{SG}} f_i\left(\frac{E_1}{E_{SG}}, \frac{E_2}{E_{SG}}, \ldots\ldots \frac{E_N}{E_{SG}}\right) = \frac{p\,a}{E_{SG}} f_i \tag{9.33}$$

Substituting Equation 9.33 into Equation 9.31 gives:

$$\sum_{i=1}^{s}\left(1 - \frac{p\,a}{E_{SG}}\frac{f_i}{\omega_i^m}\right)\left(\frac{1}{\omega_i^m}\frac{p\,a}{E_{SG}^2} f_i\right) = 0 \tag{9.34}$$

and by rearranging:

$$\frac{(p\,a)^2}{E_{SG}^3}\sum_{i=1}^{s}\left(\frac{E_{SG}}{p\,a} - \frac{f_i}{\omega_i^m}\right)\frac{f_i}{\omega_i^m} = 0 \tag{9.35}$$

which by factoring out $f_1^2$ ($f_1$ is the deflection function for the sensor under the center of the loading plate) gives:

$$f_1^2\sum_{i=1}^{s}\left(\frac{E_{SG}}{p\,a\,f_1} - \frac{f_i}{\omega_i^m\,f_1}\right)\frac{f_i}{f_1\,\omega_i^m} = 0 \tag{9.36}$$

which can be rearranged as:

$$\sum_{i=1}^{s}\left(\frac{E_{SG}\,f_i}{p\,a\,f_1^2\,\omega_i{}^m}\right) - \sum_{i=1}^{s}\left(\frac{f_i}{\omega_i{}^m\,f_1}\right)^2 = 0 \tag{9.37}$$

which allows solving for $E_{SG}$

$$E_{SG} = \frac{\sum_{i=1}^{s} \left( \frac{f_i}{\omega_i{}^m f_1} \right)^2}{\sum_{i=1}^{s} \left( \frac{f_i}{p\, a\, f_1{}^2\, \omega_i{}^m} \right)} = p\, a\, f_1 \frac{\sum_{i=1}^{s} \left( \frac{f_i}{\omega_i{}^m f_1} \right)^2}{\sum_{i=1}^{s} \left( \frac{f_i}{\omega_i{}^m f_1} \right)} \tag{9.38}$$

A numerical example of this procedure follows.

**Example 9.9**

The radial distances and deflection measurements given in Table 9.7 were obtained using an FWD with a plate radius of 10.3 cm and a contact pressure of 600 kPa. The layer thicknesses and the assumed values for the Poison ratio are given Table 9.8. Compute the layer moduli.

**ANSWER**

In developing the database of deflection functions $f_i$, a set of layer moduli is analyzed, and surface deflections are calculated at the identified radial distances using layer-elastic analysis. The values in Table 9.9 were computed using the software package EVERSTRESS, described in Chapter 7. In doing so, the subgrade modulus was selected arbitrarily equal to 140 MPa, and the moduli of the asphalt concrete and base layers were selected as multiples of the subgrade modulus (e.g., 5, 10, 20 for the asphalt concrete layer and 1, 2, 4 for the base layer). The combinations of layer thicknesses input and the resulting surface deflection predictions are shown in Table 9.9.

**Table 9.7**
Deflection Measurements for Example 9.9

| Sensor, s | 1 | 2 | 3 | 4 | 5 | 6 | 7 |
|---|---|---|---|---|---|---|---|
| Offset (cm) | 0 | 10 | 20 | 30 | 40 | 50 | 100 |
| Deflection ($\mu$m) | 189 | 159 | 114 | 91 | 74 | 62 | 34 |

**Table 9.8**
Layer Properties for Example 9.9

| Layer, $k$ | 1 | 2 | 3 |
|---|---|---|---|
| Thickness (cm) | 15 | 40 | $\infty$ |
| Poison's ratio, $\mu$ | 0.35 | 0.45 | 0.40 |

**Table 9.9**
Predicted Deflections for Example 9.9

| Modular Ratios | | Predicted Deflections $\omega_i^p$ ($\mu$m) | | | | | | |
|---|---|---|---|---|---|---|---|---|
| $E_1/E_{SG}$ | $E_2/E_{SG}$ | 1 | 2 | 3 | 4 | 5 | 6 | 7 |
| 5 | 1 | 353.2 | 285.9 | 184.4 | 134.6 | 101.5 | 79.6 | 37.2 |
| 5 | 2 | 280 | 220.9 | 137.2 | 104.8 | 84.8 | 71.3 | 39.2 |
| 5 | 4 | 227 | 174.7 | 105.1 | 85.1 | 73.4 | 65 | 40 |
| 10 | 1 | 269.9 | 231.5 | 169.4 | 131.4 | 103 | 82.2 | 37.6 |
| 10 | 2 | 216.1 | 182.1 | 130.2 | 103.6 | 85 | 71.6 | 39.1 |
| 10 | 4 | 173.2 | 143.4 | 100.8 | 83.5 | 72.1 | 63.8 | 39.6 |
| 20 | 1 | 209.6 | 187.9 | 150.6 | 123.7 | 101.4 | 83.5 | 38.6 |
| 20 | 2 | 172.1 | 157.6 | 120.6 | 100.3 | 84.3 | 71.9 | 39.2 |
| 20 | 4 | 139.2 | 122 | 95.5 | 81.3 | 70.8 | 62.7 | 39.1 |

The deflection functions $f_i$ are computed from Equation 9.33 as:

$$f_i = \omega_i^p \frac{E_{SG}}{p\ a} = \omega_i^p \frac{140000}{600\ 0.103} = \omega_i^p\ 2.2654$$

and tabulated in Table 9.10.

The values of the ratio $f_i/\omega_i^m f_1$ are tabulated in Table 9.11. They allow calculation of the subgrade moduli values (Equation 9.38), shown in Table 9.12. These subgrade values allow adjusting the predicted deflections $\omega_i^p$ by multiplying the $f_i$ functions by $\frac{p\,a}{E_{SG}}$. Finally, errors are computed between these adjusted deflections and the measured deflections, as shown in Table 9.12. The minimum value of the sum of the squares of these errors gives a back-calculated subgrade modulus as 160.46 MPa, and corresponding moduli ratios 2 and 10 for the base and asphalt concrete layers, respectively.

It should be noted that in implementing this procedure, sufficient layer moduli combinations need to be analyzed to ensure that the minimum error used to identify the back-calculated subgrade modulus does not represent a local minimum. Furthermore, care should be taken in identifying the presence of a subgrade of finite depth, that is, the presence of a stiff bedrock layer within the depth of influence of the applied loads. Failure to do so typically results in overestimation of the moduli of the pavement layers. For suggestions on the most recent software, see the Web site for this book, www.wiley.com/go/pavement.

**Table 9.10**
Computing Deflection Functions $f_i$ for Example 9.9

| Modular Ratios | | Deflection Functions, $f_i$ | | | | | | |
|---|---|---|---|---|---|---|---|---|
| $E_1/E_{SG}$ | $E_2/E_{SG}$ | 1 | 2 | 3 | 4 | 5 | 6 | 7 |
| 5 | 1 | 800.1294 | 647.6699 | 417.7346 | 304.9191 | 229.9353 | 180.3236 | 84.2718 |
| 5 | 2 | 634.3042 | 500.4207 | 310.8091 | 237.4110 | 192.1036 | 161.5210 | 88.8026 |
| 5 | 4 | 514.2395 | 395.7605 | 238.0906 | 192.7832 | 166.2783 | 147.2492 | 90.6149 |
| 10 | 1 | 611.4239 | 524.4337 | 383.7540 | 297.6699 | 233.3333 | 186.2136 | 85.1780 |
| 10 | 2 | 489.5469 | 412.5243 | 294.9515 | 234.6926 | 192.5566 | 162.2006 | 88.5761 |
| 10 | 4 | 392.3625 | 324.8544 | 228.3495 | 189.1586 | 163.3333 | 144.5307 | 89.7087 |
| 20 | 1 | 474.8220 | 425.6634 | 341.1650 | 280.2265 | 229.7087 | 189.1586 | 87.4434 |
| 20 | 2 | 389.8706 | 357.0227 | 273.2039 | 227.2168 | 190.9709 | 162.8803 | 88.8026 |
| 20 | 4 | 315.3398 | 276.3754 | 216.3430 | 184.1748 | 160.3883 | 142.0388 | 88.5761 |

**Table 9.11**
Computing $\frac{f_i}{\omega_i^m f_1}$ for Example 9.9

| Modular Ratios | | Values of $\frac{f_i}{\omega_i^m f_i}$ | | | | | | |
|---|---|---|---|---|---|---|---|---|
| $E_1/E_{SG}$ | $E_2/E_{SG}$ | 1 | 2 | 3 | 4 | 5 | 6 | 7 |
| 5 | 1 | 0.0053 | 0.0051 | 0.0046 | 0.0042 | 0.0039 | 0.0036 | 0.0031 |
| 5 | 2 | 0.0053 | 0.0050 | 0.0043 | 0.0041 | 0.0041 | 0.0041 | 0.0041 |
| 5 | 4 | 0.0053 | 0.0048 | 0.0041 | 0.0041 | 0.0044 | 0.0046 | 0.0052 |
| 10 | 1 | 0.0053 | 0.0054 | 0.0055 | 0.0053 | 0.0052 | 0.0049 | 0.0041 |
| 10 | 2 | 0.0053 | 0.0053 | 0.0053 | 0.0053 | 0.0053 | 0.0053 | 0.0053 |
| 10 | 4 | 0.0053 | 0.0052 | 0.0051 | 0.0053 | 0.0056 | 0.0059 | 0.0067 |
| 20 | 1 | 0.0053 | 0.0056 | 0.0063 | 0.0065 | 0.0065 | 0.0064 | 0.0054 |
| 20 | 2 | 0.0053 | 0.0058 | 0.0061 | 0.0064 | 0.0066 | 0.0067 | 0.0067 |
| 20 | 4 | 0.0053 | 0.0055 | 0.0060 | 0.0064 | 0.0069 | 0.0073 | 0.0083 |

**Table 9.12**
Computing $E_{SG}$, Adjusted Deflections and Sum of Squared Errors for Example 9.9

| Modular Ratios | | | | | | | | | | | |
|---|---|---|---|---|---|---|---|---|---|---|---|
| $E_1/E_{SG}$ | $E_2/E_{SG}$ | $E_{SG}$ (MPa) | Adjusted Deflections $\omega_i^p$ ($\mu$m) for Computed $E_{SG}$ | | | | | | | $\Sigma\ e^2$ |
| 5 | 1 | 216.48 | 228.42 | 184.89 | 119.25 | 87.05 | 65.64 | 51.48 | 24.06 | 0.20 |
| 5 | 2 | 175.34 | 223.56 | 176.38 | 109.55 | 83.68 | 67.71 | 56.93 | 31.30 | 0.07 |
| 5 | 4 | 148.88 | 213.46 | 164.28 | 98.83 | 80.02 | 69.02 | 61.12 | 37.61 | 0.07 |
| 10 | 1 | 194.23 | 194.55 | 166.87 | 122.11 | 94.71 | 74.24 | 59.25 | 27.10 | 0.05 |
| 10 | 2 | 160.46 | 188.55 | 158.88 | 113.60 | 90.39 | 74.16 | 62.47 | 34.11 | 0.00 |
| 10 | 4 | 136.98 | 177.02 | 146.56 | 103.02 | 85.34 | 73.69 | 65.21 | 40.47 | 0.06 |
| 20 | 1 | 177.70 | 165.13 | 148.03 | 118.65 | 97.45 | 79.89 | 65.78 | 30.41 | 0.05 |
| 20 | 2 | 151.25 | 159.30 | 145.88 | 111.63 | 92.84 | 78.03 | 66.55 | 36.28 | 0.05 |
| 20 | 4 | 129.84 | 150.10 | 131.55 | 102.98 | 87.66 | 76.34 | 67.61 | 42.16 | 0.15 |

### 9.3.5 Back-Calculating Rigid Pavement Elastic Moduli

Back-calculation of the elastic moduli of rigid pavements is carried out using deflection measurements obtained on the surface of slabs through a FWD. Several techniques are utilized for this purpose, and software for implementing them (ELCON and ILLI-BACK). ILLI-BACK utilizes a closed-form approach for computing the slab and subgrade layer moduli, requiring no iterative search routines.[17,18] The basis of this technique is theoretical relationships between a function of the deflections measured and the radius of the relative stiffness of the slab $\ell$. The latter is defined next for liquid and rigid foundations, respectively:

$$\ell = \left( \frac{E\,h^3}{12\,(1-\mu^2)\,k} \right)^{1/4} \tag{9.39}$$

$$\ell = \left( \frac{E\,h^3\,(1-\mu_s^2)}{6\,(1-\mu^2)\,E_s} \right)^{1/3} \tag{9.40}$$

where $E$ and $\mu$ are the elastic modulus and the Poison's ratio of the portland concrete, $E_s$ and $\mu_s$ are the elastic modulus and the Poison's ratio of the subgrade, and $h$ is the thickness of the slab. The function used is the area of the deflection bowl divided by the peak deflection, denoted by *AREA*. For the four deflection measurements typically available at distances of 0, 12, 24, and 36 inches (0.3, 0.6, and 0.9 m) from the point of load application, *AREA* is given by:

$$AREA = 6\left[1 + 2\left(\frac{\varpi_1}{\varpi_0}\right) + 2\left(\frac{\varpi_2}{\varpi_0}\right) + \left(\frac{\varpi_3}{\varpi_0}\right)\right] \tag{9.41}$$

where $\omega_0$, $\omega_1$, $\omega_2$, and $\omega_3$ denote these deflection measurements in inches, respectively. Note that *AREA*, as given by Equation 9.41, is in inches as well. The relationship between *AREA* and the radius of relative stiffness $\ell$ is given in Figure 9.30.

Once the radius of relative stiffness $\ell$ has been obtained, Figures 9.31 and 9.32 are used for computing the normalized deflection $d_i$ for liquid and solid foundations, respectively. The normalized deflection is defined next for liquid and solid foundations, respectively.

$$d_i = \frac{\omega_i k \ell^2}{P} \tag{9.42a}$$

$$d_i = \frac{\omega_i E_s \ell}{2\,(1-\mu_s^2)\,P} \tag{9.42b}$$

Equation 9.42a allows calculation of the modulus of subgrade reaction $k$ of the liquid foundation, since $d_i$, $\ell$, and the load $P$ applied by the FWD are known. Similarly, Equation 9.42b allows calculation of the subgrade modulus $E_s$, assuming a value for the subgrade Poison's ratio $\mu_s$. These calculations are repeated for each deflection

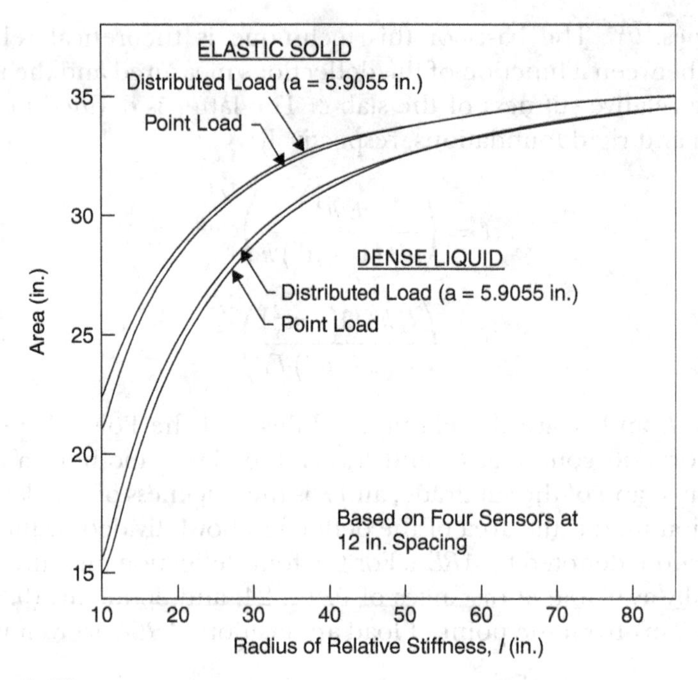


**Figure 9.30**
Relationship between *AREA* and Radius of Relative Stiffness (Ref. 18 Reproduced under License)

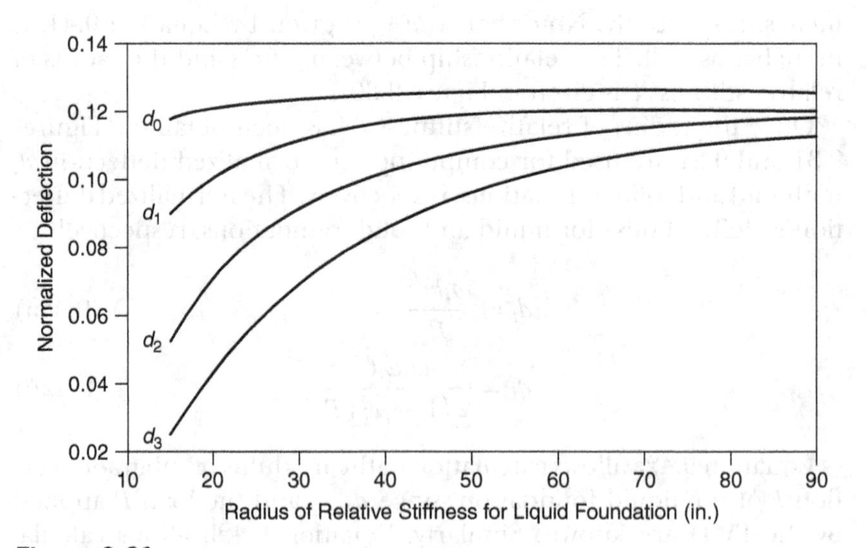


**Figure 9.31**
Relationship between Radius of Relative Stiffness and Normalized Deflection; Liquid Foundation (Ref. 17 Reproduced by Author's Permission)

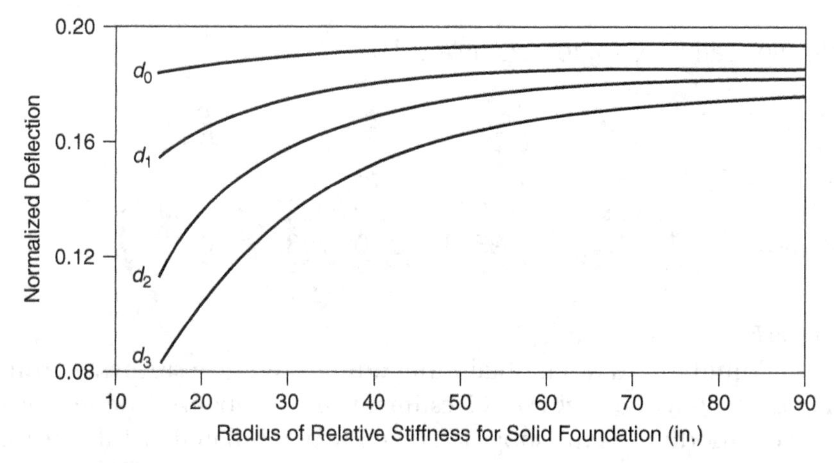


**Figure 9.32**
Relationship between Radius of Relative Stiffness and Normalized Deflection; Solid Foundation (Ref. 17 Reproduced by Author's Permission)

value available, and the average of the resulting $k$ or $E_s$ value is reported. Finally, for the solid foundation, Equation 9.40 allows calculation of the elastic modulus of the portland concrete, assuming a value for the portland concrete Poisson's ratio (typically 0.15).

**Example 9.10**

The deflection measurements given in Table 9.13 were obtained with an FWD with a loading plate of 0.15 m applying a load of 40 kN on a portland concrete layer 0.25 m thick. Back-calculate the elastic properties assuming a liquid foundation and then a solid foundation, assuming that $\mu_s$ is 0.45 and $\mu$ is 0.15.

**Table 9.13**
Pavement Deflections for Example 9.10

| Sensor, s | 0 | 1 | 2 | 3 |
|---|---|---|---|---|
| Offset (m) | 0 | 0.3 | 0.6 | 0.9 |
| Deflection ($\mu$m) | 75 | 70 | 58 | 48 |

**ANSWER**

Compute the deflections ratios $\omega_1/\omega_0$, as shown in Table 9.14. Compute *AREA* using Equation 9.41.

$$AREA = 6(1 + 2 \times 0.9333 + 2 \times 0.7733 + 0.6400) = 30.32\,\text{in.}$$

**Table 9.14**
Computing Deflection ratios $\omega_i/\omega_0$ for Example 9.10

| Sensor number | 0 | 1 | 2 | 3 |
|---|---|---|---|---|
| Offset (m) | 0 | 0.3 | 0.6 | 0.9 |
| Deflection ($\mu$m) | 75 | 70 | 58 | 48 |
| Deflection (in) | 0.00296 | 0.00276 | 0.00228 | 0.00189 |
| Deflection ratios $\omega_i/\omega_0$ | 1 | 0.93333 | 0.77333 | 0.64 |

Liquid Foundation

For a liquid foundation, obtain the radius of relative stiffness $\ell$ from Figure 9.30 as 38 inches and estimate the normalized deflections and the modulus of the subgrade reaction as tabulated in Table 9.15.

Finally, compute the elastic modulus of the Portland concrete from Equation 9.39 as:

$$E = \frac{\ell^4 12\left(1-\mu^2\right)k}{h^3} = \frac{38^4 12\left(1-0.15^2\right)265}{9.84^3}$$
$$= 6,801,400 \text{ lbs/in.}^2 (46,895 \text{ MPa})$$

Solid Foundation

For a solid foundation, obtain the radius of relative stiffness $\ell$ from Figure 9.30 as 28 inches and estimate the normalized deflections and the modulus of the subgrade reaction as tabulated in Table 9.16.

**Table 9.15**
Computing Modulus of Subgrade Reaction for Example 9.10

| Sensor number | 0 | 1 | 2 | 3 | |
|---|---|---|---|---|---|
| Estimate $d_i$ from Figure 9.31 | 0.122 | 0.113 | 0.1 | 0.083 | Average |
| Predict $k$ (pci) from Equation 9.42a | 257.52 | 255.56 | 272.95 | 273.74 | 265 (73.7 MN/m$^3$) |

**Table 9.16**
Computing Subgrade Elastic Modulus for Example 9.10

| Sensor number | 0 | 1 | 2 | 3 | |
|---|---|---|---|---|---|
| Estimate $d_i$ from Figure 9.32 | 0.182 | 0.163 | 0.14 | 0.125 | Average |
| Predict $E_s$ (lbs/in.$^2$) Equation 9.42b | 31,600 | 32,555 | 33,228 | 34,725 | 33,027 (228 MPa) |

Finally, compute the elastic modulus of the portland concrete from Equation 9.40 as:

$$E = \frac{\ell^3}{h^3}\frac{6\left(1-\mu^2\right)E_s}{\left(1-\mu_s^2\right)} = \frac{28^3}{9.84^3}\frac{6\left(1-0.15^2\right)33,027}{\left(1-0.45^2\right)}$$
$$= 5,596,281 \text{lbs/in.}^2 (38,586 \text{ MPa})$$

The discrepancy between the portland concrete elastic moduli computed for liquid and solid foundations is due to the different assumptions involved in these two foundation models. For suggestions on the most recent software, see the Web site for this book, www.wiley.com/go/pavement.

## 9.4 Surface Distress

### 9.4.1 Definitions—Introduction

This component of pavement evaluation involves collection of data related to the condition of the pavement surface, defined by the variety of pavement distresses present. Distresses are defined as the manifestations of construction defects, as well as the damaging effects of the traffic, the environment, and their interaction. They encompass a broad variety of cracks and surface distortions.

Data is typically collected manually through condition surveys performed by a technical crew riding on a slow-moving vehicle, hence the term "windshield surveys." For each distress type present, the severity and extent are recorded on special forms. These relate to the seriousness of the distress and the area of the pavement surface being affected by it, respectively. Often, spot maintenance issues, such as patches, are also identified and recorded. Effort have been made to automate the distress data collection process. The PASCO Corporation[1] manufactures a commercially available system that is equipped with vertical-aiming video cameras, as well as other sensors that collect additional data elements, ranging from roughness to rutting. Such systems have however, been used for collecting distress records for research purposes, rather than routine network surveys. Their main limitation is that they require manual image postprocessing to identify the distresses present. Image processing software capable of automating the latter have not become commercially viable yet.

Analyzing distress data is forensic in nature, offering clues to the cause of the problems present, as well as guidance in selecting applicable future pavement 4-R treatments. As such, distress evaluation is of primary interest to the engineer/manager of the roadway system. In fact, some State High Agencies use distress as the primary indicator of rehabilitation need, rather than delaying action until the ensuing roughness becomes significant.

The most widely accepted methodology for identifying distresses and judging their extent and severity is the distress identification manual developed for the Strategic Highway Research Program.[10] The following provides an overview of the distresses encountered in flexible and rigid pavements, as well as their likely causes.

### 9.4.2 Flexible Pavement Distresses

The variety of distresses encountered in asphalt concrete pavements is schematically shown in Figure 9.33. They are grouped into three main categories: cracking, surface deformation, and surface defects. A brief description of each distress and its likely causes follows. Additional information on distress identification, their causes, and remedies can be found in references 10 and 30.

#### CRACKING

Cracking appears in various forms that allow identification of its causes. Some are fatigue related, caused by the accumulation of fatigue damage from successive vehicle axles, and they appear in the wheel-paths having an interconnected polygonal pattern resembling alligator skin, or are located longitudinally along the wheel-path. They are believed to originate at the bottom and the top of the asphalt concrete layer, respectively. Some longitudinal cracks occur away from the wheel-paths; for example, along the joint between two lanes, caused by poor joint compaction during construction, or in the middle of a lane, caused by thermal stresses or subgrade settlement. Similarly, edge cracking is traffic-related; it is caused by weaker subgrade due to the lack of confinement near the edge of a driving lane next to a nonpaved shoulder. Transverse cracking is cold-temperature-related, and it occurs at regular intervals as a result of thermal stress exceeding the tensile strength of the asphalt concrete. Block cracking, resembles alligator cracking, except that the interconnected patterns are larger, and rectangular covering areas are larger than the wheel-paths. This type of cracking is cold-temperature-related, and is caused by binder hardening on

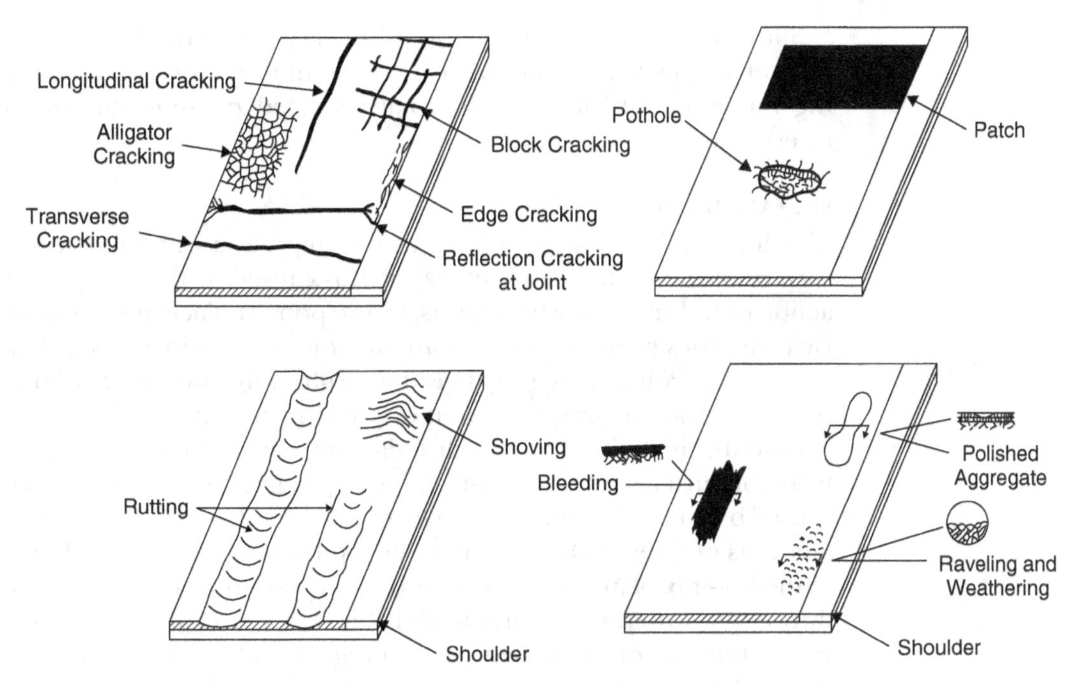


**Figure 9.33**
Schematic of Flexible Pavement Distresses. Adapted from SHRP 1993, SHRP-P-338

low-volume roads. Some of the crack types just described reappear through to the surface of an asphalt concrete overlay and, therefore, are called *reflection cracks*. Crack severity is quantified as:

- Low: mean crack opening smaller than 6 mm
- Moderate: mean crack opening between 6 and 19 mm
- High: mean crack opening larger than 19 mm

**SURFACE DEFORMATION**

Rutting is defined as longitudinal depressions in the wheel-paths caused by the compaction or plastic deformation of the asphalt concrete and the granular layers/subgrade under the action of axle loads. It is not to be confused with depressions formed in the wheel-paths by the abrasive action of passenger car studded tires. These occur predominantly in car lanes and can be distinguished by the narrower wheel base of cars that cause them. Rutting is measured using a 1.2 m straightedge placed over each wheel-path.[10]

Shoving is defined as the longitudinal displacement of a localized area of the pavement surface caused by the braking or accelerating forces of vehicles, and is usually located on hills, curves, or intersections.

### SURFACE DEFECTS

Bleeding, or flushing, is defined as the migration of binder to the surface of the asphalt concrete layer. It is caused by the compactive action of traffic in the wheel-paths, where poor in-place mix volumetric properties result in substandard air voids (i.e., values lower than 3% to 4%). Polished aggregate is the result of the abrasive action of tires on surface aggregates, often occurring near intersections.

Raveling is defined as the dislodgement and loss of aggregates from the surface of the asphalt concrete, progressing downward. It is caused by poor adhesion between aggregates and binder due to large amounts of fines in the aggregate stockpiles, poor aggregate drying in the hot-mix plant, or desegregation and poor in-place compaction during construction. A variation of this distress is called stripping, where the loss of bond between aggregate and binder is initiated at the bottom of the asphalt concrete, and progresses upward. Its cause is the chemical incompatibility of some aggregate-binder combinations and inadequate drainage.

A summary of these distress types is presented in Table 9.17, along with the units used for quantifying the extent and the levels used in judging severity. Some of these distresses are controlled through structural design (Chapter 11), while others are prevented through material selection (Chapters 4, 5, and 6) and drainage provisions (Chapter 10).

### RIGID PAVEMENT DISTRESSES

The variety of distresses encountered in portland concrete pavements is schematically shown in Figure 9.34. They are grouped into four main categories: cracking, surface defects, joint deficiencies, and miscellaneous. Obviously, joint deficiencies apply to jointed portland concrete pavements only. A brief description of each distress and the likely causes follows. Additional information can be found in references 9 and 10.

### CRACKING

Cracking appears in various forms that allow identification of its causes. Some are fatigue related, caused by the accumulation of

**Table 9.17**
Distresses in Asphalt Concrete Pavements (Ref. 10)

| Distress Type | Measurement of Extent | Severity Levels |
|---|---|---|
| *Cracking* | | |
| Fatigue (Alligator) | $m^2$ | Low, Medium, High |
| Block | $m^2$ | Low, Medium, High |
| Edge | linear m | Low, Medium, High |
| Wheel-path, longitudinal | linear m | Low, Medium, High |
| Nonwheel-path, longitudinal | linear m | Low, Medium, High |
| Transverse | linear m, number | Low, Medium, High |
| *Reflection Cracking at Joints* | | |
| Transverse | linear m, number | Low, Medium, High |
| Longitudinal | linear m | Low, Medium, High |
| *Patching and Potholes* | | |
| Patch/patch deterioration | $m^2$, number | Low, Medium, High |
| Potholes | $m^2$, number | Low, Medium, High |
| *Surface Deformation* | | |
| Rutting | mm | — |
| Shoving | $m^2$, number | — |
| *Surface Defects* | | |
| Bleeding/Flushing | $m^2$ | Low, Medium, High |
| Polished aggregates | $m^2$ | — |
| Raveling/weathering | $m^2$ | Low, Medium, High |
| *Miscellaneous* | | |
| Lane-shoulder drop-off | mm | — |
| Water bleeding and pumping | linear m, number | Low, Medium, High |

— = Either not applicable or not specified.

fatigue damage from successive vehicle axles, such as corner cracks. Other cracks, either longitudinal or transverse, can be caused by traffic, the environment, or poor construction (e.g., seized joints can cause transverse cracks at midslab). The combination of slab warping under thermal gradients and load may result in transverse cracks. Longitudinal and transverse crack severity is quantified as:

- Low: mean crack opening smaller than 3 mm
- Moderate: mean crack opening between 6 and 6 mm
- High: mean crack opening larger than 6 mm

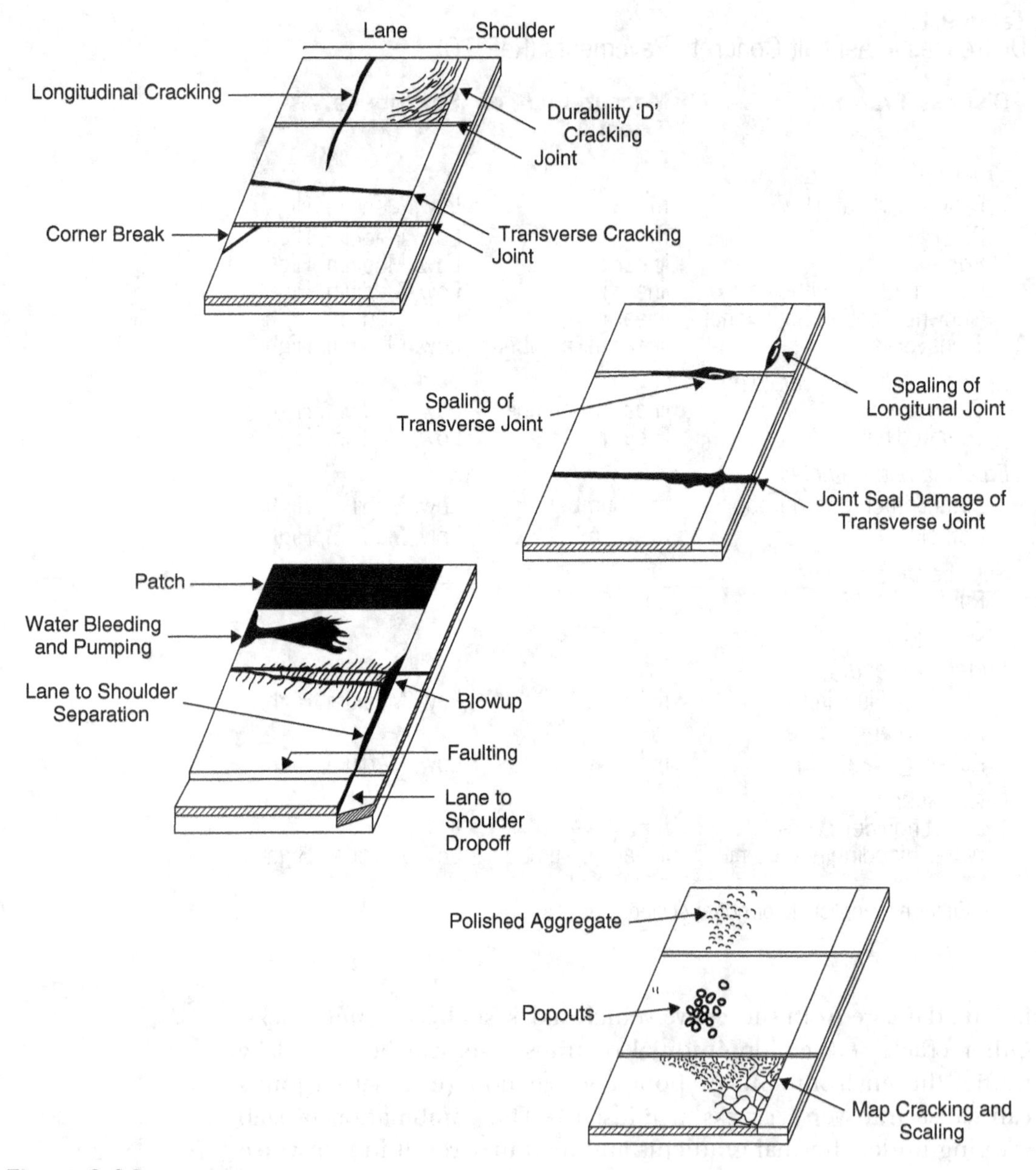


**Figure 9.34**
Schematic of Rigid Pavement Distresses. Adapted from SHRP 1993, SHRP-P-338

Another form of cracking is the so-called durability, or "D-cracking," which consists of closely spaced cracks parallel to transverse or longitudinal joints curving in a crescent shape around slab corners. These are caused by the effect of freeze/thaw cycles on porous aggregates of poor durability under high-moisture conditions.

**SURFACE DEFECTS**

Map cracking consists of interconnected cracks that extend only into the upper surface of the slab. It may be caused by poor construction (e.g., dirty aggregates resulting in fines that migrate to the surface during concrete finishing) or the action of deicing salts. Spalling is the result of dislodgement of surface blocks created by map cracking. A variation of this distress is popouts, involving the dislodgement of small pieces (25 to 100 mm) of the surface without evident map cracking. Polished aggregates are the result of the abrasive action of traffic on nondurable aggregates.

**JOINT DEFICIENCIES**

Seals of transverse/longitudinal joints can be damaged from a variety of causes, (e.g., splitting or debonding due to age hardening) and result in moisture and foreign object accumulation into the joint. Spalling is the breaking, chipping, or fraying of slab edges within 0.6 m of transverse/longitudinal joints, and it is caused by either lack of lateral support along a joint edge or by joints that do not allow slab expansion due to the presence of foreign objects (e.g., deicing sand).

**MISCELLANEOUS**

Faulting is a serious form of distress commonly found in older plain-jointed portland concrete pavements without dowel bar reinforcement. It is manifested as a slight settlement of the leading edge of each slab in the direction of travel. It is caused by the lack of vertical load transfer between slabs across joints, which creates sudden increases in pore pressure in wet subgrades, which in turn produces migration of fines and settlement under the leading edge of each slab. Where the sudden pore pressure buildup is accompanied by squirting of water and fines through the joint, the distress is refereed to as pumping. Blowups are localized heaving of the pavement surface at transverse joint or cracks, often accompanied by the shattering of the concrete in this area. They are caused by joints that do not allow slab expansion and, as a result, cannot dissipate compressive stresses.

A summary of these distresses is presented in Tables 9.18 and 9.19, for jointed and continuously reinforced portland concrete pavements, respectively. These tables also list the units used for quantifying distress extent and the levels used in judging their severity. Some of these distresses are controlled through structural design and joint/shoulder design (Chapter 8), while others are

**Table 9.18**
Distresses in Jointed Portland Concrete Pavements (Ref. 10)

| Distress Type | Unit of Measurement | Severity Levels |
|---|---|---|
| *Cracking* | | |
| Corner breaks | number | Low, Medium, High |
| Durability (D-cracking) | $m^2$, number linear | Low, Medium, High |
| Longitudinal | m | Low, Medium, High |
| Transverse | linear m, number | Low, Medium, High |
| *Joint Deficiencies* | | |
| Joint seal damage of transverse joints | number | Low, Medium, High |
| Joint seal damage of longitudinal joints | linear m, number | Low, Medium, High |
| Spalling of longitudinal joints | linear m | Low, Medium, High |
| Spalling of transverse joints | linear m, number | Low, Medium, High |
| *Surface Defects* | | |
| Map cracking | $m^2$, number | — |
| Scaling | $m^2$, number | — |
| Polished aggregate | $m^2$ | — |
| Popouts | $m^2$/number | — |
| *Miscellaneous* | | |
| Blowups | number | — |
| Faulting of transverse joints/cracks | mm | — |
| Lane-shoulder drop-off | mm | — |
| Lane-shoulder separation | mm | — |
| Patch/patch deterioration | $m^2$, number | Low, Medium, High |
| Water bleeding and pumping | linear m, number | — |

— = Either not applicable or not specified.

**Table 9.19**
Distresses in Continuously Reinforced Portland Concrete Pavements (Ref. 10)

| Distress Type | Unit of Measurement | Severity Levels |
|---|---|---|
| *Cracking* | | |
| Durability D-cracking | $m^2$, number linear | Low, Medium, High |
| Longitudinal | m | Low, Medium, High |
| Transverse | linear m, number | Low, Medium, High |
| *Surface Defects* | | |
| Map cracking and scaling | $m^2$ | Low, Medium, High |
| Scaling | $m^2$, number | — |
| Polished aggregate | $m^2$ | — |
| Popouts | $m^2$, number | — |
| *Miscellaneous* | | |
| Blowups | number | — |
| Transverse construction joint deterioration | number | Low, Medium, High |
| Lane-shoulder drop-off | mm | — |
| Lane-shoulder separation | mm | — |
| Patch/patch deterioration | $m^2$, number | Low, Medium, High |
| Punchouts | number | Low, Medium, High |
| Spalling of longitudinal joints | linear m | Low, Medium, High |
| Water bleeding and pumping | linear m, number | — |
| Longitudinal joint seal damage | linear m, number | — |

— = Either not applicable or not specified.

prevented through material selection (Chapter 6) and drainage provisions (Chapter 10).

### 9.4.3 Summarizing Pavement Distresses into an Index

Although the methodology for collecting distress data is well established, there is no widely accepted practice for summarizing them into an index that reflects pavement condition; each highway agency seems to be doing this differently. One method that has gained some acceptance is the Pavement Condition Index (*PCI*)[28] developed by the U. S. Army Corp of Engineers and subsequently approved as an ASTM Standard.[40] The calculation of the *PCI*, ranging from 100% to 0%, is explained next; it is carried out in a similar fashion for flexible and rigid pavements.

A pavement section is divided into a number of uniform sample units (i.e., an area of $225 \pm 90$ contiguous $m^2$ or $20 \pm 8$ contiguous slabs for flexible and rigid pavements, respectively). The following calculations are conducted for each sample unit. For each distress and severity level present, the area/length affected is added, then divided by the area of the sample unit, which expressed in percent is referred to as *distress density*.

Subsequently, deduct values are computed for each distress density, using a series of charts (Figures 9.35, 9.36, 9.37). These deduct values need to be processed to compute the maximum corrected deduct value (*max CDV*). The correction is necessary to ensure that the sum of the deduct values does not exceed 100%. If fewer than one of the deduct values is larger than 2%, the *max CDV* is equal to the sum of the individual deduct values. Otherwise, the *max CDV* is computed trough an iterative process, as follows. The deduct values are arranged in decreasing order. The maximum number of allowed deduct values *m*, which cannot exceed 10, is given as a function of the highest deduct value (*HDV*) (i.e., the first in the decreasing order list).

$$m = 1 + \frac{9}{98}(100 - HDV) \leq 10 \tag{9.43}$$

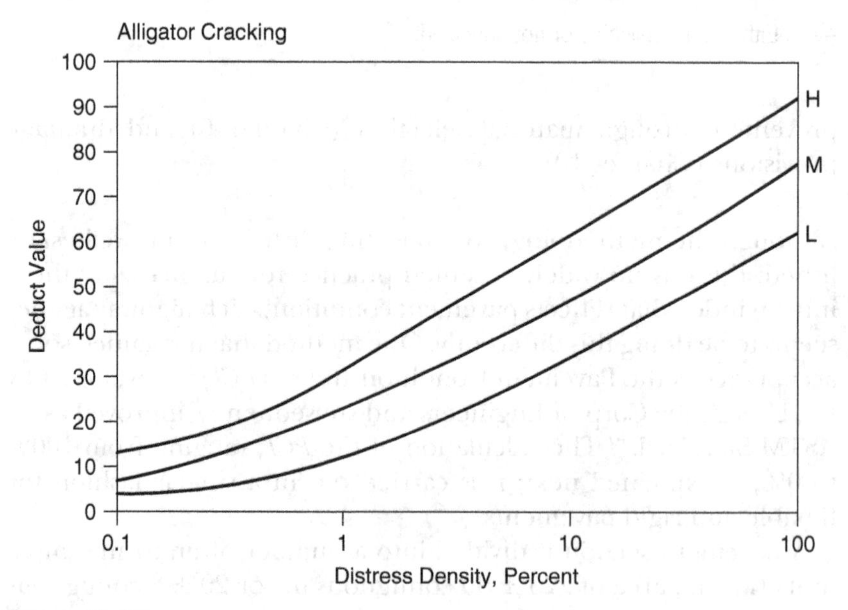


**Figure 9.35**
Example of Flexible Pavement Distress Deduct Values; Fatigue Cracking (Ref. 40)

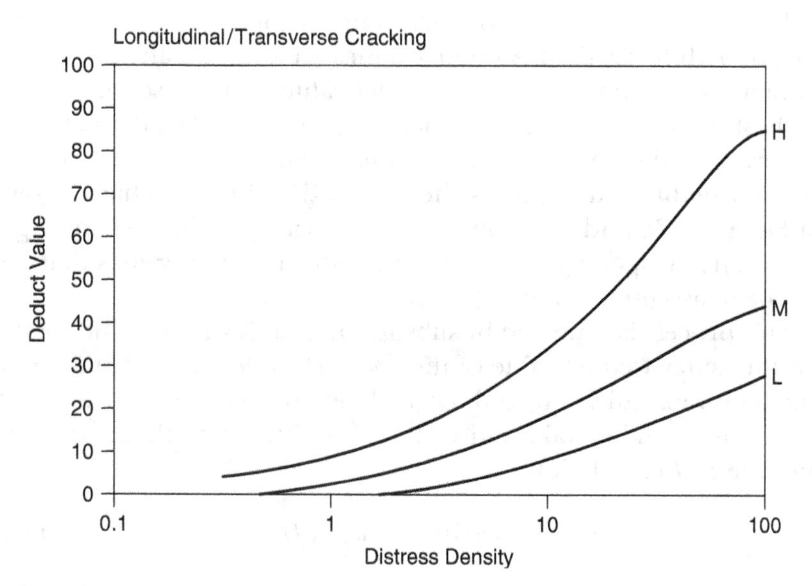


**Figure 9.36**
Example of Flexible Pavement Distress Deduct Values; Longitudinal Cracking (Ref. 40)

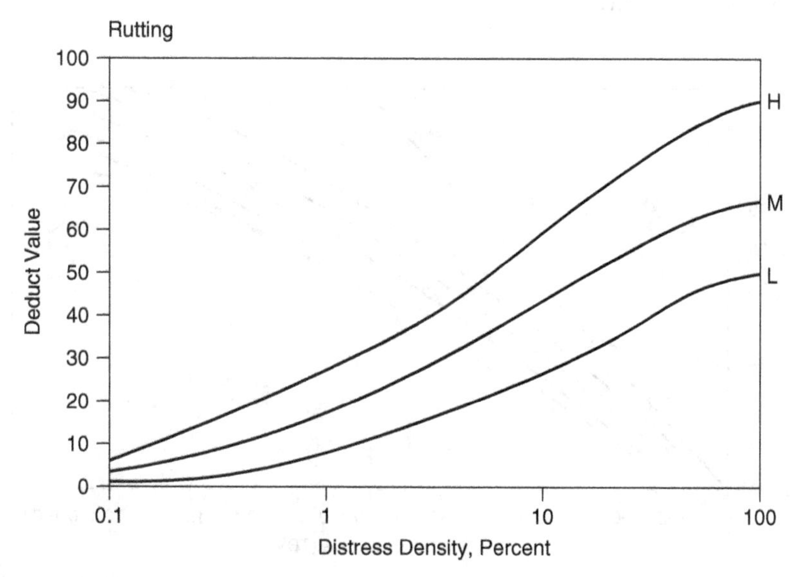


**Figure 9.37**
Example of Flexible Pavement Distress Deduct Values; Rutting (Ref. 40)

The allowed number of deduct values is computed as the integer part of $m$. If fewer than $m$ deduct values are present, all of them are summed to compute the total deduct value. Otherwise, only the $m$ highest deduct values are summed plus the $(m+1)_{th}$ deduct value factored by the real part of $m$. Thus established, the sum of the allowed deduct values, gives the *max CDV* using the charts shown in Figures 9.38 and 9.39, which are applicable to flexible and rigid pavements, respectively ($q$ is the number of deduct values that has not been assigned a value of 2%).

This process is repeated by substituting, successively, values of 2% for the actual deduct value of the $(m+1)_{th}$ deduct value, the $(m)_{th}$ deduct value, and so on, and recalculating the *max CDV*. The overall maximum of these values gives the *max CDV* value that is entered into the *PCI* calculation.

$$PCI = 100 - \max\ CDV \tag{9.44}$$

The pavement section *PCI* is computed by averaging the *PCI* values of the number of pavement sample units surveyed. The latter is selected on the basis of statistical considerations (i.e., the variation in *PCI* between sampling units and the desired confidence level). An example of the procedure for calculating the PCI for a sample unit is given next.

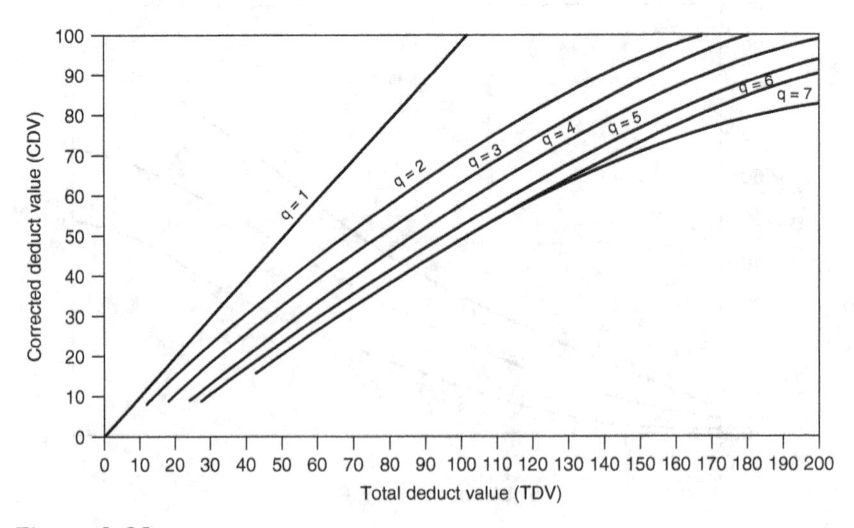


**Figure 9.38**
Computing Maximum Corrected Deduct Values; Flexible Pavements (Ref. 40)

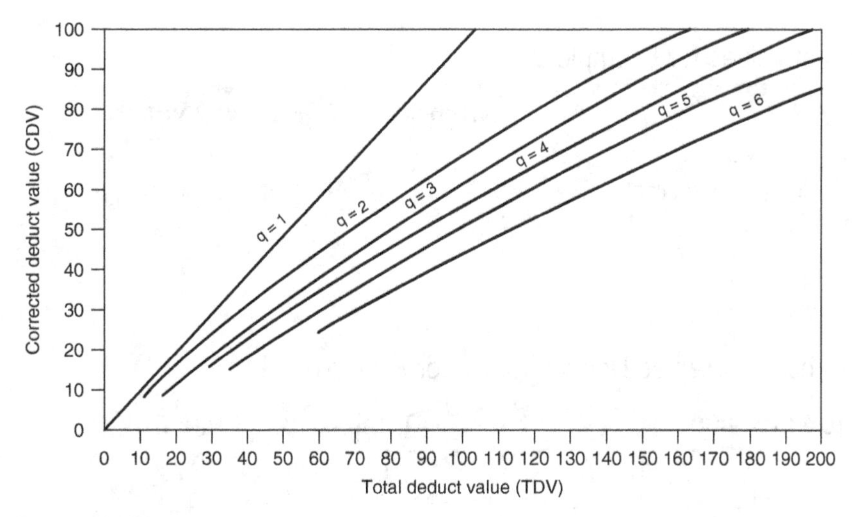


**Figure 9.39**
Computing Maximum Corrected Deduct Values; Rigid Pavements (Ref. 40)

**Example 9.11**

A pavement condition survey yielded the distress measurements given in Table 9.20 for a 250 $m^2$ (3.6 × 70 m) sample unit of a flexible pavement section. Compute the *PCI* for this pavement sample unit.

**ANSWER**

Compute the density of the distresses present by dividing their extent by the area of the pavement sample unit. For each distress, obtain the deduct values from Figures 9.35, 9.36 and 9.37, as tabulated in Table 9.21.

**Table 9.20**
Distress Data for Example 9.11

| Distress | Extent: Area/Length-Affected |
| --- | --- |
| Low-severity fatigue cracking | 25 $m^2$ |
| Medium-severity longitudinal cracking | 35 m |
| Low-severity rutting* | 15 $m^2$ |

*Low, medium, and high severity are defined by reference 40 as rut depths of 6–13 mm, 13–25 mm, and >25 mm, respectively.

**Table 9.21**
Computing Deduct Values for Example 9.11

| Distress | Density | Deduct Value |
|---|---|---|
| Low-severity fatigue cracking | 10% | 34% |
| Medium-severity longitudinal cracking | 14% | 24% |
| Low-severity rutting | 6% | 23% |

**Table 9.22**
Computing Maximum Corrected Deduct Values for Example 9.11

| Iteration | Deduct Value | | | Total | *q* | *max CDV* |
|---|---|---|---|---|---|---|
| 1 | 34% | 24% | 23% | 81% | 3 | 54% |
| 2 | 34% | 24% | 2% | 60% | 2 | 48% |
| 3 | 34% | 2% | 2% | 38% | 1 | 42% |

The highest deduct value *HDV* is 34, which gives the following value for the maximum number of distresses allowed.

$$m = 1 + \frac{9}{98}(100 - HDV) = 1 + \frac{9}{98}(100 - 34) = 7.06$$

Clearly, all three distresses present can be considered. Table 9.22 shows the sum of the distress deduct values for the successive iterative steps being taken, just as described. Note that for each iteration, the value of the *max CDV* is computed from the sum of the distresses and their number, using Figure 9.38.

Overall, 54% is the maximum of the *max CDV* values, which gives:

$$PCI = 100 - \max\ CDV = 100 - 54 = 46\%$$

which indicates a pavement sample unit of fair condition.

## 9.5 Safety

### 9.5.1 Definitions

Surface friction, alternatively referred to as skid resistance, affects the braking ability of vehicles, hence it is the main safety-related attribute of pavement surface condition. In elementary physics terms, friction

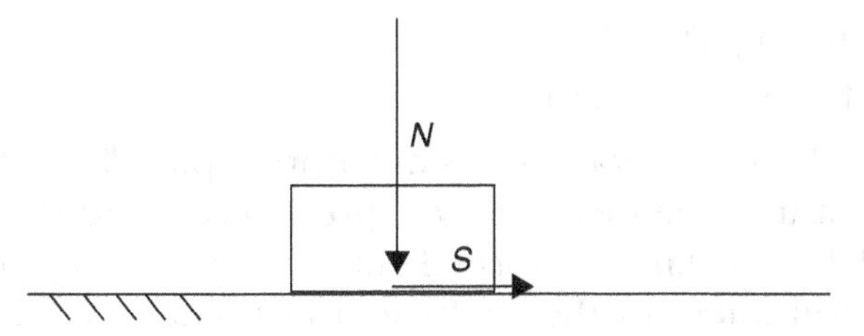


**Figure 9.40**
Defining Friction

can be indexed by the coefficient of friction $f$, defined as the ratio of the shear force $S$, generated as the result of slippage between two bodies, divided by the normal force $N$ exerted between them (Figure 9.40).

$$f = \frac{S}{N} \tag{9.45}$$

Pavement skid resistance depends on four factors: the texture of the pavement surface, the tread of a tire, the presence of water at the interface between the two, and the amount of slippage between them. Pavement texture is defined as surface irregularities with wavelengths too short to be perceived as roughness. Texture is differentiated into microtexture (wavelengths in the range of 1 $\mu$m to 0.5 mm) and macrotexture (wavelengths in the range of 0.5 mm to 50 mm).[45] Aggressive tire treads and lack of surface water improve skid resistance. Regardless of the other three factors, the slip speed between tire and pavement surface, defined as the relative speed between the contact of a tire and the pavement surface, markedly affects skid resistance. Before the brakes are applied, the slip speed is zero, whereas when the wheels are locked and the vehicle is still in motion, the slip speed equals the instantaneous speed of the vehicle. As described next, a combination of pavement surface friction and texture measurements is necessary to fully characterize skid resistance as a function of slip speed.

### 9.5.2 Measuring Pavement Friction

There are four basic types of devices for measuring pavement surface friction in the field. They involve one of the following wheel arrangements:

- Locked wheel
- Fixed slip speed

- Variable slip speed
- Side force in yaw mode

All these devices operate on a similar principle, whereby the shear force $S$ measured between the test tire imprint and the pavement is divided by the known normal force $N$ carried by the tire, to give a measurement of the coefficient of friction $f_m$, according to Equation 9.45. In the locked wheel device, the slip speed of the tire equals the speed of the towing vehicle (e.g., the ASTM E-274 trailer commonly used in the United States, and shown in Figure 9.41). Fixed slip speed devices use a reduction mechanism to force a constant difference in speeds between the test wheel and the towing vehicle (e.g., the Runway Friction Tester used in the United States), while variable slip devices go through a range of speed differences between the test wheel and the towing vehicle (e.g., the Japanese Komatsu Skid Tester). Side force in yaw mode devices use a test wheel oriented at a constant angle to the axis of travel, referred to as the yaw angle, without breaking (e.g., the British MuMeter).

In addition to these full-scale tire-based friction testers, a pendulum-based device is in use in the laboratory. Referred to as the British Pendulum Tester (BPT), it consists of a rubber shoe suspended from a pendulum mechanism (Figure 4.14). As the pendulum swings, the rubber shoe comes in contact with the surface being tested, and the associated energy loss is translated into the shear force $S$ exerted to the shoe, which gives a measurement of the coefficient of friction $f_m$. A comprehensive list of the various commercially available pavement friction measuring devices is given in references 5 and 13.

It is important to normalize all the friction measurements $f_m$ obtained with these devices to the same slip speed $S$, selected to be 60 km/h. This is done through the following relationship:[5]

$$f(60) = A + B\mathrm{f}_m \, \mathrm{e}^{\frac{S-60}{S_P}} + C\,TX \tag{9.46}$$

where $A$, $B$, and $C$ are device-specific calibration constants; $TX$ denotes pavement texture measurements (mm); and $S_p$ is a speed constant, which is a function of pavement texture, as explained in the following section. Table 9.23 lists some of the calibration constants associated with commonly used friction measuring devices.

**Figure 9.41**
Friction Tester (Courtesy of Dynatest Inc.)

**Table 9.23**
Calibration Constants for Selected Pavement Friction Measuring Devices (Ref. 5)

| Device | *S*(km/h) | *A* | *B* | *C* |
|---|---|---|---|---|
| ASTM E274 (treadless tire) | 65 | 0.045 | 0.925 | 0 |
| ASTM E274 (treaded tire) | 65 | −0.023 | 0.607 | 0.098 |
| Komatsu Skid Trailer | 10 | 0.042 | 0.0849 | 0 |
| British Pendulum Tester (BPT) | 10 | 0.056 | 0.008 | 0 |

Establishing the coefficient of friction $f(60)$ at the reference speed of 60 km/h allows calculation of the slip speed $f(S)$ at any slip speed $s$ through the following exponential relationship.

$$f(S) = f(60)\ e^{\frac{60-S}{S_p}} \tag{9.47}$$

### 9.5.3 Measuring Pavement Texture

Pavement texture measurements are essential in fully capturing the relationship between friction and slip speed. Microtexture (i.e., irregularities with wavelengths in the range of 1 $\mu$m to 0.5 mm) arises from irregularities in the surface of exposed aggregates coming in contact with vehicle tires. It is a significant contributor to skid resistance at low slip speeds.[13] Advances are being made in developing photographic methods for capturing aggregate geometric properties in the laboratory (i.e., shape, angularity, and texture), as described in Chapter 4.[25] Eventually, such methods should allow relating in-situ microtexture to skid resistance.

Pavement macrotexture (i.e., irregularities with wavelengths in the range of 0.5 mm to 50 mm) is traditionally quantified indirectly through the so-called sand-patch test.[24] It consists of spreading a specific volume of either Ottawa sand (i.e., passing and retained by Imperial sieve sizes No. 50 and 100, respectively) or spherical glass beads (i.e., passing and retained by Imperial sieve sizes No. 60 and 80, respectively) over a roughly circular area on the pavement, estimating the area covered through four diameter measurements and computing the mean texture depth (*MTD*) as the ratio of the volume divided by the area. The *MTD* is related to the amplitude of the pavement profile irregularities within the specified macrotexture wavelength range.

The recent advent of high-resolution inertial profilometers has allowed direct measurements of pavement profile elevations at the macrotexture level. The information is summarized into a single index, referred to as the mean profile depth (*MPD*).[5,8] The procedure used for this purpose consists of:

1. Dividing the profile into 0.1 m segments.
2. Removing the slope of each segment through linear regression.
3. Calculating the absolute value of the magnitude of the highest peak/trough encountered in each half-segment.
4. Averaging the mean calculated values for all segments to obtain the *MPD*.

Pavement macrotexture, indexed through the *MTD* or the *MPD* (mm), allows predicting the speed constant $S_P$ (km/h), through the following relationships:

$$S_P = 113.6\ MTD - 11.6 \tag{9.48}$$

or:

$$S_P = 89.7\ MPD + 14.2 \quad (9.49)$$

### 9.5.4 Index Summarizing Pavement Friction

The preceding discussion demonstrated that pavement texture measurements are needed (i.e., *MTD*/*MPD* or the speed constant $S_p$) in conjunction with pavement friction measurements (i.e., $f_m$), to fully quantify the relationship between the coefficient of friction and slip speed. Accordingly, a pair of numbers has been proposed to index pavement friction, namely $(S_p, f(60))$, referred to as the International Friction Index (*IFI*).[5] An example follows to demonstrate how friction and texture measurements are utilized to fully describe the relationship between friction and slip speed.

**Example 9.12**

A coefficient of friction of 0.45 was measured with an ASTM E-274 tester equipped with treadless tires. The macrotexture of this pavement, in terms of the *MPD*, was determined to be 0.8 mm. Calculate the value of the normalized coefficient of friction at 60 km/h, give the *IFI* and plot the coefficient of friction as a function of slip speed.

**ANSWER**

Utilizing Equation 9.49 gives the following value for the speed constant:

$$S_P = 89.7\ 0.8 + 14.2 = 85.9\ \text{km/h}$$

The treadless ASTM E-274 friction tester is run at 60 km/h, and, according to Table 9.23, its calibration constants *A*, *B* and *C* have values of 0.045, 0.925, and 0, respectively. Substituting these values into Equation 9.46 gives the 60 km/h normalized coefficient of friction for that pavement as:

$$f(60) = 0.045 + 0.925\ 0.45\ \ e^{\frac{65-60}{89}} + 0 = 0.49$$

Hence, the *IFI* for this pavement is given by quoting $(S_p, f(60)) =$ (89,0.49).

Finally, Equation 9.47 allows plotting the coefficient of friction as a function of slip speed (Figure 9.42) using:

$$f(S) = f(60)\ \ e^{\frac{60-S}{89}}$$

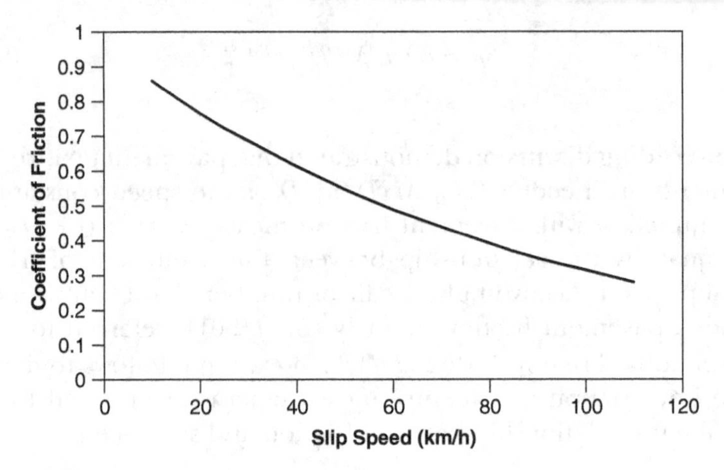


**Figure 9.42**
Coefficient of Friction versus Slip Speed for Example 9.12

## References

[1] PASCO Corporation (1987). 1 for 3 PASCO Road Survey System: From Theory to Implementation, Mitsubishi International Co. New York.

[2] American Association of State Highway Officials (1960). The AASHO Road Test, Pavement Research, Report 5, Washington, DC.

[3] Asphalt Institute (2000). *Asphalt Overlays for Highway and Street Rehabilitation*, MS-17, 2nd ed., Lexington, KY.

[4] Bendat, J. S., and A.G. Piersol (1971). *Random Data: Analysis and Measurement Procedures,* Wiley Interscience, John Wiley & Sons, Inc., New York.

[5] ASTM (1999). ''Calculating International Friction Index of a Pavement Surface,'' ASTM Standard Practice E-1960, *ASTM Book of Standards*, Volume 04.03, West Conshohocken, PA.

[6] ASTM (1999). ''Calculating Pavement Macrotexture Profile Depth,'' ASTM Standard Practice E-1845, *ASTM Book of Standards*, Volume 04.03, West Conshohocken, PA.

[7] Carey, W. N. and P. E. Irick (1960). ''The Pavement Serviceability-Performance Concept,'' *Highway Research Bulletin* 250.

[8] ISO (1998). "Characterization of Pavement Texture Using Surface Profiles—Part 1: Determination of Mean Profile Depth," ISO Standard 13473, International Standards Organization, Geneva, Switzerland.

[9] PCI (2002). *Design and Control of Concrete Mixes*, 14th ed., Portland Cement Association, Skokie, IL.

[10] SHRP (1993). *Distress Identification Manual for the Long-Term Pavement Performance Project*, Strategic Highway Research Program, National Research Council, SHRP-P-338.

[11] Gillespie, T. D., M. W. Sayers, and L. Segel (1980). "Calibration of Response-Type Road Roughness Measuring System," NCHRP Report 228, National Academy of Sciences, Washington, DC.

[12] Haas, R., W. R. Hudson, and J. Zaniewski (1994). *Modern Pavement Management*, Krieger Publishing Company, Malabar, FL.

[13] Henry, J. J. (2000). Evaluation of Pavement Friction Characteristics, NCHRP Synthesis 291, National Academy of Sciences, Washington DC.

[14] Herr, J. H., J. W. Hall, T. D. White, and W. Johnson (1995). "Continuous Deflection Basin Measurements and Back-calculation Under Rolling Wheel Load Using Scanning Laser Technology," ASCE Transportation Congress, San Diego, October 22–26.

[15] Highway Performance Monitoring System (HPMS) Field Manual, Federal Highway Administration Order M5600.1B, 1993.

[16] Huft, D. L., (1984). South Dakota "Profilometer," *Transportation Research Record*, 1000, National Research Council, Washington DC, pp. 1–8.

[17] Ioannides, A. M., E. J., Barenberg, and J. A. Larry (April 1989). "Interpretation of Falling-Weight Deflectometer Results Using Principles of Dimensional Analysis," 4th International Conference on Concrete Pavement Design and Rehabilitation, Purdue University, West Lafayette, Indiana.

[18] Ioannides, A. M., (January 1990). "Dimensional Analysis in NDT Rigid Pavement Evaluation," *ASCE Journal of Transportation Engineering*, Vol. 116. No. 1 pp. 23–36.

[19] ISO (1985). "Evaluation of Human Exposure to Whole-Body Vibration—Part 1: General Requirements," Standard 2631, International Standard Organization, Geneva, Switzerland.

[20] Janof, M. S. J. B. Nick, P. S. Davit and G. F. Hayhoe (1985). "Pavement Roughness and Rideability," NCHRP Report 275, National Academy of Sciences Washington, DC.

[21] Janof, M. S. (1988). "Pavement Roughness and Rideability Field Evaluation," NCHRP Report 308, National Academy of Sciences Washington, DC.

[22] Karamihas, S. M., T. D. Gillespie, R. W. Perrera and S. D. Kohn, (1999). "Guidelines for Longitudinal Pavement Profile Measurement," NCHRP Report 434, National Academy of Sciences Washington, DC.

[23] Kreyszig, E., (1993). *Advanced Engineering Mathematics*, 7th ed., John Wiley & Sons, Inc., New York.

[24] Masad, E., Olcott, D., White, T., and Tashman, L. (2001). Correlation of Fine Aggregate Imaging Shape Indices with Asphalt Mixture Performance, *Journal of the Transportation Research Board*, Record No. 1757, pp. 148–156.

[25] ASTM (1999). "Measuring Pavement Macrotexture Depth Using a Volumetric Technique," ASTM Standard Test Method E-965, *ASTM Book of Standards*, Volume 04.03, West Conshohocken, PA.

[26] ASTM (1999). "Measuring Surface Friction Properties Using the British Pendulum Tester," ASTM Standard Test Method E-303, *ASTM Book of Standards*, Volume 04.03, West Conshohocken, PA.

[27] Paterson, W. D. O. (1986). "International Roughness Index: Relationship to Other Measures of Roughness and Ride Quality," Transportation Research Record 1084, Transportation Research Board, Washington DC.

[28] *PAVER Asphalt Distress Manual* (1997). U.S. Army Construction Engineering Laboratories, TR 97/104 and TR 97/105, Champaign, IL.

[29] Perrera R. W., and S. D. Kohn, (July 2001). "Pavement Smoothness Measurement and Analysis: State of the Knowledge," Final Report for NCHRP Study 20-51(01), National Academy of Sciences Washington, DC.

[30] Roberts, F. L., Kandhal, P. S., Brown, E. R., Lee, D. Y., and Kennedy, T. W. (1996). "Hot-Mix Asphalt Materials, Mixture Design, and Construction," 2nd ed., National Asphalt Paving Association Education Foundation, Lanham, MD.

[31] Sayers M. W., and S. M. Karamichas (1998). *The Little Book of Profiling*, University of Michigan, Ann Arbor, MI.

[32] Sayers, M. W., T. D. Gillespie, and C. A. V. Queiroz (1986). "The International Road Roughness Experiment," World Bank Technical Paper 45, the World Bank, Washington, DC.

[33] Sayers, M. W., T. D. Gillespie, and D. W. O. Paterson (1986). "Guidelines for Conducting and Calibrating Road Roughness Measurements," Technical Paper No. 46, the World Bank, Washington, DC.

[34] Scullion, T., J. Uzan, and M. Paredes (January 1990). "Modulus: A Microcomputer-Based Backcalculation System," TRB Paper 890386, Washington DC.

[35] SHRP (1993). "SHRP Procedure for Temperature Correction of Maximum Deflections," SHRP P-654, Strategic Highway Research Program, National Research Council, Washington DC.

[36] Spangler E. B., and W. J. Kelly (1966). GMR Road Profilometer—A Method for Measuring Road Profile, *Highway Research Record* 121, pp. 27–54.

[37] ASTM (1998). "Standard Guide for Conducting Subjective Pavement Ride Quality Ratings," American Society for Testing of Materials, *ASTM Book of Standards*, Volume 04.03, E1927-98, West Conshohocken, PA.

[38] ASTM (1998). "Standard Practice for Computing International Roughness Index of Roads from Longitudinal Profile Measurements," American Society for Testing of Materials, *ASTM Book of Standards*, Volume 04.03, E1926-98, West Conshohocken, PA.

[39] ASTM (1998). "Standard Practice for Computing Ride Number of Roads from Longitudinal Profile Measurements Made by an Inertial Profile Measuring Device," American Society for Testing of Materials, *ASTM Book of Standards*, Volume 04.03, E1489-98, West Conshohocken, PA.

[40] ASTM (2000). "Standard Practice for Roads and Parking Lots Pavement Condition Index Surveys, American Society for Testing of Materials," *ASTM Book of Standards*, Volume 04.03, D6433-99, West Conshohocken, PA.

[41] ASTM (1987). "Standard Test Method for Measuring Pavement Deflections with a Falling Weight Type Impulse Load Device,"

*ASTM Book of Standards*, Volume 04.03, D-4694-87 West Conshohocken, PA.

[42] ASTM (1995). Test Method for Measuring Road Roughness By Static Level Method; American Society for Testing of Materials, *ASTM Book of Standards*, Volume 04.03, E1364-95, West Conshohocken.

[43] Timoshenko, S. P., and J. N. Goodier, *Theory of Elasticity*, 3rd ed., McGraw-Hill Inc., New York.

[44] Yang, C. Y. (1986). *Random Vibration of Structures*, John Wiley & Sons Interscience, New York.

[45] World Road Association (PIARC) (1987). Report of the Committee on Surface Characteristics, XVIII World Road Congress, Brussels, Belgium.

[46] Yoder, E. J., and M. W. Witczak, (1975). *Principles of Pavement Design*, 2nd ed., John Wiley & Sons Inc., New York.

## Problems

9.1 Synthesize a pavement profile by superimposing random elevations ranging between –0.01 and 0.01 m to two in-phase sinusoidal waves with amplitudes of 0.03 and 0.02 and wavelengths of 3 and 5 meters, respectively. Plot the pavement profile for a distance of 32 meters using 0.25 m increments. Plot the trace of a rolling straightedge (RSE) with a base length of 4 m (assume that the transport wheels and the tracing wheels are small enough to neglect their dimensions).

9.2 Filter the profile generated in problem 1 using a low-pass MA filter with a base length of 1.0 m. Plot the results.

9.3 Filter the profile generated in problem 1 using a high-pass MA filter with a base length of 2.0 m. Plot the results.

9.4 Compute and plot the power spectral density (PSD) of the artificial pavement profile given in Table 9.24. What are the dominant wavelengths and corresponding amplitudes?

9.5 The pavement profile shown in Table 9.25 was obtained with an inertial profilometer at intervals of 0.1394 m. Compute its *IRI* using commercially available software. What is the corresponding pavement serviceability (i.e., *PSI*)?

**Table 9.24**
Profile Elevation Data for Problem 9.4

| Distance (m) | Elevation (m) | Distance (m) | Elevation (m) |
|---|---|---|---|
| 0 | 10.0000 | 3.2 | 10.0165 |
| 0.2 | 10.0219 | 3.4 | 10.0229 |
| 0.4 | 10.0264 | 3.6 | 10.0101 |
| 0.6 | 10.0216 | 3.8 | 9.9918 |
| 0.8 | 10.0163 | 4 | 9.9835 |
| 1 | 10.0074 | 4.2 | 9.9788 |
| 1.2 | 9.9901 | 4.4 | 9.9726 |
| 1.4 | 9.9783 | 4.6 | 9.9783 |
| 1.6 | 9.9818 | 4.8 | 9.9994 |
| 1.8 | 9.9917 | 5 | 10.0219 |
| 2 | 9.9936 | 5.2 | 10.0272 |
| 2.2 | 9.9913 | 5.4 | 10.0218 |
| 2.4 | 9.9994 | 5.6 | 10.0169 |
| 2.6 | 10.0081 | 5.8 | 10.0076 |
| 2.8 | 10.0077 | 6 | 9.9895 |
| 3 | 10.0071 | 6.2 | 9.9788 |

9.6 The FWD measurements given in Table 9.26 were obtained on a flexible pavement using a plate radius of 10.3 cm and a contact pressure of 500 kPa. Estimate the value of the subgrade modulus. The layer thicknesses and the assumed values for the Poisson's ratio are given in Table 9.27.

9.7 The FWD measurements given in Table 9.28 were obtained on a portland concrete slab 0.30 m thick under a load of 40 kN and a plate radius of 15 cm. Determine the modulus of subgrade reaction and the elastic modulus of the slab assuming a liquid foundation and portland concrete Poisson's ratio, $\mu$ of 0.15.

9.8 For the data given in the previous question, determine the elastic moduli of the subgrade and the slab, assuming a solid foundation and a subgrade Poisson's ratio of 0.40.

9.9 A distress survey conducted on 250 $m^2$ of flexible pavement surface produced the results given in Table 9.29. Compute the *PCI*.

**Table 9.25**
Profile Elevation Data for Problem 9.5

| Distance (m) | Elevation (m) | Distance (m) | Elevation (m) | Distance (m) | Elevation (m) |
|---|---|---|---|---|---|
| 0.0000 | −0.02024 | 2.9274 | −0.012725 | 5.8548 | −0.002845 |
| 0.1394 | −0.02022 | 3.0668 | −0.011760 | 5.9942 | −0.002718 |
| 0.2788 | −0.01984 | 3.2062 | −0.011608 | 6.1336 | −0.002565 |
| 0.4182 | −0.01910 | 3.3456 | −0.010541 | 6.2730 | −0.001905 |
| 0.5576 | −0.01895 | 3.4850 | −0.010058 | 6.4124 | −0.001549 |
| 0.6970 | −0.01910 | 3.6244 | −0.010109 | 6.5518 | −0.001219 |
| 0.8364 | −0.01875 | 3.7638 | −0.009398 | | |
| 0.9758 | −0.01783 | 3.9032 | −0.008915 | | |
| 1.1152 | −0.01768 | 4.0426 | −0.008509 | | |
| 1.2546 | −0.01788 | 4.1820 | −0.008255 | | |
| 1.3940 | −0.01793 | 4.3214 | −0.007823 | | |
| 1.5334 | −0.01699 | 4.4608 | −0.007264 | | |
| 1.6728 | −0.01648 | 4.6002 | −0.006528 | | |
| 1.8122 | −0.01636 | 4.7396 | −0.006426 | | |
| 1.9516 | −0.01638 | 4.8790 | −0.005791 | | |
| 2.0910 | −0.01549 | 5.0184 | −0.005232 | | |
| 2.2304 | −0.01443 | 5.1578 | −0.004750 | | |
| 2.3698 | −0.01422 | 5.2972 | −0.004293 | | |
| 2.5092 | −0.01458 | 5.4366 | −0.004064 | | |
| 2.6486 | −0.01443 | 5.5760 | −0.003810 | | |
| 2.7880 | −0.01356 | 5.7154 | −0.003226 | | |

**Table 9.26**
Deflection Data for Problem 9.6

| Sensor, $s$ | 1 | 2 | 3 | 4 | 5 | 6 | 7 |
|---|---|---|---|---|---|---|---|
| Offset (cm) | 0 | 20 | 40 | 60 | 80 | 100 | 120 |
| Deflection ($\mu$m) | 280 | 235 | 190 | 175 | 85 | 60 | 30 |

**Table 9.27**
Layer Data for Problem 9.6

| Layer, $k$ | 1 | 2 | 3 |
|---|---|---|---|
| Thickness (cm) | 20 | 50 | $\infty$ |
| Poisson's ratio, $\mu$ | 0.33 | 0.45 | 0.45 |

**Table 9.28**
Deflection Data for Problem 9.7

| Sensor number | 0 | 1 | 2 | 3 |
|---|---|---|---|---|
| Offset (m) | 0 | 0.3 | 0.6 | 0.9 |
| Deflection ($\mu$m) | 98 | 80 | 75 | 60 |

**Table 9.29**
Distress Data for Problem 9.9

| Distress | Extent: Area/Length Affected |
|---|---|
| Low-severity fatigue cracking | 15 $m^2$ |
| High-severity longitudinal cracking | 10 m |
| Medium-low-severity rutting | 8 $m^2$ |

9.10 A coefficient of friction of 0.35 was measured with an ASTM E-274 tester equipped with treaded tires. The sand-patch test macrotexture of this pavement (i.e., *MPD*), was 0.7 mm. Calculate the value of the normalized coefficient of friction at 60 km/h, give the *IFI* and plot the coefficient of friction as a function of slip speed.

# 10 Environmental Effects on Pavements

## 10.1 Introduction

Pavements are exposed to the environment, which has a significant effect on their performance. The two main environmental factors of concern are the presence of water/ice in the pavement layers and the subgrade, and the variation of temperature throughout the year. These two factors interact with each other, such as during the freezing of pore water in frost-susceptible subgrades, which results in heaving. Furthermore, they interact with traffic loads, for example during springthaw conditions, when the base or subgrade layers can be sufficiently weakened by the presence of pore water to fail under the action of heavy axles (often, secondary roads need to be posted with lower load limits under these conditions). Another example of the interaction of traffic and environmental factors is the problem of pumping in jointed portland concrete pavements. This consists of rapid movement of base/subgrade pore water and fines near and through the joints under the high pressure being built by the rapid movement of truck axles, resulting in erosion and settlement of the downstream slab edge, called *faulting*, as described in Chapter 9.

The importance of adequate design provisions to control the effect of these environmental factors cannot be overemphasized. Proper drainage and the ability to predict pavement temperatures are paramount in ensuring proper structural behavior of the pavement layers over time. As discussed next, the latter is important to both asphalt concretes and portland concretes.

It should be noted that a number of pavement environmental problems are prevented through the proper selection of materials. Good examples are the selection of asphalt binder PG grades to prevent transverse cold-temperature-induced cracking by prescribing sufficient strength at the lowest temperature expected (see Chapter 5). Another example is the use of antistripping agents, such as lime, to control water from eroding the bond between binder and aggregates in asphalt concretes. Nevertheless, it should be understood that no structural layer thicknesses, nor material selection, can compensate for the lack of proper drainage in pavements.

## 10.2 Water in Pavements

### 10.2.1 Drainage

**PRINCIPLES**

The effect of water and the need for drainage are two of the most often overlooked aspects of pavement design and construction. Drainage follows Darcy's law, expressed as either:

$$Q = k\,i\,A \tag{10.1a}$$

or:

$$q = k\,i \tag{10.1b}$$

where $Q$ is the water discharge volume per unit time ($m^3$/hour), $q$ is the water discharge rate (m/hour), $k$ is the hydraulic permeability (m/hour), $i$ is the hydraulic gradient (total hydraulic head loss divided by the distance over which it is lost), and $A$ is the cross-sectional area ($m^2$) of the material discharging water.

The coefficient of permeability of granular media is largely a function of their gradation and especially, the amount of fines present. For soils, it varies broadly, ranging from upward of 36 m/hour (2,832 feet/day) for uniformly graded gravels to the practically impermeable of 36 $10^{-6}$ m/hour (0.0028 feet/day) for silts and clays.[15] Permeability is measured either in the lab or in-situ, through constant-head or falling-head permeameters. The permeability of manufactured layers, such as pavement bases and subbases, can be estimated using the following empirical relationship:[20]

$$k = 6.214\ 10^5\ D_{10}{}^{1.478}\ n^{6.654}/P_{200}{}^{0.597} \tag{10.2}$$

where $k$ is the permeability coefficient in feet/day (1 ft/day = 0.0127 m/hour), $D_{10}$, (inches) is the 10th percentile of the grain size distribution, $P_{200}$ is the percent passing the No. 200 sieve, and $n$ is the porosity.

**Example 10.1**

Estimate the permeability coefficient of a base layer with $D_{10}$ of 0.0165 (0.42 mm, which is the opening size of the No. 40 sieve), 2% passing the sieve No. 200 (0.075 mm), and a porosity of 0.3.

**ANSWER**

Substituting the specified values into Equation 10.2, gives:

$$\begin{aligned} k &= 6.214\ 10^5\ 0.0165^{1.478}\ 0.3^{6.654}/2^{0.597} \\ &= 0.316 \text{ feet/day or } 0.004 \text{ m/hour} \end{aligned}$$

It is noted that, often, in-situ permeability is governed by fissures or cracks, rather than the gradation and porosity of an intact layer.

### 10.2.2 Sources

There are three sources of water in the pavement layers: groundwater seepage, capillary action, and precipitation. Groundwater seepage is a problem where the water table rises to intersect the pavement layers, as may be the case on roadway cuts (Figure 10.1). Drainage through the pavement layers cannot accommodate seepage from groundwater sources. Instead, the water table needs to be lowered below the pavement layers through longitudinal trench drains and removed through properly designed and constructed perforated pipes, as shown in Figure 10.1.

The amount of groundwater to be removed is a function of the permeability of the subgrade and the ground/pavement geometry in a particular situation. The solution involves plotting flow nets, computing the amount of flow to be removed, and calculating the diameter of the pipes that can accommodate it. Guidelines for these techniques can be found in the literature.[8,20] Alternatively, flow

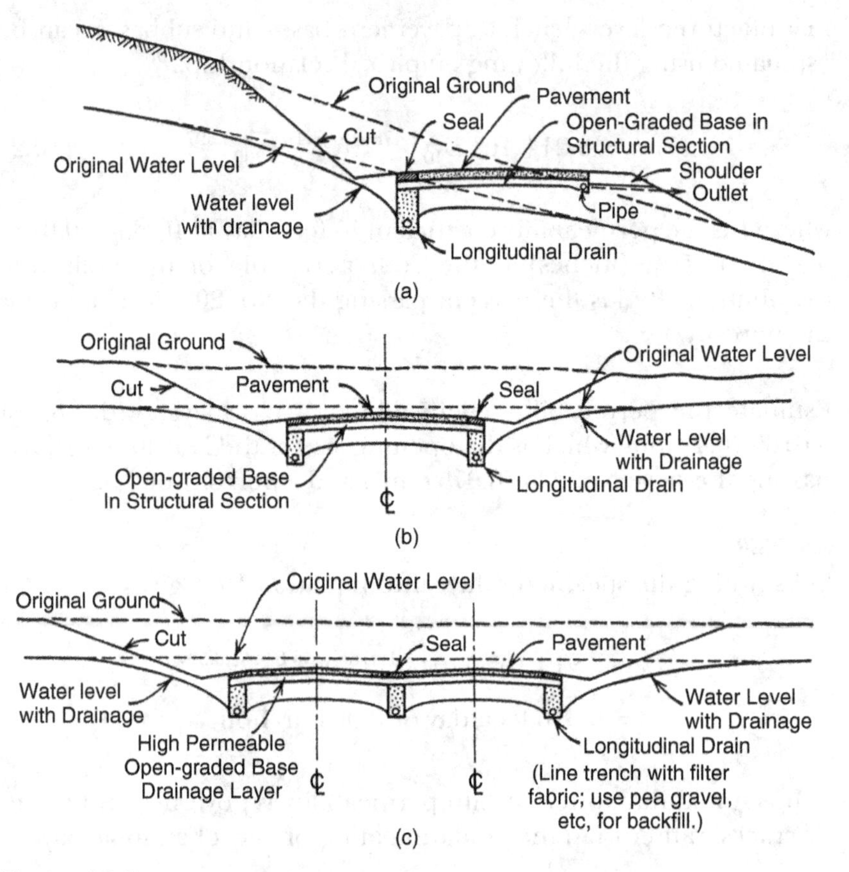


**Figure 10.1**
Drainage of Ground Water Seepage under Pavement (Ref. 8)

computations can be obtained using commercially available seepage software that use numerical techniques to provide the distribution of hydraulic heads in the flow region.

Capillary action may result in saturated conditions above the water table. Capillary pressures are the result of water surface tension in the interconnected voids of fine-grained subgrades. The actual capillary rise above the water table, $h_c$ (meters), can be computed as inversely proportional to the effective diameter of the soil pores, d (mm):[15]

$$h_c = \frac{0.03}{d} \tag{10.3}$$

In computing the effective diameter of the soil pores, a common assumption is that it is equal to 20% of their $D_{10}$ grain size (i.e., 10th percentile).

**Example 10.2**

A silty subgrade has a $D_{10}$ of 0.075 mm (the size opening of the No. 200 sieve). Compute the potential height of capillary rise.

**ANSWER**

The effective diameter of the soil pores for this subgrade is approximately 0.015 mm (0.2 × 0.075) as a result, the capillary rise height is computed from Equation 10.3 as:

$$h_c = \frac{0.03}{0.015} = 2.0 \text{ m}$$

The main source of water in the pavement layers is precipitation. This is often overlooked by designers, who believe that either the pavement surface is impermeable or that the subgrade is permeable, hence any amount of precipitation that may enter the structure is automatically removed. This is far from the truth. Pavement surfaces are permeable, whether intentionally (e.g., open-graded asphalt concretes) or by virtue of their macrostructural cracks or joints. Furthermore, even granular subgrades that are thought as permeable are far from it, when compacted near their optimum water content. This point is well pressed by Cedergren, stating, "Most of the world's pavements are so leaky that far more water soaks in than can drain away into the subsoil."[3] For a flexible pavement, the structural implications of a saturated base layer are well demonstrated by Figure 10.2, which shows that the incompressibility of water prevents the dissipation of loading stresses with depth, as assumed by layered elastic theory, thus damaging the subgrade.

There are two design considerations in handling precipitation and drainage:[9]

- Drainage rates need to be larger than infiltration rates, to prevent pavement layer saturation.
- If pavement layers become saturated, they need to be drained within a prescribed time period to prevent traffic-associated or frost-associated damage.

These are discussed next in detail.

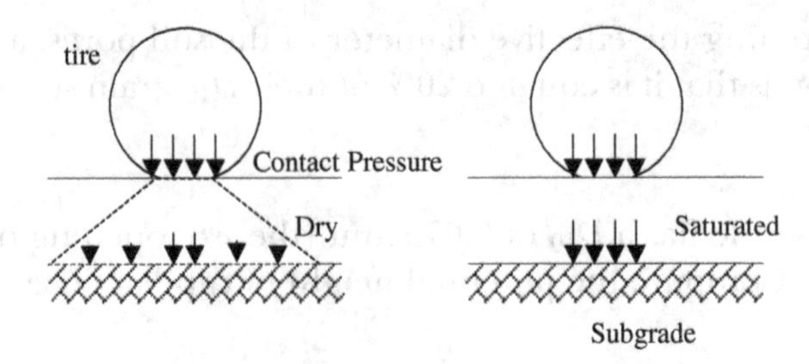


**Figure 10.2**
Stress Distribution in Dry and Saturated Pavement Layers

### 10.2.3 Infiltration Rates

A natural limit to the precipitation infiltration rate into pavements is the permeability of the surface. Considering that the hydraulic gradient for surface infiltration is unity, Darcy's law (i.e., Equation 10.1b) allows computing the maximum possible infiltration rate as:

$$q = k\,1 \tag{10.4}$$

where $k$ is the permeability of the pavement surface. In practice, however, infiltration rates cannot exceed a fraction of the rainfall rate. This fraction, denoted by $C$, accounts for runoff and evaporation. Recommended $C$ values range from 0.5 to 0.67 for rigid pavements and 0.33 to 0.50 for flexible pavements.[7] Hence, the infiltration rate $q$ (m/hour) cannot exceed:

$$q = C\,R \tag{10.5}$$

where $R$ is the design rainfall rate (m/hour), which is considered to be the maximum rate of the one-hour duration one-year frequency rainfall. Figure 10.3 gives representative values for this rainfall, in inches per hour,[6] (1 in/hour = 0.61 m/day).

More detailed rainfall rates for a particular locale can be obtained from the web site of the Hydrometerological Design Studies Center of the National Oceanographic and Atmospheric Administration (NOAA); www.ncda,noaa.gov/ox/documentlibrary/rainfall.html Hence, the design rate of infiltration is the lesser of the rates given by Equations 10.4 and 10.5.

A more elaborate approach than the one just described, estimates the infiltration rate $q$(m/day) considering the actual extent of

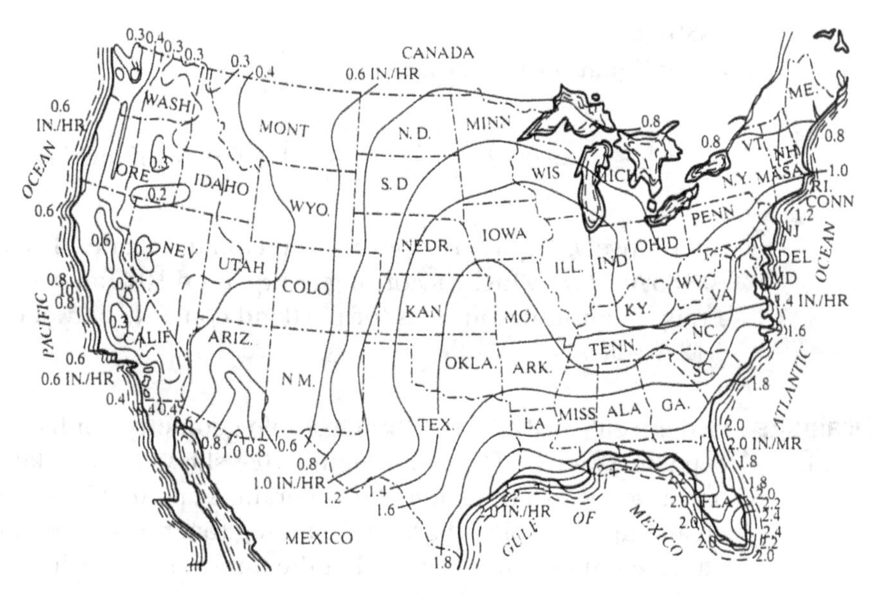


**Figure 10.3**
Representative Precipitation Rates (inches/hour) (Ref. 6)

surface cracks and/or joints present:[21]

$$q = I_c\left(\frac{N_c}{W} + \frac{W_c}{W\ C_s}\right) + k_p \tag{10.6}$$

where $W$ is the width of the permeable base (m), $N_c$ is the number of longitudinal cracks, $C_s$ is the spacing of the transverse joints/cracks (m), $W_c$ is the length of the transverse joints/cracks (m), $I_c$ is the infiltration rate per unit length of crack, typically equal to $0.223\,\text{m}^3/\text{day/m}$, and $K_p$ is the infiltration rate/permeability of the intact pavement surface (m/day).

**Example 10.3**

A two-lane rigid pavement surface is 9.6 m wide and has transverse joints spaced at 4.0 m intervals. A longitudinal construction joint is located in the middle the surface. In addition, there is a single longitudinal crack along the full length of the slab. Compute the infiltration rate according to reference 21 and compare it to the simplified approach described in reference 6, assuming it is located in eastern Washington state. Consider the infiltration rate of intact concrete to be negligible.

**ANSWER**

Using Equation 10.6 gives:

$$q = 0.223\left(\frac{2}{9.6} + \frac{9.6}{9.6\ 4}\right) + 0 = 0.102\ \text{m/day}$$

Alternatively, Figure 10.3 gives a precipitation rate of 0.3 in/hour (0.183 m/day), which, according to Equation 10.5, allows estimating a range in infiltration rates for portland concrete between 0.09 and 0.12 m/day.

### 10.2.4 Drainage Rates/Times

Cedergren[8] analyzed the ''unfavorable geometry'' of base layers, in expelling water infiltration. Figure 10.4 shows horizontal drainage base layer, with a thickness $h$ and a drainage path of length $b$, resting on an impermeable subgrade. Assuming that the layer is maintained saturated in the middle, and that the uppermost flow line exits near the bottom of the layer, the total hydraulic head loss is $h$.

Assuming layer homogeneity, flow nets were drawn for a variety of layer thickness and lane width combinations. The discharge volume per unit time $Q$ for a unit width perpendicular to the paper was computed applying Darcy's law (Equation 10.1a):

$$Q = k\,\frac{n_f}{n_d}\,h \qquad (10.7)$$

where $n_f$ is the number of flow channels and $n_d$ is the number of equipotential drops. It was observed that their ratio is proportional to the layer thickness over drainage lane width ratio $h/b$. Hence,

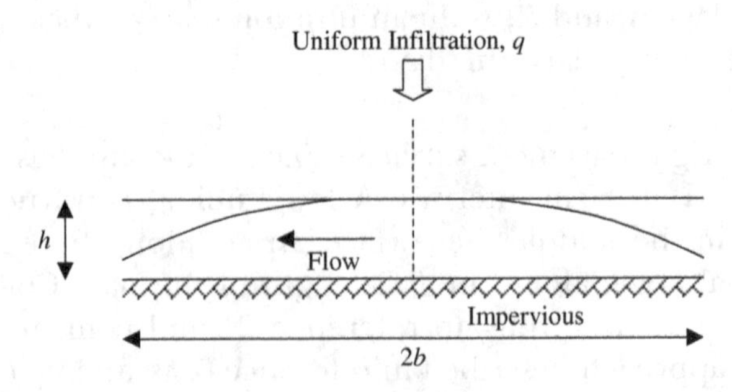


**Figure 10.4**
Drainage of Base Layer under Uniform Infiltration (Ref. 8)

Equation 10.7 can be approximated as:

$$Q = k\,\frac{h}{b}\,h \text{ or } Q = k\,\frac{h^2}{b} \tag{10.8}$$

It was furthermore noted that to maintain steady state conditions, the water infiltration rate $q$ per unit width perpendicular to the paper needs to be equal to $Q/b$. Hence, Equation 10.8 can be written as:

$$Q = q\,b = k\,\frac{h^2}{b} \text{ or } q = k\,\frac{h^2}{b^2} \tag{10.9}$$

This suggests that the capability of base layers to drain infiltration is proportional to the square of their thickness-over-width ratio. The implications of this are shown next through an example.

**Example 10.4**

A gravel layer resting on a horizontal impermeable subgrade has a coefficient of permeability of 3.6 m/hour (1 m/hour = 78.74 ft/day); it is 0.20 m thick and has a maximum drainage path of 3.6 m (width of 7.2 m). Compute the maximum rate of infiltration that this layer can accommodate, as well as the speed of movement of the pore water, given an effective porosity $n_e$ of 0.30 (the difference between $n_e$ and conventional porosity $n$ is that the volume of voids retaining water due to capillary forces are excluded).

**ANSWER**

Equation 10.9 gives the maximum water infiltration rate that can be drained through this layer as:

$$q = k\left(\frac{h}{b}\right)^2 = q = k\left(\frac{0.2}{3.6}\right)^2 = k\;0.0031$$

Substituting in the coefficient of permeability specified gives a discharge rate of 0.011 m/hour, which is the maximum amount of steady infiltration the particular layer can conduct and discharge through the side. Since only the pores conduct water, the actual speed of pore water movement is obtained by dividing the discharge speed by the effective porosity, that is $0.011/0.3 = 0.037$ m/hour. This speed suggest that a drop of water will take $3.6/0.037 = 97$ hours, or 4 days, to traverse the full length of the drainage path under steady state conditions.

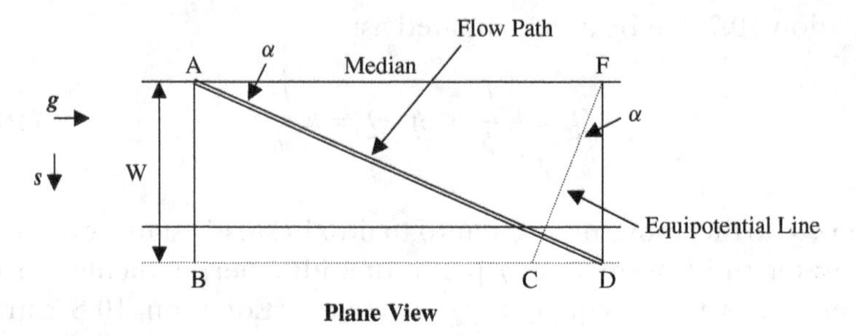


**Figure 10.5**
Drainage Geometry for Establishing Resultant Gradient (Ref. 8)

For a saturated drainage layer on an impermeable subgrade, the hydraulic gradient is simply its slope. This needs to be computed on the basis of its longitudinal and transverse slopes (i.e.,the grade and cross-slope), denoted by $g$ and $s$, respectively. Consider the geometry shown in Figure 10.5, where a flow path $AD$ is shown perpendicular to equipotential line $CF$. Let us determine the length of this flow path, which forms an angle $\alpha$ to the road center line:

$$BD = AF = \frac{W}{\tan a} \tag{10.10}$$

$$CD = W \tan a \tag{10.11}$$

where $W$ is the width of the lane plus the shoulder. Hence:

$$BC = BD - CD = W\left(\frac{1}{\tan a} - \tan a\right) \tag{10.12}$$

The differences in elevation from A to F and from A to C, denoted by $\Delta h_{AF}$, and $\Delta h_{AC}$, respectively, need to be equal, given that $CF$ is an equipotential line and there is no pressure differential between the two points. Hence:

$$\Delta h_{AF} = gAF = g\frac{W}{\tan a} \tag{10.13}$$

$$\Delta h_{AC} = \Delta h_{AB} + \Delta h_{BC} = Ws + BCg = Ws + W\left(\frac{1}{\tan a} - \tan a\right)g \tag{10.14}$$

Equating 10.13 and 10.14, gives $\tan a = s/g$, which gives the length of the drainage path as:

$$AD = \frac{W}{\sin a} = W\frac{1}{\tan a}\left(1 + (\tan a)^2\right)^{0.5}$$
$$= W\frac{g}{s}\left(1 + \left(\frac{s}{g}\right)^2\right)^{0.5} = \frac{W}{s}\left(s^2 + g^2\right)^{0.5} \qquad (10.15)$$

The combined grade along the flow line, which is equal to the hydraulic gradient driving the flow of water out of the layer, can be computed from the length of the flow line $AD$ and the difference in elevation from A to D, $\Delta h_{AD}$, which is:

$$\Delta h_{AD} = W\ s + BD\ g = W\ s + \frac{W}{\tan a}g = W\ s + W\left(\frac{g}{s}\right)g$$
$$= W\left(s + \frac{g^2}{s}\right) = \frac{W}{s}\left(s^2 + g^2\right) \qquad (10.16)$$

which allows computing the flow path grade $g_f$ as:

$$g_f = \left(s^2 + g^2\right)^{0.5} \qquad (10.17)$$

**Example 10.5**

A homogeneous base layer 0.3 m thick resting on an impermeable subgrade has a coefficient of permeability of 5.0 m/hour. The pavement surface has a grade of – 4% and a cross-slope of – 2%, (refer to Figure 10.5). The pavement width is 4.8 m. Compute the discharge rate capability of this layer.

**ANSWER**

Use Equation 10.15 to compute the length of the drainage path as:

$$AD = \frac{4.8}{0.02}(0.02^2 + 0.04^2)^{0.5} = 10.73\ \text{m}$$

Use Equation 10.16 to compute the difference in elevation between A and D:

$$\Delta h_{AD} = \frac{4.8}{0.02}(0.02^2 + 0.04^2) = 0.48\ \text{m}$$

which gives a combined grade for the flow path of 0.0447, or 4.47%. This value is confirmed using Equation 10.17.

$$g_f = (0.02^2 + 0.04^2)^{0.5} = 0.0447$$

The maximum rate of flow discharge for this layer is computed from Darcy's law (Equation 10.1b) as:

$$q = 5\ 0.0447 = 0.2236\ \text{m/hour}$$

which suggests that, permeter width perpendicular to the flow line, this 0.3 m thick base layer can discharge water at $0.2236 \times 0.3 = 0.067\ \text{m}^3/\text{hour}$ ($1.61\ \text{m}^3/\text{day}$).

Casagrande and Shannon[5] solved the transient flow problem of the draining of a layer initially saturated. They established relationships between the degree of layer drainage $U$ and the corresponding drainage time. The degree of layer drainage is defined as the ratio of the amount of water drained since the rain stopped divided by the capacity of the drainage layer (0 for totally saturated and 100% for totally drained). The results were presented in a graph of $U$ versus standardized time (Figure 10.6). The variables used in this graph are the gradient factor $S_1$ and the time factor $t/m$, given by:

$$S_1 = \frac{LS}{h} \tag{10.18}$$

and:

$$t/m = \frac{kht}{n_e L^2} \tag{10.19}$$

where $L$ and $h$ are the length and the thickness of the drainage layer, respectively, and $S$ is the slope of the layer, the remaining variables were defined earlier. Typically, a 50% degree of drainage is used to qualify a layer as drained.[11,12]

**Example 10.6** A 0.4 m thick base layer has a slope of 0.03; it is 6.5 m long, has a coefficient of permeability of 0.36 m/hour, and an effective porosity of 0.38. Compute the time it will take to reach a degree of drainage of 50%.

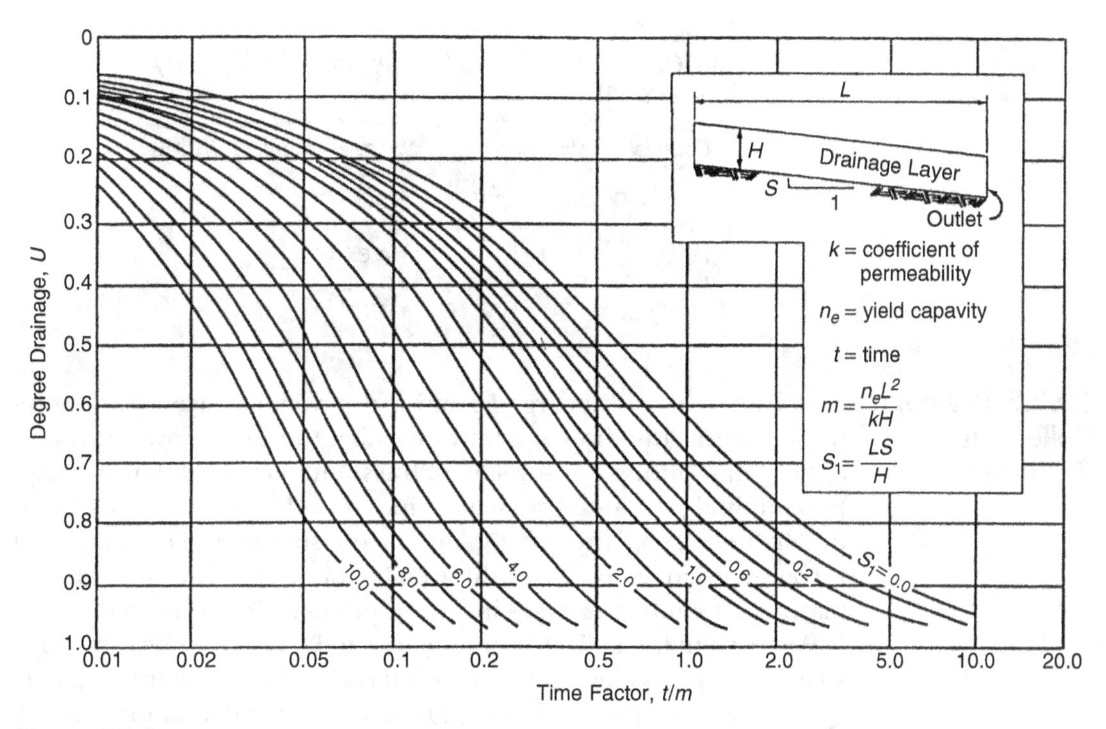


**Figure 10.6**
Degree of Drainage versus Time to Drain, (Ref. 12)

**ANSWER**

Use Equation 10.18 to compute the gradient factor $S_1$ as $6.5 \times 0.03/0.4 = 0.49$. Use Figure 10.6 with a 0.49 gradient factor to obtain a time factor $t/m$ of 0.37. Use Equation 10.19 to solve for actual time:

$$0.37 = \frac{0.36\ 0.4\ t}{0.38\ 6.5^2} \text{ or } t = 41.25 \text{ hours or } 1.72 \text{ days}$$

Needless to say, lower permeability could extend this drainage period considerably, and result in unacceptable degrees of drainage over considerably longer periods. AASHTO[11] suggests guidelines for drainage quality as a function of the length of time required for drainage that is, reaching a 50% degree of drainage (Table 10.1). According to this Table, the drainage of the layer in Example 10.6 is between good and fair.

**Table 10.1**
Guidelines for Drainage Quality, (Ref. 11 Used by Permission)

| Quality of Drainage | Water Removed Within |
|---|---|
| Excellent | 2 hours |
| Good | 1 day |
| Fair | 1 week |
| Poor | 1 month |
| Very Poor | Never |

### 10.2.5 Drainage Collection Systems

The water being discharged from the pavement layers (i.e., infiltration minus the drainage into the subgrade) is removed through either "daylighting" to the side of the shoulders or, better, through longitudinal perforated collector pipes installed in trenches located under the edge of the driving lane or the shoulder. The minimum recommended diameter for these collector pipes is 7.6 cm (3 in.) for plastic pipes and 10.2 cm (4 in.) for pipes made of other materials.

Water from the collector pipes needs to be removed by transverse solid outlet pipes placed at regular intervals. Their discharge capacity $Q$ (ft$^3$/day) is computed using Manning's formula, dating back to the 1890s:

$$Q = 86400 \frac{1.486}{n_p} A\, R^{2/3} S^{1/2} \tag{10.20}$$

where $R$ is the hydraulic radius of the pipe (ft), defined as the ratio of area $A$ of the cross-section that conducts water divided by its wetted perimeter (for a full pipe of diameter $D$, the hydraulic radius is $\pi D^2/4$ divided by $\pi D$), $S$ is the slope of the pipe, and $n_p$ is the roughness coefficient for the material of the pipe (ranges from 0.008 for smooth plastic to 0.024 for corrugated metal). Hence, the amount of lateral flow $q$ (ft$^3$/day/ft) that needs to be removed through outlet pipes spaced $L_0$ apart can be computed as:

$$q\, L_0 = 86400 \frac{1.486}{n_p} \frac{\pi\, D^2}{4} \left(\frac{D}{4}\right)^{2/3} S^{1/2} \text{ or } q\, L_0 = 40021 \frac{D^{2.6667}}{n_p} S^{1/2} \tag{10.21}$$

which, solved for $D$ gives:

$$D = \frac{1}{53.2} \left(\frac{n_p\, q\, L_0}{S^{1/2}}\right)^{0.375} \tag{10.22}$$

**Example 10.7**

Determine the required diameter of solid outlet pipes placed 152.4 m (500ft) apart capable of removing 3 $m^3$/day/m (32.3 $ft^3$/day/ft) of pavement layer drain water, given that their slope is 2% and that they are made of plastic with a $n_p$ value of 0.008.

**ANSWER**

Use Equation 10.22 to compute the diameter:

$$D = \frac{1}{53.2}\left(\frac{0.008\ 32.3\ 500}{0.02^{1/2}}\right)^{0.375} = 0.24 \text{ feet}$$

or 2.9 in. which is rounded up to 3 in. (7.6 cm).

### 10.2.6 Filters

Filters function as barriers blocking the migration of fines under the action of seepage forces. They consist of either specially graded aggregates or fabrics, the latter being referred to as *geotextiles*. Geotextiles are also used to separate structural layers and prevent segregation damage from the combined action of seepage forces and traffic loads. Aggregate filters are designed on the basis of criteria relating aspects of their gradation to that of the pavement layer being protected. The source of the criteria described here is largely a U.S. Army Corp of Engineers manual.[23]

The first criterion relates to clogging prevention, prescribing that the filter material must be fine enough compared to the adjacent pavement layer/soil to prevent migration of the latter by seepage forces (i.e., piping). It is expressed as:

$$\frac{D_{15,filter}}{D_{85,soil}} \leq 5 \tag{10.23}$$

where the subscript of grain size $D$ denotes percentile, differentiated for the filter and the soil, respectively.

The second criterion relates to high permeability by specifying that the filter material must be coarse enough compared to the adjacent pavement layer/soil to allow free removal of the water that seeps through the interface. It is expressed as:

$$\frac{D_{15,filter}}{D_{15,soil}} \geq 5 \tag{10.24}$$

The third criterion prescribes a maximum difference between the median of the two adjacent pavement layer/soil gradations to

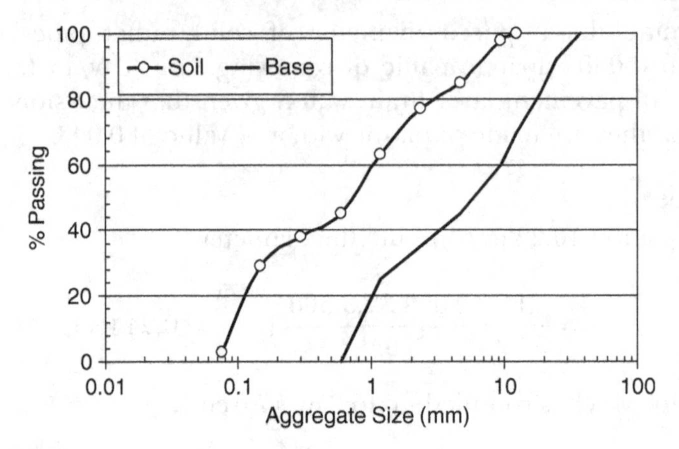


**Figure 10.7**
Gradations for Example 10.8

produce a roughly parallel arrangement between the two curves:

$$\frac{D_{50,filter}}{D_{50,soil}} \leq 25 \tag{10.25}$$

**Example 10.8** Consider a permeable base layer resting on a soil layer. Their gradations are shown in Figure 10.7. Determine if there is a need for an intermediate filter layer between the two.

**ANSWER**

Figure 10.7 allows obtaining the characteristic grain size percentiles shown in Table 10.2. The criteria suggested by Equations 10.23 to 10.25 are tested by substituting in the values from Table 10.2 as:

$$\frac{D_{15,filter}}{D_{85,soil}} = \frac{0.9}{5} = 0.18 < 5$$

$$\frac{D_{15,filter}}{D_{15,soil}} = \frac{0.9}{0.1} = 9 > 5$$

$$\frac{D_{50,filter}}{D_{50,soil}} = \frac{6.5}{0.75} = 8.66 < 25$$

which are all satisfied; hence, no intermediate filter layer is required between this base and subgrade layer.

**Table 10.2**
Characteristic Grain Sizes for Example 10.8

| Size | Soil (mm) | Base Layer (mm) |
|---|---|---|
| $D_{85}$ | 5 | 20 |
| $D_{50}$ | 0.75 | 6.5 |
| $D_{15}$ | 0.1 | 0.9 |

The design of geotextiles, woven and nonwoven, involves similar filter criteria, which are implemented by controlling the size of the geotextile openings compared to the size of the aggregates of the adjacent aggregate/soil. The size of the geotextile openings is characterized by sieving single-sized glass beads of successively increasing size through the geotextile and establishing the bead size for which fewer than 5% by weight pass.[2] This size is referred to as the apparent opening size (AOS) or the 95th percentile size ($O_{95}$). This size is selected on the basis of similar clogging and permeability criteria as those used for filter layers. These criteria are summarized next for steady state flow conditions.[13]

The clogging prevention criterion is:

- For coarse-grained soils (less than 50% passing sieve No. 200):

$$O_{95} \leq B D_{85} \tag{10.26}$$

where $B$ is a constant that depends on the coefficient of uniformity of the soil, $C_u$, (i.e., $D_{60}/D_{10}$):

- $B = 1$ for $C_u \leq 2$ or $C_u \geq 8$
- $B = 0.5/C_u$ for $2 \leq C_u \leq 4$
- $B = 8/C_u$ for $4 \leq C_u \leq 8$

- For fine-grained soils (more than 50% passing sieve No. 200):

- $O_{95} \leq D_{85}$ for woven geotextiles (10.27a)
- $O_{95} \leq 1.85$ for nonwoven geotextiles (10.27b)

The permeability criterion is:

$$k_{geotextile} \geq k_{soil} \tag{10.28}$$

where permeability is measured as described in reference 3.

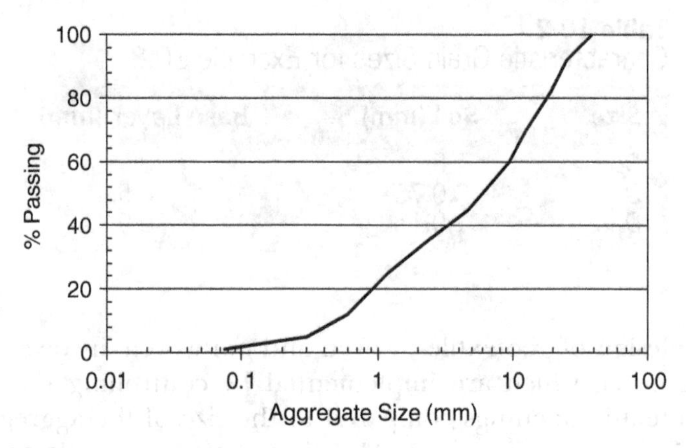


**Figure 10.8**
Gradation for Example 10.2

**Example 10.9** A geotextile is to be placed between a subgrade and a base layer. The gradation of the base layer is shown in Figure 10.8. The permeability of the subgrade is 0.036 m/hour. Determine the desired geotextile properties.

**ANSWER**

The gradation in Figure 10.8 gives a uniformity coefficient of $C_u = D_{60}/D_{10} = 9.5/0.5 = 19$, which gives a coefficient $B$ of 1.0 and a $D_{85}$ of 20 mm. Hence, Equation 10.26 suggests a geotextile with an AOS smaller than 20 mm. Its permeability needs to be higher than 0.036 m/hour.

Some of the computations described in this section were incorporated into a methodology for performing drainage analysis/design, as well as rainfall and infiltration computations.[17] The resulting model, known as the Infiltration-Design (ID) model, was subsequently incorporated into the Enhanced Integrated Climatic Model (EICM).[18] Version 3.01 of the EICM was incorporated into the NCHRP 1-37A Design Guide software[25] along with environmental data and material properties. A self-standing software alternative for performing water flow computations in pavements is DRIP.[19] It allows computing infiltration rates, drainage rates/times, as well as designing filter and drainage collection systems. See the Web site this book, www.wiley.com/go/pavement, for software sources.

## 10.3 Heat in Pavements

Exposure of pavements to changing levels of solar radiation and ambient temperature results in temperature variations at the surface, which produces temperature gradients with depth. These temperature gradients affect the behavior of both flexible and rigid pavements. In flexible pavements, temperatures affect the modulus of the asphalt concretes, which in turn affects their fatigue and plastic deformation characteristics. In rigid pavements, temperature gradients result in thermal stresses, which produce slab warping that affects their response to traffic loads, hence the accumulation of fatigue and faulting damage.

### 10.3.1 Heat Transfer

Heat flows in response to temperature gradients, following one-dimensional diffusion laws. Two related pavement layer properties are defined here, namely the coefficient of thermal conductivity and the specific heat. The coefficient of thermal conductivity, $k$, is defined as the amount of heat conducted per unit area of material per unit time in response to a unit temperature gradient, given in $W/m^2/(°C/m)$ or $BTU/hr/ft^2/(°F/ft)$. Typical $k$ values for asphalt concrete and portland concrete are 1.45 and 0.93 $W/m^2/(°C/m)$, (0.84 and 0.54 $BTU/hr/ft^2/(°F/ft)$), respectively. The mass-specific heat, $c$, is defined as the amount of heat required to raise the temperature of a unit mass of material by 1 °C. The units are J/g/°C. Typical specific heat values are given in Table 10.3. The volume-specific heat, denoted by $C$, is defined as the amount of heat required to raise the temperature of a unit volume of material by 1 °C ($J/cm^3/°C$). It is computed by multiplying the mass-specific heat by the density of the material. The combined volume-specific heat of multi-phase materials can be computed by weighing the mass-specific heat values of the individual phases according to their proportions, as described in the following example.

**Example 10.10**

Compute the volume-specific heat of a saturated subgrade under thawed and frozen conditions, given a dry density $\rho_d$ of 2.2 $g/cm^3$ and a water content $w$ of 12%. Use the typical values of the coefficients of mass-specific heat given in Table 10.3.

**Table 10.3**
Typical Values of Coefficients of Mass-Specific Heat

| Material | Coefficients of Mass-Specific Heat (J/g/°C) |
|---|---|
| Air | 1.0 |
| Water | 4.19 |
| Ice | 2.1 |
| Mineral aggregate | 0.8 |
| Soil | 0.8–1.48 |
| Asphaltic binder | 0.92 |
| Asphalt concrete | 0.55 |
| Portland concrete | 0.88 |

**ANSWER**

For thawed soil conditions, the volume-specific heat is:

$$C = \rho_d (c_{aggregate} + w\ c_{water}) = 2.2(0.80 + 0.12\ 4.19)$$
$$= 2.87 J/cm^3/^\circ C$$

where as for frozen soil conditions, assuming no volume change under freezing, it is:

$$C = \rho_d (c_{aggregate} + w c_{ice}) = 2.2\ (0.80 + 0.12\ 2.1) = 2.31\ \mathrm{J/cm^3/^\circ C}.$$

An additional relevant property is the latent heat of fusion, defined as the amount of heat released or absorbed as the soil water freezes or thaws, respectively, at a constant temperature. One Kg of water releases 334 kJ of heat when freezing at 0 °C (143.4 BTU/lb). Considering a soil with a water content $w$, and a dry density $\rho_d$ (lbs/ft$^3$), the latent heat per unit volume of soil $L$(BTU/ft$^3$) is computed using:

$$L = 1.43\ w\ \rho_d \qquad (10.29)$$

The one-dimensional diffusion law that governs the movement of heat within the pavement layers is:[18]

$$\frac{\partial^2 T}{\partial z^2} = \frac{\rho c}{k} \frac{\partial T}{\partial t} \qquad (10.30)$$

where $T$ denotes temperature, $t$ time, $z$ the depth within the layer, $\rho$ the density, $k$ the thermal conductivity, and $c$ the mass-specific heat of the material, as defined earlier (Note the similarity between this equation and Terzaghi's one-dimensional consolidation equation described in reference 15). The term $k/\rho c$ is referred to as *thermal difussivity* and has units of $m^2/hr$. Equation 10.29 can be solved using a numerical technique,[14] such as the finite difference, to advance the solution from time $t$ to time $t + 1$:

$$T_i^{t+1} = \frac{k}{\rho c}\frac{\Delta t}{\Delta z^2}\left(T_{i+1}^t + T_{i-1}^t\right) + \left(1 - 2\frac{k}{\rho c}\frac{\Delta t}{\Delta z^2}\right)T_i^t \quad (10.31)$$

where superscripts denote time and subscripts denote location, and $\Delta z$ and $\Delta t$ are the distance and time steps of the solution. This method is explained through the following example.

**Example 10.11**

An asphalt concrete layer 0.20 m thick rests on an infinite-depth subgrade. The temperature at the pavement surface is 35°C and the temperature at all the other points below it are initially 10°C. Assuming that the surface temperature remains constant and that there is no significant heat flow below a depth of 3.0 m, compute and plot their distribution with depth after 5, 10, 15, 20 and 25 hours. Given that the bulk densities of asphalt concrete and subgrade are 2.55 and 2.3 $g/cm^3$, their thermal conductivities are 1.45 and 1.0 $W/m^2/(°C/m)$, and their mass specific heat coefficients are 0.55 and 0.8 $J/g/°C$, respectively.

**ANSWER**

The heat diffusivity of the asphalt concrete and the subgrade are computed as:

$$\frac{k}{\rho c} = \frac{1.45}{2.55\ 10^6\ 0.55} = 1.03\ 10^{-6}\ m^2/sec \text{ or } 0.0037\ m^2/hr$$

and:

$$\frac{k}{\rho c} = \frac{1.00}{2.3\ 10^{-6}\ 0.8} = 0.54\ 10^{-6}\ m^2/sec \text{ or } 0.0020\ m^2/hr$$

In implementing Equation 10.30, it is important to select a depth increment, $\Delta z$, and a time increment, $\Delta t$, in such as way that the ratio $\frac{k}{\rho c}\frac{\Delta t}{\Delta z^2}$ is small (values smaller than 0.5 provide a stable solution). For this example, a depth increment of 0.05 m and a time increment of 0.1 hour (360 sec) were selected, resulting in:

$$\frac{k}{\rho c}\frac{\Delta t}{\Delta z^2} = 0.0037\ 40 = 0.148$$

$$\frac{k}{\rho c}\frac{\Delta t}{\Delta z^2} = 0.002\ 40 = 0.08$$

for the asphalt concrete and the subgrade layers, respectively. Applying Equation 10.31 is straightforward for all points except two, namely the boundary between the two layers and the lower boundary of the subgrade, below which zero heat flow is assumed. At the boundary between the two layers, Equation 10.31 is expanded to account for the different values of the ratio $\frac{k}{\rho c}\frac{\Delta t}{\Delta z^2}$ above and below the boundary:

$$T_i^{t+1} = \frac{\Delta t}{\Delta z^2}\left(\left.\frac{k}{\rho c}\right|_{subg} T_{i+1}^t + \left.\frac{k}{\rho c}\right|_{asph} T_{i-1}^t\right) + \left(1 - \frac{\Delta t}{\Delta z^2}\left.\frac{k}{\rho c}\right|_{asph} - \frac{\Delta t}{\Delta z^2}\left.\frac{k}{\rho c}\right|_{subg}\right) T_i^t$$

The lower heat-impermeable boundary is treated by visualizing a point below it, having the same temperature as the one above it (equal temperatures above and below the boundary satisfy no heat flow across). Example calculations are shown in Table 10.4, and the results are plotted in Figure 10.9 for the time intervals requested.

**Table 10.4**
Example Temperature Distribution Calculations for Example 10.11

| | Time (hrs) | | | | | |
|---|---|---|---|---|---|---|
| Depth (m) | 0 | 0.1 | 0.2 | 0.3 | 0.4 | 0.5 |
| 0 | 35 | 35 | 35 | 35 | 35 | 35 |
| 0.05 | 35 | 35 | 35 | 35 | 35 | 35 |
| 0.1 | 10 | 13.7 | 16.3048 | 18.21962 | 19.68177 | 20.83517 |
| 0.15 | 10 | 10 | 10.5476 | 11.31862 | 12.15681 | 12.98862 |
| 0.2 | 10 | 10 | 10 | 10.08104 | 10.25221 | 10.49854 |
| 0.25 | 10 | 10 | 10 | 10 | 10.01199 | 10.04659 |
| 0.3 | 10 | 10 | 10 | 10 | 10 | 10.00096 |
| 0.35 | 10 | 10 | 10 | 10 | 10 | 10 |

The example just presented demonstrates how pavement layer temperatures are computed under constant surface temperature. In reality, pavement surface temperature changes continuously, as a function of the solar radiation impacting it, the extent of daily sunshine/cloud cover, its reflectivity, and the ambient air temperature. Furthermore, the thermal properties of the pavement layers are not constant, but rather a function of their water content, as mentioned earlier. Hence, the heat flow and the moisture flow problems are interrelated. Dempsey et al.[10] developed a model, named Climatic-Materials-Structural (CMS), to solve the problem of heat flow diffusion in pavements. CMS was subsequently coupled with the ID model described earlier and incorporated into the EICM.[18] As mentioned earlier, version 3.01 of the EICM was incorporated into the NCHRP 1–37A Design Guide software[25] along with available environmental data and material properties.

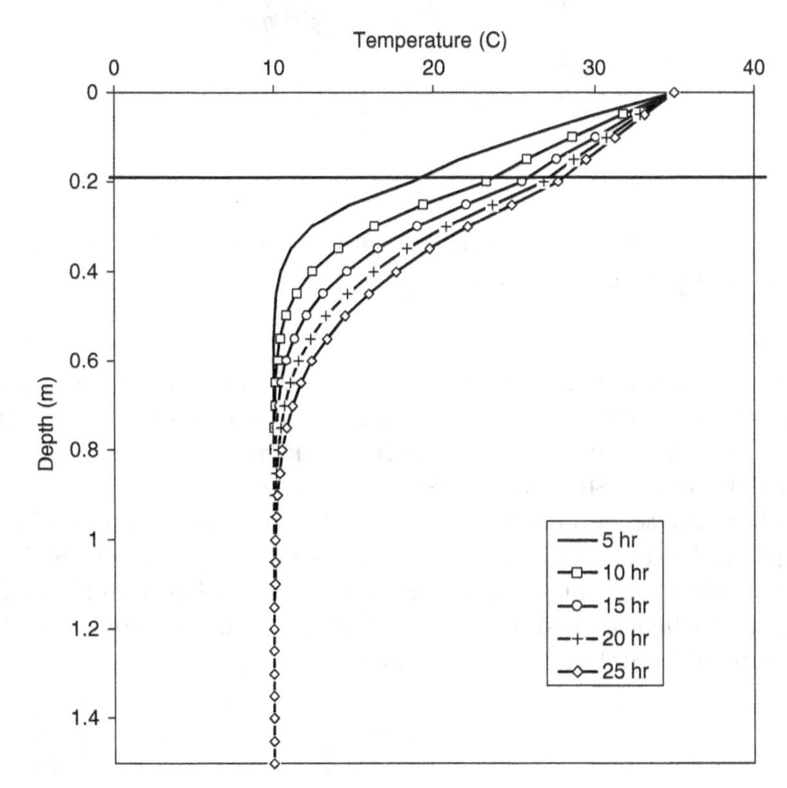


**Figure 10.9**
Answer to Example 10.11

## 10.3.2 Frost

The freezing of pore water and the melting of pore ice result in significant pavement layer volume changes, which over time and under the action of traffic loads reduce pavement serviceability. The depth of frost penetration is a function of the type of the pavement layers and the subgrade, their thermal characteristics, as well as the duration of sub-freezing temperatures, (days) and their magnitude (degrees). The last two variables are captured by the cumulative product of the number of days with mean pavement surface temperatures below freezing multiplied by the actual number of degrees below freezing, referred as the Freezing Index (*FI*).

**Table 10.5**
Temperature (°C) Data for Example 10.12

| Day | Max Temp | Min Temp |
|---|---|---|
| 1 | 5 | −12 |
| 2 | 1 | −13 |
| 3 | −3 | −18 |
| 4 | −2 | −20 |

**Example 10.12** The temperature data shown in Table 10.5 were obtained at the surface of pavement. Compute the *FI*.

**ANSWER**

The computations are shown in Table 10.6. The *FI* for these four days is −33°C-days. Similar computations can be carried out using °F. The difference is that the number of degrees below freezing are computed with reference to 32°F.

There are several empirical relationships between the *FI* and the depth of frost penetration (e.g., 9). Alternatively, the depth of frost penetration $z$(ft) can be approximately computed considering only the latent heat of fusion of the soil water. The resulting formula, developed by Stefan,[22] is expressed as:

$$z = \sqrt{\frac{48\, k\, FI}{L}} \tag{10.32}$$

**Table 10.6**
Freezing Index Computation for Example 10.12

| Day | Max Temp | Min Temp | Mean 0°C | Degrees Below 0°C | Cumulative Degree-Days |
|---|---|---|---|---|---|
| 1 | 5 | −12 | −3.5 | −3.5 | −3.5 |
| 2 | 1 | −13 | −6 | −6 | −9.5 |
| 3 | −3 | −18 | −10.5 | −10.5 | −20 |
| 4 | −2 | −20 | −11 | −11 | −33 |

where $k$ is the thermal conductivity in BTU/hr/ft$^2$/(°F/ft), $L$ is the latent heat per unit volume of soil in BTU/ft$^3$, as computed by Equation 10.29 and *FI* is in °F-days. This formula neglects the volumetric heat of the solid phase and, as a result, tends to overpredict the depth of frost in temperate climates.

**Example 10.13**

A pavement subgrade experiences winter temperatures that result in a Freezing Index of 700 °F-days per year. Estimate the depth of frost penetration, given that its coefficient of thermal conductivity is 0.43 BTU/hr/ft$^2$/(°F/ft) (0.75 W/m$^2$/(°C/m)), its water content is 15%, and its dry density is 137.34 lb/ft$^3$ (2.2 gr/cm$^3$).

**ANSWER**

Compute the latent heat per unit volume of soil using Equation 10.29:

$$L = 1.43\ 15\ 137.34 = 2,945.9\ \text{BTU/ft}^3$$

Substituting the given values into Equation 10.32 gives:

$$z = \sqrt{\frac{48\ k\ FI}{L}} = \sqrt{\frac{48\ 0.43\ 700}{2945.9}} = 2.2\ \text{ft}\ (0.67\ \text{m})$$

Berggren[4] developed a modified expression for the depth of the frost penetration, taking into account the effect of the soil volumetric heat. The resulting expression, known as the Berggren formula, reduces the depth of frost penetration computed earlier (Equation 10.32):

$$z = \lambda \sqrt{\frac{48\ k\ FI}{L}} \tag{10.33}$$

where $\lambda$ is a correction coefficient ranging from 0 to 1. Aldrich[1] developed a nomograph for estimating this correction coefficient, thus allowing solution of the Berggren formula, (Figure 10.10, after referance 24). It should be noted that this nomograph is for Imperial units. The two variables required for estimating $\lambda$ through this nomograph are the fusion parameter $\mu$ and the thermal ratio $\alpha$, defined as:

$$\mu = \frac{C_{avg}\ FI}{n\ L} \tag{10.34}$$

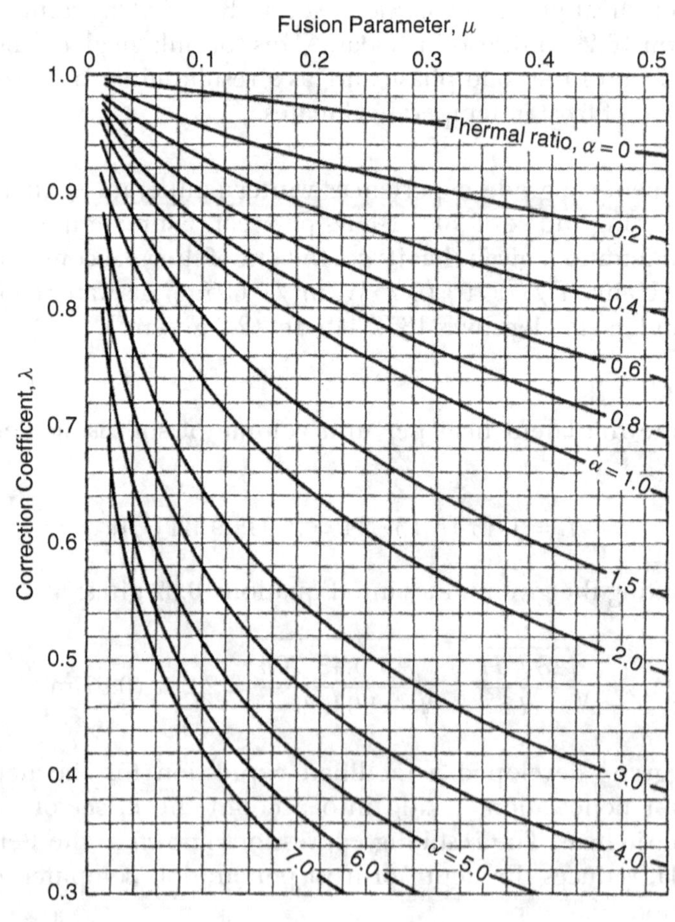


**Figure 10.10**
Nomograph for Obtaining Adjustment Factor for Berggren's Formula (Ref. 24)

and:

$$\alpha = \frac{V_0 \ n}{FI} \tag{10.35}$$

where $C_{avg}$ is the average of the volume-specific heat coefficients for the frozen and unfrozen soil (BTU/ft$^3$/°F), $n$ is the number of sub-freezing days per year, and $V_o$ is the difference between the mean annual temperature in °F and 32°F.

**Example 10.14**

A pavement subgrade experiences winter conditions consisting of 65 days of frost and 700 °F-days (389°C-days) annually with a mean annual temperature of 50°F (10°C). Compute the depth of frost penetration, given that the dry density of the soil is 131 lb/ft$^3$ (2.1 g/cm$^3$), its water content is 18%, its coefficient of thermal conductivity is 0.43 Btu/hr/ft$^2$/(°F/ft), (0.75 W/m$^2$/(°C/m), and its average volume-specific heat is 35.8 BTU/ft$^3$/°F(2.4 J/cm$^3$/°C).

**ANSWER**

Use Equation 10.29 to compute the latent heat of the soil:

$$L = 1.34 \ 18 \ 131 = 3159.7 \ \text{BTU/ft}^3 \text{ or } 117,726 \ \text{kJ/m}^3$$

Use Equations 10.34 and 10.35 to obtain the fusion parameter $\mu$ and the thermal ratio $\alpha$, respectively:

$$\mu = \frac{C_{avg} \ FI}{n \ L} = \frac{35.8 \ 700}{65 \ 3159.7} = 0.12$$

$$\alpha = \frac{V_0 \ n}{FI} = \frac{(50 - 32)65}{700} = 1.67$$

Entering these parameters into Figure 10.9 gives a value of $\lambda$ of 0.76, which allows computing the depth of frost penetration from Equation 10.33 as:

$$z = \lambda \sqrt{\frac{48 \ k \ FI}{L}} = 0.76 \sqrt{\frac{48 \ 0.43 \ 700}{3159.7}} = 1.62 \ \text{ft} \ (0.5 \ \text{m})$$

## References


[1] Aldrich, G. D, (1956). "Frost Penetration below highway and airfield pavements", Highway Research Board, Bulletin 135.

[2] ASTM (1989). "Determining Apparent Opening Size of a Geotextile," ASTM Standard D-4s751, *Concrete and Aggregates,* Vol. 04.02, Conshocken, PA.

[3] ASTM (1989). "Determining Permeability of Geotextiles," ASTM Standard D-4491, *Concrete and Aggregates,* Vol. 04.02.

[4] Berggren, W. P., (1943). "Prediction of Temperature Distribution in Frozen Soils," Transaction. AGU, Part 3, pp. 71–77.

[5] Casagrande, A., and W. L. Shannon (1952). *Base Course Drainage for Airport Pavements, Proceedings of the American Society of Civil Engineers,* Vol. 77, pp. 792–814.

[6] Cedergren, H. R., J. A. Arman, and K. H. O'Brien (1973). "Guidelines for the Design of Subsurface Drainage Systems for Highway Pavement Structural Sections," Report No. FHWA-RD-73–14, Federal Highway Administration, Washington DC.

[7] Cedergren, H. R., (1974). *Drainage of Highway and Airfield Pavements,* John Wiley & Sons, Inc., New York.

[8] Cedergren, H. R. (1974). *Seepage Drainage and Flow Nets,* John Wiley & Sons, Inc., New York 1959.

[9] U.S. Army and Air Force (1992), Pavement Design for Roads, Streets, Walks, and Open Storage Areas, TM 5-822-5/AFM 88-7, Chapter 18, pp. 18-10 to 18-13.

[10] Dempsey, B. J., W. A. Herlach, and J. Patel (1985). The Climatic Material-Structural-Pavement Analysis Program, Federal Highway Administration, Final Report, Vol. 3., FHWA-RD-84–115, Washington, DC.

[11] AASHTO (1993). *Design of Pavement Structures,* American Association of State Highway and Transportation Officials Washington, DC.

[12] FHWA (1992). "Drainage Pavement System Participant Notebook", Report No. FHWA-SA-92–008, Federal Highway Administration, Washington, DC.

13 FHWA (1989). *Geotextile Engineering Manual*, Report No. FHWA-HI-810.050, Federal Highway Administration, Washington, DC.

14 Gerald, C. F., and P. O. Wheatley (1984). *Applied Numerical Analysis*, 3rd ed. Addison-Wesley,

15 Holtz, R. D. and W. D. Kovacs (1981). *An Introduction to Geotechnical Engineering*, Prentice Hall Inc., EagleWood Cliff, NJ.

16 Lambe W. T. and R. V. Whitman, *Soil Mechanics*, (1979). SI Version, John Wiley & Sons, Inc., New York.

17 Liu, H. S. and R. L. Lytton 1985. *Environmental Effects on Pavement Drainage*, Vol. IV, Federal Highway Administration Report FHWA-DTFH-61–87-C-00057, Washington, DC.

18 Lytton, R. L., D. E. Pufhal, C. H. Michalak, H. S. Liang, and B. J. Dempsey (November 1995). *"An Integrated Model of the Climatic Effect on Pavements,"* Federal Highway Administration Report FHWA-RD-90–033, Washington, DC.

19 Mallela, J. G., Larson, T., Wyatt, J. Hall and W. Baker, (2002). *User's Guide for Drainage Requirements in Pavements* — DRIP 2.0 Microcomputer Program, Federal Highway Administration Report FHWA-DTFH61-00-F-00199, Washington, DC.

20 Moulton, L. K., (1980). ''Highway Subsurface Design'', Federal Highway Administration Report FHWA-TS-80–224, Washington, DC.

21 Ridgway, H. H., (1976). ''Infiltration of Water Through Pavement Surfaces'', Transportation Research Record 616, Transportation Research Board, Washington, DC.

22 Stefan J., (1889). *On the Theory of Ice Formation, Particularly Ice Formation in the Arctic Ocean*, (in German) S-B Wien Akad v98, 173.

23 U.S. Army Corps of Engineers (1955). ''Drainage and Erosion Control — Subsurface Drainage Facilities for Airfields,'' Part XIII, Chapter 2, *Engineering Manual*, Military Construction, Washington DC.

24 Yoder E. J. and M. W. Witczak (1975). *Principles of Pavement Design*, 2nd ed. Wiley Interscience, New York.

25 NCHRP (July 2004). ''2002 Design Guide: Design of New and Rehabilitated Pavement Structures,'' Draft Final Report, NCHRP

Study 1–37A, National Cooperative Highway Research Program, Washington DC.

## Problems

10.1 A two-lane rigid pavement surface is 14.4 m wide and has transverse joints spaced at 3.0 m intervals. A longitudinal construction joint is located in the middle the surface. In addition, there is a single longitudinal crack along the full length of the slab. Compute the infiltration rate according to reference 21 and compare it to the simplified approach described in reference 1, assuming it is located in central Texas. Consider the infiltration rate of intact concrete to be negligible.

10.2 A base layer resting on a horizontal impermeable subgrade has a coefficient of permeability of 2.0 m/hour; it is 0.40 m thick and has a maximum drainage path of 4.6 m. Compute the maximum rate of infiltration that this layer can accommodate, as well as the speed of movement of the pore water, given an effective porosity $n_e$ of 0.28.

10.3 A homogeneous base layer 0.45 m thick resting on an impermeable subgrade has a coefficient of permeability of 2.0 m/hour. The pavement surface has a grade of – 6% and a cross-slope of – 3%, (refer to Figure 10.5). The pavement width is 3.8 m. Compute the discharge rate capability of this layer.

10.4 A 0.35 m thick base layer has a slope of 0.044; it is 4.5 m long, has a coefficient of permeability of 0.2 m/hour, and an effective porosity of 0.30. Compute the time it will take to reach a degree of drainage of 50%. How is the drainage quality of this layer characterized?

10.5 Consider a permeable base layer resting on a soil layer with gradations as shown in Figure 10.11. Determine if there is a need for an intermediate filter layer between the two, and, if so, design a geotextile (i.e., provide a range of AOS) to separate the two.

10.6 An asphalt concrete layer 0.25 m thick rests on an infinite-depth subgrade. The temperature at the pavement surface

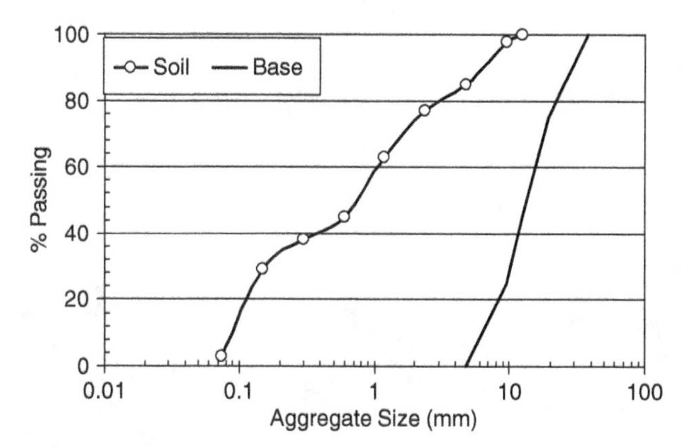


**Figure 10.11**
Gradations for Problem 10.5

is 25°C and the temperature at all the other points below it are initially 12.5°C. Assuming that the surface temperature remains constant and that there is no significant heat flow below a depth of 3.0 m, compute and plot their distribution, with depth, after 5, 10, 15, 20, and 25 hours. Given, the bulk densities of asphalt concrete and subgrade are 2.65 and 2.4 $g/cm^3$, their thermal conductivities are 1.4 and 1.1 $W/m^2/(°C/m)$, and their mass-specific heat coefficients are 0.50 and 0.85 J/g/°C, respectively. Suggestion: Use a distance increment of 0.05 m and a time increment of 0.1 hours.

10.7 A pavement subgrade experiences winter conditions consisting of 85 days of frost and 850°F-days (472°C-days) annually with a mean annual temperature of 40°F (4.5°C). Compute the depth of frost penetration, given that the dry density of the soil is 140 $lb/ft^3$ (2.24 $g/cm^3$), its water content is 12%, its coefficient of thermal conductivity is 0.43 $Btu/hr/ft^2/(°F/ft)$, (0.75 $W/m^2/(°C/m)$), and its average volume-specific heat is 35.8 $BTU/ft^3/°F$ (2.4 $J/cm^3/°C$).

# 11 Structural Design of Flexible Pavements

## 11.1 Introduction

As described in Chapter 1, flexible pavements derive their load-carrying capacity by distributing surface stresses in the underlying layers over an increasingly wide area. This layered action allows computing their structural response using relationships based on Boussinesq's solutions, as described in Chapter 7.

This chapter describes the three main methodologies available for flexible pavement design, namely the American Association of State Highway and Transportation Officials (AASHTO) 1986/1993 method[2] the Asphalt Institute (AI) design method,[3] and the method proposed by the NCHRP 1-37A Study.[24] Typically, state DOTs utilize all available design methods, including some reflecting the performance of their own pavements, and use judgment and economic considerations in selecting the final layer thickness combinations that meet their design criteria.

## 11.2 AASHTO 1986/1993 Design Method

### 11.2.1 Historical Overview

The basis of the current AASHTO flexible and rigid pavement design methods is a landmark pavement performance test conducted in the late 1950s near Ottawa, Illinois, at a cost of $27 million (1960 dollars). It was administered by the then American Association of State Highway Officials, hereafter referred to as the AASHO Road Test.[1] Its general configuration consisted of four two-lane loops, each 2 miles long, located on the future alignment of I-80. In addition, two smaller loops were constructed off this alignment for conducting special studies. Each of the large loop-lanes involved pavement test sections built of different combinations of layer thickness, both flexible and rigid. Each of these loop-lanes was assigned a particular truck configuration of fixed axle loads that drove around two eight-hour shifts per day. A variety of pavement evaluation measurements was collected at regular biweekly intervals. These included pavement roughness (summarized in the form of the slope variance measured by a device shown in Figure 9.2) and pavement distress (cracking, rutting, and so on). For each test section, data collection continued until it reached the end of its functional life—that is, a terminal serviceability value of 2.0 in terms of the *PSI*. Sections failed within a two year period, from 1958 to 1960.

The obvious limitation of the accelerated loading in this short-term experiment was that the effect of the environment was underestimated. Nevertheless, this experiment generated the first substantial database of pavement performance observations under controlled traffic. Regression analysis of this data generated the first empirical relationships between the number of axle passes to serviceability failure, structural characteristics (i.e., the *SN* defined by Equation 2.5), and axle configuration/axle load. These relationships were used in establishing the load equivalency factors (i.e., the *ESAL* factors described in chapter 2) and the first empirical pavement design equations for both flexible and rigid pavements. This early data formed the basis for the pavement design methodology adopted by AASHTO and still in use today.[2] The methodology used for the design of flexible pavements is described next.

### 11.2.2 Serviceability Loss Due to Traffic

The serviceability loss due to traffic is computed from an empirical relationship derived from AASHO Road Test data. It relates the number of cumulative *ESAL* passes to the corresponding change in pavement serviceability, $\Delta PSI$. It is expressed in the following format (Imperial units):

$$\log(W_{18}) = Z_R S_0 + 9.36 \log(SN + 1) - 0.20 + \frac{\log\left[\dfrac{\Delta PSI}{4.2 - 1.5}\right]}{0.4 + \dfrac{1094}{(SN + 1)^{5.19}}} + 2.32 \log(M_r) - 8.07 \tag{11.1}$$

where:

$W_{18}$ = the number of *ESALs* that will result in a change in serviceability of $\Delta PSI$.
$SN$ = the structural number defined by Equation 2.5.
$M_r$ = the resilient modulus of the subgrade, as defined by Equation 3.1.

The variables $Z_R$ and $S_0$ are the standard normal deviate and the standard error in predicting pavement serviceability, respectively. Values of the standard normal deviate are given in Table 11.1 for selected one-sided reliability levels. $S_0$ combines the standard errors in predicting traffic loading (i.e., *ESALs*) and in predicting performance to the end of a pavement's functional life, (e.g., *PSI* of 2.0). These two sources of uncertainty are explained through the pavement performance curve shown in Figure 11.1 and defined here:

- $w_T - N_T$ = difference between the actual and the predicted number of design life *ESALs*, which could be positive or negative.
- $N_t - W_t$ = difference between the actual and the predicted number of *ESALs* that will reduce serviceability to a terminal level, which could be positive or negative.

To compensate for these two sources of uncertainty and reliably predict layer thicknesses that will prevent failure before the design

**Table 11.1**
Values of the Standard Normal Deviate $Z_R$

| Reliability % | $Z_R$ |
|---|---|
| 80 | −0.841 |
| 85 | −1.037 |
| 90 | −1.282 |
| 95 | −1.645 |
| 99 | −2.327 |
| 99.9 | −3.090 |

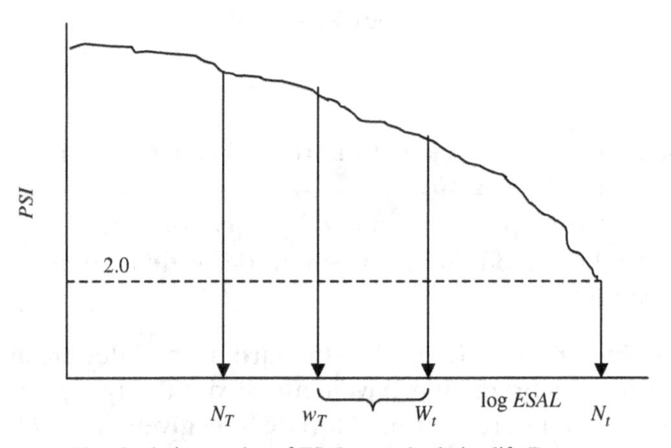


$N_T$ = the design number of *ESALs* over the design life *T*.

$w_T$ = the actual number of *ESALs* over the design life *T*.

$W_t$ = the predicted number of *ESALs* to reach terminal *PSI* (Equation 11.1).

$N_t$ = the actual number of *ESALs* to reach terminal *PSI*.

**Figure 11.1**
Pavement Design Reliability Concept (Ref. 2)

period is reached, the difference $w_T$–$W_t$ is set to a negative value (i.e., $W_t$ is selected larger than $w_T$). Their difference is set equal to the product of the standard error in predicting *PSI* (typically, between 0.25 and 0.6) multiplied by the standard normal deviate that corresponds to the desired reliability level. Note that since Equation 11.1 is in a logarithmic form, this approach results in significant increases in the number of *ESALs* being input (i.e., $W_t$). For example, for a 95% reliability, and given a standard error in predicting *PSI* of 0.5, the logarithm of *ESALs* is increased by 1.645 $\times$ 0.5 = 0.8225, which arithmetically represents a factor of $10^{0.8225}$ = 6.645.[15]

**Example 11.1**

Calculate the required layer thicknesses for a new asphalt concrete pavement on a fair draining base and subgrade (i.e., the water drains out of the pavement within a period of one week). It is estimated that the pavement structure becomes saturated less frequently than 1% of the time. The following data is also given:

- Estimated number of *ESALs* over a 12-year maximum performance period = 2.5 million.
- Subgrade resilient modulus = 30,000 lbs/in.$^2$ (206.8 MPa).
- Design reliability = 95%.
- Standard error in predicting serviceability = 0.40.
- $\Delta PSI$ = 2.2 (from 4.2 to 2.0).

**ANSWER**

Substitute the given values into Equation 11.1 and solve for *SN*.

$$6.397 = -1.645\ 0.4 + 9.36\ \log(SN+1) - 0.20 + \frac{\log\left[\dfrac{2.2}{4.2-1.5}\right]}{0.4 + \dfrac{1094}{(SN+1)^{5.19}}} + 2.32\log(30000) - 8.07$$

The value of the *SN* that satisfies this expression is 2.4. Equation 2.5 is used for decomposing *SN* into constitutive layer thicknesses. The drainage conditions described suggest drainage coefficients for the base/subgrade layers between 1.25 and 1.15 (Table 2.7). The structural layer coefficients corresponding to new layers are 0.44, 0.14, and 0.11 for the asphalt concrete, base, and subbase, respectively. Subsisting these values into Equation 2.5 gives:

$$2.4 = 0.44\ D_1 + 0.14\ 1.2\ D_2 + 0.11\ 1.2\ D_3$$

or:

$$2.4 = 0.44\ D_1 + 0.168\ D_2 + 0.132\ D_3$$

Clearly, there is no unique combination of layer thickness that satisfies this expression. A sequential method is suggested for computing the thickness of the individual layers.[2] It consists of using Equation 11.1 to compute the *SN* value of layers, one at a time, as if they were supported by a subgrade with the modulus of the underlying layer, as shown in Figure 11.2. That is, using Equation 11.1, compute $SN_1$ as if the pavement was supported by a subgrade of modulus $E_2$; and compute $SN_2$ as if the pavement was supported by a layer of modulus $E_3$. The process is explained through the steps shown in Table 11.2. The pavement layer moduli for the base and subbase are given equal to 70,000 lbs/in$^2$ (482 MPa) and 50,000 lbs/in$^2$ (344 MPa), respectively. In practice, however, the selection of layer thickness is carried out using minimum layer thickness requirements and minimum cost considerations, given the unit prices of the layers involved.

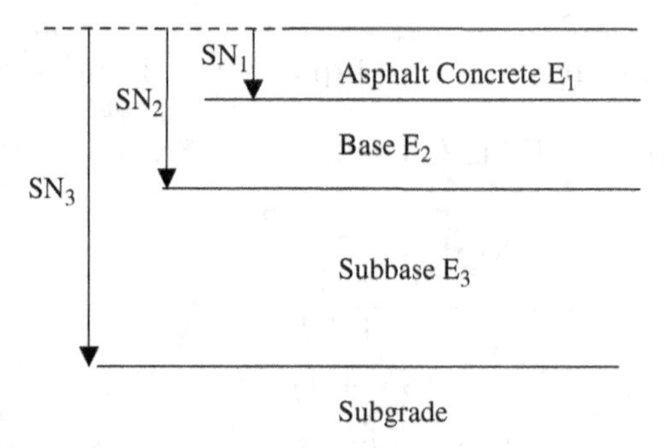


**Figure 11.2**
Schematic of the Recommended Method for Computing Layer Thicknesses from *SN*, (Ref. 2).

**Table 11.2**
Computing Layer Thicknesses from *SN* for Example 11.1

| Step | Support Modulus lbs/in.$^2$ | *SN* (Equation 11.1) | D | Rounded to |
|---|---|---|---|---|
| 1 | 70,000 | 1.75 | $D_1 = 1.75/0.44 = 3.97$ | 4 in. |
| 2 | 50,000 | 1.99 | $D_2 = (1.99 - 4\ 0.44)/0.168 = 1.36$ | 2 in. |
| 3 | 30,000 = *Mr* | 2.4 | $D_3 = (2.4 - 4\ 0.44 - 2\ 0.168)/0.132 = 2.3$ | 3 in. |

### 11.2.3 Serviceability Loss Due to Environment

The 1993 version of the AASHTO pavement design guide[2] considers pavement serviceability loss due to subgrade swelling and frost heave. The serviceability loss due to swelling, $\Delta PSI_{SW}$, is given as a function of time $t$ (years) by:

$$\Delta PSI_{SW} = 0.00335\ V_R P_s \left(1 - e^{-\theta\, t}\right) \qquad (11.2)$$

where:

$V_R$=the potential vertical rise due to swelling (inches), which is mainly a function of the Plasticity Index of the subgrade, as shown in the nomograph on Figure 11.3.

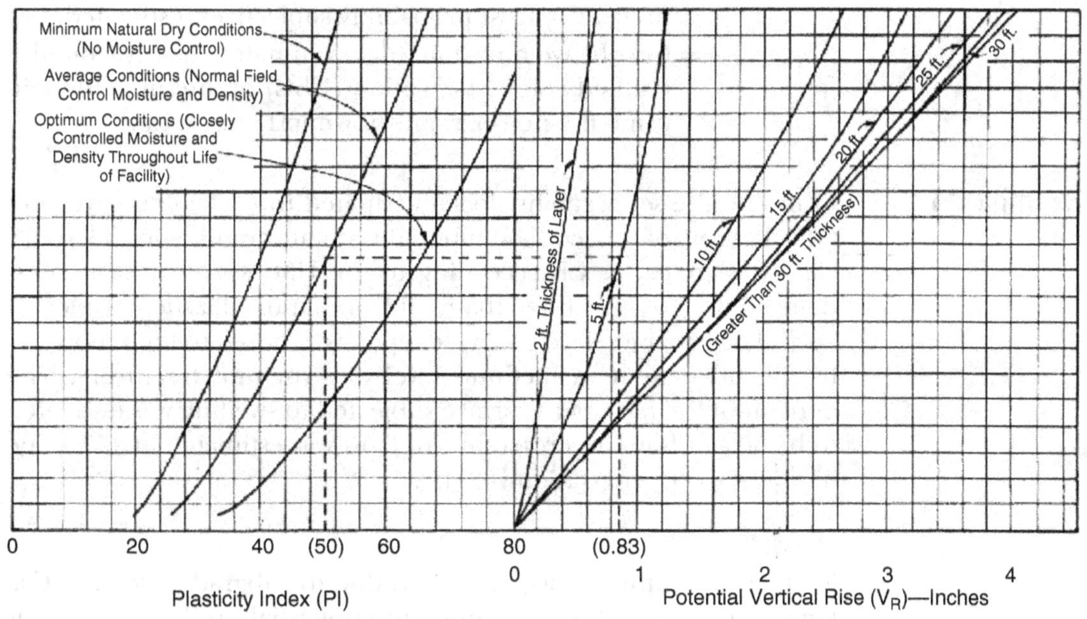


Notes:
1. This figure is predicated upon the following assumptions:
   a. The subgrade soils for the thikness shown all are passing the No. 40 mesh sieve.
   b. The subgrade soils has a uniform moisture content and Plasticity Index throughout the layer thickness for the conditions shown.
   c. A surchage pressure from 20 inc of overburden ±10 inc will have no material effect).
2. Calculations are required to determine $V_R$ for other surchage pressure

**Figure 11.3**
Chart for Estimating the Potential Subgrade Vertical Rise Due to Swelling, $V_R$ (Ref. 2 Used by Permission)

$P_S$ = the percent of the total pavement area subjected to swelling.
$\theta$ = a subgrade swelling rate constant that can be estimated from the nomograph in Figure 11.4.

The serviceability loss due to frost heave, $\Delta PSI_{FH}$, is given as a function of time $t$ (years) by:

$$\Delta PSI_{FH} = 0.01\ p_f\ \Delta PSI_{nax}\ \left(1 - e^{-0.02\,\phi\,t}\right) \tag{11.3}$$

where:

$\Delta PSI_{max}$ = the maximum serviceability loss due to frost heave estimated on the basis of drainage quality and depth of frost penetration using the nomograph shown in Figure 11.5.
$p_f$ = percent of frost probability subjectively estimated.
$\phi$ = frost heave rate (mm/day) estimated mainly from the Unified Soil Classification (USC) of the subgrade soil using the nomograph shown in Figure 11.6.

**Example 11.2** Compute the serviceability loss anticipated for a flexible pavement after 12 years of service. The subgrade is a fair-draining low-plasticity clay (designated as CL according to the USC system), having less than 60% by weight finer than 0.02 mm, and a Plasticity Index of 26%. The subgrade layer is 25 ft deep; it is exposed to high moisture levels and exhibits a medium level of structural fracturing. The percent of the pavement surface subjected to swelling was estimated to be 50%, while the probability of frost was estimated at 45%. The depth of frost penetration is 2 ft.

**ANSWER**
First, estimate the serviceability loss due to subgrade swelling. Use Figures 11.3 and 11.4 to compute the potential swelling rise, $V_R$ as 0.5 in. and the swelling rate $\theta$ as 0.15, respectively. Substituting these values into Equation 11.2, gives:

$$\Delta PSI_{SW} = 0.00335\ 0.5\ 50\ \left(1 - e^{-0.15\ 12}\right) = 0.07$$

Next, estimate the serviceability loss due to subgrade frost heave. Use Figure 11.5 and 11.6 to compute the maximum serviceability

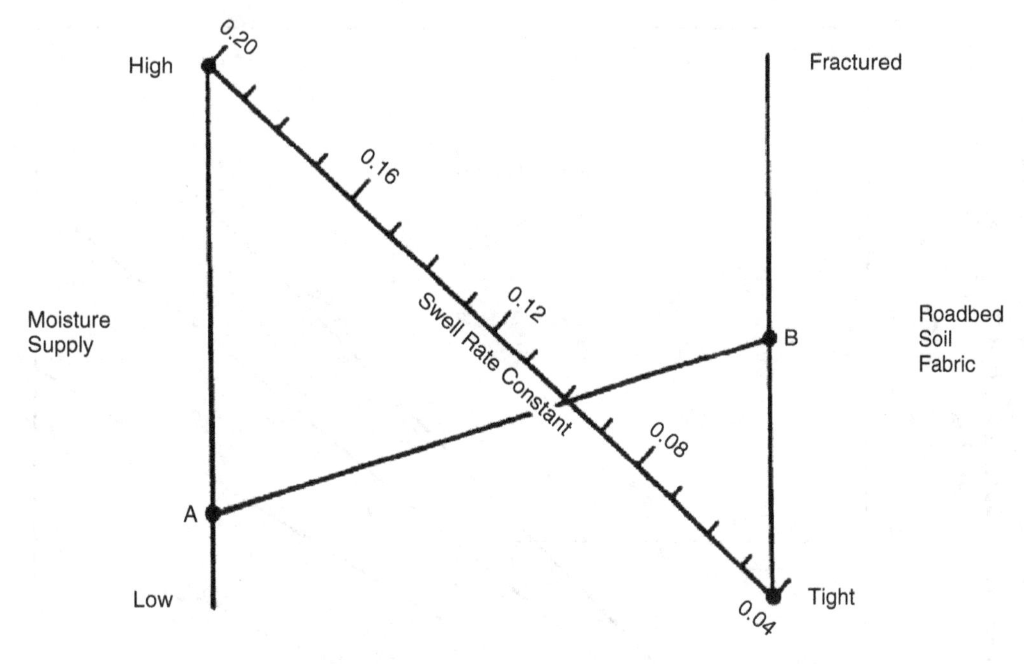


Notes: (a) Low-Moisture Supply:

Low rainfall
Good drainage

(b) High-Moisture Supply

High rainfall
Poor drainage
Vicinity of culverts, bridge abutments, inlet leads

(c) Soil Fabric Conditions (self-explanatory)

(d) Use of the Nonograph

(1) Select the appropriate moisture supply condition, which may be somewhere between low and high (such as A).

(2) Select the appropriate soil fabric (such as B). This scale must be developed by each ndividual agency.

(3) Draw a straight line between the selected points (A to B).

(4) Read swell rate constant from the diagonal axis (read 0.10).

**Figure 11.4**
Chart for Estimating the Swell Rate Constant, $\theta$ (Ref. 2 Used by Permission)

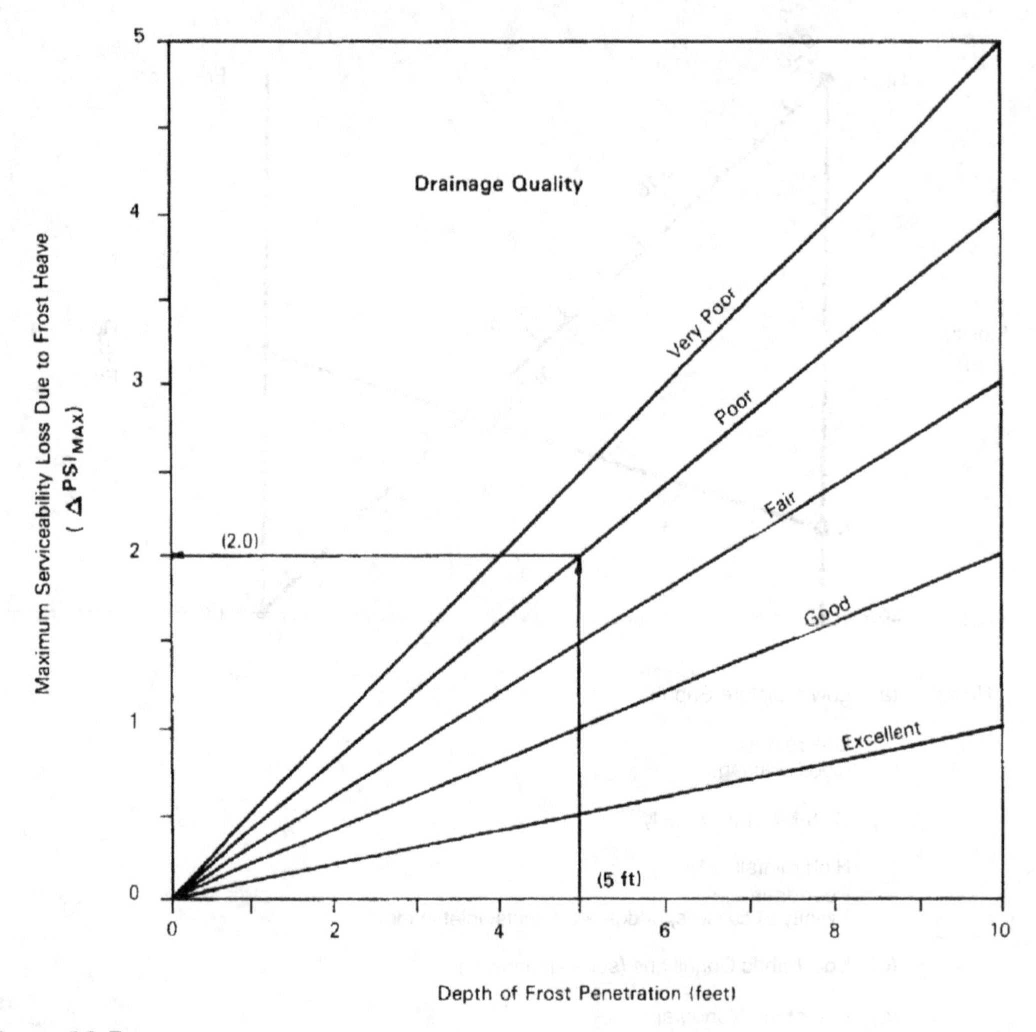


**Figure 11.5**
Chart for Estimating the Maximum Serviceability Loss Due to Frost Heave, $\Delta PSI_{max}$ (Ref. 2 Used by Permission)

loss $\Delta PSI_{max}$ as 0.6 and the frost heave rate, $\phi$, as 3 mm/day, respectively. Substituting these values into Equation 11.3, gives:

$$\Delta PSI_{FH} = 0.01\ 45\ 0.6\ \left(1 - e^{-0.02\ 3\ 12}\right) = 0.138$$

Adding the two serviceability loss components gives the total serviceability loss due to the environment, $\Delta PSI_{SWFH}$, equal to approximately 0.208.

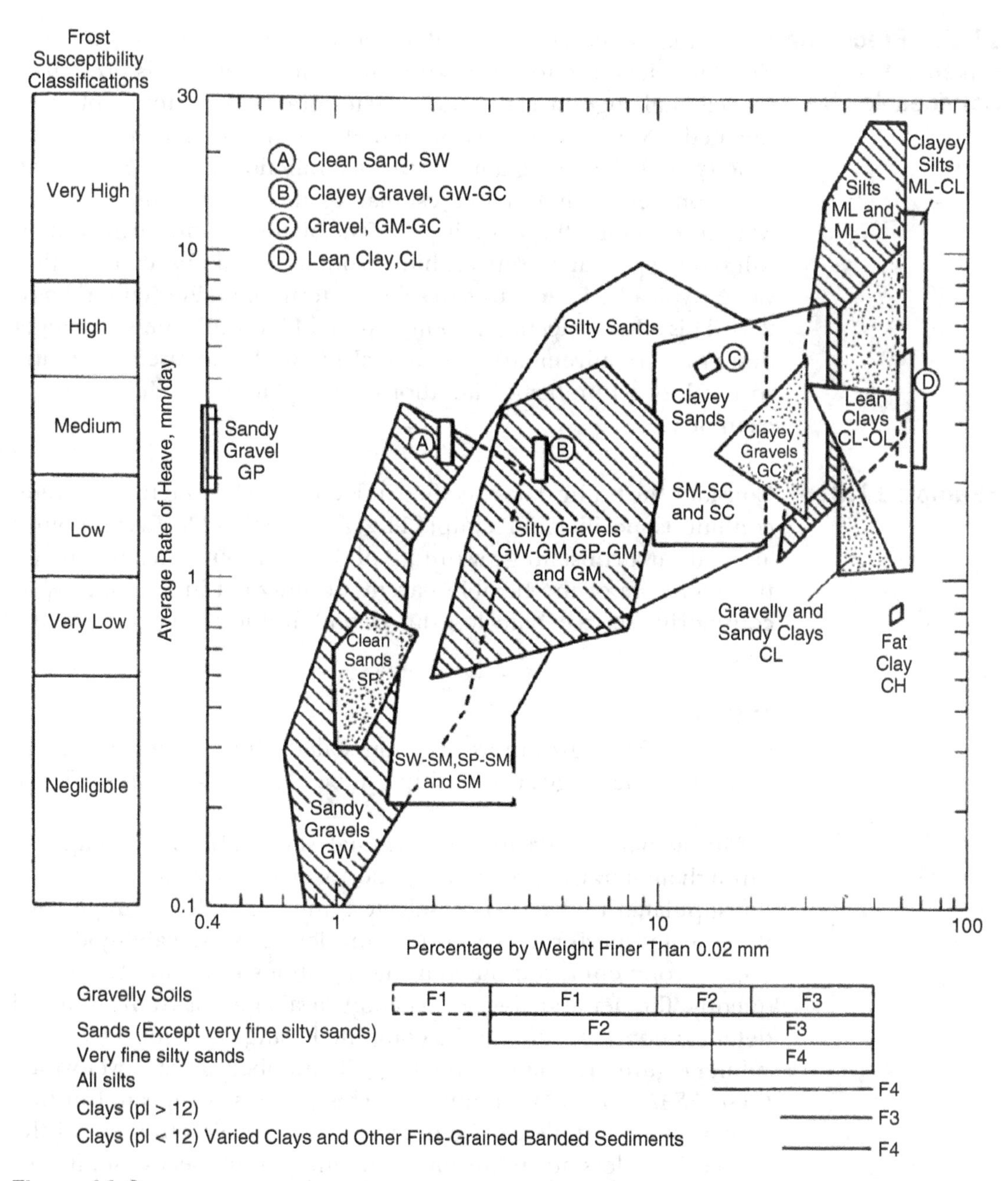


**Figure 11.6**
Chart for Estimating Frost Heave Rate, $\phi$ (Ref. 2 Used by Permission)

### 11.2.4 Predicting Pavement Serviceable Life

Predicting pavement serviceable life involves computing the serviceability loss due to traffic and the environment. This presents a special challenge, because traffic deterioration is in terms of accumulated *ESALs*, while environmental deterioration is in terms of time (years). Assuming a mathematical function for the *ESAL* accumulation versus time allows estimating the number of years that will lapse before the combined effect of traffic and environment will reduce pavement serviceability from its initial post-construction value (typically, 4.2 to 4.6) to a selected terminal value (either 2.5 or 2.0). This allows selecting a combination of layer thicknesses to meet design life requirements. The actual methodology used for doing so involves a number of iterations, as explained by the following example.

**Example 11.3**

Consider the traffic data specified in Example 11.1 and the subgrade conditions specified in Example 11.2. Assume that *ESALs* compound annually at a constant growth rate of 3%. Compute the anticipated pavement life of the flexible pavement designed in Example 11.1, considering serviceability loss due to both traffic and environmental factors.

**ANSWER**

Given the *ESAL* growth assumption stated, the accumulated *ESAL* versus pavement age relationship can be plotted as shown in Figure 11.7.

The actual life (years) to terminal serviceability is computed through an iterative procedure. A performance period shorter than the stipulated twelve years is selected, nine years for example. For the selected performance period, the loss in serviceability due to the environment is computed using Equations 11.2 and 11.3. Subsequently, the net available serviceability available for traffic-induced deterioration (*ESALs*), can be computed using Equation 11.1. Consulting Figure 11.7 allows estimating the number of years over which these *ESALs* will be accumulated. This process is repeated until a certain pavement life can be determined, for which the sum of the serviceability loss due to the environment, plus the serviceability loss due to traffic, adds to the total available, which for this example is 2.2. These steps are shown in Table 11.3.

After nine years, for example, the number of *ESALs* that correspond to a net traffic-related serviceability loss of 2.025 is computed

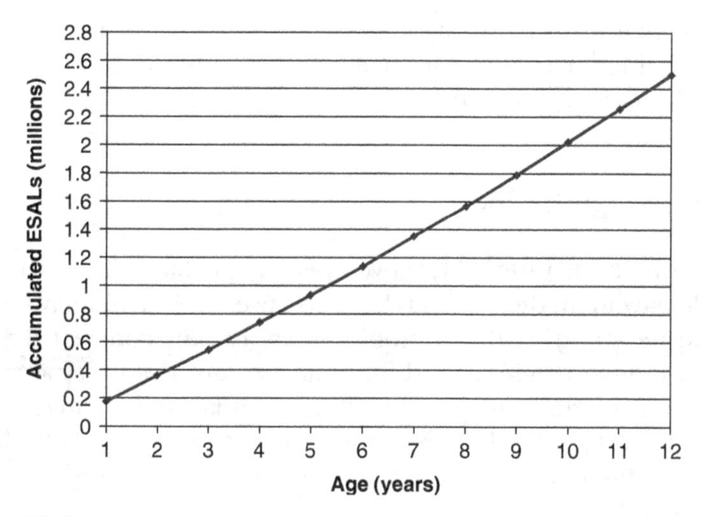


**Figure 11.7**
Accumulation of *ESALs* versus Time for Example 11.3

**Table 11.3**
Summary of Serviceability Loss Computations Due to Traffic and Environmental Factors, Example 11.3

| Iteration | Performance Period (years) | PSI Loss Due to Environment | Net PSI Available for Traffic | Period to Accumulate Corresponding ESALs (years) |
|---|---|---|---|---|
| 1 | 9 | 0.175 | 2.2–0.175 = 2.025 | 11.6 |
| 2 | 10 | 0.187 | 2.2–0.187 = 2.013 | 11.3 |
| 3 | 11 | 0.198 | 2.2–0.198 = 2.002 | 11 |

from Equation 11.1 as:

$$-1.645\ 0.4 + 9.36\ \log(2.4+1) - 0.20 + \frac{\log\left[\dfrac{2.025}{4.2-1.5}\right]}{0.4 + \dfrac{1094}{(2.4+1)^{5.19}}}$$
$$+ 2.32 \log(30000) - 8.07 = 6.3794$$

or 2,395,729 *ESALs*. The overall life for this section under the combined effect of traffic and environment is 11 years. Commercially available software can be used to perfrom these calculations, along with the other provisions of the 1993 AASHTO pavement design guide.[4] For more information on software sources, go to www.wiley.com/go/pavement.

## 11.3 Asphalt Institute Design Method

The Asphalt Institute (AI) developed a mechanistic method for flexible pavement design.[3] It is based on two criteria, namely limiting the tensile strain at the bottom of the asphalt concrete layer to prevent fatigue cracking, and limiting the compressive strain at the top of the subgrade to prevent subgrade plastic deformation that will result in rutting. The expression used to relate the number of cycles to fatigue failure $N_f$ and the asphalt concrete tensile strain $\varepsilon_t$ was adopted from work by Finn et al.:[6]

$$N_f = 0.0795\ \varepsilon_t^{-3.291}\ E^{-0.854} \tag{11.4}$$

where $E$ is the elastic modulus of the asphalt concrete layer in lbs/in$^2$. Fatigue cracking failure was defined as fatigue cracking covering 10% of the area in the wheel-paths. As described in Chapter 5, the modulus of the asphalt concrete varies with the temperature and the rate of loading; as a result, $N_f$ and the associated fatigue damage rate varies with the season and the vehicle speed.

The expression used to relate the number of cycles to rutting failure $N_r$ and the subgrade vertical compressive strain $\varepsilon_v$ is:

$$N_r = 1.365\ 10^{-9}\ \varepsilon_v{}^{-4.477} \tag{11.5}$$

Rutting failure was defined as a rut depth equal to 12.5 mm (0.5 in). To facilitate implementation of this approach, a series of layer elastic analysis computer runs were performed using different pavement layer thickness combinations, and nomographs were produced. The computer program DAMA was used for this purpose.[7] Nomographs were produced that allow solving for the asphalt concrete layer thickness, given the resilient modulus ($M_r$) of the subgrade and the *ESALs* anticipated over the life of the pavement. Different nomographs are available by base layer thicknesses and material type (untreated base and emulsified asphalt stabilized base), as well as for three distinct mean annual air temperatures (MAAT), namely 7°C, 15.5°C, and 24.4°C. An example of these nomographs is shown in Figure 11.8.

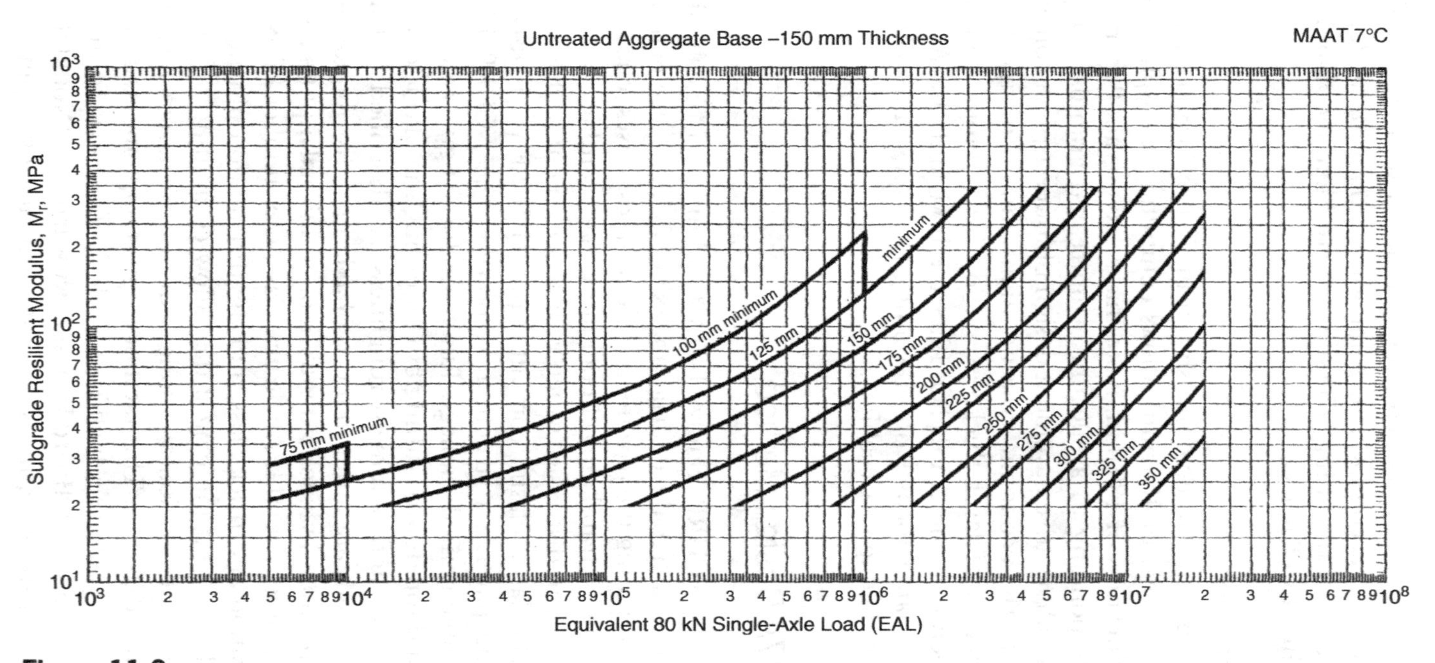


**Figure 11.8**
AI Thickness Design Nomograph; 150 mm of Untreated Aggregate Base; MAAT 7°C (after 3).

**Example 11.4** Design an asphalt concrete pavement with an untreated granular base to accommodate 3 million *ESALs* without failing in fatigue cracking or rutting. The subgrade resilient modulus is 100 MPa, and the MAAT is 7°C.

**ANSWER**

Using the nomograph shown in Figure 11.8 (i.e., selecting a base thickness of 150 mm) allows computing an asphalt concrete layer thickness of 180 mm. Note that selecting another nomograph corresponding to a 300 mm of untreated granular base would yield an alternative pavement design of lower asphalt concrete thickness. Clearly, economic considerations should dictate the best combination of layer thicknesses. No direct comparisons can be made between the AASHTO 1993 method and the AI method, since the latter focuses on preventing two types of distresses rather than predicting serviceability.

## 11.4 NCHRP 1-37A Design Method

The NCHRP Study 1-37A[24] adopted a mechanistic-empirical approach to the damage analysis of flexible pavements. This involves computing the pavement structural responses to load (i.e., stresses/strains), translating them into damage, and accumulating the damage into distresses, which reduce pavement performance over time. The layer elastic computer model JULEA[20] is used for calculating pavement structural responses. The NCHRP 1-37A approach implemented damage functions for fatigue cracking (bottom-up and top-down), rutting by computing the plastic deformation in all layers, and pavement roughness. As described in Chapter 2, traffic loads are input in terms of axle load distributions—that is, load spectra—by axle configuration. Additional input, such as the *AADTT* volume, the *MAF*s by truck class, the number of axles by configuration and truck class, and the distribution of truck traffic volume throughout the typical day, allow computing the number of axles by configuration and weight in hourly increments, as summarized in Table 2.14. This methodology is implemented into the NCHRP 1-37A software, which is available for download from the Transportation Research Board Web site, www.trb.org. (Noted that the NCHRP 1-37A software, in its current form, accepts input only in Imperial units). The following discussion provides a description of the damage functions implemented into the flexible pavement performance prediction module. It should be noted that this methodology is currently under review.[8] The outcome of this review and subsequent research is likely to result in modifications to some of the damage functions described next.

### 11.4.1 Fatigue Cracking

Fatigue damage is accumulated for estimating bottom-up alligator cracking and top-down longitudinal cracking. The expression used for computing the number of repetitions to failure $N_f$ for bottom-up and top-down cracking is a variation of the expression proposed by Finn[6] and adopted by the AI mechanistic design approach:

$$N_f = 0.00432\, k_1'\, C \left(\frac{1}{\varepsilon_t}\right)^{3.9492} \left(\frac{1}{E}\right)^{1.281} \tag{11.6}$$

where $\varepsilon_t$ is the tensile strain in the asphalt concrete layer and $E$ is the layer stiffness (lbs/in$^2$). The coefficients $C$ and $k_1$' are calibration constants. $C$ is given by:

$$C = 10^M \tag{11.7}$$

with:

$$M = 4.84 \left(\frac{V_b}{V_a + V_b} - 0.69\right) \tag{11.8}$$

where $V_b$ is the volume of binder and $V_a$ is the volume of the mix as percentages of the total mix volume. The coefficient $k_1'$ is a function of the thickness of the asphalt concrete layer $h_{ac}$ (inches). It is defined differently for bottom-up and top-down fatigue accumulation (Equations 11.9 and 11.10, respectively):

$$k_1' = \frac{1}{0.000398 + \dfrac{0.003602}{1 + e^{11.02 - 3.49\, h_{ac}}}} \tag{11.9}$$

$$k_1' = \frac{1}{0.01 + \dfrac{12}{1 + e^{15.676 - 2.8186\, h_{ac}}}} \tag{11.10}$$

Fatigue damage $FD$ (percent) is accumulated separately for bottom-up and top-down cracking, according to Miner's hypothesis[13] expressed as:

$$FD = \sum \frac{n_{i,j,k,l,m}}{N_{i,j,k,l,m}} 100 \tag{11.11}$$

where:

$n_{i,j,k,\ldots}$ = applied number of load applications at conditions $i$, $j$, $k$, $l$, $m$, $n$.

$N_{i,j,k,\ldots}$ = number of axle load applications to cracking failure under conditions $i$, $j$, $k$, $l$, $m$, where:

$i$ = month, which accounts for monthly changes in the moduli of base and subgrade due to moisture variations and asphalt concrete due to temperature variations.

$j$ = time of the day, which accounts for hourly changes in the modulus of the asphalt concrete.

$k$ = axle type (single, tandem, triple, and quad).

$l$ = load level for each axle type.

$m$ = traffic path, assuming a normally distributed lateral wheel wander.

Temperature and moisture changes are computed using the Enhanced Integrated Climatic Model[9] and weather data input from the vicinity of the area where the pavement being designed is located. The fatigue damage computations in the NCHRP 1-37A software involves a series of layered elastic analysis solutions to compute the tensile strains in the asphalt concrete layer, and the resulting number of repetitions to fatigue failure for each axle configuration and load magnitude using Equation 11.6. Subsequently, the actual number of axle passes by configuration and axle load for the particular site being analyzed is estimated from the traffic input described in chapter 2. Finally the fatigue damage accumulated versus time is computed using Equation 11.11.

The bottom-up fatigue cracking area $FC$ (percent of total lane area) is computed as:

$$FC = \frac{100}{1 + e^{c_2'(-2 + \log FD)}} \tag{11.12}$$

where *FD* is the bottom-up fatigue damage (percent) computed from Equation 11.11, and $c_2'$ is given by:

$$c_2' = -2.40874 - 39.748\ (1 + h_{ac})^{-2.856} \qquad (11.13)$$

The top-down longitudinal fatigue cracking (feet/mile) is computed as:

$$FC = \frac{10560}{1 + e^{(7.0-3.5\ \log FD)}} \qquad (11.14)$$

where *FD* is the amount of top-down fatigue damage (percent) computed from Equation 11.11. It should be noted that due to boundary problems, the linear elastic analysis yields inaccurate results near the tire-pavement surface interface. To circumvent this limitation, the NCHRP 1-37A model utilizes linear extrapolation of the strains computed deeper in the asphalt concrete layer to estimate the surface strains necessary for the top-down fatigue cracking analysis.

**Example 11.5**

A pavement section has accumulated a total of 15% bottom-up fatigue damage. Estimate its fatigue cracking, given that the asphalt concrete layer has a thickness of 20 cm (7.87 in).

**ANSWER**

Utilize Equation 11.13 to compute the coefficient $c_2'$ as:

$$c_2' = -2.40874 - 39.748\ (1 + 7.8)^{-2.856} = -2.487$$

Substituting into Equation 11.12 gives:

$$FC = \frac{100}{1 + e^{-2.487(-2+\log 15)}} = 11.41\%$$

**RUTTING DAMAGE**

The NCHRP 1-37A guide computes rutting damage by summing the plastic deformation in each pavement layer and the subgrade. The plastic deformation, *PD*, is computed by dividing each layer into a number of sublayers, computing the plastic strain in each sublayer, and adding the resulting plastic deformations through:

$$PD = \sum_{i=1}^{n} \varepsilon_p^i \, h^i \tag{11.15}$$

where, $\varepsilon_p^i$ is the plastic strain in sub-layer $i$, $h^i$ is the thickness of sub-layer $i$, and $n$ is the number of sublayers distinguished. As described next, the plastic strain $\varepsilon_p$ in each pavement layer is computed from the corresponding elastic (or resilient) vertical strain $\varepsilon_v$ using linear elastic analysis. The procedure adopted was developed by Tseng and Lytton.[19]

**PLASTIC STRAIN IN THE ASPHALT CONCRETE LAYER**

The plastic strain in the asphalt concrete layer $\varepsilon_p$ is computed as a function of the vertical elastic (resilient) strain $\varepsilon_v$ obtained from elastic layered analysis using:

$$\frac{\varepsilon_p}{\varepsilon_v} = k_1 10^{-3.4488} \; T^{1.5606} N^{0.479244} \tag{11.16}$$

where $T$ is the asphalt concrete layer temperature (°F), $N$ is the cumulative number of loading cycles experienced, and $k_1$ is a calibration factor accounting for the increased level of confinement with depth, expressed by the variable *depth* (inches):

$$k_1 = (C_1 + C_2 \; depth) \; 0.328196^{depth} \tag{11.17}$$

where:

$$C_1 = -0.1039 \; h_{ac}^2 + 2.4868 \; h_{ac} - 17.342 \tag{11.18}$$

$$C_2 = 0.0172 \; h_{ac}^2 - 1.7331 \; h_{ac} + 27.428 \tag{11.19}$$

**Example 11.6** Compute and plot the plastic strain accumulated at the middepth of an asphalt concrete layer 0.153 cm (6 in) thick at a temperature of 85°F after 1, 10, $10^2$, and $10^3$ load cycles. Also compute the plastic deformation after $10^3$ load cycles. The elastic vertical strain is $145 \; 10^{-6}$.

**ANSWER**

Compute the coefficients $C_1$ and $C_2$ using Equations 11.18 and 11.19.

$$C_1 = -0.1039\ 6^2 + 2.4868\ 6 - 17.342 = -6.1616$$

$$C_2 = 0.0172\ 6^2 - 1.7331\ 6 + 27.428 \quad = 17.6486$$

The calibration coefficient is computed from Equation 11.17.

$$k_1 = (-6.1616 + 17.6486\ 3)\ 0.328196^3 = 1.654$$

Hence, the plastic strain is computed as a function of the number of cycles $N$ using Equation 11.16, as:

$$\varepsilon_p = 145\ 10^{-6}\ 1.654\ 10^{-3.4488}\ 85^{1.5606}(N)^{0.479244}$$

The resulting plastic strain is plotted as a function of the number of cycles in Figure 11.9. The plastic deformation is obtained from Equation 11.15, considering the asphalt concrete layer as a single layer. After $10^3$ cycles, the product of the plastic strain multiplied by the thickness of the layer (6 in) gives $1.56910^{-3} \times 6 = 0.009412$ in. (0.239 mm).

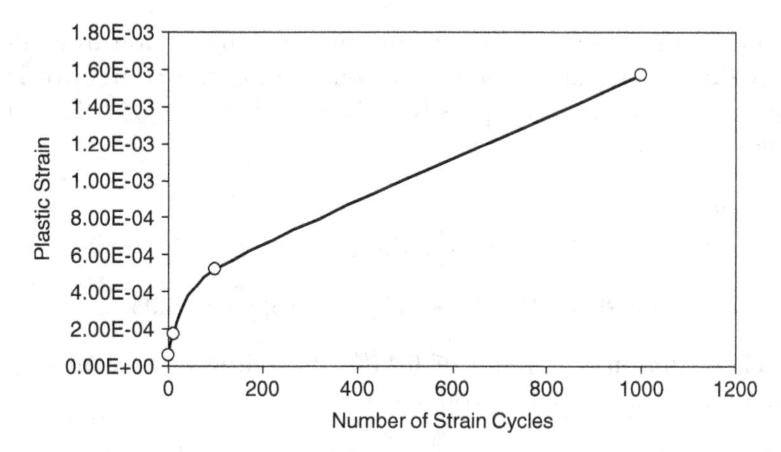


**Figure 11.9**
Relationship between Plastic Strain and Number of Strain Cycles (Example 11.6)

#### PLASTIC STRAIN IN THE UNBOUND BASE LAYERS AND SUBGRADES

NCHRP 1-37A adopted a model developed by Tseng and Lytton[19] for computing the plastic strain $\varepsilon_p$ in the unbound granular layers. It relates $\varepsilon_p$ to the vertical elastic (resilient) strain $\varepsilon_v$, calculated from layered elastic analysis using:

$$\frac{\varepsilon_p}{\varepsilon_v} = \beta_G \left(\frac{\varepsilon_0}{\varepsilon_r}\right) e^{-\left(\frac{\rho}{N}\right)^\beta} \tag{11.20}$$

where $\beta$, $\rho$, and $\varepsilon_0$ are material properties obtained from laboratory testing involving repetitive loading at resilient strain level $\varepsilon_r$, and $N$ is the number of load cycles. The procedure for developing this model was described in chapter 3. The calibration constant $\beta_G$ has the value of 1.673 for base layers and 1.35 for subgrades. The values of $\beta$ and $\rho$ are given by:

$$\log \beta = -0.6119 - 0.017638\ W_c \tag{11.21}$$

$$\rho = 10^9 \left(\frac{-4.89285}{1-(10^9)^\beta}\right)^{\frac{1}{\beta}} \tag{11.22}$$

where $W_c$ is the water content (%). The ratio $\frac{\varepsilon_0}{\varepsilon_r}$ is computed as the weighted average of the experimental measurements after 1 and $10^9$ load cycles:

$$\frac{\varepsilon_0}{\varepsilon_r} = \frac{1}{2}\left(0.15\ e^{(\rho)^\beta} + 20\ e^{\left(\frac{\rho}{10^9}\right)^\beta}\right) \tag{11.23}$$

**Example 11.7** Compute the plastic strain and the plastic deformation in a 10-in. thick (25.4 cm) granular base layer with a moisture content of 18% after 1000 cycles at a compressive strain level of 250 $10^{-6}$. Treat the base as a single layer.

#### ANSWER

Use Equation 11.21 to compute the property $\beta$:

$$\log \beta = -0.6119 - 0.017638\ 18 = -0.9294$$

which results in a $\beta$ value of 0.1177. Substituting this value into Equation 11.22 gives the property $\rho$.

$$\rho = 10^9 \left(\frac{-4.89285}{1-(10^9)^{0.1177}}\right)^{\frac{1}{0.1177}} = 0.001578 10^9$$

Equation 11.23 is used for computing the ratio $\frac{\varepsilon_0}{\varepsilon_r}$.

$$\frac{\varepsilon_0}{\varepsilon_r} = \frac{1}{2}\left(0.15\ e^{\left(0.001578\ 10^9\right)^{0.1177}} + 20\ e^{(0.001578)^{0.1177}}\right) = 31.94$$

The plastic strain is computed from Equation 11.20, using a $\beta_G$ value of 1.673:

$$\varepsilon_p = 1.673\ 31.94\ e^{-\left(\frac{0.001578\ 10^9}{1000}\right)^{0.1177}}\ 250\ 10^{-6} = 1.238\ 10^{-3}$$

which allows computing the plastic deformation of this base layer as $1.238\ 10^{-3} \times 10 = 0.01238$ in (0.31 mm).

### THERMAL (TRANSVERSE) CRACKING

The thermal cracking model adopted by the NCHRP 1-37A design approach is based on work carried out under the Strategic Highway Research Program (SHRP) contract A-005[21] and the work carried out by Witczak et al. under NCHRP study 9-19.[23] Its basic mechanism relates the thermal stresses computed from the creep compliance of the asphalt concrete to its tensile strength. The properties are determined by the Indirect Tension Test (IDT) conducted according to AASHTO Standard T322-03.[18] The IDT test is used to measure the creep compliance of the asphalt concrete in tension at various temperatures and construct its master curve following the procedure described in chapter 5. A generalized Voight-Kelvin model is fitted to the master curve, which allows expressing the creep compliance (or strain retardation modulus) of the asphalt concrete $D(\xi)$ in terms of reduced time $\xi$ as:

$$D(\xi) = D_0 + \frac{D_0\ \xi}{T_0} + \sum_{i=1}^{N} D_i(1 - e^{-\xi/T_i}) \tag{11.24}$$

where $D_i$ and $T_i$ are the constants characterizing the $N + 1$ elastic springs and dashpots, respectively (Figure 11.10).

Transforming this function into the frequency domain $\hat{D}(s)$ allows estimating the stress relaxation modulus of the asphalt concrete in the frequency domain $\hat{E}(s)$ as:

$$\hat{E}(s) = \frac{1}{s^2\ \hat{D}(s)} \tag{11.25}$$

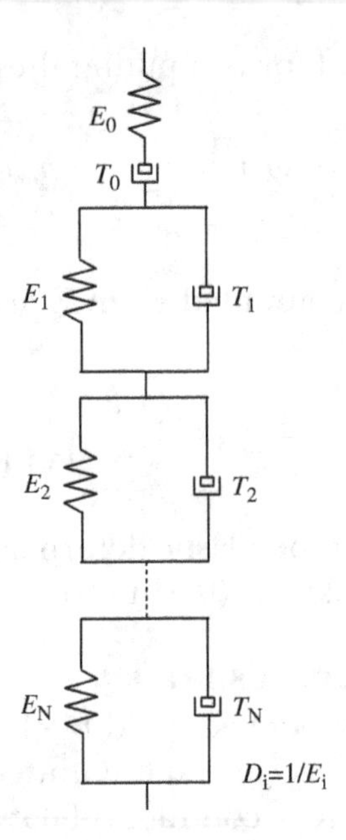


**Figure 11.10**
Generalized Voight-Kelvin Model Used to Fit Asphalt Concrete Master Curve Data

where ^ is the Laplace operator and $s$ is the frequency domain operator. A reverse Laplace transformation of function $\hat{E}(s)$ gives the asphalt concrete relaxation modulus in the time domain, denoted by $E(\xi)$. This allows computing stresses as a function of changing temperature and/or loading time using Boltzman's superposition principle (Equation 7.15). The software that was developed for handling these computations utilized a finite-difference scheme and was incorporated into the NCHRP 1-37A software.

Temperature is predicted as a function of depth in the asphalt concrete layer using the EICM model,[12] described in Chapter 10. The thermal stresses thus computed are compared to the undamaged tensile strength of the asphalt concrete. This is obtained through relatively rapid constant deformation IDT testing (loading rate of 1.27 cm/minute) conducted at various temperatures, as described in chapter 5. Thermal transverse cracking is initiated when the thermal stresses exceed the tensile strength of the asphalt concrete. Crack propagation is simulated using Paris law[16], which was adapted for cracking in viscoelastic materials,[17] and expressed as:

$$\Delta C = A\,\Delta K^n \tag{11.26}$$

where $\Delta C$ = the increase in crack length, $\Delta K$ = is the change in the stress intensity function, and $n$ and $A$ are fracture parameters. The stress intensity factor is computed using:

$$K = \sigma\left(0.45 + 1.99\ C_0^{0.56}\right) \tag{11.27}$$

where, $C_0$ is the original crack length and $\sigma$ is the stress in the asphalt concrete layer at the depth of the crack tip. The parameter $n$ is obtained by fitting an exponential relationship to the creep compliance master curve, following a technique developed by Lytton et al.[11] It relates the cracking parameter $n$ to the slope $m$ of the linear part of the log $D(\xi)$ versus log $\xi$ master curve through:

$$n = 0.8\left(1 + \frac{1}{m}\right) \tag{11.28}$$

The parameter $A$ was established through calibration using in-situ transverse cracking data:

$$\log A = 4.389 - 2.52\ \log\left(10000\ S_t\ n\right) \tag{11.29}$$

where $S_t$ is the tensile strength (lbs/in$^2$) of the asphalt concrete mix measured, as described earlier. Finally, the extent of transverse thermal cracking in asphalt concretes $AC$ (in linear feet/500 ft) is computed from the probability that the length of thermal cracks $C$ exceeds the thickness of the asphalt concrete layer $D$, expressed as:

$$AC = 353.5\,N\left(\frac{\log\ C/D}{0.769}\right) \tag{11.30}$$

where $N$ is the standard normal probability that $C$ will be larger than $D$ (i.e., 0.769 is the estimated standard deviation of the logarithm of the crack length). Computer software was developed implementing each of these steps. This software was incorporated into the NCHRP 1-37A guide software to allow predicting thermal cracking. The input to this module includes the asphalt concrete creep compliance master curve, its Poisson's ratio, its tensile strength, and the environmental data for the design location.

**Example 11.8**

The master curve of the creep compliance of an asphalt concrete at −20° C is plotted in Figures 11.11. Compute its fracture parameters and the growth of an existing 7.6 cm (3 in) deep transverse crack

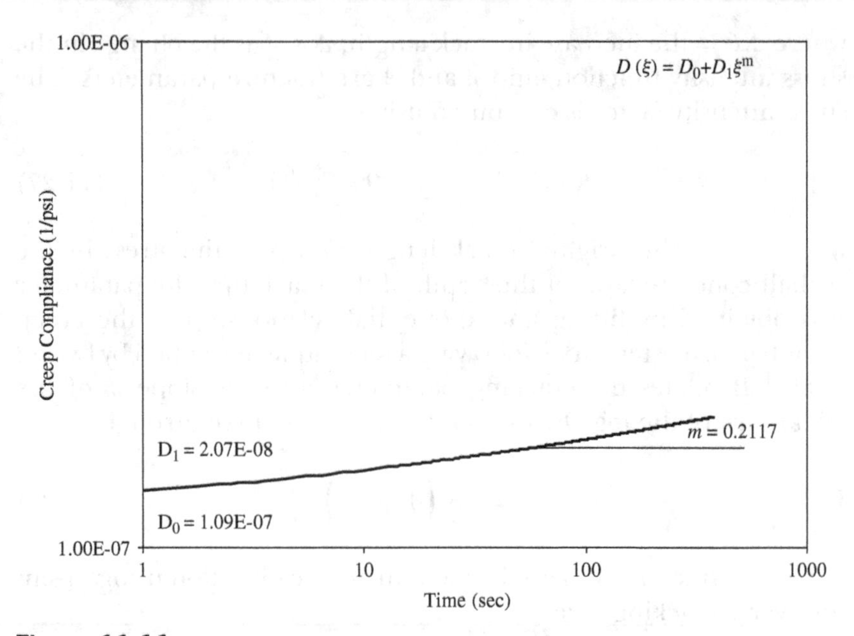


**Figure 11.11**
Obtaining the Slope *m* of the Asphalt Concrete Creep Compliance Curve; Experimental Measurements Obtained at −20°C (data after 23)

caused by an increase in stress at its tip of 310 kPa (45 lbs/in$^2$). The tensile strength of the asphalt concrete is 4.5 MPa (652 lbs/in$^2$).

**ANSWER**

Figure 11.11 yields a slope $m$ of the creep compliance exponential function equal to 0.2117. Equation 11.28 gives the fracture parameter $n$ as:

$$n = 0.8\left(1 + \frac{1}{0.2117}\right) = 4.579$$

Substituting into Equation 11.29 gives:

$$\log A = 4.389 - 2.52 \ \log\left(10000 \ 652 \ 4.579\right)$$

which gives the fracture parameter $A$ as 3.564 $10^{-15}$. The change in the stress intensity function is computed from Equation 11.27.

$$K = 45\left(0.45 + 1.99 \ 3^{0.56}\right) = 185.92$$

The resulting increase in the crack length is computed from Equation 11.26.

$$\Delta C = 3.564\ 10^{-15}\ 185.92^{4.579} = 8.77\ 10^{-5}\ \text{inches}(0.00223\ \text{mm}).$$

## 11.4.2 Roughness

The roughness model proposed by the NCHRP 1-37A guide is regression-based, using the computed distresses as the main independent variables. For asphalt concrete pavements on unbound granular bases, the expression used for predicting roughness ($IRI$ in m/km) is:

$$\begin{aligned} IRI = IRI_0 + 0.0463\ SF\ (e^{age/20} - 1) + 0.00119\ TC_T \\ + 0.1834\ COV_{RD} + 0.00384\ FC + 0.00736\ BC \\ + 0.00115\ LC_S \end{aligned} \tag{11.31}$$

where $IRI_0$ is the initial (as constructed) pavement roughness, $TC_T$ is the total length of transverse cracks, $COV_{RD}$ is the coefficient of variation in rut depth, $FC$ is the fatigue cracking in the wheel-paths, $BC$ is the area of block cracking (percent of total lane area), $age$ is the age of the section (years), and $SF$ is a site factor computed as:

$$SF = \frac{R_{SD}\ P_{0.075}\ PI}{2\ 10^4} + \frac{\ln(FI + 1)\ (P_{0.02} + 1)\ \ln(R_m + 1)}{10} \tag{11.32}$$

where $R_m$ and $R_{SD}$ is the mean and the standard deviation in annual rain fall (mm), $P_{0.075}$ and $P_{0.02}$ are the subgrade percent finer fractions for grain sizes 0.075 mm and 0.02 mm, $FI$ is the average annual Freezing Index (Chapter 10), and $PI$ is the Plasticity Index of the subgrade (Chapter 3).

## 11.4.3 Model Calibration

The pavement damage functions just described were calibrated using field performance observations from three large-scale pavement experiments: the Minnesota Road Research (MnROAD) Project,[14] the WesTrack Project,[22] and the Long-Term Pavement Performance (LTPP) Program.[10]

- The MnROAD Project is a heavily instrumented 6-mile long section of I-94 located 64 km northwest of Minneapolis/St. Paul. Instrumentation ranged from pavement strain/stress gauges to subgrade temperature/moisture gauges numbering to more than 4500. Both flexible and rigid pavements were tested under in-service traffic monitored by a WIM system.

- The WesTrack is 2.9 km long oval test track located 100 km southeast of Reno on the grounds of the Nevada Automotive Research Center. It was designed to test the performance of a number of alternative asphalt concrete mix designs and evaluate the effect of variations in structural design and materials properties (e.g., asphalt content, air void content, and aggregate gradation). Traffic was applied by means of four driverless triple-trailer trucks that applied a total of 10 million *ESALs* over a period of two years.
- The LTPP is a large-scale experiment initiated in 1986 as part of the Strategic Highway Research Program (SHRP). It involves a large number of 150 m long test sections across the United States and Canada. Experiments involve existing pavement and purpose-built pavement sections designated as general pavement sections (GPS) and special pavement sections (SPS), respectively (Tables 11.4 and 11.5). The total of number of sections was approximately 652 and 1262, respectively. These sections were exposed to in-service traffic monitored by WIM systems. Pavement data has been collected at these sections for over 20 years through four regional contracting agencies under the oversight of the FHWA. The data is assembled into a massive database, which is being periodically released to the public under the DataPave database label.[5] This database

**Table 11.4**
Identification of LTPP General Pavement Section (GPS) Experiments

| | |
|---|---|
| GPS-1 | Asphalt Concrete (AC) on Granular Base |
| GPS-2 | AC on Bound Base |
| GPS-3 | Jointed Plain Concrete Pavement |
| GPS-4 | Jointed Reinforced Concrete Pavement |
| GPS-5 | Continuously Reinforced Concrete Pavement |
| GPS-6A | Existing AC Overlay on AC Pavements |
| GPS-6B | New AC Overlay on AC Pavements |
| GPS-7A | Existing AC Overlay on Portland Cement Concrete (PCC) Pavements |
| GPS-7B | New AC Overlay on PCC Pavements |
| GPS-9 | Unbounded PCC Overlays on PCC Pavements |

**Table 11.5**
Identification of LTPP Special Pavement Section (SPS) Experiments

| | |
|---|---|
| SPS-1 | Strategic Study of Structural Factors for Flexible Pavements |
| SPS-2 | Strategic Study of Structural Factors for Rigid Pavements |
| SPS-3 | Preventative Maintenance Effective for Flexible Pavements |
| SPS-4 | Preventative Maintenance Effective for Rigid Pavements |
| SPS-5 | Rehabilitation of AC Pavements |
| SPS-6 | Rehabilitation of Jointed PCC Pavements |
| SPS-7 | Bonded PCC Overlays on Concrete Pavements |
| SPS-8 | Study of Environmental Effects in the Absence of Heavy Loads |
| SPS-9 | Validation of SHRP Asphalt Specification and Mix Design (Superpave) |

includes data on a multitude of inventory, material, traffic, environmental, and pavement evaluation variables.

It is anticipated that the flexible pavement design models described earlier will undergo further refinement as the NCHRP 1-37A pavement design approach is being evaluated.[8] Furthermore, additional model calibration will take place as individual state DOTs begin implementing this new design approach.

## References


1 AASHO (1962). AASHO Road Test, Special Report 52; Report 5, "Flexible Pavement Research," American Association of State Highway Officials, Washington, DC.

2 AASHTO (1986, 1993). *AASHTO Guide for the Design of Pavement Structures*, American Association of State Highway and Transportation Officials, *Washington, DC.*

3 AI (1981). *Asphalt Pavements for Highways and Streets*, 9th ed., Asphalt Institute Manual Series-1 (MS-1), College Park, MD.

4 DARWin Version 3.1 (1999). Applied Research Associates (ARA) Inc., Champaign IL.

5 DataPave, LTPP Database Release 20.0, December 2006, www.ltpp-products.com.

6 Finn, F. N., Saraf, C. L., Kulkrani, R., Nair, K., Smith W., and Abdulah, A., (1986). Development of Pavement Structural Subsystems, National Cooperative Highway Research Program (NCHRP) Report 291, *Washington, DC.*

[7] Hwang, D., and Witczak M. W. (1979). *Program DAMA (Chevron) User's Manual*, Department. of Civil Engineering, University of Maryland, College Park, MD.

[8] NCHRP (September 2006). Independent Review of the Recommended Mechanistic-Empirical Design Guide and Software, National Cooperative Highway Research Program (NCHRP) Research Results, Digest No. 307.

[9] Larson G., and B. J. Dempsey (1997). Integrated Climatic Model, Version 2.0, Newmark Civil Engineering Laboratory, University of Illinois at Urbana-Champaign, Report No. DTFA MN/DOT 72114.

[10] Long-Term Pavement Performance, Federal Highway Administration, Web site: www.tfhrc.gov/pavement/ltpp, accessed - Feb, 2007.

[11] Lytton, R. L., W. Shanmugham, and B. D. Garrett (1983). "Design of Asphalt Pavements for Thernal Fatigue Cracking," Research Report No. FHWA/TX-83/06 + 284-4, Texas Transportation Institute, Texas A&M University, College Station, TX.

[12] Lytton, R. L., D. E. Pufhal, C. H. Michalak, H. S. Liang, and B. J. Dempsey, (November 1993). "An Integrated Model of the Climatic Effect on Pavements," FHWA-RD-90-033, Washington DC.

[13] Miner, M. A. (1945). "Cumulative Damage in Fatigue," *Transactions of the ASCE*, Vol. 67, pp. A159–A164.

[14] Minnesota Road Research Project (MnRoad), www.mrr.dot.state.mn.us/research/MnROAD_Project/MnROADProject.asp, accessed 02.2009.

[15] Papagiannakis, A. T., M. Bracher, J. Li, and N. Jackson (2006). "Sensitivity of the NCHRP 1-37A Pavement Design to Traffic Input," *Journal of the Transportation Research Board Record*, No. 1945, pp. 49–55.

[16] Paris, P., and F. Erdogan (1963). "A Critical Analysis of Crack Propagation Laws," *Journal of Basic Engineering*, Transactions of the American Society of Mechanical Engineers, pp. 528–534.

[17] Schapery, R. A. (1973). A Theory of Crack Growth in Viscoelastic Media, ONR Contract No. N00014-68-A-0308-003, Technical Report 2, NMM 2764-73-1, Mechanics and Materials Research Center, Texas A&M University, College Station, TX.

[18] AASHTO (2003). Standard Test Method for Determining the Creep Compliance and Strength of Hot Mix Asphalt (HMA) Using the Indirect Tensile Test Device, American Association of State Highway and Transportation Officials T322-03, Washington, DC.

[19] Tseng, K., and R. Lytton (1989). Prediction of Permanent Deformation in Flexible Pavement Materials. Implications of Aggregates in the Design, Construction, and Performance of Asphalt Pavements, ASTM STP 1016, pp. 154–172, American Society for Testing of Materials, Goushohocken PA.

[20] Uzan, J., (2001) Jacob Uzan Layer Elastic Analysis (JULEA) Software.

[21] Von Quintus, H. L. (1994). "Performance Prediction Models in the Superpave Mix Design System," Strategic Highway Research Program, Report SHRP-A-699, Washington, DC.

[22] WesTrack: Accelerated Field Test of Performance-Related Specifications for Hot-Mix Asphalt Construction, FHWA Contract No. DTFH61-94-C-00004, www.westrack.com, accessed.

[23] Witczak, M. W., R. Roque, D. R. Hiltunen, and W. G. Buttlar (2000). "Modification and Re-calibration of Superpave Thermal Cracking Model," Superpave Support and Performance Models Management, Project Report NCHRP Project 9-19, Project Deliverable Task B, Washington, DC.

[24] NCHRP (July 2004). "2002 Design Guide: Design of New and Rehabilitated Pavement Structures" Draft Final Report, NCHRP Study 1-37A, National Cooperative Highway Research Program, Washington, DC.

## Problems

11.1 Calculate the required layer thicknesses for a new asphalt concrete pavement on a fair draining base and subgrade (the water drains out of the pavement within a period of two days). It is estimated that the pavement structure becomes saturated less frequently than 5% of the time. The following data is also given:

- Estimated number of *ESALs* over a 15-year maximum performance period = 3 million
- Subgrade resilient modulus = 25,000 lbs/in.$^2$ (172.4 MPa)
- Design reliability = 95%
- Standard error in predicting serviceability = 0.45
- $\Delta PSI = 2.5$ (from 4.5 to 2.0).

11.2 Compute the anticipated life of the pavement designed in problem 11.1, considering the combined effects of traffic and environment. The subgrade is a fair draining clayey sand (i.e., designated as SC according to the USC system), having less than 10% by weight finer than 0.02 mm, and a Plasticity Index of 10%. The subgrade layer is 10 ft deep; it is exposed to high moisture levels and exhibits a medium level of structural fracturing. The percent of the pavement surface subjected to swelling is estimated to be 20%, while the probability of frost is estimated at 30%. The depth of frost penetration is 4 ft.

11.3 Design an asphalt concrete pavement with an untreated granular base to accommodate 2.5 million *ESALs* without failing in fatigue cracking or rutting. The subgrade resilient modulus is 80 MPa, and the MAAT is 15.5°C. (Note: You need to obtain the proper chart from Reference 3).

11.4 A pavement section has accumulated a total of 20% bottom-up fatigue damage. Estimate its fatigue cracking, given that the asphalt concrete layer has a thickness of 15 cm (5.9 in.).

11.5 Compute and plot the plastic strain accumulated at middepth of an asphalt concrete layer 0.23 cm (8 in.) thick at a temperature of 75°F after 1, 10, 100, and 200 load cycles. Also compute the plastic deformation after 200 load cycles. The elastic vertical strain is 120 $10^{-6}$.

11.6 Compute the plastic strain and the plastic deformation in a 14-in thick (35.5 cm) granular base layer with a moisture content of 6% after 1000 cycles, at a compressive strain level of 180 $10^{-6}$. Treat the base as a single layer.

11.7 The constant describing the master curve of the creep compliance of an asphalt concrete at −20°C is given in Table 11.6. Compute its fracture parameters and the growth of an existing 5.1 cm (2 in) deep transverse crack caused by an increase in

**Table 11.6**
Constants Defining the Creep Compliance of the Asphalt Concrete for Problem 11.7

| Constant | Units | Value |
|---|---|---|
| $D_0$ | 1/lbs/in.$^2$ | 2.8 $10^{-07}$ |
| $D_1$ | 1/lbs/in.$^2$ | 5.4 $10^{-8}$ |
| $D_2$ | 1/lbs/in.$^2$ | 9.5 $10^{-08}$ |
| $D_3$ | 1/lbs/in.$^2$ | 4.0 $10^{-08}$ |
| $D_4$ | 1/lbs/in.$^2$ | 3.0 $10^{-07}$ |
| $T_0$ | sec | 13,000 |
| $T_1$ | sec | 0.9 |
| $T_2$ | sec | 1.9 |
| $T_3$ | sec | 2.8 |
| $T_4$ | sec | 3.6 |

stress at its tip of 344.7 kPa (50 lbs/in$^2$). The tensile strength of the asphalt concrete is 4.0 MPa (580 lbs/in$^2$).

11.8 Utilizing the NCHRP 1-37A Design Guide software, estimate the performance of an asphalt concrete pavement located in the southern United States (latitude = 30° N, longitude = 93° 41 W). Given:

- Layer thickness for asphalt concrete, base, and subbase of 15.24, 12.7, and 25.4 cm (6, 5, and 10 in.)

**Table 11.7**
Vehicle Classification Frequency Distribution for Problem 11.8

| FHWA Vehicle Class | Frequency (%) |
|---|---|
| 4 | 3 |
| 5 | 9.1 |
| 6 | 2.2 |
| 7 | 1.1 |
| 8 | 3.4 |
| 9 | 54.7 |
| 10 | 11 |
| 11 | 5.2 |
| 12 | 2.3 |
| 13 | 8 |

- Average annual daily truck traffic (AADTT) in the design lane for first year = 500 trucks/day.
- Traffic growth rate = 3.0% annually, compounded over the analysis period of 15 years.

The truck classification distribution is shown in Table 11.7. Assume monthly adjustment factors (*MAF*s) of 1.0 for all vehicle classes and

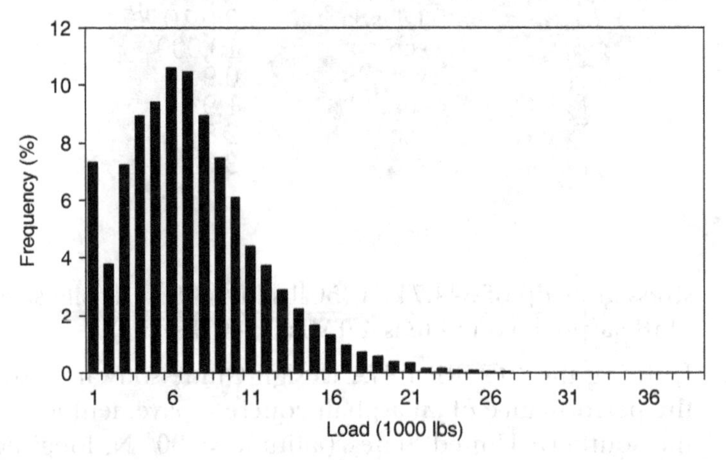


**Figure 11.12**
Single-Axle Load Frequency Distribution for Problem 11.8 (based on LTPP Annual Summary Data for Site 531002 for 1992)

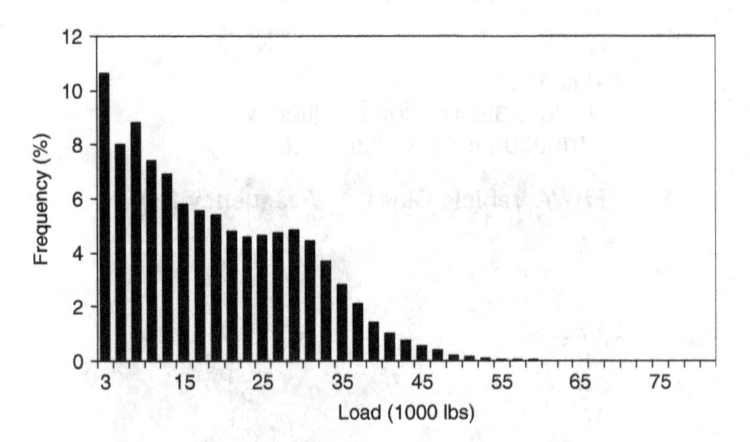


**Figure 11.13**
Tandem-Axle Load Frequency Distribution for Problem 11.8 (based on LTPP Annual Summary Data for Site 531002 for 1992)

months. Compute the number of axles by configuration, assuming that the sketches in Table 2.1 represent the vehicle type in each class. Use the axle load distributions shown in Figures 11.12 and 11.13, and assume that there are no triple- or quad-axle configurations in the traffic stream specify a tire inflation pressure of 689 kPa (100 lbs/in$^2$). Finally, the subgrade is a low-plasticity clay with a resilient modulus of 82.7 MPa (12,000 lbs/in$^2$). Present the results in the form of performance curves of the various distress versus time. Decide whether the selected pavement structural design is acceptable.

# 12 Structural Design of Rigid Pavements

## 12.1 Introduction

As described in Chapters 1 and 8, rigid pavements derive their load-carrying capacity through the combined action of plate bending and subgrade reaction. Chapter 8 described methods for computing the structural response of rigid pavements to traffic loads and environmental input. This chapter describes the most common design methodologies used for estimating the required layer thicknesses and reinforcement for the three main rigid pavement configurations:

- Jointed plain concrete pavements (JPCP)
- Jointed dowel-reinforced concrete pavements (JDRCP)
- Continuously reinforced concrete pavements (CRCP)

The chapter covers three methodologies, including the American Association of State Highway and Transportation Officials (AASHTO) 1986/1993 method[1] the Portland Cement Association (PCA) method,[16,17] and the method proposed by the NCHRP 1-37A study.[19] Typically, state DOTs utilize all available design methods, including some reflecting the performance of their own pavements, and use judgment and economic considerations in selecting the final design.

## 12.2 AASHTO 1986/1993 Design Method

### 12.2.1 Serviceability Loss Due to Traffic

The serviceability loss due to traffic is computed from an empirical relationship derived from rigid pavement performance observations made during the AASHO Road Test, which was described in Chapter 11. It relates the number of cumulative equivalent single-axle load (*ESAL*) passes (Chapter 2) to the corresponding change in pavement serviceability, $\Delta PSI$. It is expressed in the following, imperial units, format:

$$\log(W_{18}) = Z_R S_0 + 7.35\log(D+1) - 0.06 + \frac{\log\left[\dfrac{\Delta PSI}{4.5-1.5}\right]}{1+\dfrac{1.62410^7}{(D+1)^{8.46}}} + (4.22 - 0.32P_t) \times \log\frac{S_c' C_d (D^{0.75} - 1.132)}{215.63J\left[D^{0.75} - \dfrac{18.42}{(E_{c/k})^{0.25}}\right]} \qquad (12.1)$$

where:

$W_{18}$ = the number of *ESALs*.
$D$ = the portland concrete slab thickness (inches).
$P_t$ = the terminal serviceability of the section.
$S_c'$ = the modulus of rupture of the portland concrete (lbs/in$^2$), measured through beam flexure tests and third-span loading, according to AASHTO Standard T97.[2]
$C_d$ = a drainage coefficient (Table 12.1).
$J$ = a load transfer coefficient (Table 12.2).
$E_c$ = the modulus of elasticity of the portland concrete (lbs/in$^2$).
$k$ = the modulus of subgrade reaction (lbs/in$^3$).

$Z_R$ and $S_0$ are the standard normal deviate and the standard error in predicting pavement serviceability, respectively, which were described in Chapter 11.

**Example 12.1** Calculate the required slab thickness for a JPCP with asphalt shoulders on a fair-draining subgrade (i.e., the water drains out of the pavement structure within a period of one week). It is estimated that the pavement structure becomes saturated less frequently than 1% of the time. The following data are also given:

- Estimated number of *ESALs* over a 20-year maximum performance period = 3 million.
- Modulus of subgrade reaction = 150 lbs/in$^3$ (40 MPa/m).
- Design reliability = 95%.
- Standard error in predicting serviceability = 0.50.
- Modulus of rupture of portland concrete (28 days) = 425 lbs/in$^2$ (2.9 MPa).
- Elastic modulus of portland concrete = 4,000,000 lbs/in$^2$ (28,000 MPa).
- $\Delta PSI = 2.5$ (from 4.5 to 2.0) and $P_t = 2.0$.

**ANSWER**

The drainage coefficient and the load transfer coefficient for the conditions specified are obtained as 1.1 and 4.2, from Tables 12.1 and 12.2, respectively. The value of the standard normal deviate for 95% confidence is −1.645. Substituting these values into Equation 12.1 gives:

$$\log(3{,}000{,}000) = -1.645\ 0.5 + 7.35\log(D+1) - 0.06$$

$$+\frac{\log\left[\frac{2.5}{3}\right]}{1+\frac{1.624\ 10^{7}}{(D+1)^{8.46}}} + (4.22 - 0.32\ 2.0)$$

$$\times \log\frac{425\ 1.1\ (D^{0.75} - 1.132)}{215.63\ 4.2\left[D^{0.75} - \frac{18.42}{(4{,}000{,}000/150)^{0.25}}\right]}$$

which, solved for *D*, gives a value of about 12.5 in to be rounded up to 13 in (33 cm). The solution of this equation can be obtained through a nomograph[1] or, preferably, by programming it in a spreadsheet, as done for this example. Note that the 20-year period is referred to as the maximum performance period because it is computed ignoring

**Table 12.1**
Recommended Drainage Coefficient $C_d$ Values (Ref. 1 Used by Permission)

| | Percent of Time Pavement Structure Is Saturated | | | |
|---|---|---|---|---|
| **Drainage Quality** | **<1%** | **1–5%** | **5–25%** | **>25%** |
| Excellent (drainage within 2 hrs) | 1.25–1.20 | 1.20–1.15 | 1.15–1.10 | 1.10 |
| Good (drainage within 1 day) | 1.20–1.15 | 1.15–1.10 | 1.10–1.00 | 1.00 |
| Fair (drainage within 1 week) | 1.15–1.10 | 1.10–1.00 | 10.0–0.90 | 0.90 |
| Poor (drainage within 1 month) | 1.10–1.00 | 10.0–0.90 | 0.90–0.80 | 0.80 |
| Very poor (no drainage) | 1.00–0.90 | 0.90–0.80 | 0.80–0.70 | 0.70 |

**Table 12.2**
Recommended Values for the Load Transfer Coefficient $J$ (Ref. 1 Used by Permission)

| | Shoulder Material/Load Transfer Reinforcement across Joints or Transverse Cracks | | | |
|---|---|---|---|---|
| **Pavement Type** | **Asphalt/Yes** | **Asphalt/No** | **Rebar-Tied PC/Yes** | **Rebar Tied PC/No** |
| **JPCP or JDRCP** | 3.2 | 3.8–4.4 | 2.5–3.1 | 3.6–4.2 |
| **CRCP** | 2.9–3.2 | – | 2.3-2.9 | – |

the serviceability loss due to the enviroment, which is described next.

### 12.2.2 Serviceability Loss Due to Environment

The 1993 version of the AASHTO Pavement Design Guide[1] considers pavement serviceability loss due to subgrade swelling and frost heave in a fashion similar to the method used for flexible pavements (Chapter 11). The serviceability loss due to swelling, $\Delta PSI_{SW}$, is given as a function of time $t$ (years) by:

$$\Delta PSI_{SW} = 0.00335\ V_R P_s \left(1 - e^{-\theta t}\right) \tag{12.2}$$

where:

$V_R$ = the potential vertical rise due to swelling (inches), which is mainly a function of the Plasticity Index of the subgrade, presented earlier as Figure 11.3.

$P_S$ = the percent of the total pavement area subjected to swelling, which is subjectively estimated.
$\theta$ = a subgrade swelling rate constant that can be estimated from the nomograph in Figure 11.4.

The serviceability loss due to frost heave, $\Delta PSI_{FH}$, is given as a function of time $t$ (years) by:

$$\Delta PSI_{FH} = 0.01\ p_f\ \Delta PSI_{max}\left(1 - e^{-0.02\ \phi t}\right) \qquad (12.3)$$

where:

$\Delta PSI_{max}$ = the maximum serviceability loss due to frost heave estimated on the basis of drainage quality and depth of frost penetration using the nomograph shown in Figure 11.5
$p_f$ = percent of frost probability subjectively estimated
$\phi$ = frost heave rate (mm/day) estimated mainly from the Unified Soil Classification (USC) of the subgrade soil using the nomograph shown in Figure 11.6.

**Example 12.2**

Compute the serviceability loss anticipated for a rigid pavement after 15 years of service. The subgrade is a fair-draining low-plasticity clay (designated as CL according to the USC system), having less than 60% by weight finer than 0.02 mm, and a Plasticity Index of 50%. The subgrade layer is 50 ft deep; it is exposed to high moisture levels and exhibits a medium level of structural fracturing. The percent of the pavement surface subjected to swelling is estimated to be 50%, while the probability of frost is estimated at 35%. The depth of frost penetration is 5 ft.

**ANSWER**

First, estimate the serviceability loss due to subgrade swelling. Use Figures 11.3 and 11.4 to compute the potential swelling rise, $V_R$, as 2 in and the swelling rate $\theta$ as 0.15, respectively. Substituting these values into Equation 12.2 gives:

$$\Delta PSI_{SW} = 0.00335\ 2\ 50\left(1 - e^{-0.15\ \ 15}\right) = 0.30$$

Next, estimate the serviceability loss due to subgrade frost heave. Use Figures 11.4 and 11.5 to compute the maximum serviceability loss, $\Delta PSI_{max}$, as 1.5, and the frost heave rate, $\phi$, as 3 mm/day, respectively. Substituting these values into Equation 12.3 gives:

$$\Delta PSI_{FH} = 0.01\ 30\ 1.5\left(1 - e^{-0.02\ 3\ 15}\right) = 0.267$$

Adding the two serviceability loss components gives the total serviceability loss due to the environment, $\Delta PSI_{SWFH}$, equal to 0.567.

### 12.2.3 Predicting Pavement Serviceable Life

Predicting pavement serviceable life involves computing the serviceability loss due to traffic and the environment. This presents a special challenge, because traffic deterioration is in terms of accumulated *ESALs*, while environmental deterioration is in terms of time (i.e., years). Assuming a mathematical function of *ESAL* accumulation with time allows estimating the number of years that will lapse before the combined effect of traffic and environment will reduce pavement serviceability from its initial postconstruction value (typically, 4.2 to 4.6) to a selected terminal value (either 2.5 or 2.0). This allows selecting a pavement slab thickness to meet design life requirements. The actual methodology used for doing so was described earlier under Example 11.3.

**Example 12.3**

Consider the traffic data specified in Example 12.1 and the subgrade conditions specified in Example 12.2. Assume that *ESALs* compound annually at a constant growth rate of 3%. Compute the anticipated pavement life of the 13-in JPCP pavement designed in Example 12.1, considering serviceability loss due to both traffic and environmental factors.

**ANSWER**

Given the *ESAL* growth assumption just stated, the accumulated *ESAL* versus pavement age relationship is plotted in Figure 12.1.

Select a performance period shorter than the stipulated 20 years, 15 years for this example (i.e., it is assumed that at the end of this performance period the pavement section will need rehabilitation). For the selected performance period, compute the loss in serviceability due to the environment (0.567 per Example 12.2). Subtract this value from the maximum performance period serviceability loss ($4.5 - 2.0 = 2.5$) to obtain the serviceability loss available for the

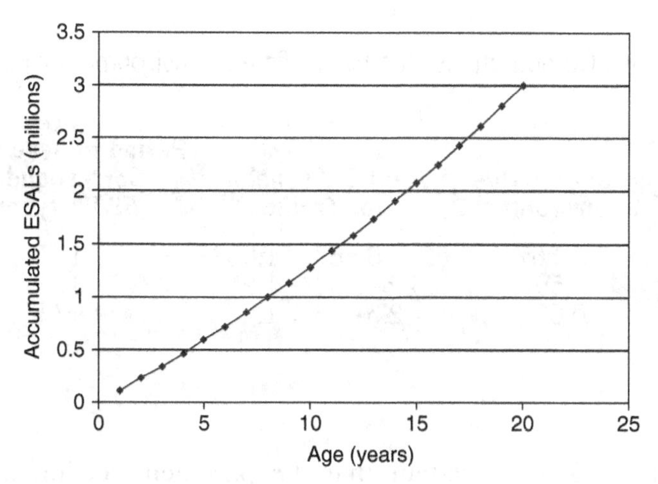


**Figure 12.1**
Cumulative *ESAL*s as a Function of Time; Example 12.3

traffic alone, $\Delta PSI_{TR}$, which after 15 years is:

$$\Delta PSI_{TR} = \Delta PSI - PSI_{SWFH} = 2.5 - 0.567 = 1.933$$

Compute the *ESALs* that will bring about this serviceability loss using Equation 12.1:

$$\log(W_t) = -1.645\ 0.5 + 7.35 \log(13 + 1) - 0.06 + \frac{\log\left[\dfrac{1.933}{3}\right]}{1 + \dfrac{1.624\ 10^7}{(13 + 1)^{8.46}}} + (4.22 - 0.32\ 2.0) \times \log \frac{425\ 4.2\ (13^{0.75} - 1.132)}{215.63\ 1.1\left[13^{0.75} - \dfrac{18.42}{(4{,}000{,}000/150)^{0.25}}\right]}$$

Solving this equation gives $W_t$ = 2,567,236 *ESALs*. This level of *ESALs* is accumulated over a period of 17.7 years, as indicated by Figure 12.1. The environmental and traffic serviceability loss calculations are repeated for another trial period, until a time period is found over which the sum of serviceability losses due to traffic and environment add to 2.5. The results are tabulated

**Table 12.3**
Summary of Serviceability Loss Computations Due to Traffic and Environmental Factors, Example 12.3

| Iteration | Performance Period (years) | PSI Loss Due to Environment | Net PSI Available for Traffic | Period to Accumulate Corresponding ESALs (years) |
|---|---|---|---|---|
| 1 | 15 | 0.567 | 2.5–0.567 = 1.933 | 17.7 |
| 2 | 16.4 | 0.59 | 2.5–0.59 = 1.91 | 17.6 |
| 3 | 17 | 0.60 | 2.5–0.6 = 1.9 | 17.5 |

in Table 12.3. It is evident that the pavement section in question will remain serviceable for approximately 17 years under the combined effect of traffic and environment. Commercially available software can be used to perfrom these calculations, along with the other provisions of the 1993 AASHTO Pavement Design Guide.[5]

### 12.2.4 Steel Reinforcement across Joints and Cracks

Joints and cracks open in response to concrete volume changes caused by temperature reductions and early-life shrinkage. Although the opening of these joints and cracks is resisted by subgrade friction, it can be fully controlled only through tiebar reinforcement, as described in Chapter 8. This is the case in longitudinal construction joints (e.g., adjacent jointed or continuously reinforced slabs cast in two stages) or the unavoidable transverse cracks in CRCPs. AASHTO[1] utilizes Equation 8.24, generalized in the form of Equation 8.25, to compute the area of tiebar reinforcement required per unit width of slab.

### 12.2.5 Steel Reinforcement in CRCPs

The method recommended by AASHTO[1] for designing CRCP steel reinforcement is based on three criteria:

- Transverse crack spacing (ranging between 3.5 and 8 ft)
- Transverse crack opening, which is not to exceed 0.04 in.
- Stresses in the steel reinforcement, which is to be limited to 75% of the ultimate yield stress

The following empirical equations were developed relating the crack spacing, $c_s$ (ft), and the crack width, $c_w$ (in), respectively, to

**Table 12.4**
Portland Concrete Shrinkage Coefficient (Ref. 1)

| Tensile Strength of Concrete at 28 Days (lbs/in$^2$) | Shrinkage Coefficient $z$ (in/in) |
|---|---|
| ≤300 | 0.0008 |
| 400 | 0.0006 |
| 500 | 0.00045 |
| 600 | 0.0003 |
| ≥700 | 0.0002 |

the percent reinforcement, $p$:

$$c_s = \frac{1.32\left(1+\dfrac{f_t}{1000}\right)^{6.70}\left(1+\dfrac{\alpha_s}{2\,\alpha_c}\right)^{1.15}(1+\Phi)^{2.19}}{\left(1+\dfrac{\sigma_W}{1000}\right)^{5.20}(1+p)^{4.6}(1+1000\,z)^{1.79}} \tag{12.4}$$

$$c_w = \frac{0.00932\left(1+\dfrac{f_t}{1000}\right)^{6.53}(1+\Phi)^{2.20}}{\left(1+\dfrac{\sigma_W}{1000}\right)^{4.91}(1+p)^{4.55}} \tag{12.5}$$

where:

$f_t$ = the tensile strength of the concrete (lbs/in$^2$), measured through indirect tension tests according to AASHTO Standard T198.[3] Normally, portland concrete tensile strength is about 86% of its modulus of rupture, $S_c'$, measured as described earlier.

$z$ = the concrete shrinkage strain at 28 days (in/in), given in Table 12.4.

$\Phi$ = is the diameter of the steel reinforcement (in.).

$\sigma_w$ = stress under a load $w$ (lbs), given by the nomograph in Figure 12.2.

The variables $\alpha_s$ and $\alpha_c$ are the coefficients of thermal expansion for the steel and the concrete, respectively. The value of $\alpha_s$ is typically 5.0 $10^{-6}$/°F. Values for $\alpha_c$ are given in Table 12.5 as a function of the type of aggregate in the portland concrete. The stress in the steel

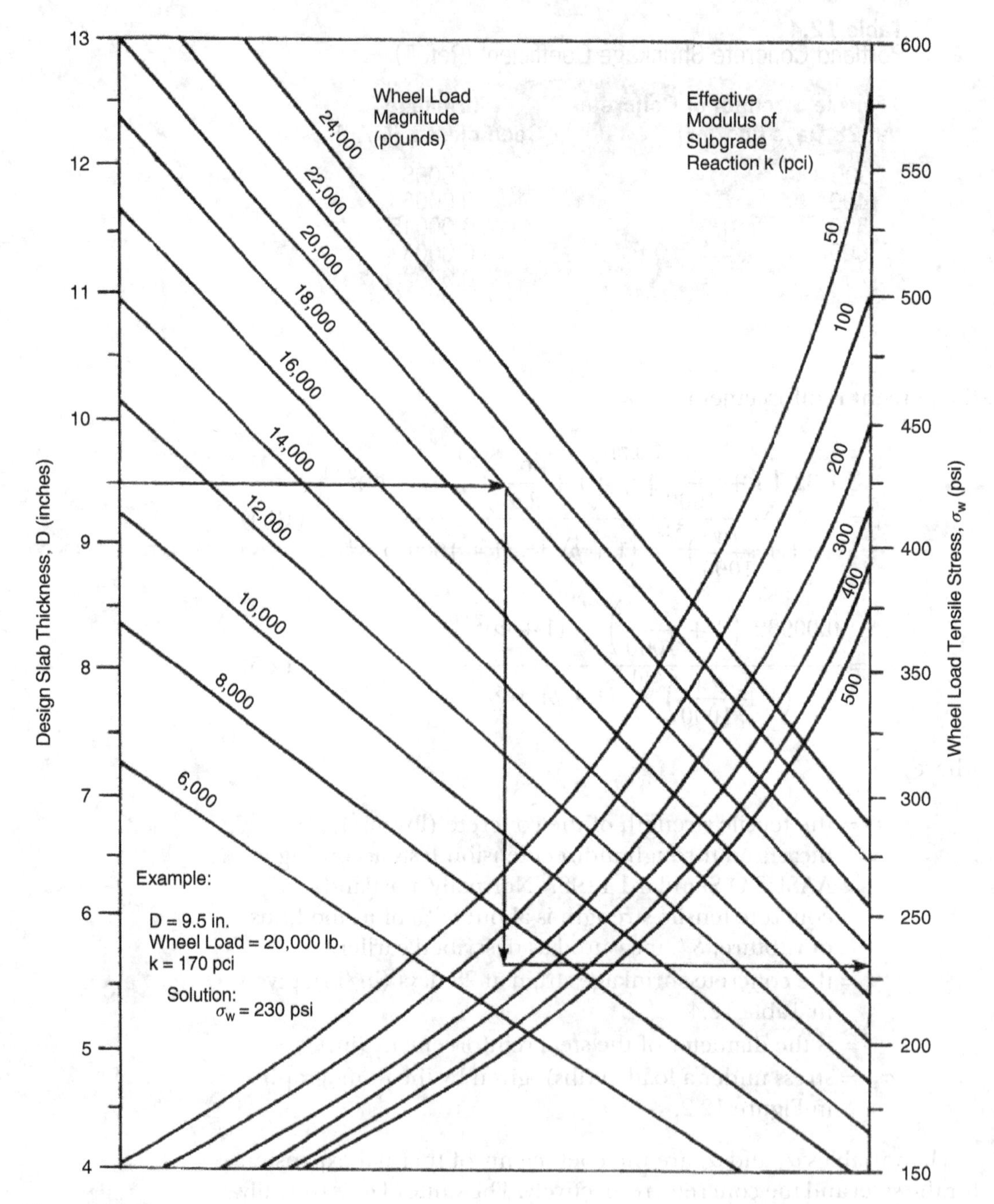


**Figure 12.2**
Computing Rigid Pavement Stresses $\sigma_W$ for Steel Reinforcement Design (Ref. 1 Used by Permission)

**Table 12.5**
Thermal Coefficient of Portland Concrete (Ref. 1 Used by Permission)

| Dominant Aggregate Type | Thermal Coefficient $\alpha_c$ ($10^{-6}$/°F) |
|---|---|
| Quartz | 6.6 |
| Sandstone | 6.5 |
| Gravel | 6.0 |
| Granite | 5.3 |
| Basalt | 4.8 |
| Limestone | 3.8 |

reinforcement, $\sigma_s$, is given by:

$$\sigma_s = \frac{47300\left(1+\frac{DT_D}{100}\right)^{0.425}\left(1+\frac{f_t}{1000}\right)^{4.09}}{\left(1+\frac{\sigma_w}{1000}\right)^{3.14}(1+1000\ z)^{0.494}\ (1+p)^{2.74}} \tag{12.6}$$

where $DT_D$ is the difference between the average high daily temperature of the month the concrete was cast, $T_H$, and the average daily low temperature during the coldest month of the year, $T_L$.

$$DT_D = T_H - T_L \tag{12.7}$$

The limiting working stresses in the steel reinforcement is taken as 75% of its ultimate yield stress. Limiting working stress values for Grade 60 steel, according to ASTM A615[4] are given in Table 12.6 as a function of bar reinforcement size.

The percent steel reinforcement that satisfies all three criteria is determined through the following steps:

1. Select a steel reinforcement bar diameter and compute the percentage of steel required to satisfy a maximum of $c_s = 8$ ft, $c_w = 0.04$ in., and $\sigma_s =$ the allowable steel stress. The maximum of these values is defined as $p_{min}$.
2. Compute $p_{max}$ as the percentage of steel that satisfies $c_s = 3.5$ ft.
3. If $p_{max}$ is larger than $p_{min}$, the design is feasible; otherwise, the steel reinforcement bar diameter should be changed and these steps repeated.

**Table 12.6**
Allowable Steel Working Stress in Portland Concrete Reinforcement (Ref. 1 Used by Permission)

| Tensile Strength of Concrete at 28 Days (lbs/in²) | Allowable Steel Stress (1000 lbs/in²) Steel Bar No. 4 1/2-in diameter | Steel Bar No. 5 5/8-in diameter | Steel Bar No. 6 3/4-in diameter |
|---|---|---|---|
| ≤300 | 65 | 57 | 54 |
| 400 | 67 | 60 | 55 |
| 500 | 67 | 61 | 56 |
| 600 | 67 | 63 | 58 |
| 700 | 67 | 65 | 59 |
| ≥800 | 67 | 67 | 60 |

Having established a feasible design, compute the number of steel reinforcement bars required per unit width for the $p_{max}$ and $p_{min}$, using:

$$N_{\min} = 0.01273\ p_{\min}\ W\ D\ /\Phi^2 \quad (12.8)$$

$$N_{\max} = 0.01273\ p_{\max}\ W\ D\ /\Phi^2 \quad (12.9)$$

where:

$N_{min}$ and $N_{max}$ = the minimum and maximum number of reinforcement bars required, respectively.

$W$ and $D$ = the width and the thickness of the slab (inches), respectively.

An integer number of dowel bars is finally selected from the numbers computed here. Note that Equations 12.4, 12.5, and 12.6 can be solved through nomographs[1] or, preferably, by programming them into a spreadsheet, as done for the next example.

**Example 12.4** Compute the amount of steel reinforcement required for a limestone aggregate CRCP slab 9.5 in. thick. Also given:

- $T_H$, and $T_L = 75°$ and 20°F, respectively
- $f_t = 550\,\text{lbs/in}^2$
- Axle load of 20,000 lbs
- Subgrade reaction modulus $k = 170\,\text{lbs/in}^3$
- $\alpha_s = 5.0\ 10^{-6}/°\text{F}$

**ANSWER**

Use Figure 12.2 to compute the tensile stress in the slab due to the 20,000 lb load, $\sigma_w$, as 230 lbs/in$^2$ Table 12.5 for a limestone aggregate gives a value of 3.8 $10^{-6}$/°F for the coefficient of thermal expansion of the portland concrete, $\alpha_c$. The tensile strength of the concrete suggests a shrinkage coefficient $z$ of 0.0004 (Table 12.4). Selecting as a trial a No. 5 steel bar gives a limiting steel stress of 62,000 lbs/in$^2$ (Table 12.6). Substituting the available values in Equations 12.4, 12.5, and 12.6 allows solving for the minimum percent steel reinforcement, $p$:

$$8 = \frac{1.32\left(1+\frac{550}{1000}\right)^{6.70}\left(1+\frac{5}{2\ 3.8}\right)^{1.15}(1+0.625)^{2.19}}{\left(1+\frac{230}{1000}\right)^{5.20}(1+p)^{4.6}(1+1000\ 0.0004)^{1.79}}$$

$$0.04 = \frac{0.00932\left(1+\frac{550}{1000}\right)^{6.53}(1+0.625)^{2.20}}{\left(1+\frac{230}{1000}\right)^{4.91}(1+p)^{4.55}}$$

$$62000 = \frac{47300\left(1+\frac{55}{100}\right)^{0.425}\left(1+\frac{550}{1000}\right)^{4.09}}{\left(1+\frac{230}{1000}\right)^{3.14}(1+1000\ 0.0004)^{0.494}(1+p)^{2.74}}$$

The calculated values for the minimum percent steel are 0.425, 0.36, and 0.46, respectively, which yield a minimum percent steel reinforcement $p_{min}$ value of 0.46. The maximum percent steel reinforcement value is obtained by substituting the minimum crack spacing value ($c_s$ = 3.5 ft) into Equation 12.4. The corresponding maximum percent steel reinforcement value is 0.7. Since $p_{max}$ is larger than $p_{min}$, the design is feasible. Utilizing Equation 12.8 and 12.9 gives the number of reinforcement bars required:

- $N_{min} = 0.01273\ 0.46\ 12\ 9.5\ /0.625^2 = 1.71$ No. 5 bars
- $N_{max} = 0.01273\ 0.7\ 12\ 9.5\ /0.625^2 = 2.6$ No. 5 bars

Finally, select two No. 5 bars per foot-width of slab, which gives a 6-in. center-to-center bar spacing.

## 12.3 PCA Design Method

The Portland Cement Association (PCA) developed a mechanistic design method for rigid highway and street pavements.[16] It is based on two design criteria associated with slab fatigue cracking and subgrade erosion. Traffic is handled in the form of the design life axle load repetitions by load increment and axle configuration (i.e., single axles and tandem axles). For a given axle load and configuration, structural response parameters associated with fatigue and erosion are computed using a finite element algorithm; and the associated fatigue and erosion life is calculated, in terms of the number of passes that would cause failure. Subsequently, the fatigue or erosion damage contributed by each axle load and configuration is computed as the ratio of the actual number of lifelong axle passes divided by the number of passes that would cause fatigue or erosion failure, respectively. Finally, these fatigue or erosion damage ratios are summed over all the applied axle loads and configurations according to Miner's hypothesis.[10] A pavement design is deemed adequate if it satisfies independently fatigue damage accumulation lower than 100% and erosion damage accumulation lower than 100%. The following sections present the actual methodology used for computing fatigue and erosion damage and the trial-and-error method used for arriving at a rigid pavement configuration and thickness. The tables and nomographs used are obtained from an SI version of this method published by the Canadian Portland Cement Association.[17]

### 12.3.1 Fatigue Damage

Slab fatigue is associated with the flexural stresses in the concrete slab under a given axle load. The flexural stress divided by the modulus of rupture of the portland concrete, $S_c'$, defines the stress ratio factor (*SRF*):

$$SRF = \frac{ES}{Sc'} \tag{12.10}$$

where *ES* (equivalent stress) is the flexural stress under a reference axle load. Tables are used to obtain *ES* as a function of slab thickness and modulus of subgrade reaction. A different table is available for

**Table 12.7**
Equivalent Stress *ES* (MPa) (Single/Tandem Axles) Rigid Pavement with Asphalt Concrete Shoulders (Ref. 17)

| Slab Thickness (mm) | Modulus of Subgrade Reaction, *k* (MPa/m) | | | | | |
|---|---|---|---|---|---|---|
| | 20 | 40 | 60 | 80 | 140 | 180 |
| 100 | 5.42/4.39 | 4.75/3.83 | 4.38/3.59 | 4.13/3.44 | 3.66/3.22 | 3.45/3.15 |
| 110 | 4.71/3.88 | 4.16/3.35 | 3.85/3.12 | 3.63/2.97 | 3.23/2.76 | 3.06/2.68 |
| 120 | 4.19/3.47 | 3.69/2.98 | 3.41/2.75 | 3.23/2.62 | 2.88/2.40 | 2.73/2.33 |
| 130 | 3.75/3.14 | 3.30/2.68 | 3.06/2.46 | 2.89/2.33 | 2.59/2.13 | 2.46/2.05 |
| 140 | 3.37/2.87 | 2.97/2.43 | 2.76/2.23 | 2.61/2.10 | 2.34/1.90 | 2.23/1.83 |
| 150 | 3.06/2.64 | 2.70/2.23 | 2.51/2.04 | 2.37/1.92 | 2.13/1.72 | 2.03/1.65 |
| 160 | 2.79/2.45 | 2.47/2.06 | 2.29/1.87 | 2.17/1.76 | 1.95/1.57 | 1.86/1.50 |
| 170 | 2.56/2.28 | 2.26/1.91 | 2.10/1.74 | 1.99/1.63 | 1.80/1.45 | 1.71/1.38 |
| 180 | 2.37/2.14 | 2.09/1.79 | 1.94/1.62 | 1.84/1.51 | 1.66/1.34 | 1.58/1.27 |
| 190 | 2.19/2.01 | 1.94/1.67 | 1.80/1.51 | 1.71/1.41 | 1.54/1.25 | 1.47/1.18 |
| 200 | 2.04/1.90 | 1.80/1.58 | 1.67/1.42 | 1.59/1.33 | 1.43/1.17 | 1.37/1.11 |
| 210 | 1.91/1.79 | 1.68/1.49 | 1.56/1.34 | 1.48/1.25 | 1.34/1.10 | 1.28/1.04 |
| 220 | 1.79/1.70 | 1.57/1.41 | 1.46/1.27 | 1.39/1.18 | 1.26/1.03 | 1.20/0.98 |
| 230 | 1.68/1.62 | 1.48/1.34 | 1.38/1.21 | 1.31/1.12 | 1.18/0.98 | 1.13/0.92 |
| 240 | 1.58/1.55 | 1.39/1.28 | 1.30/1.15 | 1.23/1.06 | 1.11/0.93 | 1.06/0.87 |
| 250 | 1.49/1.48 | 1.32/1.22 | 1.22/1.09 | 1.16/1.01 | 1.05/0.88 | 1.00/0.83 |
| 260 | 1.41/1.41 | 1.25/1.17 | 1.16/1.05 | 1.10/0.97 | 0.99/0.84 | 0.95/0.79 |
| 270 | 1.34/1.36 | 1.18/1.12 | 1.10/1.00 | 1.04/0.93 | 0.94/0.80 | 0.90/0.75 |
| 280 | 1.28/1.30 | 1.12/1.07 | 1.04/0.96 | 0.99/0.89 | 0.89/0.77 | 0.86/0.72 |
| 290 | 1.22/1.25 | 1.07/1.03 | 0.99/0.92 | 0.94/0.85 | 0.85/0.74 | 0.81/0.69 |
| 300 | 1.16/1.21 | 1.02/0.99 | 0.95/0.89 | 0.90/0.82 | 0.81/0.71 | 0.78/0.66 |
| 310 | 1.11/1.16 | 0.97/0.96 | 0.90/0.86 | 0.86/0.79 | 0.77/0.68 | 0.74/0.64 |
| 320 | 1.06/1.12 | 0.93/0.92 | 0.86/0.83 | 0.82/0.76 | 0.74/0.66 | 0.71/0.62 |
| 330 | 1.02/1.09 | 0.89/0.89 | 0.83/0.80 | 0.78/0.74 | 0.71/0.63 | 0.68/0.59 |
| 340 | 0.98/1.05 | 0.85/0.86 | 0.79/0.77 | 0.75/0.71 | 0.68/0.61 | 0.65/0.57 |
| 350 | 0.94/1.02 | 0.82/0.84 | 0.76/0.75 | 0.72/0.69 | 0.65/0.59 | 0.62/0.55 |

rigid pavements with asphalt concrete shoulders and rigid pavements with portland concrete shoulders. This is to reflect the differences in lateral support that yields different flexural stresses when the right-hand-side wheel is at the edge of the driving lane. Table 12.7, for example, gives *ES* values (MPa) for rigid pavements with asphalt concrete shoulders in metric units.[17] The two values separated by a slash correspond to single and tandem axles, respectively.

These two values, divided by the modulus of rupture of the portland concrete give *SRF* values for single and tandem axles, respectively. These *SRF* values allow computing the number of single

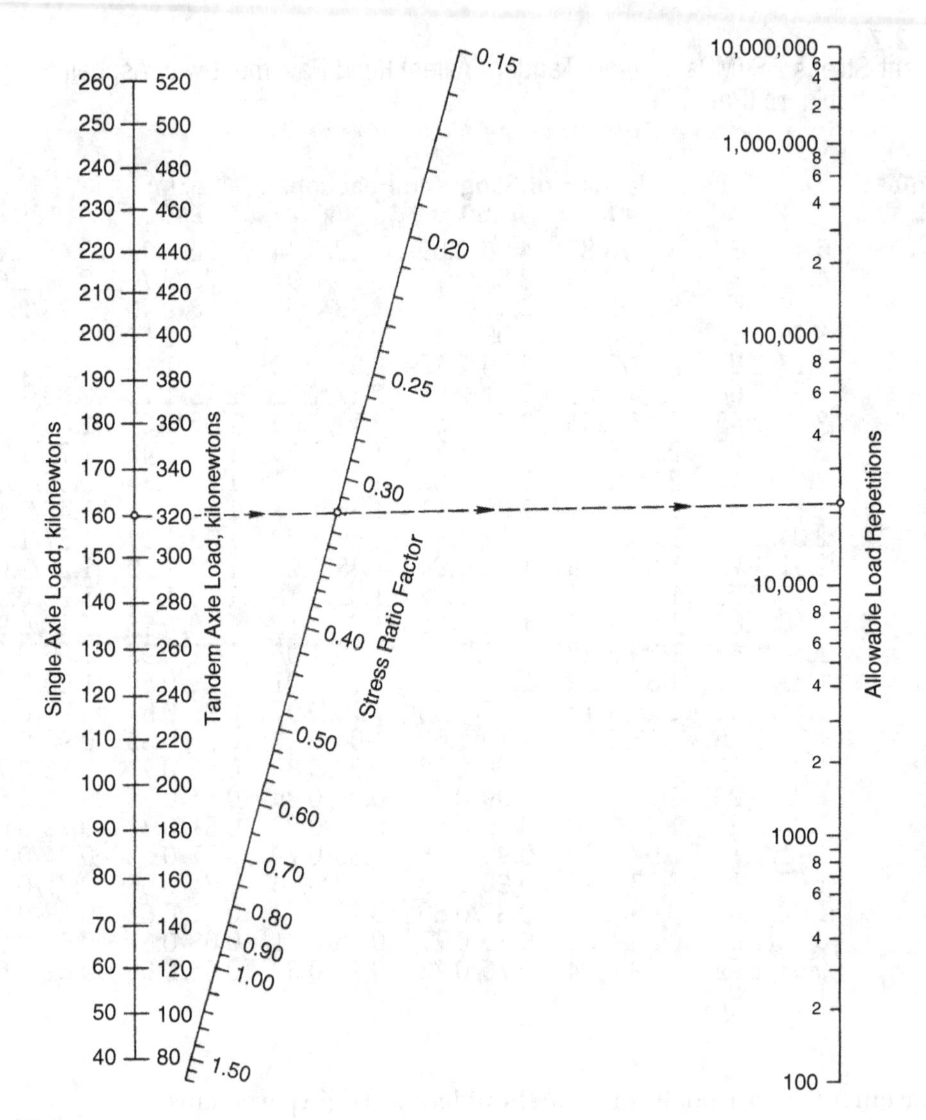


**Figure 12.3**
Nomograph for Computing Fatigue Life by the PCA Rigid Pavement Design Method (Ref. 17)

and tandem axle passes to fatigue failure for each load increment through Figure 12.3.

The following example explains the methodology used for computing fatigue damage ratios, and summing them to check whether the particular design yields an accumulated fatigue damage lower than 100%.

**Example 12.5**

Design a JPCP for a four-lane divided highway with asphalt concrete shoulders. The load frequency distribution data is given in Table 12.8, in the form of the number of axles per 1000 heavy trucks (FHWA classes 4 to 13). Noted that, from a fatigue damage point of view, only the heavy axle loads are of interest. Furthermore, some of the axle loads listed in this table are over the legal limit, which explains their low expected frequency. Also given:

- Load safety factor (LSF) = 1.00
- Modulus of subgrade reaction $k = 40\,\text{MPa/m}$
- Portland concrete modulus of rupture = 4.5 MPa
- Average daily traffic (ADT) over design life = 15,670 vehicles/day
- Directional split = 50/50
- Percent of heavy trucks (FHWA classes 4 to 13) = 19%
- Percent of trucks in the right-hand-side lane = 80%
- Design life = 20 years

**ANSWER**

Compute the total number of heavy trucks in the design lane over the 20 year design life:

$$15,670\ 365\ 20\ 0.50\ 0.80\ 0.19 = 8,693,716 \text{ heavy trucks}$$

Use the load frequency distribution given in Table 12.8 to compute the actual number of axle passes, by configuration, as shown in Table 12.9 (e.g., $5042 = 0.58 \times 8693.716$). Select a trial slab thickness of 240 mm to test for fatigue damage. For 240 mm, Table 12.7 gives *ES* values of 1.39 MPa and 1.28 MPa for single and tandem axles, respectively. Using Equation 12.10 and a modulus of rupture of 4.5 MPa allows computing *SRF* values of $1.39/4.5 = 0.31$ and $1.28/4.5 = 0.28$, respectively. Entering these *SRF*s into Figure 12.3, allows computing the number of axle passes to fatigue failure. In doing so, it was conservatively elected to use the upper limit of the load interval. Finally, the damage ratios caused by each axle load interval are computed and summed for all load intervals and

**Table 12.8**
Load Frequency Distribution (i.e., Number of Axles per 1000 Heavy Trucks) Example 12.5

| Load Range (kN) | Single Axles<br>Number of Axles/1000 Heavy Trucks |
|---|---|
| 125–133 | 0.58 |
| 115–125 | 1.35 |
| 107–115 | 2.77 |
| 97.8–107 | 5.92 |
| 88.8–97.8 | 9.83 |
| 80–88.8 | 21.67 |
| **Load Range (kN)** | **Tandem Axles<br>Number of Axles/1000 Heavy Trucks** |
| 213–231 | 1.96 |
| 195–213 | 3.94 |
| 178–195 | 11.48 |
| 160–178 | 34.27 |
| 142–160 | 81.42 |
| 125–142 | 85.54 |

**Table 12.9**
Actual Number of Heavy Truck Axles by Load and Configuration Example 12.5

| Axle Load (kN) | Single Axles<br>Actual Number of Axle Passes |
|---|---|
| 125–133 | 5042 |
| 115–125 | 11737 |
| 107–115 | 24082 |
| 97.8–107 | 51467 |
| 88.8–97.8 | 85459 |
| 80–88.8 | 188393 |
| **Axle load (kN)** | **Tandem Axles<br>Actual Number of Axle Passes** |
| 213–231 | 17040 |
| 195–213 | 34253 |
| 178–195 | 99804 |
| 160–178 | 297934 |
| 142–160 | 707842 |
| 125–142 | 743660 |

axle configurations. The results, tabulated in Table 12.10, show that 240 mm is perhaps too thick for the traffic loads specified. Before testing a lesser slab thickness for fatigue, however, it is prudent to make sure that the 240 mm slab can pass the erosion test as well, as described next.

### 12.3.2 Erosion Damage

Subgrade erosion is associated with the work being input into the subgrade from a given axle load. The latter is computed as the product of the subgrade deflection at the edge of a slab multiplied by the subgrade stress and divided by the radius of the relative stiffness of the slab (i.e., the variable $\ell$ given earlier by Equation 8.21), defined as the erosion factor (*EF*). *EF* is obtained from tables as a function of the slab thickness. Four different tables are available, depending on the type of shoulder (asphalt concrete versus portland concrete) and the vertical load transfer mechanism between slabs (JPCPs versus JDRCPs). The latter applies to CRCPs as well. An example is given in Table 12.11 for a JPCP with an asphalt concrete shoulder. The two values separated by a slash correspond to single and tandem axles, respectively. The remaining tables can be found in reference 17. The values of *EF* established allow estimating the number of axle load repetitions to erosion failure using Figure 12.4.

As for fatigue, erosion ratios are computed for each axle load and configuration as the ratios of the actual number of passes divided by the number of passes to erosion failure. Finally, these ratios are summed to check if the particular design provides a cumulative erosion lower than 100%. An example of the erosion calculations follows.

**Example 12.6**

Evaluate whether the slab design described in Example 12.5 meets erosion requirements.

**ANSWER**

For the 240 mm JPCP pavement with asphalt shoulders on a subgrade with 40 MPa/m reaction modulus, Table 12.11 gives *EF* values of 2.83 and 3.05 for single and tandem axles, respectively. The number

**Table 12.10**
Fatigue Damage Ratios by Heavy Truck Axle Load and Configuration Example 12.5

| Single Axles | | | |
|---|---|---|---|
| **Axle Load (kN)** | **Number of Passes to Failure (Figure. 12.7)** | **Actual Number of Axle Passes** | **Damage Ratio (%)** |
| 125–133 | 500,000 | 5042 | 1.0084 |
| 115–125 | 4,000,000 | 11737 | 0.293 |
| 107–115 | — | 24082 | 0 |
| 97.8–107 | — | 51467 | 0 |
| 88.8–97.8 | — | 85459 | 0 |
| 80–88.8 | — | 188393 | 0 |
| **Tandem Axles** | | | |
| **Axle Load (kN)** | **Number of Passes to Failure (Figure. 12.7)** | **Actual Number of Axle Passes** | |
| 213–231 | — | 17040 | 0 |
| 195–213 | — | 34253 | 0 |
| 178–195 | — | 99804 | 0 |
| 160–178 | — | 297934 | 0 |
| 142–160 | — | 707842 | 0 |
| 125–142 | — | 743660 | 0 |
| | | Total Accumulated Fatigue Damage: | 1.3% |

— = Practically infinite axle passes to fatigue failure.

of axle load passes to erosion failure and the erosion factors are given in Table 12.12.

The total accumulated erosion damage for this trial is 48.8%. In conclusion, the critical distress for this design is clearly erosion. Another erosion check should be tried—for example, 230 mm slab, to see if some slab thickness can be economized without raising the accumulated erosion damage higher than 100%.

The PCA methodology described was implemented into a proprietary software package named PCAPAV.[14] Note that the truck traffic volume in this software package is input in terms of the bidirectional daily heavy truck traffic in the design lane (for examples 12.5 and 12.6, this is 15,670 0.80 0.19 = 2,382 trucks).

## 12.4 NCHRP 1-37A Design Method

The NCHRP Study 1-37A[19] adopted a mechanistic-empirical rigid pavement design approach. It involves analytical computations of

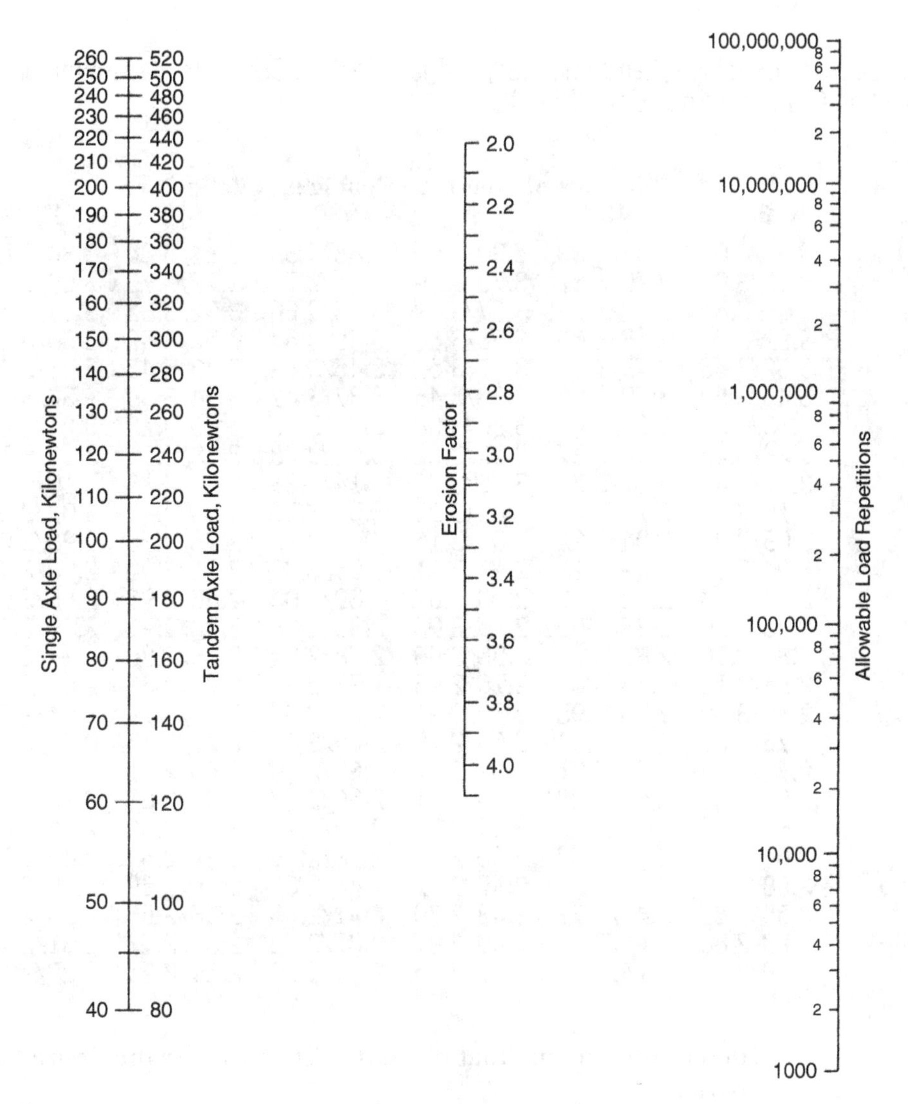


**Figure 12.4**
Nomograph for Computing Erosion Life by the PCA Rigid Pavement Design Method (Ref. 17)

pavement structural response to load and environmental input, and incorporation of these parameters into mechanistic-empirical damage functions to predict the accumulation of pavement distresses versus time. Design is based on successive trials, whereby the performance of selected structural pavement configurations is evaluated to

**Table 12.11**
Erosion Factors *EF* (single/tandem Axles) —Plain Jointed Concrete Pavement (JPCP) with Asphalt Concrete Shoulders (Ref. 17)

| Slab Thickness (mm) | Modulus of Subgrade Reaction, *k* (MPa/m) 20 | 40 | 60 | 80 | 140 | 180 |
|---|---|---|---|---|---|---|
| 100 | 3.94/4.00 | 3.92/3.93 | 3.90/3.90 | 3.88/3.88 | 3.84/3.84 | 3.80/3.82 |
| 110 | 3.82/3.90 | 3.79/3.82 | 3.78/3.79 | 3.76/3.76 | 3.72/3.72 | 3.69/3.70 |
| 120 | 3.71/3.81 | 3.68/3.73 | 3.67/3.69 | 3.65/3.66 | 3.62/3.62 | 3.59/3.59 |
| 130 | 3.61/3.73 | 3.58/3.65 | 3.56/3.60 | 3.55/3.57 | 3.52/3.52 | 3.50/3.49 |
| 140 | 3.52/3.66 | 3.49/3.57 | 3.47/3.52 | 3.46/3.49 | 3.43/3.43 | 3.41/3.41 |
| 150 | 3.43/3.59 | 3.40/3.50 | 3.38/3.45 | 3.37/3.42 | 3.34/3.36 | 3.32/3.33 |
| 160 | 3.35/3.53 | 3.32/3.43 | 3.30/3.38 | 3.29/3.35 | 3.26/3.28 | 3.24/3.26 |
| 170 | 3.28/3.48 | 3.24/3.37 | 3.22/3.32 | 3.21/3.28 | 3.18/3.22 | 3.17/3.19 |
| 180 | 3.21/3.42 | 3.17/3.32 | 3.15/3.26 | 3.14/3.23 | 3.11/3.16 | 3.10/3.13 |
| 190 | 3.15/3.37 | 3.11/3.27 | 3.08/3.21 | 3.07/3.17 | 3.04/3.10 | 3.03/3.07 |
| 200 | 3.09/3.33 | 3.04/3.22 | 3.02/3.16 | 3.01/3.12 | 2.98/3.05 | 2.96/3.01 |
| 210 | 3.04/3.28 | 2.99/3.17 | 2.96/3.11 | 2.95/3.07 | 2.92/3.00 | 2.90/2.96 |
| 220 | 2.98/3.24 | 2.93/3.13 | 2.90/3.07 | 2.89/3.03 | 2.86/2.95 | 2.85/2.92 |
| 230 | 2.93/3.20 | 2.88/3.09 | 2.85/3.03 | 2.83/2.98 | 2.80/2.91 | 2.79/2.87 |
| 240 | 2.89/3.16 | 2.83/3.05 | 2.80/2.99 | 2.78/2.94 | 2.75/2.86 | 2.74/2.83 |
| 250 | 2.84/3.13 | 2.78/3.01 | 2.75/2.95 | 2.73/2.91 | 2.70/2.82 | 2.69/2.79 |
| 260 | 2.80/3.09 | 2.73/2.98 | 2.70/2.91 | 2.69/2.87 | 2.65/2.79 | 2.64/2.75 |
| 270 | 2.76/3.06 | 2.69/2.94 | 2.66/2.88 | 2.64/2.83 | 2.61/2.75 | 2.59/2.71 |
| 280 | 2.72/3.03 | 2.65/2.91 | 2.62/2.84 | 2.60/2.80 | 2.56/2.71 | 2.55/2.68 |
| 290 | 2.68/3.00 | 2.61/2.88 | 2.58/2.81 | 2.56/2.77 | 2.52/2.68 | 2.50/2.64 |
| 300 | 2.65/2.97 | 2.57/2.85 | 2.54/2.78 | 2.52/2.74 | 2.48/2.65 | 2.46/2.61 |
| 310 | 2.61/2.94 | 2.54/2.82 | 2.50/2.75 | 2.48/2.71 | 2.44/2.62 | 2.42/2.58 |
| 320 | 2.58/2.91 | 2.50/2.79 | 2.47/2.72 | 2.44/2.68 | 2.40/2.59 | 2.38/2.55 |
| 330 | 2.55/2.89 | 2.47/2.77 | 2.43/2.70 | 2.41/2.65 | 2.36/2.56 | 2.35/2.52 |
| 340 | 2.52/2.86 | 2.44/2.74 | 2.40/2.67 | 2.37/2.62 | 2.33/2.53 | 2.31/2.49 |
| 350 | 2.49/2.84 | 2.14/2.71 | 2.37/2.65 | 2.34/2.60 | 2.29/2.51 | 2.28/2.47 |

determine the one that prevents distresses over the desired analysis period.

The NCHRP 1-37A rigid design approach distinguishes two broad categories of portland concrete pavement structures, namely jointed (JCP) and continuously reinforced (CRCP). JCPs can be either undoweled or doweled—that is, transmit vertical forces between adjacent slabs through aggregate interlock or through dowel action across the joints, respectively. This design approach considers the following damage mechanisms:

- Fatigue transverse cracking, both bottom-up and top-down, for JCP only

**Table 12.12**
Erosion Damage Ratios by Heavy Truck Axle Load and Configuration Example 12.6

| Single Axles | | | |
|---|---|---|---|
| Axle Load (kN) | Number of Passes to Failure (Figure. 12.8) | Actual Number of Axle Passes | Damage Ratio (%) |
| 125–133 | 900,000 | 5042 | 0.56 |
| 115–125 | 1,100,000 | 11737 | 1.07 |
| 107–115 | 2,000,000 | 24082 | 1.20 |
| 97.8–107 | 3,050,000 | 51467 | 1.69 |
| 88.8–97.8 | 6,000,000 | 85459 | 1.42 |
| 80–88.8 | 10,000,000 | 188393 | 1.88 |
| **Tandem Axles** | | | |
| Axle Load (kN) | Number of Passes to Failure (Figure. 12.8) | Actual Number of Axle Passes | |
| 213–231 | 600,000 | 17040 | 2.84 |
| 195–213 | 1,000,000 | 34253 | 3.43 |
| 178–195 | 1,800,000 | 99804 | 5.54 |
| 160–178 | 3,000,000 | 297934 | 9.93 |
| 142–160 | 6,000,000 | 707842 | 11.80 |
| 125–142 | 10,000,000 | 743660.5 | 7.44 |
| | | Total Accumulated Erosion Damage: | 48.8% |

- Joint faulting, for JCP only
- Punchouts, for CRCP only
- Roughness, for both JCP and CRCP

Pavement structural response is computed using the Finite Element computer model ISLAB2000.[7] Environmental input to the model is computed through the Enhanced Integrated Climatic Model (EICM),[8] described in Chapter 10. Temperature profiles are computed at 11 equally spaced depth increments into slabs at hourly intervals, based on climatic data from the weather station nearest to the pavement design location. The effect of relative humidity is translated into an equivalent temperature gradient. Stresses are computed for a limited combination of axle locations and slab conditions, including:

- Slab curling due to temperature/moisture gradients
- Loss of subgrade support due to curling
- Slab-to-slab interaction

This output is fed into a neural-network algorithm for estimating the critical structural response parameters to be input into the pavement damage functions. These analysis components are implemented into the NCHRP 1-37A software. The following provides a description of the damage functions implemented into the rigid pavement analysis method adopted by NCHRP Study 1-37A.[19] Noted that this methodology is currently under review.[6] The outcome of this review and subsequent research is likely to result in modifications to some of the damage functions described here. Also note that the NCHRP 1-37A software, in its current form, accepts input only in imperial units.

### 12.4.1 Fatigue Transverse Cracking

The total percentage of slab cracking, *TCRACK*, in jointed portland concrete pavements is determined by adding the percentage of top-down and bottom-up cracking, while subtracting their product. This is expressed as:

$$TCRACK = \left(CRK_p + CRK_q - CRK_p CRK_q\right)\ 100\% \qquad (12.11)$$

where $CRK_p$, $CRK_q$ = predicted percentage of bottom-up and top-down cracking, respectively.

Each specific cracking type is modeled as a function of its respective fatigue damage using:

$$CRK_{p,q} = \frac{1}{1 + FD_{p,q}^{-1.68}} \qquad (12.12)$$

where:

$$FD_{p,q} = \sum \frac{n_{i,j,k,l,m,n}}{N_{i,j,k,l,m,n}} \qquad (12.13)$$

and:

$FD_{p,q}$ = total fatigue damage for bottom-up or top-down cracking, accumulated according to Miner's hypothesis[10]

$n_{i,\,j,\,k,\,l,\,m,\,n}$, = applied number of load applications at condition *i, j, k, l, m, n,*

$N_{i,j,k,l,m}$ = number of load applications to failure, (i.e., 50% slab cracking) under conditions *i, j, k, l, m, n*, where:

$i$ = age, which accounts for changes in portland concrete modulus of rupture, layer bond condition and the deterioration of shoulder load transfer efficiency (*LTE*)
$j$ = month, which accounts for changes in base and effective modulus of subgrade reaction
$k$ = axle type (single, tandem, tridem, and quad)
$l$ = load level for each axle type
$m$ = temperature difference between top and bottom of slab
$n$ = traffic path, assuming a normally distributed lateral wheel wander

The number of load application to failure (50% of the slabs cracked), $N_{i,j,k,l,m,n}$, under conditions *i, j, k, l, m, n* is given by:

$$\log\left(N_{i,j,k,l,m,n}\right) = 2.0\left(\frac{MR_i}{\sigma_{i,j,k,l,m,n}}\right)^{1.22} + 0.4371 \qquad (12.14)$$

where:

$MR_i$ = modulus of rupture of portland concrete at age $i$ ($\text{lb/in}^2$) (Note that this is the same property as the one denoted by $S_c'$ earlier.)
$\sigma_{i,j,k,,l,m,n}$ = critical stresses under load conditions *i, j, k, l, m, n* ($\text{lb/in}^2$)

In performing these fatigue calculations, the locations of calculated stresses are crucial. These differ as a function of axle type, lateral placement, temperature difference between top and bottom of slab, subgrade support, and so on. As mentioned earlier, a neural network approach was used to establish critical stress locations and their magnitudes. This approach drastically reduces software running time.

**Example 12.7** Table 12.13 gives the number of single-axle passes by axle load for a particular hour of a particular month (note that only heavier axle loads are listed, as lighter ones cause insignificant fatigue damage). This table also lists the strain levels predicted at the top and bottom of a portland concrete slab for a particular wheel-path. The modulus of rupture for the portland concrete is equal to 525 lbs/in$^2$ (3.6 MPa). Compute the fatigue damage accumulated from these axles in this time period.

**ANSWER**

For each axle load level, fatigue life is computed using Equation 12.14; the corresponding fatigue damage is computed using Equation 12.13. The calculations are presented in Table 12.14.

Summing each of the top-down and bottom-up fatigue damage components gives accumulated damage of 0.258% and 0.335%, respectively. The corresponding percent of slab cracking is computed using Equation 12.12 as 4.468 $10^{-5}$ and 6.962 $10^{-5}$, respectively. Finally, the combined cracking is computed using

**Table 12.13**
Number of Single-Axle Load Cycles and Corresponding Stresses Example 12.7

| Load Level (lbs) | Number of Passes | Stress at Top of Slab (lbs/in$^2$) | Stress at Bottom of Slab (lbs/in$^2$) |
|---|---|---|---|
| 15500 | 85 | 120 | 131 |
| 16500 | 76 | 135 | 140 |
| 17500 | 45 | 155 | 165 |
| 18500 | 40 | 175 | 189 |
| 19500 | 35 | 190 | 201 |
| 20500 | 20 | 210 | 220 |
| 21500 | 16 | 240 | 253 |
| 22500 | 14 | 260 | 271 |
| 23500 | 10 | 275 | 288 |
| 24500 | 9 | 290 | 301 |
| 25500 | 6 | 310 | 318 |
| 26500 | 5 | 340 | 343 |
| 27500 | 3 | 351 | 360 |
| 28500 | 2 | 360 | 369 |

**Table 12.14**
Fatigue Damage Calculations Example 12.7

| Load Level (lbs) | Number of Repetitions to Top-Down Fatigue Failure (Equation. 12.14) | Number of Repetitions to Bottom-Up Fatigue Failure (Equation. 12.14) | Top-Down Fatigue Damage (Equation. 12.13) | Bottom-Up Fatigue Damage (Equation. 12.13) |
|---|---|---|---|---|
| 15500 | 3.49727E+12 | 2.07E+11 | 0.000 | 0.000 |
| 16500 | 83806197416 | 2.94E+10 | 0.000 | 0.000 |
| 17500 | 1980675894 | 4.43E+08 | 0.000 | 0.000 |
| 18500 | 119545217 | 24705954 | 0.000 | 0.000 |
| 19500 | 22292799 | 7754567 | 0.000 | 0.000 |
| 20500 | 3579272 | 1645556 | 0.001 | 0.001 |
| 21500 | 430791 | 204331 | 0.004 | 0.008 |
| 22500 | 141496 | 82866 | 0.010 | 0.017 |
| 23500 | 69021 | 39609 | 0.014 | 0.025 |
| 24500 | 36544 | 23967 | 0.025 | 0.038 |
| 25500 | 17395 | 13305 | 0.035 | 0.046 |
| 26500 | 6840 | 6293 | 0.068 | 0.074 |
| 27500 | 5077 | 4038 | 0.051 | 0.064 |
| 28500 | 4038 | 3252 | 0.050 | 0.062 |
| | Total Accumulated Fatigue Damage (%) | | 0.258 | 0.335 |

Equation 12.11, as:

$$TCRACK = \left(4.468\ 10^{-5} + 6.962\ 10^{-5} - 4.468\ 10^{-5}\ 6.962\ 10^{-5}\right)\ 100\%$$
$$= 0.011\%$$

This means that loading during this period will cause 0.011% of slabs to crack in fatigue.

### 12.4.2 Faulting

Faulting in jointed portland concrete pavements is computed using an incremental approach, whereby the faulting increments by month $i, \Delta Fault_i$, are summed to compute the total faulting after $m$ months, $Fault_m$, (in):

$$Fault_m = \sum_{i=1}^{m} \Delta Fault_i \qquad (12.15)$$

For each month, the faulting increment, $\Delta Fault_i$, is assumed proportional to the energy dissipated in deforming the slab support, expressed as:

$$\Delta Fault_i = C_{34}\ (FAULTMAX_{\ i-1} - Fault_{i-1})^2\ DE_i \tag{12.16}$$

where $Fault_{i-1}$ is the accumulated mean faulting up to the previous month $i-1$, $FAULTMAX_{i-1}$ is the maximum mean faulting for the previous month, $i-1$, and $DE_i$ is the differential energy of subgrade deformation. Assuming a liquid foundation (Figure 8.2), allows computing the energy input into the subgrade as the product of the modulus of the subgrade reaction multiplied by the square of the deflection. Hence, the differential energy is expressed as:

$$DE = \frac{1}{2} k\ (w_l^2 - w_{ul}^2) \tag{12.17}$$

where $w_l$ and $w_{ul}$ are the surface vertical deflections at the loaded and unloaded edges of the joint between two slabs (i.e., the "leave" and the "approach" slabs, respectively). $DE$ depends on the load transfer efficiency ($LTE$) of the joint, defined as:

$$LTE = \frac{w_{ul}}{w_l} 100\% \tag{12.18}$$

$LTE$ ranges from 0% for no vertical load transfer to 100% for perfect load transfer between adjacent slabs. Substituting Equation 12.18 into Equation 12.7 gives:

$$DE = \frac{1}{2} k\ (w_l + w_{ul})\,(w_l - w_{ul}) = \frac{1}{2} k\ (w_l + w_{ul})^2 \frac{1 - \dfrac{LTE}{100}}{1 + \dfrac{LTE}{100}} \tag{12.19}$$

Hence, $LTE$ affects significantly the subgrade deformation energy, which drives faulting (the sum $w_l + w_{ul}$ equals the deflection of the free end of a slab and, therefore, is not relevant to faulting). Three sources contribute to $LTE$, namely aggregate interlock, dowels (if present), and base/subgrade reaction. The methodology used for analytically estimating each of these $LTE$ components is described later.

In Equation 12.16, the constant $C_{34}$ is a function of the freezing ratio, $FR$, which is defined as the percentage of time the temperature at the top of the base layer is below freezing; it is expressed as:

$$C_{34} = C_3 + C_4\ FR^{0.25} \tag{12.20}$$

with $C_3 = 0.001725$ and $C_4 = 0.0008$.

The maximum mean transverse joint faulting for a particular month is computed using:

$$FAULTMAX_i = FAULTMAX_0 + C_7 \sum_{j=1}^{m} DE_j \log\left(1 + C_5 5.0^{EROD}\right)^{C_6} \tag{12.21}$$

where $FAULTMAX_0$ is the initial (i.e., the first month pavement is open to traffic) maximum mean transverse joint faulting, computed as:

$$FAULTMAX_0 = C_{12}\ \delta_{curling} \left[\log\left(1 + C_5\ 5.0^{EROD}\right) \times \log\left(\frac{P_{200}\ WetDays}{P_s}\right)\right]^{C_6} \tag{12.22}$$

where:

$\delta_{curling}$ = maximum mean monthly slab corner upward deflection (in), due to temperature and moisture gradients.

$EROD$ = index of the erosion potential of the slab support,[12] ranging from 1 for heavily stabilized bases with geotextiles and drainage to 5 for no base layer at all (i.e., slab on subgrade).

$P_s$ = overburden stress on subgrade (lb/in$^2$) computed as the sum of the layer thicknesses above the subgrade multiplied by their density.

$P_{200}$ = percent of subgrade grain sizes passing sieve No. 200.

$WetDays$ = average annual number of wet days (daily rainfall higher than 0.1 in).

The constant $C_{12}$ is also a function of the freezing ratio:

$$C_{12} = C_1 + C_2\ FR^{0.25} \tag{12.23}$$

with:

$C_1 = 1.29$
$C_2 = 1.1$
$C_5 = 250$
$C_6 = 0.4$
$C_7 = 1.2$

Critical structural response parameters are computed under the combined action of traffic loads and temperature/moisture gradients. The distribution of temperature and moisture within slabs is computed from the geographical location of a pavement design site using the Enhanced Integrated Climatic Model (EICM),[8] which is incorporated into the NCHRP 1-37A Design Guide software.[19] For each calendar month, an effective temperature gradient is estimated, representing the warping condition of slabs as a result of the temperature and moisture distribution versus depth. This equivalent temperature gradient is defined as:

$$\Delta T_m = \Delta T_{t,m} - \Delta T_{b,m} + \Delta T_{sh,m} + \Delta T_{PCW} \tag{12.24}$$

where:

$\Delta T_m$ = effective temperature differential for month $m$.
$\Delta T_{t,m}$ = mean nighttime temperature (from 8 P.M. to 8 A.M.) for month $m$ at the top of the slab.
$\Delta T_{b,m}$ = mean nighttime temperature (8 P.M. to 8 A.M.) for month $m$ at the bottom of the slab.
$\Delta T_{sh,m}$ = equivalent temperature differential due to reversible shrinkage for month $m$ (for old concrete shrinkage is fully developed).
$\Delta T_{PCW}$ = equivalent temperature differential due to permanent curl/warp.

The particular night-time period is selected because it results in wider joint openings and a concave upper slab shape, which exacerbate faulting damage.

In addition to $DE$ (Equation 12.19), two critical structural response parameters are computed, namely the dimensionless shear stress at the slab edge, $\tau$, and the maximum dowel bearing stress, $\sigma_b$ (they are normalized by dividing by the product $k\ell$):

$$\tau = \frac{J_{agg}}{h}(w_l - w_{ul}) \tag{12.25}$$

$$\sigma_b = \frac{J_d}{d}(w_l - w_{ul}) \tag{12.26}$$

where:

$J_{agg}$ = a dimensionless aggregate interlock stiffness variable, computed as explained later.
$J_d$ = a dimensionless load transfer stiffness variable due to dowel bar action, computed as explained later.
$h$ = slab thickness (in).
$d$ = dowel diameter (in).
$k$ = modulus of subgrade reaction (lb/in$^3$).
$\ell$ = radius of relative stiffness, defined earlier (in) (Equation 8.21).

The following sections describe the methodology used for estimating each of the three *LTE* components identified: those due to aggregate interlock, dowel bar action, and subgrade action.

### COMPUTING LTE FROM AGGREGATE INTERLOCK, LTE$_{AGG}$

The *LTE* across joints is affected by the opening width of the joint, *jw*, which is computed from the linear decrease in slab length in response to a uniform temperature decrease and post-construction shrinkage, as described earlier (Equation 8.27), and written as:

$$jw = 12000\ L\ \beta\ \left(a_c(T_{constr} - T_{mean}) + \varepsilon_{sh,m}\right) \tag{12.27}$$

where:

$jw$ = joint opening (0.001 in).
$L$ = joint spacing (ft).
$\beta$ = friction coefficient between the base/subgrade and the portland concrete slab (assumed equal to 0.65 for stabilized base and 0.85 for granular base/subgrade).

$\alpha_t$ = portland concrete coefficient of thermal expansion (/°F).
$T_{mean}$ = mean monthly nighttime middepth temperature (°F).
$T_{constr}$ = portland concrete postcuring temperature (°F).
$\varepsilon_{sh,m}$ = portland concrete slab mean shrinkage strain for month $m$.

The dimensionless initial (first month the pavement is open to traffic) joint shear capacity, $s_0$, is given by:

$$s_0 = 0.05\,h \exp\left(-0.032jw\right) \tag{12.28}$$

where $h$ is the thickness of the portland concrete slab (in). The following expression was adopted[18] for computing the dimensionless joint transverse stiffness from aggregate interlock, $J_{agg}$.

$$\log\left(J_{agg}\right) = -3.19626 + 16.09737 \quad \exp\left\{-2.7183^{-\left(\frac{s_0 - 0.35}{0.38}\right)}\right\} \tag{12.29}$$

Finally, the $LTE$ due to aggregate interlock, $LTE_{agg}$, is given by:

$$LTE_{agg} = \frac{100}{1 + 1.2\,J_{agg}^{-0.849}} \tag{12.30}$$

For subsequent monthly increments, the dimensionless shear joint capacity, $s$, is computed as:

$$s = s_0 - \Delta s \tag{12.31}$$

where $\Delta s$ is the loss in shear joint capacity, computed as:

$$\Delta s = \begin{cases} 0 & \text{for } jw < 0.001\,h \\ \sum_j \dfrac{0.005}{1 + (jw/h)^{-5.7}} \left(\dfrac{n_j}{10^6}\right)\left(\dfrac{\tau_j}{\tau_{ref}}\right) & \text{for } 0.001\,h < jw < 3.8\,h \\ \sum_j \dfrac{0.0068}{1 + 6(jw/h - 3)^{-1.98}} \left(\dfrac{n_j}{10^6}\right)\left(\dfrac{\tau_j}{\tau_{ref}}\right) & \text{for } jw > 3.8\,h \end{cases} \tag{12.32}$$

where:

$n_j$ = number of load applications for the current increment by load group $j$.

$\tau_j$ = shear stress on the transverse crack for load group $j$ (Equation 12.25).

$\tau_{ref}$ = dimensionless reference shear stress given by:

$$\tau_{ref} = 111.1 \exp\left[-\exp\left(0.9988 \exp\left\{-0.1089 \log J_{agg}\right\}\right)\right] \quad (12.33)$$

**Example 12.8**

Compute the $LTE$ of the joints of a 12 in thick JPCP pavement with joints spaced 10 ft apart under a mean January temperature of $-15°$F, given that it was cured at $+45°$F and it is experiencing mean shrinkage strain for the month equal to $2\ 10^{-4}$. Also given is the coefficient of friction between the slab and the subgrade equal to 0.65 and that the coefficient of thermal expansion of the concrete is equal to $5\ 10^{-6}/°$F.

**ANSWER**

Use Equation 12.27 to compute the joint opening as:

$$jw = 12000\ 10\ 0.65\ (5\ 10^{-6}(45+15) + 0.0002) = 39\ 10^{-3}\ \text{in}$$

Use Equation 12.28 to compute the initial joint shear capacity, $s_0$:

$$s_0 = 0.05\ 12 \exp\left(-0.032\ \ 39\right) = 0.17225$$

Use Equation 12.29 to compute the dimensionless joint transverse stiffness from aggregate interlock, $J_{agg}$:

$$\log\left(J_{agg}\right) = -3.19626 + 16.09737\ \exp\left\{-2.7183^{-}\left(\frac{0.17225 - 0.35}{0.38}\right)\right\}$$

$$= 0.06536,$$

which gives a $J_{agg}$ value of 1.162. Substituting this value into Equation 12.30 gives:

$$LTE_{agg} = \frac{100}{1 + 1.2\ \ 1.162^{-0.849}} = 48.64\%$$

**Example 12.9** As a continuation to the previous example, compute the loss in $LTE_{agg}$ at the end of the month of January, resulting from 100,000 axle passes of a single-axle configuration/load that causes a vertical deflection at the edge of the loaded slab $w_l = 10\ 10^{-3}$ in

**ANSWER**

For the month of January, the $LTE_{agg}$ was calculated as 48.64%, which, given a deflection $w_l$ of 10 $10^{-3}$ in, allows computing a deflection $w_{ul}$ of 4.9 $10^{-3}$ in. This gives the dimensionless shear stress $\tau$, using Equation 12.25.

$$\tau = \frac{1.162}{12}(10 - 4.9) = 0.49$$

Use Equation 12.33 to compute the reference shear stress.

$$\tau_{ref} = 111.1\ \exp\left[-\exp\left(0.9988\ \exp\{-0.1089 \log\ 1.162\}\right)\right] = 7.5$$

Considering that 3.8 12 = 45.6 is larger than $jw$ = 39, and that there is only one axle configuration/load to be considered, allows computation of the loss in shear stress $\Delta s$, using Equation 12.32, as:

$$\Delta s = \frac{0.005}{1 + (39/12)^{-5.7}}\left(\frac{100000}{10^6}\right)\left(\frac{0.49}{7.5}\right) = 3.2610^{-5}$$

which is to be subtracted from the dimensionless shear stress for the following month (Equation 12.31).

**COMPUTING LTE FROM DOWELS, $LTE_{DOWEL}$**

*LTE* from dowel action is computed on the basis of the load transfer stiffness variable $J_d$. Initially (the first month the pavement is open to traffic), it is given by:

$$J_d = J_0 = \frac{120\ d^2}{h} \qquad (12.34)$$

where:

$h$ = slab thickness (in).
$d$ = dowel diameter (in).

In subsequent months, this stiffness is computed considering the cumulative damage of the portland concrete supporting the dowels:

$$J_d = J_d^* + (J_0 - J_d^*)\exp(-DAM_{dowel}) \tag{12.35}$$

where $J_d^*$ = critical dowel stiffness, given by:

$$J_d^* = Min\left(118,\ Max\left[165\frac{d^2}{h} - 19.8120,\ 0.4\right]\right) \tag{12.36}$$

$DAM_{dowel}$ equals cumulative damage of a doweled joint, which depends on dowel bearing stress, (Equation 12.26) and the number of load repetitions. For the first month the pavement is open to traffic, $DAM_{dowel}$ is set equal to zero. For subsequent months, it is computed as:

$$DAM_{dowel} = C_8 \sum_j \left(\frac{n_j}{10^6}\right)\left(\frac{\sigma_{b_j} k\,\ell}{f_c'}\right) \tag{12.37}$$

where:

$j$ = signifies summation over axle configuration/load for the current month.
$\sigma_b$ = the dimensionless bearing stress in the portland concrete (Equation 12.26).
$C_8$ = a coefficient equal to 400.
$f_c'$ = the portland concrete compressive strength (lbs/in$^2$).

Finally, the $LTE$ due to dowel action, $LTE_{dowel}$, is computed using:

$$LTE_{dowel} = \frac{100}{1 + 1.2\,J_d^{-0.849}} \tag{12.38}$$

**Example 12.10**

Compute the $LTE$ due to the action of dowels, 1 in in diameter, set 1 ft apart, for the first and second months a pavement section is open to traffic. During the first month, 50,000 axle passes were experienced of a configuration/load that produced deflections at the loaded edge of the slabs $w_l = 10\ 10^{-3}$ in. Given that the thickness of the slab is 12 in, the modulus of subgrade reaction is 100 lbs/in$^3$, the elastic properties of the concrete are 4 $10^6$ lbs/in$^2$, and 0.15 and its compressive strength is 3500 lbs/in$^2$

**ANSWER**

During the first month, the load transfer stiffness variable $J_d$ is computed from Equation 12.34:

$$J_d = J_0 = \frac{120\ 1^2}{12} = 10$$

which allows computing $LTE_{dowel}$ using Equation 12.38:

$$LTE_{dowel} = \frac{100}{1 + 1.2\ 10^{-0.849}} = 85.5\%$$

which allows computing the $w_{ul}$ for the current month (Equation 12.18) as 0.01 0.85 = 0.0085 in. This, in turn, allows computing the dimensionless bearing stress of the portland concrete (Equation 12.26):

$$\sigma_b = \frac{10}{1}(0.01 - 0.0085) = 0.0145$$

Use Equation 8.21 to compute the radius of relative stiffness.

$$\ell = \left(\frac{4000000\ 12^3}{12\,(1 - 0.15^2)\ 100}\right)^{1/4} = 49.3 \text{ in.}$$

Substituting these values into Equation 12.37 gives the dowel support damage at the end of the first month:

$$DAM_{\ dowel} = 400 \sum_j \left(\frac{50000}{10^6}\right)\left(\frac{0.0145\ 100\ 49.3}{3500}\right) = 0.408$$

which allows computing the $J_d$ for the following month (Equation 12.35), after computing $J_d^*$ as 0.4 (Equation 12.36).

$$J_d = 0.4 + (10 - 0.4)\exp(-0.408) = 6.78$$

As a result, the $LTE_{dowel}$ for the following month will be:

$$LTE_{dowel} = \frac{100}{1 + 1.2\ 6.78^{-0.849}} = 80.9\%$$

**COMPUTING LTE FROM BASE/SUBGRADE REACTION, LTE$_{BASE}$**

The contribution of the base/subgrade reaction to the $LTE$ ($LTE_{base}$) is assumed to be a constant value equal to 20%, 30%, and 40% for granular base, asphalt-treated, or concrete-treated material, respectively.

**COMPUTING THE COMPOSITE LTE FOR A TRANSVERSE JOINT, LTE$_{JOINT}$**

After computing each of the individual *LTE* factors, namely $LTE_{agg}$, $LTE_{dowel}$, and $LTE_{base}$, the combined joint load transfer efficiency, $LTE_{joint}$, is computed as follows:

$$LTE_{joint} = 100\left(1 - \left\{1 - \frac{LTE_{dowel}}{100}\right\}\left\{1 - \frac{LTE_{agg}}{100}\right\}\left\{1 - \frac{LTE_{base}}{100}\right\}\right) \tag{12.39}$$

### 12.4.3 Punchouts

Punchouts in CRCPs are the result of the formation of longitudinal top-down fatigue cracks spanning two adjacent transverse cracks. They typically occur near the edge of a driving lane, when the upper surface of the pavement is concave and the load transfer between slab and shoulder is poor. Computing the occurrence of punchouts involves a number of steps, some of which are similar to the ones described earlier under the faulting model (e.g., computation of equivalent monthly temperature/moisture gradients and calculation of the *LTE* across joints). The majority of the steps unique to this model are described next.

Crack spacing is important because it affects crack width, which in turn affects *LTE* across the crack. The mean transverse crack spacing $\overline{L}$ (in) is calculated using:[15]

$$\overline{L} = \frac{f_t - C\sigma_0\left(1 - \frac{2\zeta}{h}\right)}{\frac{f}{2} + \frac{u_m\, p}{c_1 d}} \tag{12.40}$$

where:

$f_t$ = portland concrete tensile strength at 28 days (lb/in$^2$).
$f$ = coefficient of friction between slab and supporting layer.
$u_m$ = maximum bond stress between steel bars and concrete (lb/in$^2$).

$p$ = ratio of steel reinforcement area divided by slab cross sectional area (percent).
$d$ = reinforcing steel bar diameter (in).
$h$ = slab thickness (in).
$\zeta$ = depth to steel reinforcement location (in).
$C$ = Bradbury's curling/warping stress coefficient (Figure 8.6), computed for the typical lane width of 144 in.

The bond slip coefficient $c_1$ is given by:

$$c_1 = 0.577 - 9.499\ 10^{-9} \frac{\ln \varepsilon_{tot-\xi}}{\left(\varepsilon_{tot-\xi}\right)^2} + 0.00502\ \overline{L}\ \ln\left(\overline{L}\right) \qquad (12.41)$$

where:

$\varepsilon_{tot-\xi}$ = total strain (in/in) at the depth of the steel reinforcement, caused by temperature gradient and shrinkage for the current month.
$\sigma_0$ = Westergard's nominal stress factor:

$$\sigma_0 = \frac{E\ \Delta\varepsilon_{tot}}{2\,(1-\mu)} \qquad (12.42)$$

with:

$\varepsilon_{tot}$ = the equivalent unrestrained curling/warping strain difference between top and bottom of slab (in/in) as a result of the temperature/moisture gradient for the current month.

The corresponding average crack width, $\overline{cw}$, (0.001 in) at the level of steel reinforcement is computed as:

$$\overline{cw} = CC\ \overline{L}\ \left(\varepsilon_{shr} + \alpha_t\ \Delta T_\xi - \frac{c_2\ f_\sigma}{E}\right)\ 1000 \qquad (12.43)$$

where:

$\varepsilon_{shr}$ = unrestrained concrete drying shrinkage at the depth of the reinforcement (in/in $10^{-6}$).

$\alpha_t$ = portland concrete coefficient of thermal expansion (/°F).

$\Delta T_\zeta$ = the difference in portland concrete temperature between the monthly mean temperature and its "set" temperature at the depth of the steel $\zeta$ (°F).

$f_\sigma$ = maximum longitudinal tensile stress in the portland concrete at the steel level (lb/in$^2$), which has an upper limit of $f_t$.

$CC$ = calibration constant, with a default value of 1.0.

$c_2$ = another bond slip coefficient, computed as:

$$c_2 = a + \frac{b}{k_1} + \frac{c}{\bar{L}^2} \tag{12.44}$$

where $a$, $b$, and $c$ are regression functions of $\varepsilon_{tot-\xi}$:

$$a = 0.7606 + 1772.5\,\varepsilon_{tot-\xi} - 2\ \ 10^6 \varepsilon^2_{tot-\xi} \tag{12.45a}$$

$$b = 9\ \ 10^8 \varepsilon_{tot-\xi} + 149486 \tag{12.45b}$$

$$c = 3\ \ 10^9 \varepsilon^2_{tot-\xi} - 5\ \ 10^6 \varepsilon_{tot-\xi} + 2020.4 \tag{12.45c}$$

and $k_1$ is a function of the portland concrete compressive strength, $f_c'$, at 28 days (lbs/in$^2$).

$$k_1 = 117.2\, f_c' \tag{12.46}$$

The average crack width, $\overline{cw}$, allows computing the dimensionless shear transfer capacity of the crack due to aggregate interlock, $s$. This is done using Equation 12.28, by substituting the crack width computed for the joint width, $jw$. The combined crack stiffness $J_c$ from aggregate interlock and pavement shoulder action is computed using:

$$\log(J_c) = -2.2\ \exp\left\{-2.718^{-\frac{J_s+11.26}{7.56}}\right\} - 28.85\exp\left\{-2.718^{-\frac{s-0.35}{0.38}}\right\}$$
$$+49.8\exp\left\{-2.718^{-\frac{J_s+11.26}{7.56}}\right\}\ \exp\left\{-2.718^{-\frac{s-0.35}{0.38}}\right\} \tag{12.47}$$

where $J_s$ is the stiffness contribution of the lane-shoulder joint, which ranges from 0.04 to 4.0 for unpaved/asphalt concrete to tied portland concrete, respectively. Note that this expression could be readily simplified, but it is presented in the exact form given in the literature.[19] The *LTE* of the crack, $LTE_c$, due to the contribution of aggregate interlock and the action of the reinforcing steel, *R*, is given by:

$$LTE_c = \frac{100}{1 + 10^{\left(\frac{0.214 - 0.183\frac{a}{\ell} - \log(J_c) - R}{1.18}\right)}} \tag{12.48}$$

where:

$p$ = percent of longitudinal reinforcement.
$a$ = typical radius of a circular tire imprint (in.).
$\ell$ = radius of relative stiffness (in.) (this is computed for each analysis increment to reflect the time-dependent modulus of the subgrade).
$R$ = factor accounting for the load transfer provided by the steel reinforcement, calculated as:

$$R = 2.5\,p - 1.25 \tag{12.49}$$

Finally, the overall *LTE* of the crack, $LTE_{tot}$, is computed by combining $LTE_c$ and the contribution of the base layer, $LTE_{base}$, as done earlier (Equation 12.39).

**Example 12.11** Compute the *LTE* across the cracks of CRCP slab in the first month that it is open to traffic. The slab is 12 in. thick with asphalt-concrete shoulders. Consider aggregate interlock, steel reinforcement action, and shoulder action. The continuous reinforcement consists of 0.75 in. diameter steel bars spaced 6 in. center to center. Given:

$\varepsilon_{shr} = 250\ 10^{-6}$ in/in
$\varepsilon_{tot-\xi} = 50\ 10^{-6}$ in/in
$\Delta T_{\zeta} = 60°$F

The mean material-related values for the particular monthly increment are given as:

$E = 4\ 10^{6}$ lbs/in$^2$
$k = 110$ lbs/in$^3$

Also given:

$\zeta = 6\,\text{in.}$
$\alpha_t = 5.0\ 10^{-6}/°\text{F}$
$\mu = 0.15$
$f_c' = 4000\,\text{lbs/in}^2$
$f_t = 380\,\text{lbs/in}^2$
$f = 0.85$
$u_m = 340\,\text{lbs/in}^2$
$f_\sigma = 200\,\text{lbs/in}^2$

**ANSWER**

The reinforcement is in the middle of the slab ($\zeta = h/2$) and, as a result, the second term of the nominator in Equation 12.40 reduces to zero.

The percentage of steel reinforcement $p$ is:

$$\frac{2\frac{\pi\ 0.75^2}{4}}{12\ 12}100 = 0.61\%$$

The bond slip coefficient $c_1$ is computed from Equation 12.41, assuming a seed value for $\overline{L}$ of 85 in, as:

$$c_1 = 0.577 - 9.499\ 10^{-9}\frac{\ln(50\ 10^{-6})}{(50\ 10^{-6})^2} + 0.00502\ 85\ \ln(85) = 40.07$$

As a result, the average length of cracks, $\overline{L}$, is computed as:

$$\overline{L} = \frac{380}{\frac{0.85}{2} + \frac{200\ 0.61}{40.07\ 0.75}} = 84.72\ \text{in}$$

The variables used in computing the bond coefficient $c_2$ (Equations 12.45 and 12.46) are:

$a = 0.7606 + 1772.5\ 50\ 10^{-6} - 2\ 10^6\ (50\ 10^{-6})^2 = 0.844$
$b = 9\ 10^8\ 100\ 10^{-6} + 149486 = 1.9410^5$
$c = 3\ 10^9\ (100\ 10^{-6})^2 - 5\ 10^6\ 100\ 10^{-6} + 2020.4 = 1.7810^3$
$k_1 = 117.2\ 4000 = 468800$

Subsequently, the bond coefficient $c_2$ is computed as:

$$c_2 = 0.844 + \frac{1.94\ 10^5}{468800} + \frac{1.78\ 10^3}{804.59^2} = 9.79$$

The corresponding mean crack width is given by Equation 12.43.

$$\overline{cw} = 84.72\ \left(250\ 10^{-6} + 50\ 10^{-6}60 - \frac{9.79\ 200}{4\ 10^6}\right)\ 1000$$
$$= 5.14\ (0.001\ \text{in})$$

Substituting this mean crack width for the joint opening $jw$ in Equation 12.28 gives:

$$s_0 = 0.05\ \ 12\ \exp\left(-0.032\ 5.14\right) = 0.509$$

for the first month the pavement is open to traffic. Substituting this dimensionless shear transfer coefficient and a lane-shoulder joint stiffness coefficient $J_s$ of 0.04 for asphalt-concrete shoulders into Equation 12.47 gives:

$$\log\left(J_c\right) = -2.2\ \exp\left\{-2.718^{-\frac{0.04+11.26}{7.56}}\right\} - 28.85\exp\left\{-2.718^{-\frac{0.509-0.35}{0.38}}\right\}$$
$$+49.8\exp\left\{-2.718^{-\frac{0.04+11.26}{7.56}}\right\}\ \exp\left\{-2.718^{-\frac{0.509-0.35}{0.38}}\right\} = 3.909$$

Finally, the factor $R$ for the shear transfer effect of the reinforcing steel is computed using Equation 12.49.

$$R = 2.5\ 0.61 - 1.25 = 0.275$$

The radius of relative stiffness is computed as:

$$\ell = \left(\frac{4000000\ 12^3}{12\left(1 - 0.15^2\right)\ 110}\right)^{1/4} = 48.1\ \text{in.}$$

Substituting these values into Equation 11.48, along with a value $a$ for the typical radius of a tire imprint of trucks of 6 in, gives the $LTE$ of the crack as:

$$LTE_c = \frac{100}{1 + 10^{\left(\frac{0.214-0.183\frac{6}{48.1}-3.909-0.275}{1.18}\right)}} = 99.96\%$$

The shear transfer capacity across CRCP cracks is reduced in subsequent months according to Equation 12.31. The shear transfer capacity loss, $\Delta s$, is given by:

$$\Delta s = \begin{cases} \sum_j \left(\frac{0.005}{1+(\frac{c\overline{w}}{h})^{-5.7}}\right)\left(\frac{n_j}{10^6}\right)\left(\frac{\tau_j}{\tau_{ref}}\right) ESR & \text{for } \frac{c\overline{w}}{h} < 3.7 \\ \sum_j \left(\frac{0.068}{1+6(\frac{c\overline{w}}{h}-3)^{-1.98}}\right)\left(\frac{n_j}{10^6}\right)\left(\frac{\tau_j}{\tau_{ref}}\right) ESR & \text{for } \frac{c\overline{w}}{h} > 3.7 \end{cases} \tag{12.50}$$

where the reference shear stresses, $\tau_{ref}$, is computed as:

$$\tau_{ref} = 111.1 \exp\left(-2.718^{x'}\right) \tag{12.51}$$

with:

$$x\prime = 0.9988 \exp\left(-0.1089 \ln(J_c)\right)$$

and *ESR* is the equivalent shear ratio used to account for lateral wheel-path wander:

$$ESR = a + \frac{b\,\ell}{L} + c\frac{LTE_c}{100} \tag{12.52}$$

where:

$$a = 0.0026\overline{D}^2 - 0.1779\overline{D} + 3.2206 \tag{12.53a}$$

$$b = 0.1309 \ln(\overline{D}) - 0.4627 \tag{12.53b}$$

$$c = 0.5798 \ln(\overline{D}) - 2.061 \tag{12.53c}$$

and $\overline{D}$ is the average distance between the lane edge and the center of the wheel-path.

Having established the load transfer characteristics between the slabs created by the inherent cracking of the continuously reinforced slabs allows structural analysis of stresses under load using the neural network approach. Fatigue damage, *FD*, is computed using a similar approach to cracking fatigue damage accumulation (Equations 12.13 and 12.14). Finally, punchouts, *PO*, (number per mile), are computed using:

$$PO = \frac{106.3}{1 + 4.0\ FD^{-0.4}} \tag{12.54}$$

where *FD* is the fatigue damage accumulated

### 12.4.4 Roughness

Pavement roughness in the NCHRP 1-37A Design Guide is in terms of the International Roughness Index (*IRI*), described in Chapter 9. The relationships used for predicting *IRI* are empirical, having independent variables related to the local conditions, the pavement age, and the distresses present. They are described next for jointed and continuously reinforced portland concrete pavements.

**IRI MODEL FOR JOINTED PAVEMENTS**

The *IRI* model for jointed portland concrete pavements is expressed in terms of the distresses present, namely fatigue cracking, faulting, and spalling, in addition to the initial (post-construction) roughness, $IRI_i$ (in/mi). Spalling, described in Chapter 9, is predicted on the basis of the following empirical expression:

$$SPALL = \left[\frac{AGE}{AGE + 0.01}\right]\left[\frac{100}{1 + 1.005^{(-12\,AGE + SCF)}}\right] \quad (12.55)$$

where:

$SPALL$ = percent of joints with medium-high-severity spalling.
$AGE$ = pavement age (years).
$SCF$ = scaling factor based on site, design, and climate-related variables, given by:

$$SCF = -1400 + 350\,AIR\%\ (0.5 + PREFORM) + 1.36\,f_c'$$
$$-0.2\,FTCYC\,AGE + 43\,h - 536\,WC\_RATIO \quad (12.56)$$

where:

$AIR\ \%$ = air content of the Portland concrete (%).
$WC\_RATIO$ = water/cement ratio of the portland concrete.
$FTCYC$ = average annual number of freeze/thaw cycles.
$PREFORM$ = either 1.0 or 0 for sealed and non-sealed joints, respectively.

Finally, the *IRI* (in/mi) for a particular month is given by:

$$IRI = IRI_i + 0.0823\,CRK + 0.4417\,SPALL$$
$$+1.4929\,TFAULT + 25.24\,SF \quad (12.57)$$

where:

$CRK$ = percent slabs with transverse cracks.
$SPALL$ = percent of joints with medium-high-severity spalling.
$TFAULT$ = total cumulative joint faulting per mile (in.) (the sum of the $Fault_m$ values computed according to Equation 12.15).
$SF$ = site factor, defined as:

$$SF = AGE\ (1 + 0.5556\ FI)\ \frac{1 + P_{200}}{10^6} \qquad (12.58)$$

where:

$FI$ = freezing index (°F-days).
$P_{200}$ = percent subgrade material passing the No. 200 sieve.

**_IRI_ MODEL FOR CONTINUOUSLY REINFORCED PAVEMENTS**

For CRCPs, pavement roughness, ($IRI$ in/mi), is expressed as a function of initial (i.e., post construction) $IRI$, age as reflected by the site factor $SF$ (Equation 12.56), and punchouts, $PO$, as computed earlier. The model used[19] is given by:

$$IRI_m = IRI_i + 3.15\ PO + 28.35\ SF \qquad (12.59)$$

**Example 12.12**

A jointed portland concrete pavement 10 in thick with unsealed joints is 45 years old and is experiencing the following distresses:

- $CRK = 12\%$
- $TFAULT = 33$ in/mile

Its postconstruction roughness, $IRI_i$, was 75 in/mi. Also given, the following portland concrete properties:

- $AIR\% = 3$
- $WC\_RATIO = 0.5$
- $f_c' = 3500$ lbs/in$^2$

and environmental/subgrade properties:

- $FI = 800$°F-days
- $FTCYC = 200$ freeze/thaw cycles per year
- $P_{200} = 15\%$

**ANSWER**

Use Equation 12.58 to compute the site factor:

$$SF = 45\ (1 + 0.5556\ 800)\ \frac{1 + 15}{10^6} = 0.3207$$

Use Equation 12.56 to compute the scaling factor:

$$SCF = -1400 + 350\ 3\ \ (0.5) + 1.36\ 3500 - 0.2\ 200\ 45$$
$$+43\ 10\ \ - 536\ 0.50 = 2247$$

Use Equation 12.55 to compute the extent of spalling:

$$SPALL = \left[\frac{45}{45 + 0.01}\right]\left[\frac{100}{1 + 1.005^{(-12\ 45+2247)}}\right] = 2.01\%\ \text{of joints}$$

Use Equation 12.57 to compute the present roughness:

$$IRI = 75 + 0.0823\ 12 + 0.4417\ 2.01 + 1.4929\ 33 + 25.24\ 0.3207$$
$$= 134.23\ \ \text{in/mi}(2.11\ \ \text{m/km})$$

### 12.4.5 Model Calibration

The pavement damage functions just described were calibrated using field performance observations from two large-scale pavement experiments described earlier, namely the Minnesota Road Research (MnROAD) Project,[11] and the Long-Term Pavement Performance (LTPP) Program.[9] It is anticipated that these rigid pavement design models will undergo further refinement as the NCHRP 1-37A pavement design approach is being evaluated.[6] Furthermore, additional model calibration will take place as individual state DOTs begin implementing the new design approach.

## References

1 AASHTO (1986, 1993). *AASHTO Guide for the Design of Pavement Structures*, American Association of State Highway and Transportation Officials, Washington, DC.

[2] AASHTO (1997). AASHTO Standard T97-97, "Flexural Strength of Concrete (Using Simple Beam with Third-Point Loading)," American Association of State Highway and Transportation Officials, Washington, DC.

[3] AASHTO (2002). AASHTO Standard T198-02, "Splitting Tensile Strength of Cylindrical Concrete Samples," American Association of State Highway and Transportation Officials, Washington, DC.

[4] ASTM (2007). A615, Standard Specification for Deformed and Plain Billet-Steel for Concrete Reinforcement, American Society for Testing of Materials, *Book of Standards*, Vol. 01.04, Conshocken, PA.

[5] DARWin, Version 3.1 (1999). Applied Research Associates (ARA) Inc., Champaign IL.

[6] NCHRP (July 2006). Independent Review of the Recommended Mechanistic-Empirical Design Guide and Software, National Cooperative Highway Research Program (NCHRP) Research Results Digest, No. 307, Wasington, DC.

[7] Khazanovich, L., H. T. Yu, S. Rao, K. Galasova, E. Shats, and R. Jones (2000). *ISLAB 2000-Finite Element Analysis Program for Rigid and Composite Pavements User's Guide*, ERES Division of ARA Inc., Champaign, IL.

[8] Larson G., and B. J. Dempsey (1997). Integrated Climatic Model, Version 2.0, Newmark Civil Engineering Laboratory, University of Illinois at Urbana-Champaign, Report No. DTFA MN/DOT 72114.

[9] Long-Term Pavement Performance, Federal Highway Administration, www.tfhrc.gov/pavement/ltpp/datapave.htm, accessed 02, 2007.

[10] Miner, M. A. (1945). Cumulative Damage in Fatigue, *Transactions of the ASCE*, Vol. 67, pp. A159–A164, (1945).

[11] Minnesota Road Research Project (MnRoad), www.mrr.dot.state.mn.us/research/MnROAD_Project/MnROADProject.asp, accessed 02, 2007.

[12] Mohamed, A. R., and W. Hansen (1997). "Effect of Non-Linear Temperature Gradient on Curling Stresses in Concrete Pavements," *TRB Record* 1568, Washington DC, pp. 65– 71.

[13] PIARC (1987). "Combating Concrete Pavement Slab Pumping by Interface Drainage and Use of Low-Erodibility Materials: State of the Art and Recommendations," Permanent International Association of Road Congress, Paris, FR.

[14] PCAPAV (1985–1986). Thickness Design of Highway and Street Pavements, Software MC003X, by the Portland Cement Association.

[15] Reis, E. E. Jr., J. D. Mozer, A. C. Bianchinni, and C. E. Kesler (1965). "Causes and Control of Cracking in Concrete-Reinforced with High-Strength Steel Bars—A Review of Research", *Engineering Experimental Station Bulletin* 479, University of Illinois, Urbana-Champaign.

[16] PCI (1984/1995). "Thickness Design for Concrete Highway and Street Pavements," Portland Cement Association (imperial units), Skokie, IL.

[17] CPCA (1984/1995). "Thickness Design for Concrete Highway and Street Pavements," Canadian Portland Cement Association (SI units), Ottawa, Ont.

[18] Zolinger, D. G., N. Buch, D. Xin, and J. Soares (February 1998). "Performance of CRCP Pavements," Report Volumes 1 to 7, FHWA Contract DTFH61-90-C-00073, U.S. Department of Transportation, Washington, DC.

[19] NCHRP (July 2004). "2002 Design Guide: Design of New and Rehabilitated Pavement Structures," Draft Final Report, NCHRP Study 1-37A, National Cooperative Highway Research Program, Washington, DC.

## Problems

12.1 Calculate the required slab thickness for a JDRCP with tied portland concrete shoulders on a good-draining subgrade (the water drains out of the pavement structure within a period of one day). It is estimated that the pavement structure becomes saturated less frequently than 5% of the time. The following data are also given:

- Estimated number of *ESAL*s over a 30-year maximum performance period = 5 million
- Modulus of subgrade reaction = 140 lbs/in$^3$ (37.3 MPa/m)

- Design reliability = 95%
- Standard error in predicting serviceability = 0.40
- Modulus of rupture of portland concrete (28 days) = 525 lbs/in$^2$ (3.6 MPa)
- Elastic modulus of portland concrete = 4,000,000 lbs/in$^2$ (28,000 MPa)
- $\Delta PSI = 2.5$ (from 4.5 to 2.0) and $P_t = 2.0$

12.2 Compute the environment-related serviceability loss anticipated for a rigid pavement after 30 years of service. The subgrade is a poorly draining low-plasticity clay (designated as CL according to the USC system), having less than 70% by weight finer than 0.02 mm, and a Plasticity Index of 30%. The subgrade layer is 8 ft deep; it is exposed to high moisture levels and exhibits a medium level of structural fracturing. The percent of the pavement surface subjected to swelling is estimated to be 40%, while the probability of frost is estimated at 50%. The depth of frost penetration is 2 ft.

12.3 Consider the traffic data specified in problem 12.1 and the subgrade conditions specified in problem 12.2. Assume that *ESALs* compound annually at a constant growth rate of 3%. Compute the anticipated pavement life the JDRCP slab designed in problem 12.1, considering serviceability loss due to both traffic and environmental factors.

12.4 Using the AASHTO method,[1] compute the amount of steel reinforcement required for a CRCP slab 12 in thick with limestone aggregate. Also given:

- $T_H$ and $T_L = 80$ and 15°F, respectively
- $f_t = 500$ lbs/in$^2$
- Axle load of 22,000 lbs
- Subgrade reaction modulus $k = 150$ lbs/in$^3$
- $\alpha_s = 5.0\ 10^{-6}$/°F

12.5 Using the PCA method,[17] design a JDRCP for a four-lane divided highway with tied portland concrete shoulders. Use the load frequency distribution data given in Table 12.8. Also given:

- Load safety factor (LSF) = 1.10

- Modulus of subgrade reaction $k = 60$ MPa/m
- Portland concrete modulus of rupture = 4.0 MPa
- Average daily traffic (ADT) over design life = 20,000 vehicles/day
- Directional split = 50/50
- Percent of heavy trucks (FHWA classes 4 to 13) = 15%
- Percent of trucks in the right-hand-side lane = 75%
- Design life = 25 years

  Note that the proper charts from Ref. 17 need to be obtained.

12.6 Compute the *LTE* of the joints of a 10 in thick plain jointed portland concrete pavement with joints spaced 8 ft apart under mean January temperature of −25°F, given that it was cured at + 55°F and it is experiencing mean shrinkage strain for the month of $1.5\ 10^{-4}$. Also given that the coefficient of friction between slab and subgrade is 0.85, and that the coefficient of thermal expansion of the concrete is $5\ 10^{-6}/°F$.

12.7 As a continuation to the previous problem, compute the loss in $LTE_{agg}$ at the end of the month of January, resulting from 250,000 axle passes of a single-axle configuration/load that causes a vertical deflection at the edge of the loaded slab, $w_l = 12\ 10^{-3}$ in.

12.8 Compute the *LTE* due to the action of dowels, 1.25 in in diameter, set 1 ft apart, for the first and second months a pavement section is open to traffic. During the first month, 100,000 axle passes were experienced of a configuration/load that produced deflections at the loaded edge of the slabs, $w_l = 15\ 10^{-3}$ in. Given that the thickness of the slab is 14 in, the modulus of subgrade reaction is 120 lbs/in$^3$, the elastic properties of the concrete are $4\ 10^6$ lbs/in$^2$ and 0.15, and its compressive strength is 4200 lbs/in$^2$.

12.9 Compute the *LTE* across the cracks of CRCP slab in the first month it is open to traffic. The slab is 14 in thick with asphalt-concrete shoulders. Consider aggregate interlock, steel reinforcement action, and shoulder action. The continuous reinforcement consists of 0.75 in diameter steel bars spaced 6 in center to center. Given:

- $\varepsilon_{shr} = 300\ 10^{-6}$ in/in
- $\varepsilon_{tot-\xi} = 50\ 10^{-6}$ in/in
- $\Delta T_{\zeta} = 60°F$

The mean material-related values for the particular monthly increment are given as:

- $E = 4\ 10^{6}$ lbs/in$^2$
- $k = 110$ lbs/in$^3$

Also given:

- $\zeta = 6$ in
- $\alpha_t = 5.0\ 10^{-6}/°F$
- $\mu = 0.15$
- $f_c' = 4000$ lbs/in$^2$
- $f = 0.85$
- $u_m = 340$ lbs/in$^2$
- $f_\sigma = 200$ lbs/in$^2$

12.10 A jointed portland concrete pavement 12 in thick with unsealed joints is 35 years old and is experiencing the following distresses:

- $CRK = 22\%$
- $TFAULT = 25$ in/mile

Its postconstruction roughness, $IRI_i$, was 65 in/mi. Also given the following portland concrete properties:

- $AIR\% = 7$
- $WC\_RATIO = 0.45$
- $f_c' = 4000$ lbs/in$^2$

and environmental/subgrade properties:

- $FI = 1000°F$-days
- $FTCYC = 100$ freeze/thaw cycles per year
- $P_{200} = 20\%$

12.11 Utilizing the NCHRP 1-37A Design Guide software, estimate the performance of a plain jointed portland concrete pavement located in the western United States (latitude = 47° 33′ N, longitude = 117° 41′ W). Given:

- Slab thickness $h = 14$ in.
- Unbound base thickness = 20 in.
- Joint spacing = 10 ft
- Average annual daily truck traffic (AADTT) in the design lane for first year = 800 trucks/day
- Traffic growth rate = 3.0% annually, compounded over the analysis period of 40 years

The vehicle classification distribution is as shown in Table 11.7. Assume monthly adjustment factors (*MAF*s) of 1.0 for all vehicle classes and months. Compute the number of axles by configuration, assuming that the sketches in Table 2.1 represent the vehicle type in each class. Use the axle load distributions shown in Figures 11.12 and 11.13, and assume that there are no tridem or quad-axle configurations in the traffic stream. Finally, assume the subgrade is a low-plasticity clay with a Resilient Modulus of 82.7 MPa (12,000 lbs/in$^2$). The tire inflation pressure is equal to 689 kPa (100 lbs/in$^2$). Present the results in the form of performance curves of the various distress versus time. Determine whether the selected pavement structural design is acceptable.

# 13 Pavement Rehabilitation

## 13.1 Introduction

Pavement rehabilitation refers to the broad range of treatments for repair, rehabilitation, restoration and replacement of pavements, colloquially referred as the 4-Rs. Historically, these treatments excluded routine maintenance activities (e.g., pothole filling), which did not qualify for federal fund sharing. This is no longer the case, provided that agencies can document that maintenance is cost-effective in preserving pavements between more comprehensive 4-R treatments.[10,11] Restoration refers to a variety of surface treatments such as crack filling and coating, while resurfacing includes chip sealing and overlaying with either asphalt concrete or portland concrete. Recycling consists mainly of incorporating reclaimed asphalt pavement (RAP) into new asphalt concrete. Reconstruction is, in essence, new construction for which design methods were described earlier in this book (Chapters 11 and 12).

Typically, rehabilitation is triggered using pavement distress criteria (e.g., reaching threshold values in particular distresses) or some type of aggregate distress index reflecting the type of distresses present and their extent/severity (e.g., the *PCI* index described in Chapter 9). Some jurisdictions utilize pavement roughness or a combined index of distress and roughness as an alternative rehabilitation-triggering criterion. In addition, there are circumstances where pavement rehabilitation is triggered by safety rather than distress/roughness considerations (e.g., the need to increase pavement texture in order to improve skid resistance).

The range of feasible 4-R treatments depends on pavement condition, as defined by the type of distresses present and their extent/severity, as well as the degree of structural strengthening necessary. Ideally, the best among the feasible 4-R treatments should be selected on the basis of life-cycle cost analysis (LCCA) described in Chapter 14. General 4-R treatment guidelines for a selected set of circumstances are shown in Tables 13.1, 13.2, and 13.3 for asphalt concrete pavements, jointed portland concrete pavements, (either JPCP or JDRCP), and continuously reinforced portland concrete pavements (CRCP), respectively. These tables provide estimates of the expected service lives of these treatments. Their actual service lives may vary significantly from these estimates, depending on the particular traffic/environmental circumstances and construction quality.

The focus of this chapter is on overlays and their design methodologies. Regardless of the design methodology and the materials involved, the existing pavement surface needs to be prepared to repair distresses that, if left untreated, may compromise the performance of the overlay. This is especially true in portland concrete pavements, where the surface needs to be treated to control reflection cracking. This involves either fracturing of the entire surface or keeping the surface intact while effecting full-depth repairs/slab replacement at particularly deteriorated locations.[9] Fracturing an existing portland concrete pavement in preparation for an overlay is cost-effective where the existing portland concrete surface is badly distressed. It is carried out by either rubbilizing (i.e., mechanically fracturing slabs into pieces smaller than 0.3 m), cracking/seating (i.e., breaking slabs into 0.3 to 1.0 m pieces), or breaking/seating (i.e., breaking the slabs into 0.3 to 1.0 m pieces and ensuring that the bond between concrete and rebar reinforcement is also broken). The fracturing method is dictated by portland concrete type: cracking/seating is suited for JPCPs; breaking/seating is suited to JDRPCs; and rubbilizing suits all portland concrete types, including CRCPs. Overlay design requires special care in identifying the limits of the pavement sections to be overlaid, so they have uniform age, surface condition, traffic level, and structural characteristics. The overlay design methods described next are attributed to the

**Table 13.1**
Guidelines for Flexible Pavement Rehabilitation Treatments and Estimated Service Lives

***Fatigue Cracking***

| *Severity* | *Extent (% Length)* | *Asphalt Layer Thickness* | *Typical Treatments* | *Estimated Service Life* |
|---|---|---|---|---|
| Low | 1–10% | <200 mm | Patch or chip sealing | 5 |
| Medium | 11–25% | >200 mm | 50 mm overlay | 12 |
| Medium | 11–25% | <200 mm | 75 mm overlay | 12 |
| Medium | 25–50% | >200 mm | 75 mm overlay | 12 |
| Medium | 25–50% | <200 mm | 150 mm overlay | 12 |
| High | >50% | >200 mm | 100 mm overlay | 12 |
| High | >50% | <200 mm | Reconstruction | 15 |

***Transverse Cracking***

| *Severity* | *Extent (Spacing)* | *Typical Treatments* | *Estimated Service Life* |
|---|---|---|---|
| Low | >15 m | Rout and Seal | 8 |
| Medium | 10–15 m | Clean and Seal | 5 |
| Medium | 10–15 m | Clean and slurry seal | 3 |
| High | <10 m | Clean/seal, and 50 mm overlay | 10 |

***Rutting***

| *Severity* | *Extent (% Length)* | *Typical Treatments* | *Service Life* |
|---|---|---|---|
| <6 mm | >25% | Do nothing | — |
| 6–12 mm | >25% | Grinding | 7 |
| 12–25 mm | >25% | Grinding | 4 |
| >25 mm | >50% | Mill and inlay 50 mm | 10 |

1993 edition of the AASHTO Pavement Design Guide,[1] the Asphalt Institute,[8] and the NCHRP Study 1-37A.[13]

## 13.2 AASHTO (1993) Flexible Pavement Overlay Design Method

AASHTO[1] recommends an empirical flexible pavement overlay thickness design method that relies on the effective structural number($SN$) of an existing pavement, denoted by $SN_{EFF}$. Three

**Table 13.2**
Guidelines for Jointed Rigid Pavement Rehabilitation Treatments and Estimated Service Lives

| Faulting Severity | Extent (% Length) | LTE* | Typical Treatments | Service Life |
|---|---|---|---|---|
| Low | >25% | | Do nothing | — |
| Medium | >25% | >70% | Grinding | 5 |
| Medium | >25% | <70% | Load transfer restoration and grinding | 15 |
| Medium | >25% | >70% | Grinding | 6 |
| Medium | >25% | <70% | Load transfer restoration and grinding | 10 |
| High | >50% | <70% | Crack/seat or rubbilize and 125 mm overlay | 12 |
| High | >50% | <70% | Reconstruction | 35 |

*Load Transfer Efficiency

**Panel and Corner Cracking**

| Severity | Extent (% Length) | Typical Treatments | Service Life |
|---|---|---|---|
| Low | 1–10 % | Panel replacement | 7 |
| Medium | 11–25% | AC overlay | 12 |
| Medium | 11–25% | Unbonded PC overlay | 35 |
| High | >25% | Reconstruction | 35 |

**Table 13.3**
Guidelines for Continuously Reinforced Rigid Pavement Rehabilitation Treatments and Estimated Service Lives

| Punchouts Severity | Extent | Typical Treatments | Estimated Service Life |
|---|---|---|---|
| Medium | <3 / km | Full-depth patching | 10 |
| Medium | 3–6 / km | Full-depth patching | 5 |
| Medium | >6 / km | AC overlay | 12 |
| High | >15 / km | Reconstruction | 35 |

alternatives methods are described for obtaining $SN_{EFF}$, based on either:

- Estimating structural layer coefficients $a_i$ for Equation 2.5 from qualitative surface distress criteria.

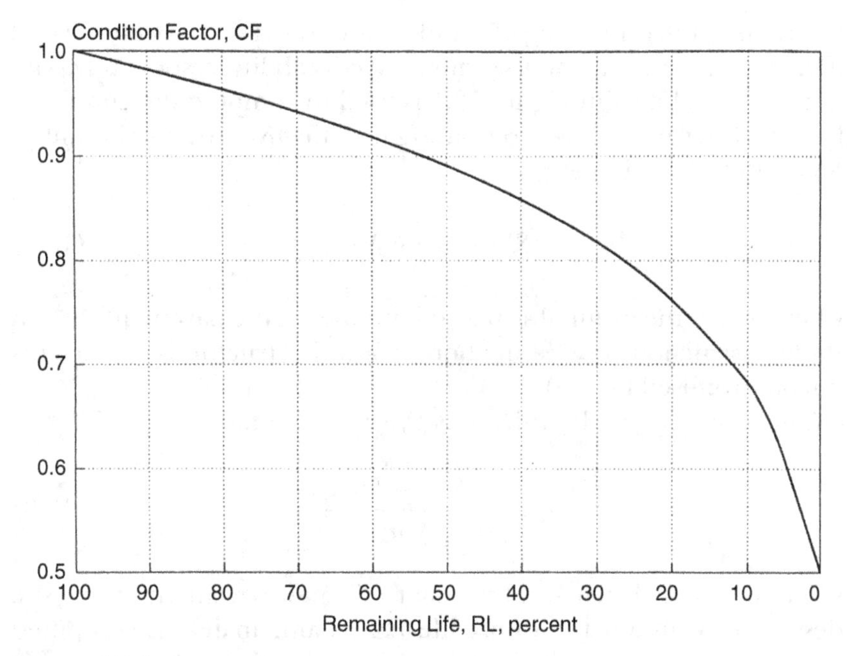


**Figure 13.1**
Nomograph for Estimating the Condition Factor *CF* (Ref. 1 Used by Permission)

- Discounting the original (i.e., post-construction) structural number using a condition factor based on the remaining *ESAL*-life to reach terminal serviceability.
- Estimating structural layer coefficients $a_i$ for Equation 2.5 based on the in-situ effective moduli of the pavement layers, as determined by nondestructive testing.

The last two methods are preferable and are described next in detail, followed by the recommended approach for computing the required asphalt concrete overlay thickness.

### 13.2.1 $SN_{EFF}$ from ESAL-Life Calculations

The remaining *ESAL*-life of an existing flexible pavement section is defined as the percent difference between the number of *ESAL*s estimated to reach a terminal serviceability of 1.5 and the number of *ESAL*s accumulated up to the present, expressed as:

$$RL = 100\left(1 - \frac{N_P}{N_{1.5}}\right) \tag{13.1}$$

where $RL$ is the remaining life, and $N_p$ and $N_{1.5}$ are the accumulated $ESALs$ to the present and to terminal serviceability, respectively. The nomograph shown in Figure 13.1 is used to compute the condition factor ($CF$), which allows computing the effective structural number $SN_{EFF}$ of this section as:

$$SN_{EFF} = CF\ SN_0 \tag{13.2}$$

where $SN_0$ is the original structural number of the pavement section post-construction (i.e., Equation 2.5 with structural coefficients uncompromised by age).

The required overlay thickness $D_{OL}$ is computed using:

$$D_{OL} = \frac{SN_F - SN_{EFF}}{a_1} \tag{13.3}$$

where $SN_F$ is the structural number of a new pavement at this site designed to sustain the future-life $ESALs$ anticipated as computed using the procedure described in Chapter 11. Typically, an $a_1$ value of 0.44 is used for new asphalt concrete layers.

**Example 13.1**

Determine the thickness of the asphalt concrete overlay required for an existing pavement with asphalt concrete, unbound base and subbase layer thicknesses of 18 cm, 15 cm and 20 cm (7, 5.9, and 7.9 in.), respectively. The section has experienced 3.5 million $ESALs$ to date; its serviceability is 2.5 in the $PSI$ scale; and it is estimated that if left untreated, it could sustain another 1.5 million $ESALs$ before its $PSI$ reduces to a terminal value of 1.5. It is estimated that the overlaid pavement needs to have a structural number of 4.8 to sustain future traffic. The pavement layers drain within one day if they become saturated, which happens 10% of the time.

**ANSWER**

Draining within one day characterizes the drainage as "good" (Table 10.1), which, combined with 10% of time saturation, allows estimating drainage coefficients for the base and the subbase layers of 1.1 (Table 2.7). Hence, the original structural number of the section postconstruction can be computed from Equation 2.5 as:

$$SN_0 = 0.44\ 7 + 0.14\ 1.1\ 5.9 + 0.11\ 1.1\ 7.9 = 4.94$$

The remaining life factor is computed from Equation 13.1 as:

$$RL = 100\left(1 - \frac{3.5}{5.0}\right) = 30\%$$

Using a 30% remaining life, Figure 13.1 gives a condition factor($CF$) of 0.82, which allows computing the effective structural number $SN_{EFF}$ from Equation 13.2 as:

$$SN_{EFF} = CF\ SN_0 = 0.82\ 4.94 = 4.05$$

Given a required future structural number of 4.8, the overlay thickness can be computed from Equation 13.3 as:

$$D_{OL} = \frac{SN_F - SN_{EFF}}{a_1} = \frac{4.8 - 4.05}{0.44} = 1.7 \text{ in to be rounded up 2 in (5 cm)}$$

### 13.2.2 SN$_{EFF}$ from Non-destructive Testing Data

AASHTO[1] describes an alternative method for establishing the structural coefficients that give the effective structural number $SN_{EFF}$. It is based on obtaining the resilient modulus of the subgrade $M_r$ from pavement surface deflection measurements, and using this value to compute an effective modulus of the pavement layers above the subgrade $E_p$. The subgrade resilient modulus is obtained from an expression derived from Equation 7.6 by setting the Poisson's ratio equal to 0.5:

$$M_r = \frac{0.24\ P}{w_r\ r} \qquad (13.4)$$

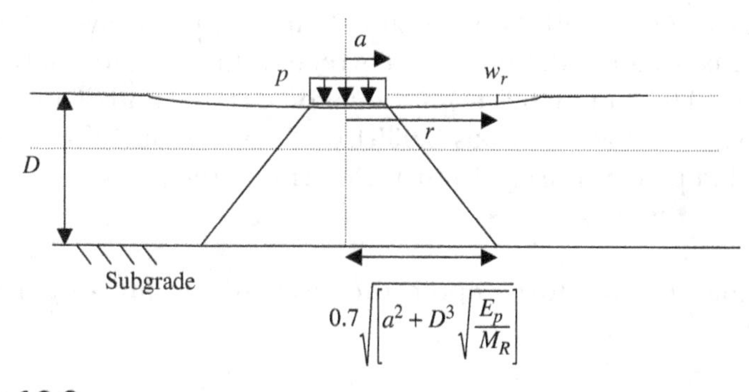


**Figure 13.2**
Radial Offset for Subgrade-only Contribution to Surface Deflection

where $w_r$ is the deflection at a radial offset $r$ from the centerline of the load of magnitude $P$. This radial offset should be sufficiently large to ensure that only vertical strains within the subgrade contribute to the surface deflection being measured (Figure 13.2). As a result, the deflection beyond this offset is not temperature-sensitive.

The effective modulus of the layers above the subgrade $E_p$ is subsequently computed through:

$$w_0 = 1.5\, p\, a \left\{ \frac{1}{M_r \sqrt{1 + \left( \frac{D}{a} \sqrt[3]{\frac{E_P}{M_r}} \right)^2}} + \frac{1 - \frac{1}{\sqrt{1 + \left( \frac{D}{a} \right)^2}}}{E_p} \right\} \tag{13.5}$$

where $w_0$ is the pavement surface deflection at the centerline of the loading plates the rest of the variables are defined in Figure 13.2. Note that the $w_0$ deflection must be adjusted to correspond to the reference temperature of 68°F (20°C). A procedure similar to the one described in Chapter 9 can be used for this purpose. Solving Equation 13.5 for $E_p$ allows determining $SN_{EFF}$ using the nomograph shown in Figure 13.3. An example of this procedure follows.

**Example 13.2**

A flexible pavement has asphalt concrete and base/subbase layer thicknesses of 15 and 25 cm (6 and 10 in), respectively. FWD measurements yielded deflections, corrected to 68°F, of 0.015 and 0.008 in, (0.038 and 0.02 cm) at radial offsets of 0 and 20 in, respectively. The load applied was 8,000 lbs (35.5 kN), and the radius of the loaded plate was 6 in (15 cm). Determine the $SN_{EFF}$.

**ANSWER**

Use Equation 13.4 to compute the resilient modulus of the subgrade:

$$M_r = \frac{0.24\, P}{w_r r} = \frac{0.24\; 8000}{0.008\; 20} = 12,000\ \mathrm{lb/in^2}\ (82.7\ \mathrm{MPa})$$

Given a contact area of $\pi 6^2 = 113.1\ \text{in}^2$, the contact pressure $p$ on the FWD plate is $70.7\ \text{lb/in}^2$ (487 kPa). Use Equation 13.5 to solve for $E_p$:

$$0.015 = 1.5\ 70.7\ 6\left\{\frac{1}{12000\sqrt{1+\left(\frac{16}{6}\sqrt[3]{\frac{E_P}{12000}}\right)^2}} + \frac{1-\frac{1}{\sqrt{1+\left(\frac{16}{6}\right)^2}}}{E_p}\right\}$$

This gives an $E_p$ value of approximately $27{,}500\ \text{lbs/in}^2$ (206.8 MPa). Entering this value, along with a $D$ value of 16 in into Figure 13.3 gives an $SN_{EFF}$ value of 2.2. Consideration of the structural number required to accommodate future traffic allows computation of the required overlay thickness, as in the previous example.

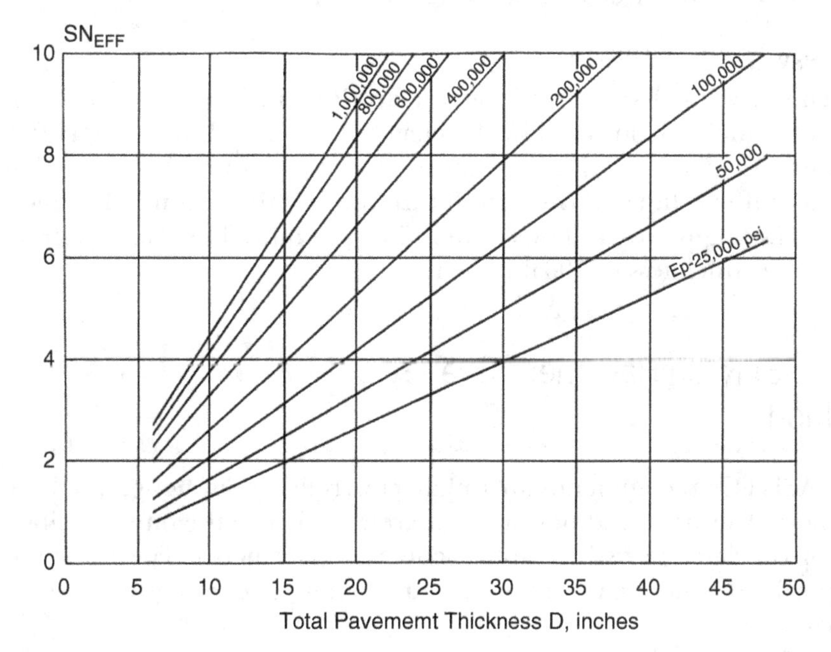


**Figure 13.3**
Nomograph for Estimating $SN_{EFF}$ from the Estimated Subgrade Modulus $E_p$ (Ref. 1 Used by Permission)

It should be noted that the methods just described for determining flexible pavement overlay thickness represent a significant simplification of the approach described in the 1986 edition of the AASHTO guide, which based $SN_{EFF}$ estimation on a combination of remaining *ESAL*-life and structural considerations.

## 13.3 Asphalt Institute Flexible Pavement Overlay Design Method

This method bases flexible pavement overlay thickness on the representative rebound deflection (*RRD*) of a section, obtained as explained earlier (Chapter 9). Asphalt concrete overlay thickness is estimated as a function of the future anticipated *ESALs* over the life of the overlaid section, using the nomograph shown in Figure 13.4.[8] Its use is explained through the following example.

**Example 13.3** Design an overlay for the pavement deflection measurements and particulars described earlier in Examples 9.6, 9.7, and 9.8.

**ANSWER**

The answer to Example 9.7 suggests an *RRD* of 1.735 mm (0.0068 in), which makes it structurally deficient (as described in Example 9.8, the *SAI* of this section was 2.0 on a scale of 0 to 10). Given that the traffic estimated over the future life of this section is 1 million *ESALs*, Figure 13.4 allows computing a required asphalt concrete overlay thickness of 100 mm (4 in).

## 13.4 AASHTO (1993) Rigid Pavement Overlay Design Method

AASHTO[1] recommends an asphalt concrete overlay design methodology over fractured portland concrete similar to the one described for overlaying existing asphalt concrete pavements. The fractured portland concrete is simply treated as a high-strength granular base layer.

The method AASHTO[1] recommends for designing asphalt concrete overlays over intact portland concrete pavements relies on the effective slab thickness of the existing pavement, denoted by $D_{EFF}$.

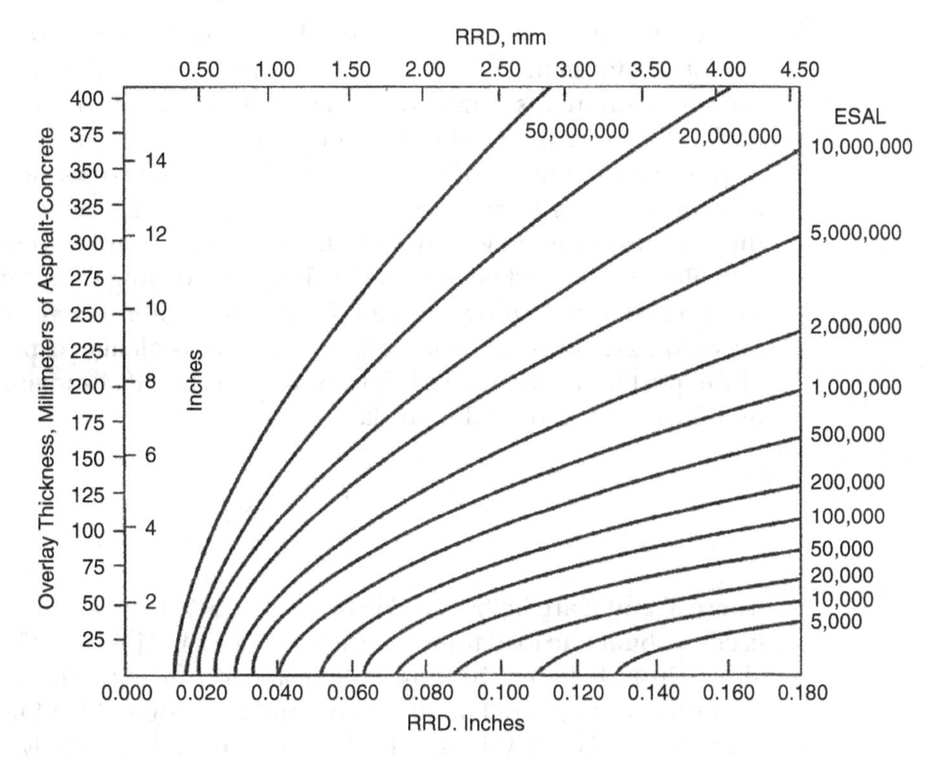


**Figure 13.4**
Nomograph for Estimating Overlay Thickness from the *RRD* (Ref. 8)

Two alternatives methods are described for obtaining $D_{EFF}$, based on either:

- Qualitative surface distress criteria
- Discounting the original (post-construction) slab thickness using a condition factor based on the remaining *ESAL*-life to reach terminal serviceability

The latter is preferable, and it is in essence identical to the method described earlier for computing the $SN_{EFF}$ for existing asphalt concrete pavements. The remaining life (*RL*) is computed using Equation 13.1. and a condition factor (*CF*) is obtained using Figure 13.1, while Equation 13.2 is written as:

$$D_{EFF} = CF\ D \qquad (13.6)$$

where $D$ is the as-built thickness of the portland concrete slab. Note that no method is described for obtaining $D_{EFF}$ from deflection measurements. Instead, FWD measurements are used for back-calculating the modulus of subgrade reaction $k$ and the modulus of the existing portland concrete slab $E$. The back-calculation methodology was described in detail in Chapter 9. It is suggested that the $k$ values obtained with this method are roughly twice as large as the values that would be obtained with a plate loading test, and need to be adjusted accordingly.[1] In addition, where indirect tension data on extracted cores are not available, the back-calculated modulus of the portland concrete slab $E$ is used for indirectly estimating the modulus of rupture of the portland concrete, $S_c'$ using:

$$S_C' = 43.5 \frac{E}{10^6} + 488.5 \tag{13.7}$$

where $S_c'$ and $E$ are in lbs/in$^2$. Furthermore, FWD measurements are used to obtain the load transfer efficiency ($LTE$) of joints. $LTE$ values above 70%, between 70% and 50%, and below 50% characterize load transfer as good, fair, and poor, and are assigned load transfer coefficients (i.e., $J$ values) of 3.2, 3.5, and 4, respectively. Thus, FWD measurements provide essential input data in determining the thickness of the future portland concrete slab, $D_F$, following the methodology described in Chapter 12 (Equation 12.1).

The following describes the methodology used for computing the overlay thickness for the most common types of overlays on portland concrete pavements, including asphalt concrete and portland concrete, the latter being either bonded or unbonded. Bonded portland concrete overlays are constructed after a thin coat of cement grout or liquid epoxy is placed on the existing portland concrete pavement ahead of the paver.[4] Unbonded portland concrete overlays are constructed after a separation/debonding interlayer is placed on the existing portland concrete pavement surface.[5] This interlayer is typically thin (25 to 50 mm) and consists of unbound aggregates or asphalt concrete.

## ASPHALT CONCRETE OVERLAYS

Asphalt concrete overlays on intact portland concrete pavements need to be constructed after the original portland concrete surface has been properly repaired. This includes full-depth repair of cracked slabs and spalled joints and drainage improvement, followed by a tack coat prior to overlaying. The thickness of the asphalt concrete overlay $D_{OL}$ required is computed using:

$$D_{OL} = A\ (D_F - D_{EFF}) \tag{13.8}$$

where $A$ is a function of structural deficiency obtained from:

$$A = 2.2233 + 0.0099\ (D_F - D_{EFF})^2 - 0.1534\ (D_F - D_{EFF}) \tag{13.9}$$

with the slab thicknesses in inches. The typical range in such overlays is between 75 and 150 mm (3 to 6 in).

### 13.4.1 Bonded Portland Concrete Overlays

Bonded portland concrete overlays are effective where the existing surface has relatively low distress, which can be practically repaired before the overlay. The procedure recommended for determining the overlay thickness of a bonded portland concrete overlay is a variation of the flexible pavement overlay approach just described. The main difference is that instead of using Equation 13.8, the portland concrete overlay thickness is computed from:

$$D_{OL} = D_F - D_{EFF} \tag{13.10}$$

The typical range in this type of overlay is between 50 and 100 mm (2 to 4 in).

### 13.4.2 Unbonded Portland Concrete Overlays

Unbonded portland concrete overlays are effective where the existing surface is badly deteriorated, and repairing it would be too expensive. Even so, a basic treatment of the surface and improvement of drainage are advisable. The overlay thickness is computed from:

$$D_{OL} = \sqrt{D_F - D_{EFF}} \tag{13.11}$$

**Example 13.4**

Design an asphalt concrete overlay for a jointed dowel-reinforced rigid pavement 8 in thick (0.20 m). FWD measurements on this pavement yield a modulus of subgrade reaction of 140 lbs/in$^3$ (38 MPa/m) and a portland concrete elastic modulus of 3,500,000 lbs/in$^2$ (24.8 MPa). In addition, FWD measurements across the joints yield a $LTE$ of 75%. Additional information given includes 6.5 million *ESALs* accumulated to present and 8 million *ESALs* estimated to reduce *PSI* to 1.5. The number of future *ESALs* that needs to be carried over the design life of the overlaid section is 7 million. Consider only traffic-related serviceability loss of 2.5,(4.5 to 2.0). The standard error in predicting serviceability is 0.5, and the desired reliability is 85%. The drainage condition for this section is considered good, with pavement layer saturation less frequent than 10%.

**ANSWER**

Use Equation 12.1 to compute the required thickness of a portland concrete layer to carry future traffic, $D_F$. Compute the necessary input from the information given. For 85% confidence, the value of the standard normal deviate $Z_R$ is − 1.037. The adjusted modulus of the subgrade reaction is 70 lbs/in$^3$. The load transfer coefficient $J$ is 3.2, and the drainage coefficient $C_d$ is 1.05 (Table 12.1). The modulus of rupture of the portland concrete, $S_c'$, is estimated from Equation 13.7 as:

$$S_C' = 43.5\ 3.5 + 488.5 = 640 \text{ lbs/in}^2 \text{ (4412 kPa)}$$

Substituting these values into Equation 12.1 allows solving for the required thickness of the future pavement:

$$\log(7\ 10^6) = -1.037\ 0.5 + 7.35\log(D+1) - 0.06 + \frac{\log\left[\dfrac{2.5}{4.5-1.5}\right]}{1+\dfrac{1.624\ 10^7}{(D+1)^{8.46}}}$$

$$+\,(4.22 - 0.32\ 2)\log\frac{640\ 1.05\ (D^{0.75} - 1.132)}{215.63\ 3.2\left[D^{0.75} - \dfrac{18.42}{(3.5\ 10^6/70)^{0.25}}\right]}$$

which gives a value of $D_F$ of 9.5 in (0.24 m).

Compute the remaining life from Equation 13.1 as:

$$RL = 100\left(1 - \frac{6.5}{8}\right) = 18.75\%$$

which gives a condition factor of 0.75 (Figure 13.1) and results in a $D_{EFF}$ value of $0.75 \times 0.2 = 0.15$ m (5.9 in) (Equation 13.6). The correction factor $A$ is computed from Equation 13.9.

$$A = 2.2233 + 0.0099\ (9.5 - 5.9)^2 - 0.1534\ (9.5 - 5.9) = 1.8$$

Finally, the asphalt concrete overlay thickness is computed using Equation 13.8.

$$D_{OL} = 1.8\ (9.5 - 5.9) = 6.4 \text{ in } (0.165 \text{ m})$$

## 13.5 NCHRP 1-37A Overlay Design Method

The NCHRP 1-37A study[13] describes a mechanistic-empirical approach to the design of asphalt concrete and portland concrete overlays. The approach expands on the mechanistic response and empirical damage function design approach presented earlier for designing new pavements (Chapters 11 and 12). A key part of this overlay design approach is determining the properties of the existing pavement layers. Three levels of detail are described for this purpose, namely non-destructive testing (NDT), correlation to index properties such as the California Bearing Ratio, and empirical relationships to surface distress. Only level 1 design involving NDT will be described next, given the widely accepted use of this technology in designing overlays. The discussion treats separately asphalt concrete overlays and portland concrete overlays. Each overlay type is discussed separately as applied over existing flexible pavements, fractured portland concrete pavements, and intact portland concrete pavements. The actual overlay alternative to be selected in each circumstance should be decided on the basis of life-cycle cost analysis, as described in Chapter 14. The procedures described are under review by another NCHRP Study.[11] The outcome of this review and subsequent research is likely to result in modifications to some of the design procedures described here.

### 13.5.1 Asphalt Concrete Overlays

Asphalt concrete is an effective overlay material for asphalt concrete pavements, as well as portland concrete pavements. Asphalt concrete overlaying of portland concrete requires considerable preparation of the existing surface to avoid reflection cracking, as described earlier.

**ASPHALT CONCRETE OVERLAYS OVER FLEXIBLE PAVEMENTS**

The design of asphalt concrete overlays on asphalt concrete pavements involves determining the in-situ properties of the existing pavement layers, including their in-situ elastic layer moduli, the amount of fatigue damage, and the amount of plastic deformation experienced prior to the overlay. As described next, the fatigue damage in the asphalt concrete is estimated from the differences in dynamic moduli between the existing and the as-built layer, which is computed from extracted cores. The plastic deformation of the pavement layers is measured directly through trenching.

The pavement layer elastic moduli are computed from NDT data through back-calculation techniques, as described in Chapter 9. The in-situ moduli of the granular layers need to be adjusted to account for the differences between field testing and laboratory conditions. Typically, factors of 0.67 and 0.40 are used in converting back-calculated moduli to resilient moduli ($M_r$) for granular base and subgrade layers, respectively. The dynamic modulus of the "damaged" asphalt concrete layer, back-calculated from NDT data and denoted by $E^*_{dam}$, is used in combination with the "undamaged" (i.e., as-built) dynamic modulus, denoted by $E^*$, to estimate the amount of fatigue damage accumulated. The master curve for $E^*$ is reconstructed from field core data (i.e., volumetric, gradation, and viscosity-temperature susceptibility data of the extracted binder, as described in Chapter 5). It allows estimating $E^*$ for the loading rate/temperature conditions (i.e., reduced time) prevailing during NDT. Knowing the $E^*_{dam}$ and $E^*$ values under the same temperature/loading condition, allows computing the fatigue damage in the asphalt concrete $d_{AC}$ prior to the overlay by solving the following expression:

$$E^*_{dam} = E_{\min} + \frac{E^* - E_{\min}}{1 + e^{-0.3+5 \log d_{AC}}} \tag{13.12}$$

where $E_{min}$ is the minimum dynamic modulus for the asphalt concrete, and the moduli are in lbs/in$^2$. Note that Equation 13.12 is also used to compute further reductions in the modulus of the asphalt concrete layer, as fatigue damage keeps accumulating throughout the life of the overlaid section.

**Example 13.5**

Compute the fatigue damage accumulated in an existing asphalt concrete pavement, given that its $E^*$ and $E^*_{dam}$, computed as just described, have values of 450,000 lb/in$^2$ and 375,000 lb/in$^2$ (3.1 GPa and 2.5 GPa), respectively, and the minimum $E^*_{min}$ is 200,000 lb/in$^2$ (1.37 GPa).

**ANSWER**

Substitute given moduli values into Equation 13.12 and solve it for $d_{AC}$:

$$375000 = 200000 + \frac{450000 - 200000}{1 + e^{-0.3+5\log d_{AC}}}$$

which gives a $d_{AC}$ value of 0.77, or 77%.

Successive trial performance simulations involving different overlay thicknesses are carried out to establish the minimum feasible thickness that can accommodate future design life traffic. The criterion for the latter is predicting distresses below selected critical threshold values. The software developed by the NCHRP 1-37A study is used for implementing this approach. In doing so, the properties of the existing pavement layers, and their structural condition, obtained as described previously need to be specified.

**ASPHALT CONCRETE OVERLAYS OVER FRACTURED PORTLAND CONCRETE PAVEMENTS**

Asphalt concrete overlays over fractured portland concrete pavements require characterization of the underlying granular layers and the fractured portland concrete slabs. This involves determining the moduli of the pavement layers, as well as the plastic deformation in the granular layers. The latter is obtained directly through trenching. The moduli are computed through back-calculation using NDT data, as described in Chapter 9. The effective modulus of the fractured portland concrete slab is computed as a function of the variability in the fracturing process. Table 13.4 shows the coefficient of variation in the back-calculated slab piece moduli as a function of the quality of slab fracturing and the corresponding effective slab modulus. The latter was obtained as the 25th percentile of the observed values.[9] Using these effective moduli requires ensuring that no more than 5%

**Table 13.4**
Effective Fractured Portland Concrete Moduli for Asphalt Concrete Overlay Design (Ref. 1)

| Control of Fracturing Process | Coefficient of Variation in Slab Modulus | Effective Slab Modulus (GPa) |
|---|---|---|
| Good to excellent | 25% | 4.1 |
| Fair to good | 40% | 3.1 |
| Poor to fair | 60% | 2.0 |

of the slab pieces have a modulus higher than 6.9 GPa (1 million lbs/in$^2$).

### ASPHALT CONCRETE OVERLAYS OVER INTACT PORTLAND CONCRETE PAVEMENTS

Asphalt concrete overlays over intact portland concrete pavements are feasible following proper treatment of the existing surface to limit overlay reflection cracking, as described earlier. An integral part of the design is characterizing the modulus of the underlying granular layers and the structural condition of the existing portland concrete layer. The moduli of the underlying granular layers and the effective modulus of the portland concrete are obtained from back-calculation of NDT data, following the procedures described in Chapter 9. In addition, characterization of the structural condition of the portland concrete layer requires its in-situ elastic modulus/Poisson's ratio, modulus of rupture, and current fatigue damage level. It is recommended to obtain the in-situ portland concrete elastic modulus/Poisson's ratio and the modulus of rupture from field-extracted cylindrical cores and beams, respectively. The laboratory testing procedures to be followed on these samples are described in references 7, 2, respectively. Finally, the amount of fatigue damage in the existing portland concrete $d_{PCC}$ is computed from its effective modulus, $E_{PCC}$, (i.e., back-calculated from NDT) and its as-built modulus, $E$, by solving the following equation:

$$E_{PCC} = 1600000 + \frac{E - 1600000}{1 + e^{-5+5\,d_{PCC}}} \tag{13.13}$$

The moduli in Equation 13.13 are in lbs/in$^2$ and 1.6 million lbs/in$^2$ (11.03 MPa) is a representative modulus for rubbilized portland concrete (i.e., a lower-limit value for an intact slab). Equation 13.13 is also used to reduce the modulus of the portland concrete, with increasing fatigue damage in simulating the performance of the asphalt concrete overlay.

### 13.5.2 Portland Concrete Overlays

The NCHRP Study 1-37A[13] includes design procedures for portland concrete overlays over existing flexible pavements, as well as rigid pavements, including JPCP/JDRCPs and CRCPs. The former, are similar to analyze/design as new portland concrete pavements with an asphalt concrete base, will not be further discussed here (see Chapter 12). Special provisions for constructing ultrathin portland concrete overlays over asphalt concrete pavements can be found in the literature (e.g., reference 6). The following describes the design particulars of portland concrete overlays over existing portland concrete pavements.

Portland concrete overlays can be either bonded or unbonded. Bonded portland concrete overlays involve no separation/debonding layer, hence can only match the configuration of the existing rigid pavement. This means that jointed portland concrete overlays are feasible only over an existing jointed portland concrete pavement with the same joint configuration and spacing. Similarly, CRCP overlays are feasible only over existing CRCP pavements. The reason is that bonding between the old and the new surfaces precludes relative movement of the two layers of slabs. On the other hand, unbonded portland concrete overlays involve a separation/debonding layer, hence allow relative movement of the two layers of slabs. As a result, there is no reason for matching the configuration of the portland concrete overlay to that of the existing pavement.

The analysis/design approach for portland concrete overlays is similar to the approach used for designing new rigid pavements. There are several differences, namely the need to describe the pavement layers and their structural properties, which includes the existing layers, the new layer, and the interface layer (if one is present). The moduli for the existing pavement layers are best obtained from back-calculation of NDT data and extracted samples, as described earlier. The damage functions for the overlaid structure are similar to the those described in Chapter 12. There are some differences in accounting for past damage in the existing layers (e.g., the fatigue damage involves an additional term to the one shown in Equation 12.13 to account for past damage in the existing layers).

## References


1. AASHTO (1993). AASHTO Guide for the Design of Pavement Structures, American Association of State Highway and Transportation Officials, Washington, DC.
2. AASHTO (1997). AASHTO Standard T97-97, "Flexural Strength of Concrete (Using Simple Beam with Third-Point Loading)," American Association of State Highway and Transportation Officials, Washington, DC.
3. American Concrete Pavement Association (1993). "Guidelines for Full-Depth Repair," *Technical Bulletin* TB-002P, Skokie IL.
4. American Concrete Pavement Association (1993). "Guidelines for Bonded Concrete Overlays," *Technical Bulletin* TB-007P, Skokie IL.
5. American Concrete Pavement Association (1993). "Guidelines for Unbonded Concrete Overlays," *Technical Bulletin* TB-009P, Skokie IL.
6. American Concrete Pavement Association (1993). "Ultra-thin Whitetopping Technical," Bulletin RP-273P, Skokie, IL.
7. ASTM (2002). "ASTM Standard Test Method for Static Modulus of Elasticity and Poisson's Ratio of Concrete in Compression," American Society for Testing of Materials, C469-02e1.
8. AI (2000). *Asphalt Overlays for Highway and Street Rehabilitation*, 2nd ed., Manual Series MS-17, Asphalt Institute, Lexington, KY.
9. NAPA (1993). "Guidelines for the Use of HMA Overlays to Rehabilitate PCC Pavements," National Asphalt Paving Association, IS-117, Lanham MD.
10. Hot-Mix Asphalt Pavement Evaluation and Rehabilitation, NHI Course 131063, October 2001.
11. NCHRP (September 2006). Independent Review of the Recommended Mechanistic-Empirical Design Guide and Software, National Cooperative Highway Research Program *Research Results Digest*, No. 307, Washington, DC.
12. Portland Concrete Pavement Evaluation and Rehabilitation, NHI Course 131062, October 2001.

[13] NCHRP (July 2006). "2002 Design Guide: Design of New and Rehabilitated Pavement Structures," Draft Final Report, NCHRP Study 1-37A, National Cooperative Highway Research Program, Woshington, DC.

## Problems

13.1 A flexible pavement has asphalt concrete and base/subbase layer thicknesses of 12 and 30 cm, respectively. FWD measurements yielded deflections, corrected to 68°F, of 0.010 and 0.007 in at radial offsets of 0 and 20 in, respectively. The load applied was 10,000 lbs, and the radius of the loaded plate was 6 in. Determine the $SN_{EFF}$ and the required overlay thickness. The section has experienced 5.0 million *ESALs* to-date (*PSI* to 2.5) and it is estimated that it could sustain another 3.0 million *ESALs* before its serviceability is reduced to 1.5.

13.2 Determine the thickness of the asphalt concrete overlay required for an existing flexible pavement with asphalt concrete, unbound base, and subbase layer thicknesses of 22 cm, 18 cm, and 15 cm, respectively. The section has experienced 4.5 million *ESALs* to date; its serviceability is 3.0 in the *PSI* scale; and it is estimated that, if left untreated, it could sustain another 1 million *ESALs* before its *PSI* reduces to a terminal value of 2.3. It is estimated that the overlaid pavement needs to have a structural number of 4.8 to sustain future traffic. The pavement layers drain within one week if they become saturated, which happens 5% of the time.

13.3 Design an asphalt concrete overlay for a jointed dowel-reinforced rigid pavement 8 in thick. FWD measurements on this pavement yield a modulus of subgrade reaction of 110 lbs/in$^3$ and a portland concrete elastic modulus of 3,000,000 lbs/in$^2$. In addition, FWD measurements across the joints yield an *LTE* of 65%. Additional information includes 4.5 million *ESALs* accumulated to present and 5 million *ESALs* estimated to reduce *PSI* to 1.5. The number of future *ESALs* that needs to be carried over the design life of the overlaid section is 6.5 million. Consider only traffic-related serviceability loss of 2.5 (4.5 to 2.0). The standard error in predicting serviceability is 0.45, and the desired reliability is 95%. The drainage condition for this section is considered good, with pavement layer saturation less frequent than 5%.

13.4 Compute the fatigue damage accumulated in an existing asphalt concrete pavement, given that its $E^*$ and $E^*_{dam}$ computed as described previously have values of 400,000 lb/in$^2$ and 325,000 lb/in$^2$, respectively, and the minimum $E^*_{min}$ is 190,000 lb/in$^2$

# 14 Economic Analysis of Pavement Project Alternatives

## 14.1 Introduction

As described in Chapter 1, the economic analysis of alternative pavement designs is an essential component of the pavement design process. The decision to construct new pavement alternatives A or B, for example (Figure 14.1), requires the ability to predict their performance and quantify their economic implications. Similarly, decisions for routine 4-R activities for existing pavements require economic analysis to ensure the best utilization of available funds.

Pavement 4-R activity decisions are made at two levels, the network level and at the project level. At the network level, the needs of all the roadway pavement sections in a jurisdiction compete for the limited budget available. At this level, the need for pavement 4-R is established on the basis of pavement surface condition (e.g., roughness, distress, friction) rather than economic considerations. For this purpose, particular surface condition threshold or trigger values are set, depending on road functional classification (e.g., a *PSI* value of 2.0, a rut depth of 0.10 cm, or a friction coefficient of 0.35).

Pavement sections identified at the network level as "due" for a 4-R treatment are assembled into projects and further analyzed at the project level. At the project level, feasible 4-R alternative

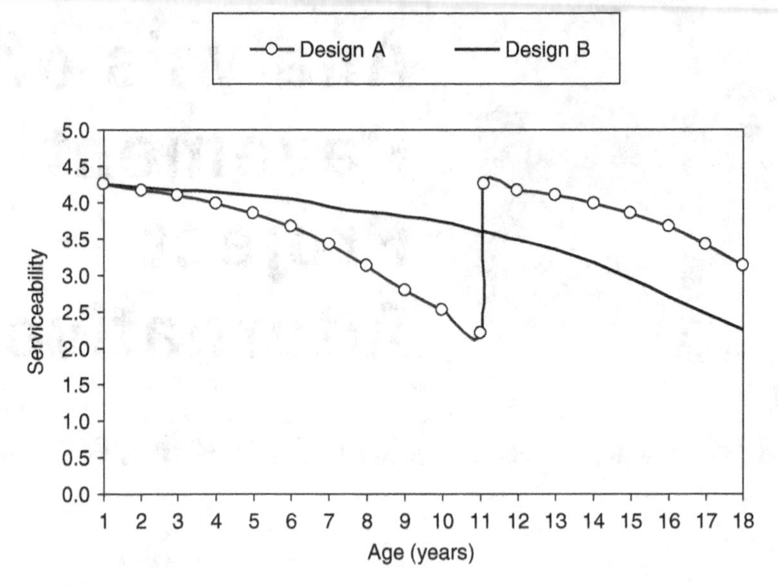


**Figure 14.1**
Examples of Alternative Pavement Designs and Their Performance

treatments are compared on a project-by-project basis to define the fiscally optimum alternative for each project over a selected analysis period. This process is known as life-cycle cost analysis (LCCA). Federal directives[25] define LCCA as "a process for evaluating the total economic worth of one or more projects, that is investments, by analyzing both initial costs as well as discounted future costs, such as maintenance and user costs over the life of the project." Hence, LCCA involves economic comparisons of alternative project 4-R treatments, including the financial implications to both the agency and the user, in a fashion that maximizes the effectiveness of the pavement budget allocation throughout a jurisdiction. For practical purposes, user benefits are calculated as savings in user costs from some arbitrary benchmark of pavement condition (e.g., pavement with a *PSI* of 2.0). The actual process used varies between highway authorities; however, the basic elements involved are common and will be presented in the first half of this chapter. The second half deals with a review of the literature on quantifying the cost components involved in LCCA, including those of a roadway agency and those of the road user.

It should be noted that routine pavement 4-R activities encompass major work covering the entire length of a pavement section, with the intention of restoring it to its original as-constructed condition. They are not to be confused with as-needed spot pavement maintenance (e.g., crack filling, patching, and so on) that is intended to slow pavement deterioration and, hence, delay the need for a routine pavement 4-R treatment. As-needed pavement maintenance is typically funded from the operating budget of a roadway agency. By nature, it is more difficult to predict and quantify, being budgeted in conjunction with other activities, such as grass cutting and snow removal. Hence, the economic implications of maintenance are typically ignored for the purpose of conducting project-level pavement LCCA.

## 14.2 Overview of Time Value of Money Concepts

Pavement LCCA involves comparisons of streams of money over an analysis period, hence requires an understanding of time value of money concepts. This section provides an overview of rates and methods for economic analysis.

### 14.2.1 Inflation, Market, and Borrowing Rates

Clearly, the worth of sums of money varies with time as a function of market interest rates and inflation. The following example explains this concept.

**Example 14.1**

A highway department considers purchasing a piece of equipment costing \$1,000,000 today. How much money would it have available for this purpose, if it decided to wait for one year, assuming that it could earn a market interest rate of $i = 5\%$ per year on its investments and that inflation runs at $f = 3\%$ per year?

**ANSWER**

Clearly, \$1,000,000, after one year, would have accumulated to \$1,050,000 of "actual" money, but would have less buying power due to inflation, that is \$1,050,000/$(1 + f)$ in terms of "real" (i.e., inflation-free) money. Hence, if the real interest rate is defined as $i'$, the following is true:

$$1 + i' = \frac{1 + i}{1 + f} \tag{14.1}$$

which can be written as:

$$i' = \frac{i - f}{1 + f} \tag{14.2}$$

which yields a real interest rate of:

$$i' = \frac{0.05 - 0.03}{1 + 0.03} = \frac{0.02}{1.03} = 0.0194 = 1.94\%$$

Hence, although the actual money available after one year is \$1,050,000, its real purchasing value is only \$1,019,400.

Several important conclusions can be drawn form this example:

- Unless $i$ is larger than $f$, keeping money in savings would make no financial sense.
- Banks need to charge a rate higher than $i$ to their borrowing customers (e.g., mortgage holders, credit card users, and so on) to remain profitable.
- Economic analyses conducted in actual money terms using the market interest rate or real money terms using with the real interest rate are equivalent.

The last point is further demonstrated by another example.

**Example 14.2** A highway department purchased right-of way land in 1951 at \$1/m$^2$, and it estimated that it had a year 2000 market value of \$40/m$^2$. What is the inflation- free rate of return on its investment if inflation runs at an average of 3% annually over that period (assume end-of-year transactions)?

**ANSWER IN TERMS OF ACTUAL MONEY**

The market interest rate $i$ is estimated as:

$$(1 + i)^{50} = \frac{40}{1} \rightarrow i = 7.66\%$$

which suggests (Equation 14.2) a real interest rate $i'$ of:

$$i' = \frac{0.0766 - 0.03}{1 + 0.03} = 4.52\%$$

**ANSWER IN TERMS OF REAL MONEY**

Translate first the $40 million dollars for year 2000 into equivalent 1951 dollars:

$$\frac{\$40}{(1+0.03)^{50}} = \$9.12 \text{ million}$$

which allows calculating a real interest rate of:

$$(1+i')^{50} = \frac{9.12}{1} \rightarrow i' = 4.52\%$$

The point is that care should be taken in utilizing either actual money and the market interest rate or real money and the real (inflation-free) interest rate, without mixing the two. In pavement LCCA analysis, it is customary to use real money and real interest rate in comparing pavement project alternatives. Therefore, the remainder of the discussion in this chapter uses real money and the real interest rate. For this type of analysis, the real interest rate is also referred to as the *discount rate*, because it discounts inflation.

Selecting an appropriate discount rate over an analysis period requires assuming representative values for the market interest rate and the inflation rate over that period. Clearly, these rates depend on market forces and are, at best, difficult to predict over the decades involved in pavement LCCA. Inflation rates can, to some degree, be predicted from historic data on consumer prices. The Consumer Price Index (CPI) (Table 14.1) is a summary statistic reflecting the relative annual prices of goods and services. It can be used as a guideline for extrapolating future inflation trends, as explained in the following example.

**Example 14.3**

Predict the average annual inflation rate expected over the next 20 years.

**ANSWER**

An estimate of future inflation trends can be obtained by studying the historic inflation data shown in Table 14.1 for the past 20 years. Between 1981 and 2000, inflation averaged 3.57% annually, which

**Table 14.1**
CPI Summary Data: Urban Conditions (U.S. Department of Labor Bureau of Labor Statistics)

| Year | CPI | Change % | Year | CPI | Change % | Year | CPI | Change % |
|---|---|---|---|---|---|---|---|---|
| 1960 | 29.6 | 1.4 | 1975 | 53.8 | 6.9 | 1990 | 130.7 | 6.1 |
| 1961 | 29.9 | 0.7 | 1976 | 56.9 | 4.9 | 1991 | 136.2 | 3.1 |
| 1962 | 30.2 | 1.3 | 1977 | 60.6 | 6.7 | 1992 | 140.3 | 2.9 |
| 1963 | 30.6 | 1.6 | 1978 | 65.2 | 9 | 1993 | 144.5 | 2.7 |
| 1964 | 31 | 1 | 1979 | 72.6 | 13.3 | 1994 | 148.2 | 2.7 |
| 1965 | 31.5 | 1.9 | 1980 | 82.4 | 12.5 | 1995 | 152.4 | 2.5 |
| 1966 | 32.4 | 3.5 | 1981 | 90.9 | 8.9 | 1996 | 156.9 | 3.3 |
| 1967 | 33.4 | 3 | 1982 | 96.5 | 3.8 | 1997 | 160.5 | 1.7 |
| 1968 | 34.8 | 4.7 | 1983 | 99.6 | 3.8 | 1998 | 163 | 1.6 |
| 1969 | 36.7 | 6.2 | 1984 | 103.9 | 3.9 | 1999 | 166.6 | 2.7 |
| 1970 | 38.8 | 5.6 | 1985 | 107.6 | 3.8 | 2000 | 172.2 | 3.4 |
| 1971 | 40.5 | 3.3 | 1986 | 109.6 | 1.1 | | | |
| 1972 | 41.8 | 3.4 | 1987 | 113.6 | 4.4 | | | |
| 1973 | 44.4 | 8.7 | 1988 | 118.3 | 4.4 | | | |
| 1974 | 49.3 | 12.3 | 1989 | 124 | 4.6 | | | |

is a reasonable estimate of average annual inflation over the next 20 years. It should be noted, however, that inflation rates in individual years can vary considerably from the mean.

Market interest rates are more difficult to predict, as they largely depend on the type of investment involved. They range from the rates earned at guarantied conventional bank accounts or Treasury bills (T-bills) to the rates earned from the stock market or business ventures. A good gauge for the interest rates that can be earned in a secure investment (i.e., an investment where the principal amount is guarantied) is the T-bill auctions conducted by the U.S. Department of the Treasury. Table 14.2 lists average annual yields of six-month T-bills.

Figure 14.2 demonstrates the high correlation between annual inflation rates and six-month T-bill rates. It also shows the variation of the discount rate calculated from this data through Equation 14.2. This figure suggests that over the past 20 years discount rates ranged from 0.5% to 6%, with an average of about 3%. This information

**Table 14.2**
Annual Treasury Bill Rates (U.S. Department of the Treasury)

| Year | Average % Rate for six-Month T-Bills | Year | Average % Rate for six-Month T-Bills |
|---|---|---|---|
| 1979 | 8.68 | 1990 | 7.25 |
| 1980 | 11.06 | 1991 | 5.21 |
| 1981 | 13.52 | 1992 | 3.46 |
| 1982 | 10.44 | 1993 | 3.25 |
| 1983 | 8.93 | 1994 | 4.88 |
| 1984 | 9.59 | 1995 | 5.50 |
| 1985 | 7.46 | 1996 | 5.12 |
| 1986 | 5.97 | 1997 | 5.17 |
| 1987 | 6.15 | 1998 | 4.78 |
| 1988 | 7.22 | 1999 | 4.92 |
| 1989 | 7.94 | 2000 | 5.84 |

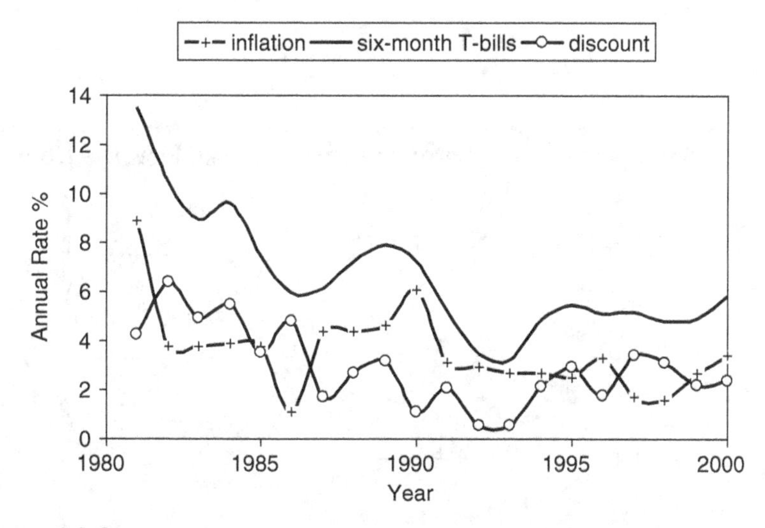


**Figure 14.2**
Historic Data on Inflation, T-Bill, and Discount Rates

provides guidelines for establishing future trends in real interest rates for conducting LCCA of alternative pavement designs. Often, the selection of the range in real interest rates to be used is a matter of policy by individual DOTs.

### 14.2.2 Basic Time Value Relationships

The relationships most commonly used for moving isolated sums of money or uniform series of sums of money in time are summarized in Table 14.3. Note that these relationships were derived assuming that isolated sums occur at the beginning of a period (e.g., year), while uniform sums occur at the end of each period. Furthermore, the interest rate and the time period used for the cash flow diagrams must correspond. The derivation of these relationships can be found in any introductory economics textbook (e.g., White et al.[68]). A few examples are shown next to explain their application. It is important to note that drawing the cash flow diagram for each of these problems greatly facilitates their solution.

**Example 14.4** A state DOT is planning on rehabilitating an asphalt concrete pavement using a 5 cm asphalt concrete overlay, which is projected

**Table 14.3**
Time Value of Money Formulas

| Amounts | Notation | Formula | Cash Flow Diagram |
|---|---|---|---|
| Compound | (F/P, i, n) | $F = P(1+i)^n$ | P, F |
| Present Worth | (P/F, i, n) | $P = F(1+i)^{-n}$ | P, F |
| Series Compound | (F/A, i, n) | $F = A\frac{(1+i)^n - 1}{i}$ | A, F |
| Sinking Fund | (A/F, i, n) | $A = F\frac{i}{(1+i)^n - 1}$ | A, F |
| Capital Recovery | (A/P, i, n) | $A = P\frac{i(1+i)^n}{(1+i)^n - 1}$ | P, A |
| Series Present Worth | (P/A, i, n) | $P = A\frac{(1+i)^n - 1}{i(1+i)^n}$ | P, A |

to last 12 years. Given that the overlay costs $20,000 per lane-km, calculate its annualized costs (assume a 4% discount rate).

**ANSWER**

The cash flow diagram for this problem is:

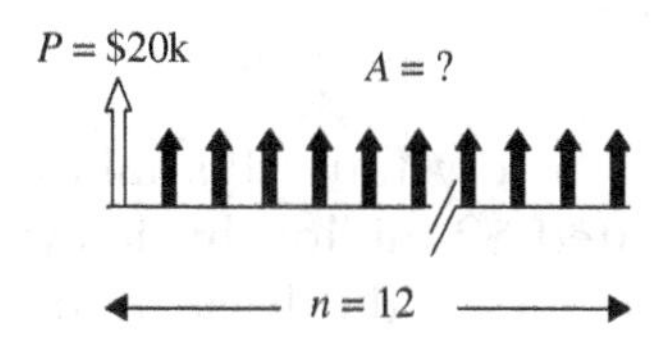


which, according to the Capital Recovery formula (Table 14.3), gives an annual cost of:

$$A = P\frac{i(1+i)^n}{(1+i)^n - 1} = \$20\frac{0.04(1.04)^{12}}{1.04^{12} - 1} = \$2.13 \text{ thousands, or } \$2{,}130$$

Notice that the timing of the amounts was assumed to comply with the convention made in deriving these formulas; that is, isolated sums occur at the beginning of a period, while series of uniform payments occur at the end of each period.

This example gives the lane-km cost of a particular treatment. The cost of such pavement 4-R treatments constitutes a DOT expenditure. Detailed information on the unit costs of other treatments is given later in this chapter.

**Example 14.5**

A state is considering issuing a five year bond for the purpose of financing a roadway construction program. The bond is to be repaid by an increase in the state gasoline tax, which will generate an average annual income of $20 million. What is the amount the state DOT can hope to raise at present, if the bond pays an inflation-free interest rate of 3.5% annually?

**ANSWER**

The cash flow diagram is identical to the one corresponding to the Series Present Worth formula (Table 14.3), where the uniform annual amount is $20 million over five years. This gives:

$$P = A\frac{(1+i)^n - 1}{i(1+i)^n} = 20\frac{1.035^5 - 1}{0.035(1.035)^5} = \$90.3 \text{ million}$$

Hence, $90.3 million can be spent at present for the construction program. This example indicates one of the alternative mechanisms for raising capital for special roadway projects.

**Example 14.6** A roadway is being overlaid every 20 years. Every time this takes place, the traveling public incurs an estimated $2 million in delay costs, due to the necessary lane closures. Estimate the least monetary benefit the public must realize annually from the improved pavement condition to offset this delay cost (assume a discount rate of 4%).

**ANSWER**

The cash flow diagram is plotted assuming that the present worth of the traveling public benefit from the improved pavement condition is equal to the delay cost–that is, $2 million.

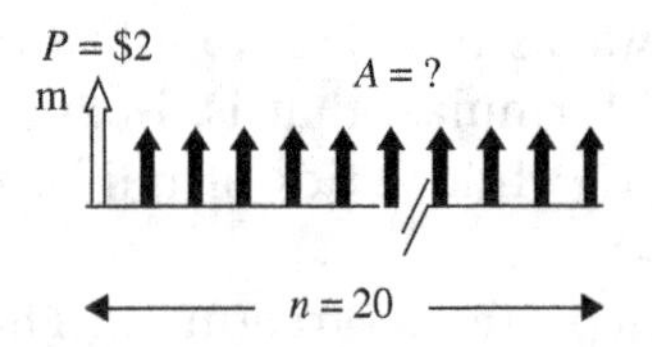


Applying the Capital Recovery formula (Table 14.3) gives: $A = P\frac{i(1+i)^n}{(1+i)^n - 1} = \$2\frac{0.04(1.04)^{20}}{1.04^{20} - 1} = \$0.147$million, or $147,160 of least annual benefit need to be generated from the pavement condition improvement to compensate for public delays during construction.

The last example introduces the concept of traveling public (i.e., user) costs and benefits, which will be the subject of much of the later discussion on pavement LCCA. In this context, benefits are quantified as savings in user costs due to the pavement condition improvement resulting from the 4-R activity. Improvement is referenced to a benchmark pavement condition, such as the one reflected by the minimum acceptable Present Serviceability Index (*PSI*), described in Chapter 9.

### 14.2.3 Spreadsheet Functions Implementing These Relationships

PC spreadsheets include a number of functions that facilitate the performance of some of these calculations. Excel, for example, includes a number of functions related to uniform payments A, given a present value $P$, a future value $F$, a discount rate $i$, and number of periods $n$:

- $FV(i, n, A, type)$, which gives the value $F/A$, for $type = 0$
- $PMT(i, n, P, F, type)$, which gives the value $A/P$, for $F = 0$ and $type = 0$
- $PV(i, n, A, F, type)$, which gives the value of $P/A$, for $F = 0$ and $type = 0$

where *type* is a binary variable indicating the convention made for the timing of the payments (0 for end of period and 1 for beginning of period).

## 14.3 Methods for Economic Comparison of Alternatives

The relationships presented so far allow reducing streams of costs and benefits to either present worth or equivalent annual amounts. Analyzing the economic implications of pavement project alternatives requires application of one of the following methods of economic comparison:

- Net present worth
- Net annualized worth
- Benefit-cost ratio
- Increment benefit-cost ratio
- Rate of return
- Incremental rate of return

Each of these methods is explained in turn, and examples are given of their applicability in comparing alternative pavement designs and 4-R treatments.

### 14.3.1 Net Present Worth (NPW)

The NPW method consists of translating streams of benefits and costs into present worth (i.e., time zero) and subtracting them to calculate the net present worth of benefits minus costs. It can be used to determine the feasibility of a single alternative or to compare two or more alternatives, whereby the alternative with the largest NPW is best. It can also be used to compare two alternatives that have the same benefits, which is referred to as a *fixed output comparison.* For the latter, the alternative with the lowest present worth of cost is best.

**Example 14.7** The stream of benefits and costs for two alternative pavement designs, A and B, that have the same service life are shown here in the form of cash flow diagrams in millions:

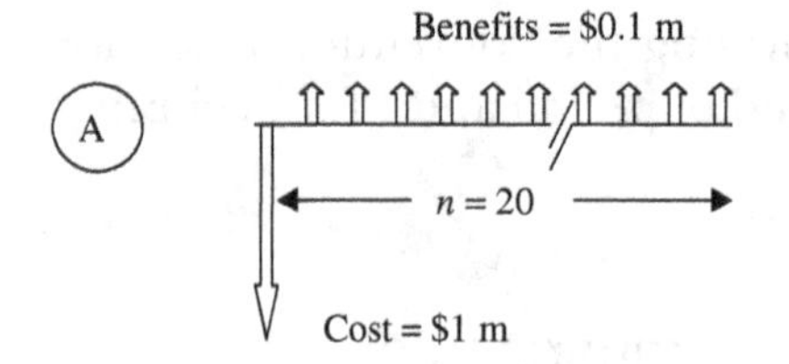


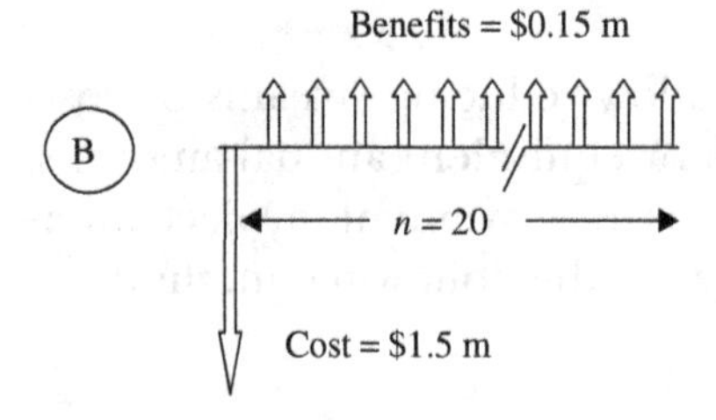


As stated earlier, agency costs incur during construction, while user benefits can be visualized as cost savings accrued annually by users from a pavement condition that is better than the benchmark pavement condition. Determine the best alternative, assuming a discount rate of 4%.

**ANSWER**

The NPW for alternative A is calculated as:

$$0.1\frac{1.04^{20}-1}{0.04(1.04)^{20}}-1=\$0.359\ \text{million}$$

Alternative B is calculated as:

$$0.15\frac{1.04^{20}-1}{0.04(1.04)^{20}} - 1.5 = \$0.538 \text{ millions}$$

which suggests that alternative B is best.

Two important observations can be made from this example:

- NPW comparisons are valid only when the length of the service life of the alternatives is identical. Since this is not true for the wide range of pavement 4-R alternatives, the NPW method is not commonly used for pavement LCCA.
- If the comparison just shown was treated as a fixed output problem—that is, the difference in user benefits between the two alternatives was ignored—alternative A would be best, since it requires a lower capital expenditure to construct.

## 14.3.2 Net Annualized Worth (NAW)

The NAW method consists of translating streams of benefits and costs into equivalent annual amounts and subtracting them to calculate the net annual worth of benefits minus costs. It can be used to determine the feasibility of a single alternative or to compare two or more alternatives, whereby the alternative with the largest NAW is best. Similar to the NPW method, NAW can be used to compare two alternatives that have the same benefits, which is referred to as a *fixed output comparison*. Since annual costs and benefits are compared, there is no requirement that the alternatives have the same service lives.

### Example 14.8

Solve Example 14.7 using the NAW method.

**ANSWER**

The solution involves translating the capital (i.e., construction) costs into equivalent annual terms and subtracting annual benefits minus annual costs.

The equivalent annual worth of alternative A is:

$$\$0.1 - \$1\frac{0.04(1.04)^{20}}{1.04^{20}-1} = \$0.0264 \text{ million}$$

The equivalent annual worth of alternative B is:

$$\$0.15 - \$1.5\frac{0.04(1.04)^{20}}{1.04^{20} - 1} = \$0.0396 \text{ million}$$

The NAW is $0.0132 million which, as expected, suggests that alternative B is best.

The NAW is the method of choice for pavement LCCA that treats the problem as one of fixed output, by ignoring the differences in user benefits between pavement alternatives.

### 14.3.3 Benefit over Cost Ratio (BCR)

The BCR method consists of translating streams of benefits and costs into either present worth or equivalent annual amounts and dividing them to calculate the benefit-over-cost ratio. It can be used to determine the feasibility of a single alternative or to compare two or more alternatives, whereby the alternative with the largest BCR over 1.00 is best. Obviously, it is not feasible to use the BCR for evaluating fixed output pavement alternatives, since benefits are not considered.

**Example 14.9**

Compare the following two alternatives:

| Alternative | Present Worth of Benefit | Present Worth of Cost |
|---|---|---|
| A | $700,000 | $500,000 |
| B | $1,200,000 | $1,000,000 |

**ANSWER**

The benefit-cost ratios for these alternatives are:

$$\frac{700,000}{500,000} = 1.4 \text{ and } \frac{1,200,000}{1,000,000} = 1.2, \text{ respectively,}$$

which suggests that alternative A is best.

Notice that the NPW method would not allow a conclusive comparison of these two alternatives, since their net present worth is identical (i.e., $200,000). Conversely, there are problems where the BCR between alternatives is identical, such as, two alternatives that yield benefits twice as large as their costs. Such problems are best analyzed considering the incremental benefits and costs between alternatives. This can be done in either incremental net terms (i.e., subtracting the difference in benefits minus the difference in costs between alternatives) or, most commonly, in incremental benefit/cost ratio terms, as explained next.

### 14.3.4 Incremental Benefit over Cost Ratio (IBCR)

The IBCR method consists of translating streams of benefits and costs into either present worth or equivalent annual amounts and comparing the difference in benefits divided by the difference in costs between two alternatives. If the ratio is larger than 1.00, the higher-capital cost alternative is better. The method can be used to compare more than two alternatives, by arranging them in order of increasing capital cost, and comparing them two at a time. The better of the two alternatives in the first paired comparison competes with the next alternative, until the best overall alternative is established. The ICBR approach is well suited to define the optimum level of expenditure, given a number of alternatives.

**Example 14.10**

Determine which of the following four alternative pavement treatments, arranged in order of increasing annual cost, is best.

| Alternative | Annual Benefit | Annual Costs |
|---|---|---|
| A | \$12,000 | \$8,000 |
| B | \$44,000 | \$20,000 |
| C | \$55,000 | \$40,000 |
| D | \$190,000 | \$100,000 |

**ANSWER**

First, calculate the incremental ratios of benefits and costs between alternatives A and B.

$$\frac{44,000 - 12,000}{20,000 - 8,000} = 2.67 > 1$$

suggesting that alternative B is better than A. Subsequently, calculate the incremental ratios of benefits and costs between alternatives B and C.

$$\frac{55,000 - 40,000}{40,000 - 20,000} = 0.75 < 1$$

suggesting that alternative B is better than C.

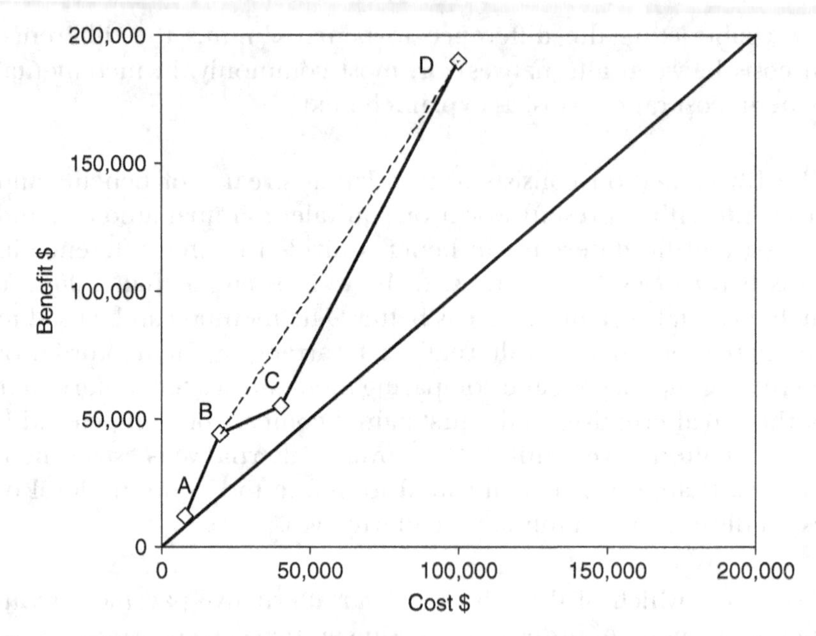


**Figure 14.3**
Graphical Solution of IBCR Problem 14.10

Finally, calculate the incremental ratios of benefits and costs between alternatives B and D.

$$\frac{190,000 - 44,000}{100,000 - 40,000} = 2.435 > 1$$

suggesting that alternative D is best.

Note that an effective alternative for obtaining an answer to this problem is to plot benefits versus costs (Figure 14.3). If benefits and costs are plotted using equal scales for the *y*-and *x*-axis, respectively, the slope of the line connecting any pair of solutions represents their ICBR. This representations allows, for example, determining at a glance that alternative C is no better than B, hence it should be excluded from subsequent comparisons.

### 14.3.5 Rate of Return (RR)

The RR method consists of establishing the interest rate that renders the present worth of benefits equal to the present worth of costs (i.e., the break-even interest rate). The RR is a good indicator of the economic feasibility of a particular alternative, by comparing the RR with estimates of the discount rate over the analysis period (Figure 14.2). Clearly, if the RR is. larger than the discount rate, the project is feasible, otherwise, it is not.

**Example 14.11**

A contractor is considering the purchase of an earth-moving piece of equipment that costs \$835,000, has a service life of 12 years, and is expected to yield sums of net annual income, in thousands of dollars, as shown here (assume that the discount rate is 4%).

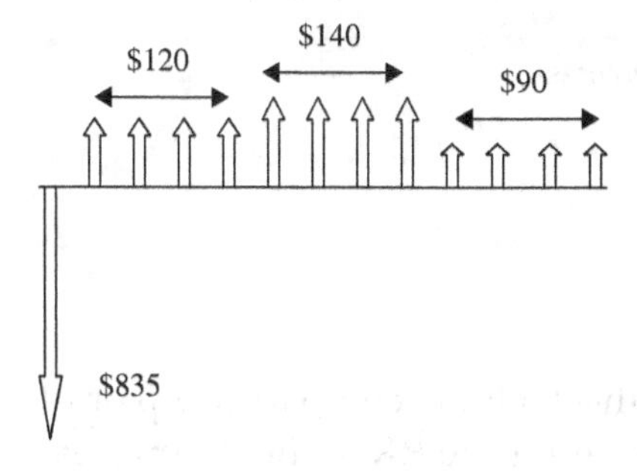


**ANSWER**

The rate of return of this investment can be found by defining the interest rate that will reduce the series of benefits to a present worth of \$835,000, or, alternatively, reduce the over all NPW to zero. This is expressed as:

$$835 = 120\frac{(1+i)^4 - 1}{i(1+i)^4} + 140\frac{(1+i)^4 - 1}{i(1+i)^4}(1+i)^{-4} + 90\frac{(1+i)^4 - 1}{i(1+i)^4}(1+i)^{-8}$$

which cannot be solved in a close form. Instead, a trial-and-error or a graphical method can be used. The graphical method is shown in Figure 14.4, yielding an RR of approximately 9.5%.

Clearly, if the RR is higher than the rate that can be obtained in the free market, the project is feasible: otherwise, it is not. The RR is often referred to as the *internal rate of return*, to differentiate it from external rates of return, where external suggests investments outside of the project being evaluated.

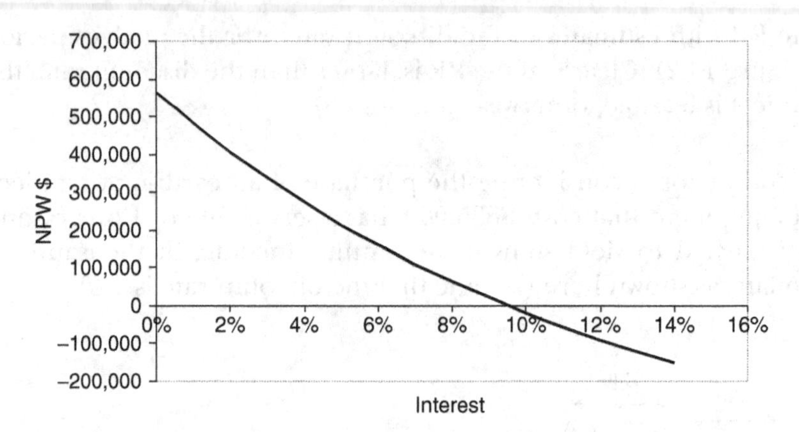


**Figure 14.4**
Determining the RR Graphically

### 14.3.6 Incremental Rate of Return (IRR)

The IRR, a derivative of the RR method, allows comparing a pair of alternatives. It is carried out by computing the RR of the differential streams of benefits and costs for the two alternatives. If the IRR is higher than market interest rates, the higher-cost alternative is better.

**Example 14.12** Two alternatives, as described by their cash flow diagrams in thousands of dollars, are shown here. Analyze them using the IRR method (assume a discount rate of 5%).

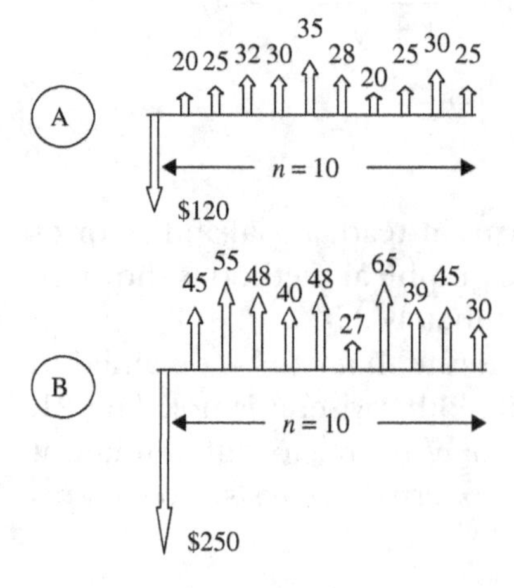

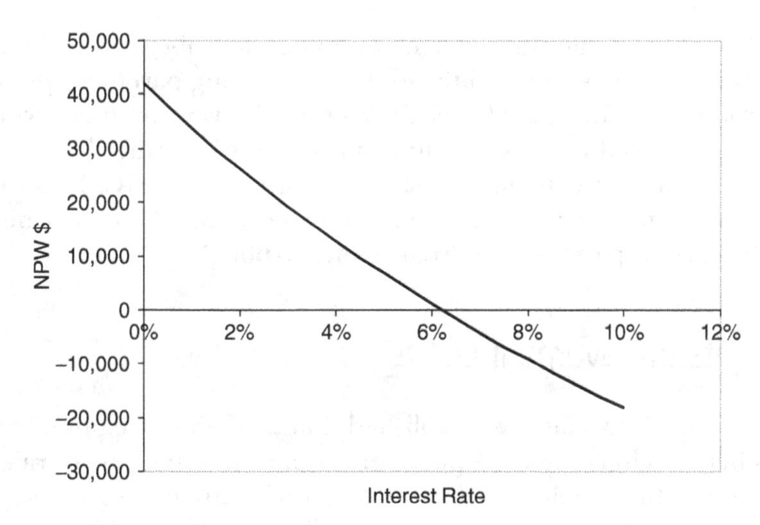


**Figure 14.5**
Graphical Solution of the IRR for Example 14.12.

**ANSWER**

Plotting the cash flow diagram of the difference between the cash flows for alternatives B and A reduces the comparison into an RR problem, which is solved graphically to yield an IRR of approximately 6% (Figure 14.5). Hence, alternative B is better than A.

B − A
25 30 16 10 13 1 45 14 15 5
$n = 10$
$130

### 14.3.7 Spreadsheet Functions Implementing These Comparisons

PC spreadsheets include a number of functions that facilitate the performance of some of these calculations. Excel, for example, includes a number of functions related to blocks of entries (either columns or rows), given a discount rate $i$:

- *NPV*($i$, *block*), which gives the NPW of the cash flow entered in a *block*.
- *IRR*(*block, guess*), which gives the RR for a cash flow entered in a *block,* given *guess,* an estimate of the rate of return.

#### 14.3.8 Summarizing Economic Evaluation Alternatives

Of the six economic evaluation methods just described, the NAW and the IBCR are most commonly used in comparing pavement project alternatives in LCCA. The NAW method is typically used, where LCCA is treated as a fixed output problem, where the differential user benefits between the alternatives are ignored. The ICBR method is most commonly used where the agency costs and the user benefits of the various pavement alternatives are explicitly considered.

## 14.4 Cost Components in Pavement LCCA

The discussion so far has established that analyzing streams of costs and benefits in comparing pavement project alternatives is straightforward. The problem is quantifying these streams of costs and benefits for the highway agency and the traveling public, respectively. The remainder of this chapter presents a review of the literature on agency and user costs.

#### 14.4.1 Agency Costs

Pavement 4-R-related agency costs include all the expenses associated with designing, financing, constructing, and rehabilitating roadway pavements. They can be grouped into the following major categories:

- Initial pavement construction costs, which typically exclude the cost of acquiring the right-of way
- Future 4-R costs (e.g., overlays)
- Administrative/engineering/mobilization costs, which are typically calculated as a fixed percentage of the costs for initial construction or 4-R
- Financing costs
- Salvage costs, which are, in essence, negative agency costs, reflecting the residual value of the layers of a pavement at the end of its design life.

Since the majority of the pavement initial construction and 4-R work is contracted out, unit cost data for the various materials and activities involved are readily available from state DOT contract bid records. An example is presented in Table 14.4, reflecting average bid prices submitted to the Washington State DOT in the 2006 construction season. Understandably, unit costs vary between regions within a state and between states. Furthermore, they need to be adjusted for inflation to reflect present conditions.

The actual cost of a pavement 4-R treatment is calculated from such unit cost data and the geometric data of the roadway, which includes:

- Section length
- Number of lanes and their width
- Shoulder width
- Pavement layer thicknesses

Representative specific gravity values are used to translate the calculated volumes of the materials to weights, prior to applying the unit prices.

**Example 14.14**

Calculate the agency cost for 1 km of a four-lane divided freeway built of flexible pavement. Given:

- Lane widths of 3.6 m
- Shoulder widths of 1.8 m
- Layer thicknesses of 16 cm of asphalt concrete and 40 cm of unbound base, respectively (specific gravities of 2.49 and 2.55)

Use the unit costs in Table 14.4. Assume that engineering and mobilization overhead add 15% and 20% to the contract cost, respectively. Ignore excavation and appurtenance costs (e.g., lane markings).

**Table 14.4**
Example of Average Unit Costs for Pavement Construction Materials (Washington State DOT; Statewide Average for third Quarter of 2006)

| Description | Unit Measure | Unit Cost |
|---|---|---|
| Roadway excavation | $m^3$ | $13.64 |
| Crushed surfacing (base/subbase) | Metric ton | $12.65 |
| Hot-mix asphalt concrete | Metric ton | $68.71* |
| Portland concrete pavement | $m^3$ | $348.8 |
| Steel reinforcement | kg | $2.40 |

*Depends on the price of crude oil.

**ANSWER**

| Item | Unit Cost | Item Cost |
|---|---|---|
| Asphalt concrete friction course | \$68.71/metric ton | $0.16 \times 1000 \times (3.6 \times 4 + 1.8 \times 4) \times 2.49 \times 68.71 =$ \$591,280 |
| Base layer | \$12.65/metric ton | $0.40 \times 1000 \times (3.6 \times 4 + 1.8 \times 4) \times 2.55 \times 12.65 =$ \$278,704 |
| | Subtotal | \$ 869,984 |
| Engineering | 15% | \$ 130,498 |
| Mobilization | 20% | \$ 173,997 |
| | Grand Total | \$1,174,478 |

Dividing the calculated total cost by the total pavement surface (i.e., 21,600 $m^2$/km) reveals a cost of about \$54/$m^2$ (\$5.052/$ft^2$) for pavement materials and their associated DOT overhead cost only.

As discussed in Chapter 13, the extent and severity of the distresses present dictate the feasible pavement 4-R treatments, which in turn dictate agency costs. As suggested by Tables 13.1, 13.2 and 13.3, the worse the pavement condition, the fewer and the more capital-intensive the feasible treatments are. For this reason, many state DOTs utilize pavement distress as the trigger for pavement rehabilitation, rather than pavement roughness. This approach allows repairing distresses early, before they become more extensive and severe thereby, reducing pavement life-cycle cost. A more thorough treatment of this subject can be found in a variety of pavement management references (e.g., 33).

### 14.4.2 User Costs

Roadway user costs can be grouped into two major categories,[1] namely vehicle operating costs and non-vehicle operating costs, each encompassing a number of components:

- Vehicle operating costs:
  - Fuel
  - Repair/maintenance, including parts and labor

- Tires
- Other (i.e., motor oil and usage-related depreciation)

- Non-vehicle operating costs:
  - Travel delays due to lane closures for pavement 4-R
  - Other (i.e., travel delays due to reduced speed caused by pavement roughness, pavement-related occupational injuries, cargo damage/packing costs, and pavement condition-related accidents).

Pavement roughness is the main pavement condition attribute affecting vehicle operating costs. Clearly, there are other factors affecting vehicle operating costs, such as roadway geometrics and traffic congestion, that are not relevant to pavement LCCA, hence need not be considered in analyzing the impact of alternative pavement project treatments to the road user. Pavement roughness also affects a number of non-vehicle operating costs, such as travel delays due to reduced speed, occupational injuries, and cargo damage costs. Travel delays from pavement 4-R activities depend on traffic conditions and lane closure practices. Surface friction is the main pavement attribute that affects traffic accidents. However, it is used more as an indicator of the need to carry out pavement 4-R, rather than as a predictor of pavement condition associated traffic accident costs.

The remainder of this chapter presents an overview of the literature on the effect of pavement roughness on user costs, as well as the procedures developed for quantifying delay costs during pavement 4-R activities.

**FUEL CONSUMPTION**

Early efforts to quantify the effect of roadway characteristics on fuel consumption date back to the 1960s [5,17,54]. These early studies agreed that there is a considerable difference in fuel consumption between paved roads and earth/gravel roads.

Winfrey[69] identified pavement surface stiffness and roughness as the main roadway attributes affecting fuel consumption, but presented only relationships between the total vehicle operating costs on paved surfaces and those on gravel surfaces.

In the 1980s, the World Bank sponsored an extensive program for establishing user cost models for developing countries.[12] Observations were made in Kenya, Brazil, the Caribbean, and India, using a number of vehicles instrumented to monitor fuel consumption. Roads included both unpaved and paved surfaces with a wide range of roughness. Fuel consumption was plotted versus pavement roughness by vehicle type and country where the observations were made (Figure 14.6). No consistent trends emerged from these results, which indicated that only the Brazilian data showed increasing fuel consumption with increasing pavement roughness for all the vehicle types monitored.

Work by Zaniewski et al.,[70] examined the relationship between fuel consumption and pavement serviceability (i.e., *PSI*), which is largely a function of pavement roughness (Equations 9.1a and 9.1b). Tests were conducted in the United States on a variety of road surfaces, including portland concrete, asphalt concrete, surface treatments and gravel. Fuel consumption was measured using a fuel meter mounted on eight different test vehicles (four cars and four trucks, ranging in size). Some differences in fuel consumption were shown between paved and gravel roads, but no statistically significant differences were found between paved surfaces of different roughness. Similar studies conducted in New Zealand[7,14] and South Africa[19,20] also reached the conclusion that pavement roughness has little effect on fuel consumption.

Work by Haugodegard et al.[34] produced regression relationships between fuel consumption and pavement roughness by combining World Bank data[12] and vehicle operating cost survey data from Norway (Figure 14.7). Fuel consumption was depicted in relative terms, with reference to an *IRI* roughness of 3 m/km.

Watanatada et al.[66,67] proposed a hybrid mechanistic-empirical approach for calculating the effect of pavement roughness on fuel consumption from vehicle rolling resistance analysis, where:

$$F_r = M\,g\,CR \tag{14.3}$$

with $F_r$ the result of the rolling resistance forces acting on the vehicle (in N), $M$ the mass of the vehicle (in kg), $g$ the acceleration of gravity (in m/sec$^2$), and $CR$ a dimensionless coefficient of rolling resistance. The following relationships were developed between $CR$ and pavement roughness ($IRI$ in m/km) for cars and trucks, respectively.

$$CR = 0.0218 + 0.00061\,IRI \tag{14.4a}$$

$$CR = 0.0139 + 0.00026\,IRI \tag{14.4b}$$

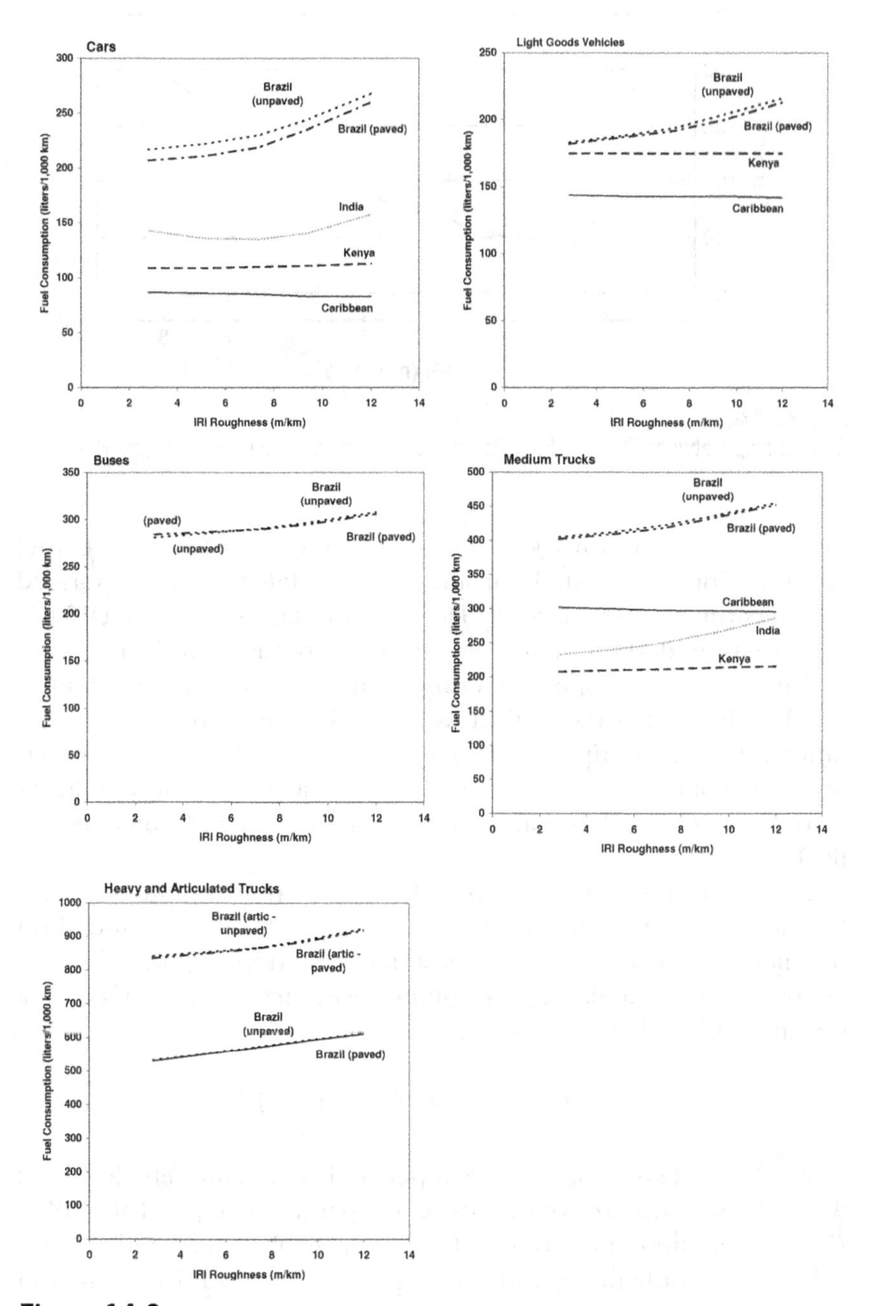


**Figure 14.6**
Fuel Consumption (liters per 1,000 km) versus Pavement Roughness (Ref. 12)

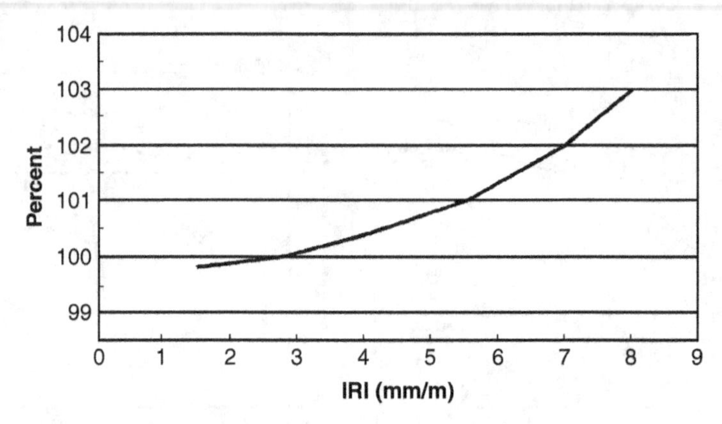


**Figure 14.7**
Relationship between Relative Fuel Consumption and Pavement Roughness (Ref. 34)

Empirical relationships were, in turn, used to relate *CR* to fuel consumption. These fuel consumption models were incorporated into version 3 of the Highway Design and Maintenance (HDM-III) software intended for analyzing alternative roadway projects.

Fancher et al.[23] proposed modifying these relationships to account for the effect of surface stiffness on rolling resistance. It was recommended to multiply *CR* by factors of 1.0, 1.2, and 1.5 for new portland concrete, old portland concrete/asphalt concrete under cold conditions and asphalt concrete under hot conditions, respectively.

The next generation of HDM-III, referred to as the Highway Development and Management program(HDM-4),[44] expanded on the mechanistic approach just described by adopting the following expression for calculating the rolling resistance force, $F_r$ (N), on a moving vehicle developed by Biggs:[4]

$$F_r = CR2(b11N_w + CR1(b12\,M + b13\,v^2)) \tag{14.5}$$

where $N_w$ is the number of wheels per vehicle; $v$ is the vehicle speed; $b11$, $b12$, and $b13$ are rolling resistance parameters, per Table 14.5; $CR1$ is a function of tire design (1.3 for cross-ply, 1.0 for radials, and 0.9 for low-profile tires); and $CR2$ is given in Table 14.6 as a function of road surface type.

**Table 14.5**
HDM-4 Representative Vehicle Rolling Resistance Model Parameters (Ref. 44)

| Vehicle Number | Type | Number of Wheels | Wheel Diameter (m) | Tire Type | CR1 | b11 | b12 | b13 |
|---|---|---|---|---|---|---|---|---|
| 1 | Motorcycle | 2 | 0.55 | Bias | 1.3 | 20.35 | 0.1164 | 0.0793 |
| 2 | Small Car | 4 | 0.60 | Radial | 1.0 | 22.20 | 0.1067 | 0.1333 |
| 3 | Medium Car | 4 | 0.60 | Radial | 1.0 | 22.20 | 0.1067 | 0.1333 |
| 4 | Large Car | 4 | 0.66 | Radial | 1.0 | 24.42 | 0.0970 | 0.1102 |
| 5 | Light Delivery Vehicle | 4 | 0.70 | Radial | 1.0 | 25.90 | 0.0914 | 0.0980 |
| 6 | Light Goods Vehicle | 4 | 0.70 | Bias | 1.3 | 25.90 | 0.0914 | 0.0980 |
| 7 | Four-Wheel Drive | 4 | 0.70 | Bias | 1.3 | 25.90 | 0.0914 | 0.0980 |
| 8 | Light Truck | 4 | 0.80 | Bias | 1.3 | 29.60 | 0.0800 | 0.0750 |
| 9 | Medium Truck | 6 | 1.05 | Bias | 1.3 | 38.85 | 0.0610 | 0.0653 |
| 10 | Heavy Truck | 10 | 1.05 | Bias | 1.3 | 38.85 | 0.0610 | 0.1088 |
| 11 | Articulated Truck | 18 | 1.05 | Bias | 1.3 | 38.85 | 0.0610 | 0.1959 |
| 12 | Minibus | 4 | 0.70 | Radial | 1.0 | 25.90 | 0.0914 | 0.0980 |
| 13 | Light Bus | 4 | 0.80 | Bias | 1.3 | 29.60 | 0.0800 | 0.0750 |
| 14 | Medium Bus | 6 | 1.05 | Bias | 1.3 | 38.85 | 0.0610 | 0.0653 |
| 15 | Heavy Bus | 10 | 1.05 | Bias | 1.3 | 38.85 | 0.0610 | 0.1088 |
| 16 | Coach | 10 | 1.05 | Bias | 1.3 | 38.85 | 0.0610 | 0.1088 |

**Table 14.6**
HDM-4 Proposed Values for CR2 (Ref. 44)

| Pavement Surface | Vehicle Mass | CR2 |
|---|---|---|
| Rigid Pavement | M ≤ 2500 kg | CR2 = 0.89 + 0.03 IRI (0.38 + 0.93 Tdsp)$^2$ |
| | M > 2500 kg | CR2 = 0.64 + 0.03 (Tdsp + IRI) |
| Flexible Pavement | M ≤ 2500 kg | CR2 = 0.89 + 0.03 IRI (0.38 + 0.93 Tdsp)$^2$ |
| | M > 2500 kg | CR2 = 0.84 + 0.03 (Tdsp + IRI) |

Note: Tdsp is the sand-patch-derived texture depth in mm.[59]

**Example 14.15** Compare the rolling resistance forces acting on an 18-wheel articulated truck operating on two different pavements with *IRI* roughnesses of 1 m/km and 5 m/km. Given, vehicle speed of 105 km/h (29.2 m/sec), bias-ply tires, *Tdsp* of 1 mm, and vehicle mass *M* of 36,287 kgr. Ignore the effect of pavement surface stiffness.

**ANSWER**

Tables 14.5 and 14.6 give:

$CR1 = 1.3$

$CR2 = 0.64 + 0.03\ (1 + 1) = 0.7$ for *IRI* of 1 m/km

$CR2 = 0.64 + 0.03\ (1 + 5) = 0.82$ for *IRI* of 5 m/km

$b11 = 38.85$

$b12 = 0.061$

$b13 = 0.1959$

which, substituted into Equation 14.5 give the rolling resistance force for roughness levels of 1 m/km and 5 m/km in the *IRI* scale as:

$$F_r = 0.70(38.85 18 + 1.3\ (0.0610\ \ 36,287\ + 0.1959\ 29.2^2)) = 2,656\text{N}$$

$$F_r = 0.82\ (38.85\ 18 + 1.3(0.0610\ \ 36,287\ + 0.1959\ 29.2^2)) = 3,111\text{N}$$

respectively. Hence, the difference in traction forces due solely to the difference in pavement roughness is 455 N.

HDM-4 continues by calculating the power required to overcome total traction forces $P_{tr}$ (kW), which include, in addition to $F_r$, aerodynamic forces, $F_a$, grade forces, $F_g$, curvature forces $F_c$, and inertial forces $F_i$ as:

$$P_{tr} = v\,(F_r + F_a + F_g + F_c + F_i)/1000 \tag{14.6}$$

Clearly, the last four force components are not relevant to pavement-oriented economic analyses. The power requirement thus calculated is translated into fuel consumption *IFC* (ml/sec), conceptually expressed as:

$$IFC = fn(P_{tr},\ P_{accs} + P_{eng}) \tag{14.7}$$

where $P_{accs}$ is the power required for engine accessories (e.g., fan belts, alternator) and $P_{eng}$ is the power required to overcome internal engine friction. The total power requirement, $P_{tot}$ is expressed as:

$$P_{tot} = \frac{P_{tr}}{edt} + P_{accs} + P_{eng} \qquad \text{for going uphill/level, i.e., } P_{tr} \geq 0 \tag{14.8a}$$

$$P_{tot} = P_{tr}\,edt + P_{accs} + P_{eng} \qquad \text{for going downhill, i.e., } P_{tr} < 0 \tag{14.8b}$$

where *edt* is the drive-train efficiency factor (Table 14.7). Clearly, the last two power components are not relevant to pavement-oriented economic analyses. Given the fuel-to-power efficiency ratio, $\xi$ (ml/kW/s), the fuel consumption can be calculated as:

$$IFC = \xi\ P_{tot} \tag{14.9}$$

where $\xi$ is factored to account for decreased efficiency at high engine rotation rates, using:

$$\xi = \xi_b\left(1 + ehp\frac{P_{tot} - P_{eng}}{P_{\max}}\right) \tag{14.10}$$

where, typical values for $\xi_b$, *ehp*, and $P_{max}$ are shown in Table 14.7 by vehicle type.

**Example 14.16**

Continue the previous example by comparing the difference in fuel consumption for the same 18-wheel articulated truck operating on *IRI* roughness levels of 1 m/km versus 5 m/km. Assume that the power required to overcome engine friction $P_{eng}$ is negligible and that the engine is running at midpower ($P_{tot} = P_{max}/2$).

**Table 14.7**
HDM-4 Fuel Consumption Parameters for Equations 14.8a and 14.9 (Ref. 44)

| Vehicle Number | Type | $\xi_b$ | *ehp* | $P_{max}$ | *edt* |
|---|---|---|---|---|---|
| 1 | Motorcycle | 0.067 | 0.25 | 15 | 0.95 |
| 2 | Small Car | 0.067 | 0.25 | 60 | 0.90 |
| 3 | Medium Car | 0.067 | 0.25 | 70 | 0.90 |
| 4 | Large Car | 0.067 | 0.25 | 90 | 0.90 |
| 5 | Light Delivery Vehicle | 0.067 | 0.25 | 60 | 0.90 |
| 6 | Light Goods Vehicle | 0.067 | 0.25 | 55 | 0.90 |
| 7 | Four-Wheel Drive | 0.057 | 0.10 | 60 | 0.90 |
| 8 | Light Truck | 0.057 | 0.10 | 75 | 0.86 |
| 9 | Medium Truck | 0.057 | 0.10 | 100 | 0.86 |
| 10 | Heavy Truck | 0.057 | 0.10 | 280 | 0.86 |
| 11 | Articulated Truck | 0.057 | 0.10 | 300 | 0.86 |
| 12 | Minibus | 0.057 | 0.10 | 60 | 0.90 |
| 13 | Light Bus | 0.057 | 0.10 | 75 | 0.86 |
| 14 | Medium Bus | 0.057 | 0.10 | 100 | 0.86 |
| 15 | Heavy Bus | 0.057 | 0.10 | 120 | 0.86 |
| 16 | Coach | 0.057 | 0.10 | 150 | 0.86 |

**ANSWER**

Table 14.7 suggests that $P_{max}$ for an articulated truck is 300 kW, hence $P_{tot}$ at midpower is 150 kW. Equation 14.6 suggests that the difference in the power requirement due to the calculated difference in rolling resistance force (i.e., 455 N calculated earlier) is:

$$\Delta P_{tot} = \Delta P_{tr} = v\ \Delta F_r/1000 = 29.2\ 455/1000 = 13.29\ \text{kW}$$

which, according to Equations 14.9 and 14.10, results in a difference in fuel consumption of:

$$\Delta IFC = \xi\ \Delta P_{tr} = \xi_b \left(1 + ehp \frac{P_{tot} - P_{eng}}{P_{\max}}\right) \Delta P_{tr}$$

$$= 0.055 \left(1 + 0.10 \frac{150}{300}\right) 13.29 = 0.76$$

That is, 0.76 ml/s, or 2.74 liters, of fuel per hour, which, assuming a constant speed of 105 km/h translates into a 0.026 liters/km. To put this difference into perspective, consider an average fuel consumption for a heavy truck of 0.40 liters/km (about 5.9 miles/gallon).

The difference in fuel cost due to pavement roughness is about 6.5%.

Recent experimental evidence confirms that, indeed, pavement roughness has an effect on the fuel consumption of heavy trucks. Preliminary observations from the WesTrack experiment,[57] involving driverless articulated instrumented trucks running at 65 km/h in a closed circuit, suggest a more significant impact of roughness on fuel consumption (an *IRI* reduction of about 10% resulted in a 4.4% decrease in fuel consumption).

### VEHICLE MAINTENANCE AND REPAIRS

Early work on quantifying user costs has demonstrated significant differences in repair/maintenance costs between earth/gravel roads and paved roads,[17] as well as between new and old pavements.[35]

Zaniewski et al.[70] studied the relationship between vehicle maintenance/repair costs and pavement serviceability (i.e., *PSI*) and developed factors for adjusting repair/maintenance costs with reference to a baseline *PSI* of 3.5 (Table 14.8). Work by Clouston[14] also agreed that vehicle maintenance costs and tire costs are significantly affected by pavement roughness (Table 14.9).

Chesher et al.[12] developed relationships between vehicle maintenance and pavement roughness from the data obtained from the World Bank study mentioned earlier. Figure 14.8 summarizes the trends in the cost of maintenance-related parts for each of the four countries from which vehicle operating cost data was collected. The corresponding vehicle maintenance-related labor cost was

**Table 14.8**
Maintenance and Repair Factors versus *PSI* (Ref. 70)

| Present Serviceability Index | Passenger Passenger Cars and Pickup Trucks | Single-Unit Trucks | Single-Trailer Trucks (2-S2, 3-S2) |
|---|---|---|---|
| 4.5 | 0.83 | 0.90 | 0.86 |
| 4.0 | 0.90 | 0.94 | 0.92 |
| 3.5 | 1.00 | 1.00 | 1.00 |
| 3.0 | 1.15 | 1.07 | 1.11 |
| 2.5 | 1.37 | 1.17 | 1.27 |
| 2.0 | 1.71 | 1.30 | 1.50 |
| 1.5 | 1.98 | 1.48 | 1.82 |
| 1.0 | 2.30 | 1.73 | 2.35 |

**Table 14.9**
Vehicle Maintenance Costs as a Function of Road Roughness (New Zealand 1984 $) (Ref. 14)

| | CARS | | | | TRUCKS (Mass > 2 tons) | | | |
|---|---|---|---|---|---|---|---|---|
| Roughness (mm/km) | Parts (¢/km) | Labor (¢/km) | Tires (¢/km) | Total (¢/km) | Parts (¢/km) | Labor (¢/km) | Tires (¢/km) | Total (¢/km) |
| 1500 | 0.70 | 0.74 | 0.27 | 1.71 | 11.0 | 11.0 | 2.0 | 23.0 |
| 2000 | 1.65 | 1.63 | 0.30 | 3.58 | 12.9 | 12.8 | 2.11 | 27.8 |
| 3000 | 3.55 | 3.11 | 0.82 | 7.48 | 16.9 | 16.2 | 2.33 | 35.4 |
| 4000 | 5.44 | 4.16 | 1.34 | 10.94 | 20.8 | 19.4 | 2.56 | 42.8 |
| 5000 | 7.33 | 4.80 | 1.87 | 14.00 | 24.7 | 22.4 | 2.78 | 49.9 |
| 6000 | 9.23 | 5.02 | 2.39 | 16.64 | 28.6 | 25.2 | 3.00 | 56.8 |
| 7000 | 11.12 | 6.05 | 2.91 | 20.08 | 32.5 | 28.7 | 3.23 | 64.4 |
| 8000 | 13.01 | 7.08 | 3.44 | 23.53 | 36.5 | 32.1 | 3.45 | 72.1 |
| 9000 | 14.91 | 8.11 | 3.96 | 26.98 | 40.4 | 35.6 | 3.68 | 79.7 |
| 10,000 | 16.80 | 9.14 | 4.48 | 30.42 | 44.3 | 39.0 | 3.90 | 87.2 |

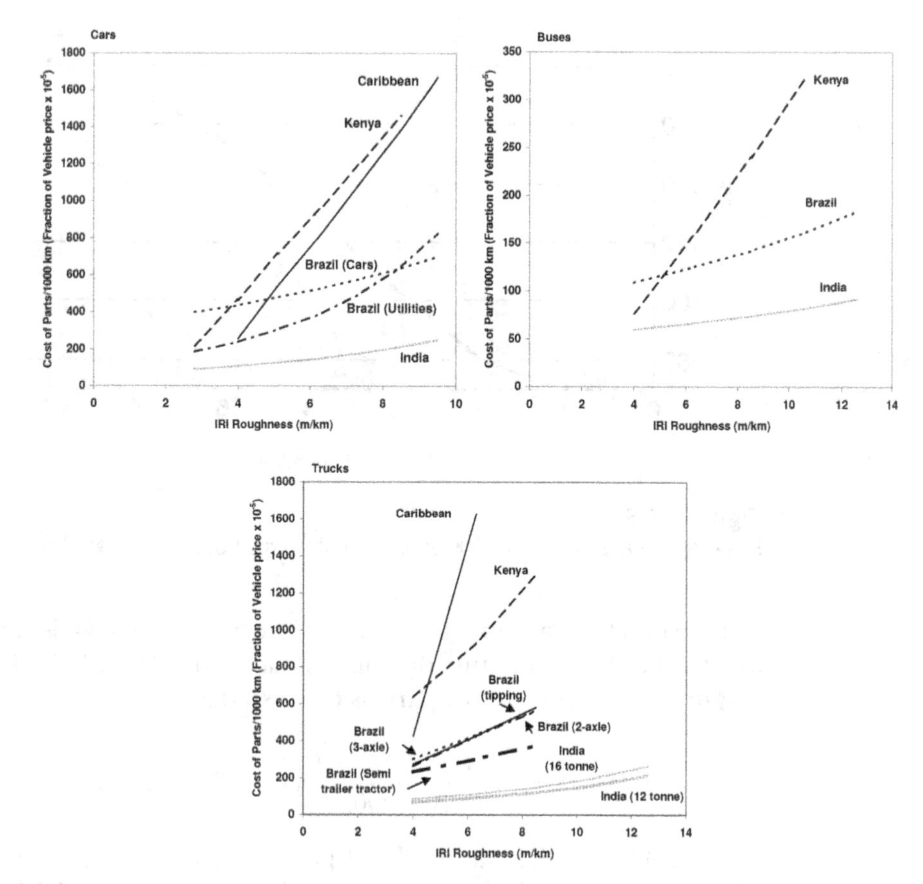


**Figure 14.8**
Cost of Parts for Vehicle Maintenance, Expressed as the Ratio of the Cost of Parts per 1,000 km Divided by the Purchase Price of a Vehicle $\times$ $10^{-5}$, versus Pavement Roughness (Ref. 12)

expressed as a function of the cost of parts. The HDM-III model[66,67] implemented the regression relationships developed from the data obtained from the World Bank Brazilian experiment.

Work by du Plessiss et al.[20] showed a significant increase in car and truck maintenance costs with increasing pavement roughness. An increase in roughness from 1.1 to 7.7 m/km, for example, resulted in a 95% increase in maintenance costs. Similar trends were reported by Haugodegard et al.[34] (Figure 14.9).

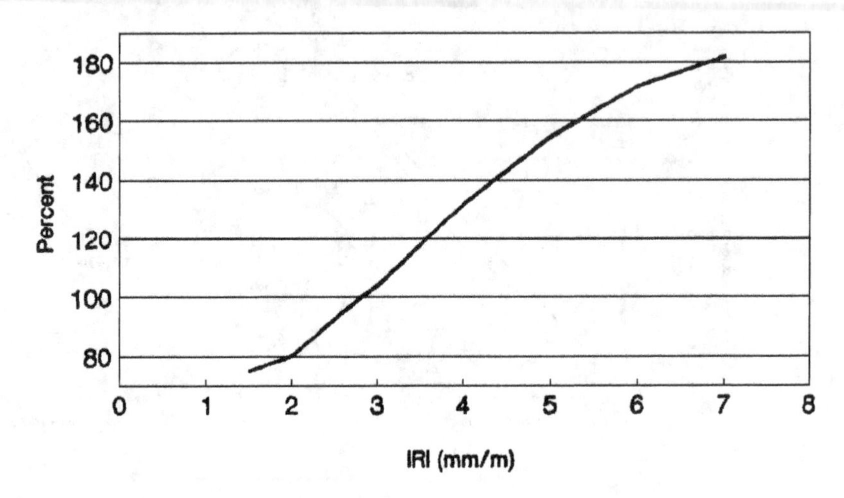


**Figure 14.9**
Relative Increase in Vehicle Maintenance Cost versus Roughness (Ref. 34).

The vehicle/maintenance cost relationships in the HDM-4 model[44] are largely based on the Brazilian data of the World Bank study (Figure 14.8). The cost of parts is expressed as:

$$PARTS = C0\left(\frac{CKM}{100,000}\right)^{kp}(1 + CIRI(RI - 3)) \qquad (14.11)$$

where *PARTS* is the standardized parts consumption (cost of parts divided by the purchase price of the vehicle in $1000s) per 1000 km; *CKM* is the cumulative odometer reading of a vehicle (km); *C*0, *CIRI*, and the exponent *kp* are calibration parameters that depend on vehicle type (Table 14.10); and *RI* is a function of pavement roughness (*IRI* in m/km):

$$RI = IRI \qquad \text{for} \quad IRI \geq IRI0 \qquad (14.12a)$$

$$RI = a0 + a1\ IRI^{a2} \qquad \text{for} \quad IRI < IRI0 \qquad (14.12b)$$

with *IRI*0 selected to have a value of 3.25 m/km, and *a*0, *a*1, and *a*2 regression constants equal to 3.0, $5.54 \times 10^{-8}$, and 13, respectively. It should be noted that Equation 14.11 suggests that a roughness level below an *IRI* of 3.25 m/km has limited effect on the cost of parts.

**Table 14.10**
Regression Constants for Calculating the Cost of Parts; Vehicles of Modern Technology (Ref. 16)

| Vehicle Number | Type | *CO* | *CIRI* | *kp* |
|---|---|---|---|---|
| 1 | Motorcycle | 0.200 | 0.230 | 0.230 |
| 2 | Small Car | 1.000 | 0.230 | 0.230 |
| 3 | Medium Car | 1.000 | 0.230 | 0.230 |
| 4 | Large Car | 1.000 | 0.230 | 0.230 |
| 5 | Light Delivery Vehicle | 1.000 | 0.230 | 0.230 |
| 6 | Light Goods Vehicle | 1.000 | 0.230 | 0.230 |
| 7 | Four Wheel Drive | 0.650 | 0.200 | 0.200 |
| 8 | Light Truck | 0.870 | 0.230 | 0.280 |
| 9 | Medium Truck | 1.100 | 0.150 | 0.280 |
| 10 | Heavy Truck | 1.100 | 0.150 | 0.280 |
| 11 | Articulated Truck | 1.100 | 0.150 | 0.280 |
| 12 | Mini-bus | 1.000 | 0.230 | 0.230 |
| 13 | Light Bus | 0.870 | 0.230 | 0.280 |
| 14 | Medium Bus | 0.519 | 0.061 | 0.483 |
| 15 | Heavy Bus | 0.519 | 0.061 | 0.483 |
| 16 | Coach | 0.519 | 0.061 | 0.483 |

The HDM-4 model calculates vehicle maintenance-related labor costs as a function of the cost of the corresponding vehicle parts. The regression relationships used for this purpose were obtained from analyzing Brazilian World Bank data[66] and were expressed as:

$$LH = COLH\ PARTS^{CLHPC}\ e^{CLHIRI\ RI} \tag{14.13}$$

where $LH$ is the number of labor hours, and $COLH$, $CLHPC$, and $CLHIRI$ are regression constants given in Table 14.11.

**Example 14.17**

Compare the cost of vehicle maintenance parts and labor for an articulated truck-tractor that has an odometer reading of 150,000 km and operates on pavements with an $IRI$ of 5 m/km, versus operating on a pavement with an $IRI$ of 1 m/km. Given, the purchase cost of the tractor is \$100,000 and the shop charges are \$60 per hour labor.

**Table 14.11**
Regression Constants for Calculating the Vehicle Maintenance Labor Cost per Equation 14.13 (Ref. 67)

| Vehicle Number | Type | *COLH* | *CLHPC* | *CLHIRI* |
|---|---|---|---|---|
| 1 | Motorcycle | 77.14 | 0.547 | 0 |
| 2 | Small Car | 77.14 | 0.547 | 0 |
| 3 | Medium Car | 77.14 | 0.547 | 0 |
| 4 | Large Car | 77.14 | 0.547 | 0 |
| 5 | Light Delivery Vehicle | 77.14 | 0.547 | 0 |
| 6 | Light Goods Vehicle | 77.14 | 0.547 | 0 |
| 7 | Four-Wheel Drive | 77.14 | 0.547 | 0 |
| 8 | Light Truck | 242.03 | 0.519 | 0 |
| 9 | Medium Truck | 242.03 | 0.519 | 0 |
| 10 | Heavy Truck | 242.03 | 0.519 | 0 |
| 11 | Articulated Truck | 652.51 | 0.519 | 0 |
| 12 | Minibus | 77.14 | 0.547 | 0 |
| 13 | Light Bus | 242.03 | 0.519 | 0 |
| 14 | Medium Bus | 293.44 | 0.517 | 0.0715 |
| 15 | Heavy Bus | 293.44 | 0.517 | 0.0715 |
| 16 | Coach | 293.44 | 0.517 | 0.0715 |

**ANSWER**

Substituting the constants for the articulated vehicle from Table 14.10 into Equation 14.11 gives:

$$PARTS = 1.10\left(\frac{150{,}000}{100{,}000}\right)^{0.28}(1 + 0.150(5 - 3))$$
$$= 1.6019 \text{ of the price of a new vehicle in } \$1000$$

That is 1.6019 × 100 = \$160.2 per 1000 km for operating on a pavement with an *IRI* of 5 m/km. The corresponding labor is calculated from Equation 14.13 as:

$$LH = 652.51\, PARTS^{0.519} e^{0} = 652.5\; 0.0016019^{0.519}$$
$$= 23.1 \text{ hours per } 1000 \text{ km}$$

which translates to labor costs of \$1386 and brings the total vehicle maintenance cost to \$1546 per 1000 km.

Repeating these calculations for an *IRI* of 1 m/km gives:

$$RI = 3.0 + 5.54 \times 10^{-8} IRI^{13} = 3.000$$

which, when substituted into Equation 14.11, gives:

$$PARTS = 1.10 \left(\frac{150{,}000}{100{,}000}\right)^{0.28} (1 + 0.150(3 - 3))$$
$$= 1.2322\% \text{ of the price of a new vehicle}$$

That is \$123 per 1000 km, and involves labor of:

$$LH = 652.51\ PARTS^{0.519} e^{0} = 652.5\ 0.001232^{0.519}$$
$$= 20.1 \text{ hours per } 1000 \text{ km},$$

which translates to a labor cost of \$1206 and brings the total vehicle maintenance cost to \$1329 per 1000 km. Hence, the difference in total vehicle maintenance cost attributed to pavement roughness is \$217 per 1000 km (\$1546 – \$1329), which amounts to about 16.3% additional parts cost for operating on the rougher pavement.

#### TIRE WEAR

Early work by Winfrey[69] suggested that road-related tire wear is hard to quantify due to the necessary repairs/replacements following tire damage from road hazards. De Weille[17] found that tire wear was mainly influenced by ambient temperature and vehicle weight. Zaniewski et al.[70] developed adjustment factors for tire wear as a function of *PSI*, using a *PSI* of 3.5 as the reference (Table 14.12).

Subsequent work by Clouston[14] suggested a significant effect of roughness on tire costs, while work by Brown[7] concluded that the km driven and the vehicle weight have a more significant impact on tire costs than pavement roughness.

The World Bank study[12] found increasing tire cost trends with increasing pavement roughness, although the rates of increase varied between the four country experiments (Figure 14.10).

The HDM-III model[66,67] implemented tire cost regression equations fitted to the World Bank data (Figure 14.10). Work by du Plessis et al.[19,20] found similar tire wear trends as in previous studies, but also noted that surface texture has an effect on tire wear.

**Table 14.12**
Tire Expense Adjustment Factors versus *PSI* (Ref. 70)

| Present Serviceability Index | Passenger Cars and Pickup Trucks[2] | Single- Trucks and Single-Trailer Trucks (2-S2, 3-S2) |
|---|---|---|
| 4.5 | 0.76 | 0.92 |
| 4.0 | 0.86 | 0.95 |
| 3.5 | 1.00 | 1.00 |
| 3.0 | 1.16 | 1.07 |
| 2.5 | 1.37 | 1.16 |
| 2.0 | 1.64 | 1.27 |
| 1.5 | 1.97 | 1.44 |
| 1.0 | 2.40 | 1.67 |

Work done by Haugodegard et al.[34] showed an increasing trend in tire wear costs with increasing pavement roughness (Figure 14.11).

HDM-4[44] calculates tire costs from their volumetric tread wear rates under prevailing operating conditions[9] and the volume of tread available.[3] The volumetric tread wear rate *TWT* ($dm^3/1000\,km$) is calculated from:

$$TWT = FLV\left(\frac{CT_C FNC\left(|CFT|\right)^2}{NFT} + \frac{CT_L FNL\left(|LFT|\right)^2}{NFT} + TWT_0\right) \tag{14.14}$$

where:

$TWT_0$ = the volumetric tread wear rate constant for a tire ($dm^3/1000\,km$) (i.e., the component of tire wear that is independent of the external forces).

*CFT*, *LFT* = the circumferential (i.e., longitudinal) and the lateral (i.e., transverse) components of the shear force (N) on the tire imprint.

*FNC*, *FNL* = factors indicating the variation in the circumferential and the lateral forces on the tire.

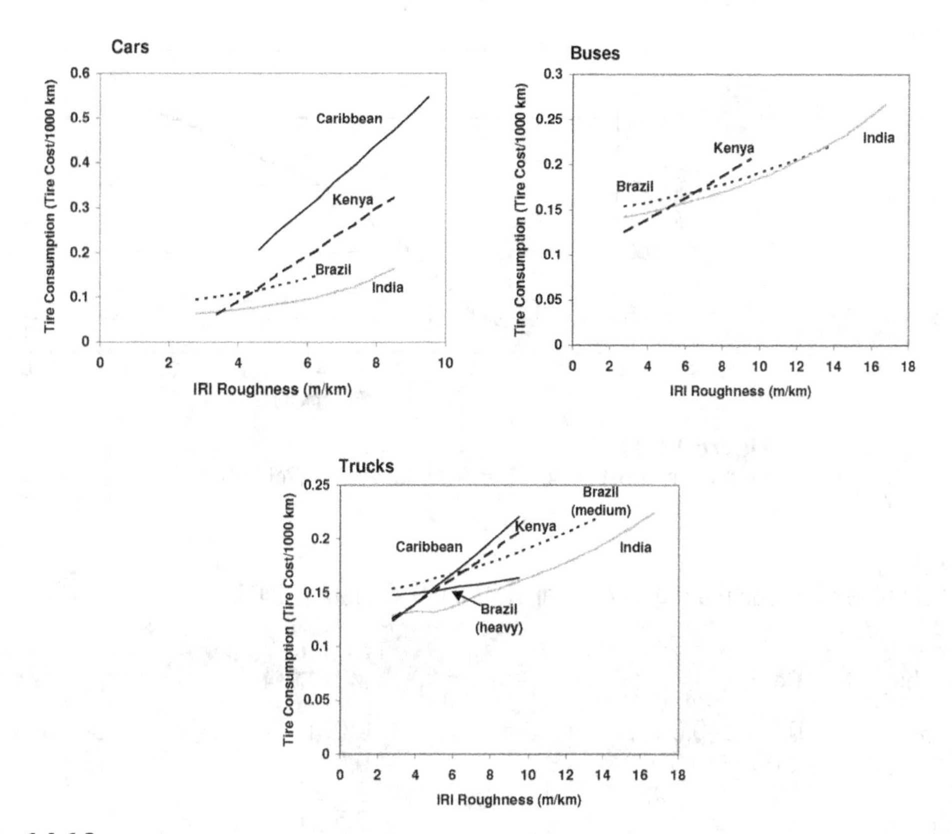


**Figure 14.10**
Tire Cost, with Reference to the Purchase Price of a New Tire, per 1,000 km versus Pavement Roughness (Ref. 12)

$NFT$ = the normal (i.e., vertical) force on the tire imprint (N).

$CT_C$ and $CT_L$ = the circumferential and lateral coefficients of tire tread wear ($dm^3/N/1000\,km$).

$FLV$ = a composite factor that reflects prevailing conditions, including pavement roughness/texture, vehicle type, weather, driving style, and so on.

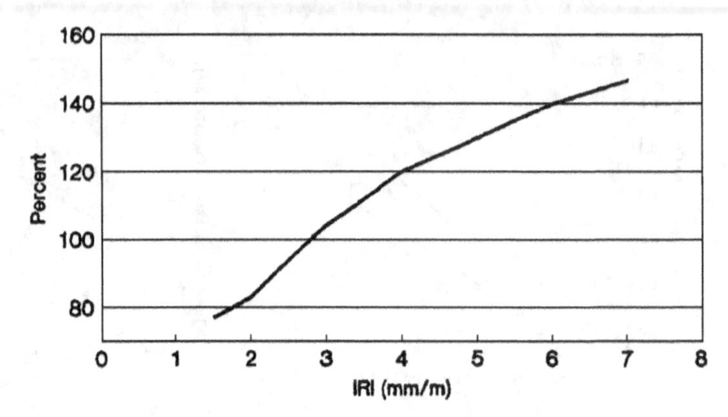


**Figure 14.11**
Relative Tire Cost versus Pavement Roughness (Ref. 34)

**Table 14.13**
Typical Values for Calculating Tire Wear Rates per Equation 14.14

| Variable | Range | Default Value in HDM-4 | Units |
|---|---|---|---|
| $TWT_0$ | 0.0005–0.0015 | 0.001 | $dm^3$/MN/m |
| $CT$ | 0.0003–0.0009 | 0.0005 | $dm^3$/MN/m |
| $FNC$ | 1.1–12 (cars);1.1–29.4 (trucks) | 2 (cars/trucks) | — |
| $FNL$ | 2.4–2.6 | 2.5 (cars/trucks) | — |

Table 14.13 summarizes the range of values to be used for these variables, as well as the default values used in HDM-4.

Ignoring the differences between tires on drive axles and nondrive axles allows calculation of *CFT* as the ratio of the total tractive force (i.e., the sum of the force components in the right hand-side of Equation 14.14) divided by the number of tires in a vehicle. The average *LFT* is calculated from the average road curvature ($^o$/km, which allows calculation of a representative turning radius $r$ per km) and the cornering speed $V_c$, assumed to be a fraction (e.g., 80%), of the vehicle speed on a tangent:

$$LFT = \frac{m\,V_c^2}{r} \tag{14.15}$$

where $m$ is the vehicle mass per wheel (kg).

**Example 14.18**

Calculate the circumferential and lateral tire forces for an 18-wheel truck with a mass of 36,000 kg, driving on tangents at speeds of 100 km/h, given that the average curvature of the road is 50°/km and that the average traction force on level terrain is 9,000 N.

**ANSWER**

*CFT* is calculated as 9,000/18 = 500 N, and *LFT* is calculated from Equation 14.15 as:

$$LFT = \frac{2000\ (0.8x27.78)^2}{\dfrac{360}{50}\dfrac{1}{2\ \pi}1000} = \frac{987{,}812}{1{,}146} = 861\ \ \text{N}$$

*FLV* is a function of pavement roughness and texture. To date, there have been no studies establishing this relationship, hence *FLV* is typically assumed to have a value of 1.00.

**Example 14.19**

Continue the previous example by calculating the rate of volumetric tire tread wear, assuming that the mass of the vehicle is distributed equally among all its 18 tires (the vertical force on each tire imprint, *NFT*, is 353.16/18 = 19.62 kN).

**ANSWER**

Substituting values for all known variables into Equation 14.14 gives:

$$TWT = 1.00\left(\frac{0.0005\ 2\ 500^2}{19620} + \frac{0.0005\ 2.5\ 861^2}{19620} + 0.001\right)$$
$$= 0.061\ \text{dm}^3/1000\ \ \text{km}$$

After obtaining the tire wear rate *TWT* the tire life in km, and, the unit tire cost per km can be calculated from the total available tread volume of a tire. The latter is computed from tire dimensions and tread data, as described by Bennett.[3] The area of the tread cross section, *AREA* ($mm^2$) is calculated as:

$$AREA = \frac{LL + LU}{2}(DE - DE_{\text{min}})\ ARUB \qquad (14.16)$$

where *LL* and *LU* are the width of the tire tread at the pavement interface (mm) when the tire is new and when it needs replacement, respectively, $DE - DE_{min}$ represents the net thickness of the tire tread

**Table 14.14**
Tire Wear Model Parameters (Ref. 3)

| Vehicle Number | Type | $DE - DE_{min}$ (mm) | LU – LL (mm) | ARUB |
|---|---|---|---|---|
| 1 | Motorcycle | 5 | 4 | 0.9 |
| 2 | Small Car | 8 | 6 | 0.85 |
| 3 | Medium Car | 8 | 6 | 0.85 |
| 4 | Large Car | 8 | 6 | 0.85 |
| 5 | Light Delivery Vehicle | 8 | 6 | 0.85 |
| 6 | Light Goods Vehicle | 8 | 6 | 0.85 |
| 7 | Four-Wheel Drive | 9 | 6 | 0.85 |
| 8 | Light Truck | 11 | 10 | 0.80 |
| 9 | Medium Truck | 15 | 10 | 0.70 |
| 10 | Heavy Truck | 17 | 10 | 0.70 |
| 11 | Articulated Truck | 17 | 10 | 0.70 |
| 12 | Minibus | 11 | 10 | 0.80 |
| 13 | Light Bus | 11 | 10 | 0.80 |
| 14 | Medium Bus | 15 | 10 | 0.70 |
| 15 | Heavy Bus | 15 | 10 | 0.70 |
| 16 | Coach | 15 | 10 | 0.70 |

available for wear, and *ARUB* is the fraction of tire tread in the tire imprint (decimal). Empirical data was analyzed and typical values for tread depth, *LL*, *DL*, and *ARUB*, were compiled for each of the 16 vehicle types considered in HDM-4 (Table 14.14). Empirical data on a number of tires were analyzed, and a regression equation was developed relating *LL* to nominal tire width:

$$LL = 1.05\ w - constant \tag{14.17}$$

where $w$ is the nominal tire width and *constant* is either 52.5 for vehicle numbers 1 to 3 or 66.7 for vehicle numbers 4 and higher.

The volume of the wearable rubber, *VOL* in $dm^3$, is in turn calculated as:

$$VOL = \frac{\pi\ DIAM\ AREA}{1{,}000{,}000} \tag{14.18}$$

where *DIAM* is the diameter of the tire (mm) given by:

$$DIAM = z + 2\ w\ a/100 \tag{14.19}$$

with $z$ = rim size mm and $a$ = aspect ratio (i.e., the ratio of nominal tire width divided by tire section height in percent, which is typically around 80).

**Example 14.20**

One of the tires of the truck described in the previous examples bears the designation 11 R22.5 (i.e., 11 in. or 27.94 cm nominal width, and a rim radius of 22.5 in. or 57.15 cm). Calculate the volume of its wearable tread and its anticipated life.

**ANSWER**

Substituting the values from Table 14.14 into Equation 14.16 gives:

$$AREA = \frac{227 + 237}{2}(17)\ 0.70 = 2{,}759\ \text{mm}^2$$

Substituting the given values into Equation 14.19 gives:

$$DIAM = 571.5 + 2\ 279.4\ 80/100 = 1019\ \text{mm}$$

which, substituted into Equation 14.18, gives a wearable tire rubber volume of:

$$VOL = \frac{\pi\ 1019\ 2{,}759}{1{,}000{,}000} = 8.83\ \text{dm}^3$$

which, combined with the wear rate calculated under Example 14.19, gives a tire life of 8.83/0.061 = 144,000 km.

Considering a purchase price for this tire of $250 results in a unit cost of $0.0017/km under the conditions specified. Obviously, from a pavement LCCA point of view, the interest is calculating differences in unit tire costs as pavement conditions change, which as mentioned earlier, in not quite feasible yet.

**OTHER VEHICLE OPERATING COST COMPONENTS**

Other vehicle operating costs include motor oil and depreciation. Motor oil costs are a function of oil consumption and the frequency of oil changes. These depend to a significant extent on factors other than pavement condition (e.g., engine condition, driving style). Hence, although motor oil cost may be important in other types of transportation economic studies, it is not typically considered in pavement LCCA.

Vehicle-usage-related depreciation is defined in terms of the life-km of a vehicle. Vehicle life-km is to be differentiated from vehicle life-year expectancy, because it is strictly related to the amount of usage of a particular vehicle. Considerable work has been done in differentiating these two sources of vehicle depreciation (e.g., references 69, 70). The unit cost for vehicle-usage-related depreciation is calculated as the ratio of the purchase price of a vehicle divided by its life-km ($/km). Pavement condition, primarily in terms of roughness, may affect the unit depreciation cost of vehicles by altering their life-km. However, the literature[5,7,12,17,34] suggests no consistent trends in vehicle unit depreciation costs in response to pavement roughness. The reason is that the life-km of vehicles is, to some extent subjective, often being decided by the cost of the maintenance/repairs required as the odometer reading of a vehicle increases. Hence, vehicle depreciation costs are often ignored in pavement LCCA.

### TRAVEL DELAY DUE TO LANE CLOSURES FOR PAVEMENT 4-R

Pavement 4-R activities require lane closures, which result in reduced roadway capacity and, as a consequence, higher travel times through work zones. The resulting travel time delays can be translated into cost by selecting the appropriate unit cost for the value of time of the traveling public. In addition to travel delay, vehicle operating costs are higher due to potential speed changes and stopping through the work zone. The configuration of the lane closure, the length of the lane or lanes closed, and the duration of the closure depend on local circumstances and vary widely. The following discussion deals with the procedures to be used in calculating travel time delay for only two simple lane closure scenarios: one where one or more lanes of a multilane divided freeway facility are closed, while some lanes remain open in each direction; and, two, where one lane is closed in a two-lane undivided highway facility, and directional traffic needs to be alternated on the remaining lane (Figure 14.12a and 14.12b, respectively). The source of the information presented next is a study by Memmott et al.[46].

#### Scenario 1

The method used for calculating lane-closure-associated delays in multilane divided highways is based on queuing (i.e., vehicles having to stop) and speed reduction (i.e., vehicles having to slow down) calculations through work zones. The latter is computed according

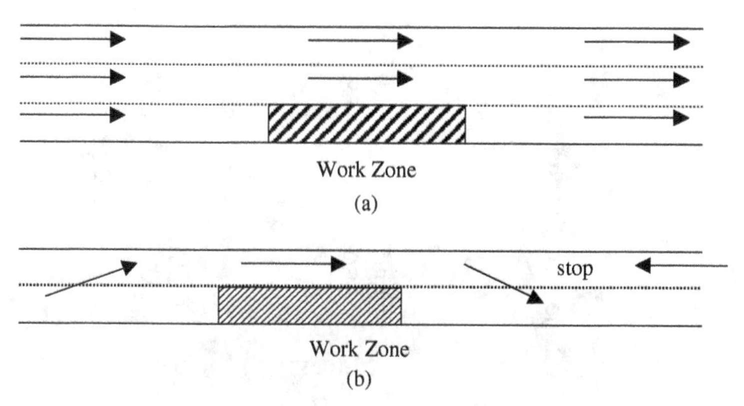


**Figure 14.12**
Schematic of Lane Closure Scenarios for Calculating Pavement 4-R Delay

to procedures described in the *Highway Capacity Manual.*[36] The simplifying assumption of this method is that vehicles arrive at a work zone uniformly distributed through time. The analysis is carried out at distinct time intervals, *ANALINT*, that are typically subsets of an hour (min). The number of vehicles stopped in the queue in interval $i$ $VEHQUE_i$ is calculated by:

$$VEHQUE_i = VEHQUE_{i-1} + (ARRIVAL_i - CAPWZ_i)\frac{ANALINT}{60} \tag{14.20}$$

where $VEHQUE_{i-1}$ = the number of vehicles in the queue in the previous time interval, $ARRIVAL_i$ = the arrival rate for interval $i$ in passenger car equivalents/hr (pce/hr), and $CAPWZ_i$ = the capacity through the work zone, calculated as:

$$CAPWZ = (a0 - a1\ CERF)\ WZLANES \tag{14.21}$$

where the coefficients $a0$ and $a1$ are given in Table 14.15, $CERF$ = the capacity estimate risk factor (the probability that the estimated capacity will be lower than or equal to the actual capacity), and $WZLANES$ = the number of open lanes through the work zone.

**Table 14.15**
Restricted Capacity Coefficients During Work Zone Activity Hours (Ref. 46)

| | **Intercept a0 for Equation 14.21 Number of Open Lanes through Work Zone per Direction WZLANES** | | | | |
|---|---|---|---|---|---|
| Number of Lanes per Direction | 1 | 2 | 3 | 4 | 5 |
| 2 | 1460 | | | | |
| 3 | 1370 | 1600 | | | |
| 4 | 1200 | 1580 | 1560 | | |
| 5 | 1200 | 1460 | 1500 | 1550 | |
| 6 | 1200 | 1400 | 1500 | 1550 | 1580 |
| | Slope a1 for Equation 14.21 | | | | |
| | 1 | 2 | 3 | 4 | 5 |
| 2 | 2.13 | | | | |
| 3 | 4.05 | 1.81 | | | |
| 4 | 0.00 | 1.60 | 0.57 | | |
| 5 | 0.00 | 1.46 | 0.00 | 0.00 | |
| 6 | 0.00 | 0.00 | 0.00 | 0.00 | 0.00 |

The length of the queue is therefore calculated as:

$$QUELEN_i = \frac{VEHQUE_i \; VEHSPC}{NLAPP} \tag{14.22}$$

where $VEHSPC$ = the average vehicle spacing in m/pce (average of 12m) and $NLAPP$ = the number of open lanes approaching the work zone.

**Example 14.21**

Calculate the queue length accumulated over a 10-minute interval, given the lane closure situation shown in Figure 14.12a, the vehicle arrival rate is 3,100 pce/hr, the fact that there are 10 vehicles stopped at the beginning of the time interval, and that the desired capacity estimate risk factor is 60%.

**ANSWER**

Equation 14.21 gives: $CAPWZ = (1600 - 1.81\ 60)\ 2 = 2{,}982$ pce/hr

The number of vehicles at the end of the interval is calculated from Equation 14.20 as:

$$VEHQUE_i = 10 + (3{,}100 - 2{,}982)\frac{10}{60}$$
$$= 29.7 \text{ rounded to } 30 \text{ pce}$$

The associated queue length is given from Equation 14.22 as:

$$QUELEN_i = \frac{30\ 12}{3} = 120 \text{ m}$$

The total time delay $DQUE_i$ (vehicle-sec) is subsequently calculated by:

$$DQUE_i = \frac{(VEHQUE_{i-1} + VEHQUE_i)}{2}\ 60\ ANALINT \tag{14.23a}$$

when the queue persists throughout the time interval $i$ or:

$$DQUE_i = \frac{VEHQUE_{i-1}^2}{2\ (CAPWZ_i - ARRIVAL_i)}\ 60\ ANALINT \tag{14.23b}$$

when the queue dissipates during the time interval $i$.

**Example 14.22** Continue the previous example by calculating the total delay involved and the associated cost, assuming single-vehicle occupancy and a constant unit cost for the travel time equal to $10/hr.

**ANSWER**

Substituting the calculated values into Equation 14.23a gives:

$$DQUE_i = \frac{(10+30)}{2} 60\ 10 = 12{,}000 \text{ vehicle-sec or } 3.33 \text{ hours}$$

which translates to $33.33 for the time interval of the 10 minutes analyzed.

Memmott et al.[46] proposed calculating capacity-constrained speeds through work zones from simplified speed versus volume over capacity relationship, such as the one shown in Figure 14.13. Under these conditions, the length of the road experiencing reduced vehicle speeds $WZEL_i$ (m) is, in general, different from the length of the work zone $WZLEN$ (m), and is expressed as:

$$WZEL_i = WZLEN + 320 \tag{14.24a}$$

when $WZLEN \leq 160$ or $\frac{ARRIVAL_i}{CAPWZ_i} > 1$ or:

$$WZEL_i = 160 + (WZLEN + 160)\left(\frac{ARRIVAL_i}{CAPWZ_i}\right) \tag{14.24b}$$

otherwise.

Subsequently, the delay $DELAYWZ_i$ associated with driving through the length of the road experiencing reduced vehicle speeds is:

$$DELAYWZ_i = WZEL_i\left(\frac{1}{VELWZ} - \frac{1}{VELAPP}\right) \tag{14.25}$$

where $VELAPP$ is the vehicle speed upstream from the influence of the work zone (m/s) and $VELWZ$ is the lowest of the speed imposed by the posted speed limit and that imposed by capacity constraints through the work zone.

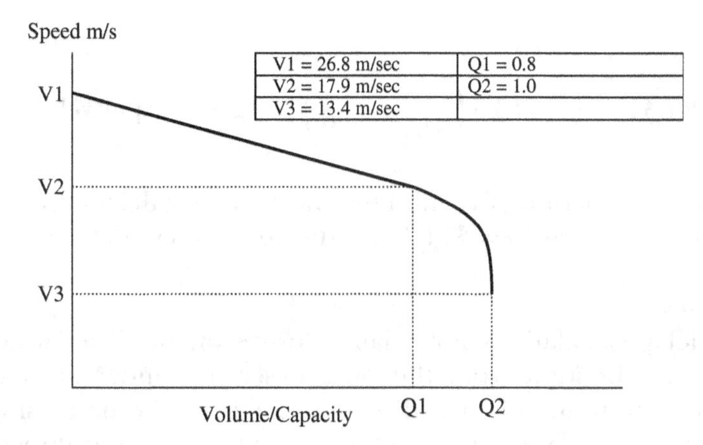


**Figure 14.13**
Speed versus Volume Relationships (Ref. 46)

**Example 14.23**

Calculate the delay and the associated cost for a single-lane per direction closure of 500 m in length on a six-lane divided freeway (Figure 14.12a). Given that lane capacity upstream from the work zone is 2000 pce/hr, the capacity of the work zone is 2,982 pce/hr (Example 14.21), the arrival rate is 2,500 pce/hr in the 10-minute interval to be analyzed, and no lower speed limit is posted through the work zone. Assume that all vehicles are single occupancy, that there is no lower speed limit posted in the approach to the work zone, and that the unit cost of travel time is \$10/hr.

**ANSWER**

Equation 14.24b gives the length of the road influenced by the work zone as:

$$WZEL_i = 160 + (500 + 160)\left(\frac{2{,}500}{2{,}982}\right) = 713.3 \text{ m}$$

For the approach to the work zone and the work zone, the lane traffic volumes are calculated as 2,500/3 = 833.3 and 2,500/2 = 1250 pce/hr, respectively, while the lane capacities (i.e., value of $Q2$ in Figure 14.13) are 2,000 and 2,982/2 = 1,491 pce/hr, respectively. The corresponding volume-over-capacity ratios are 0.416 and 0.838, which give vehicle speeds of 22.2 m/sec and 16 m/sec, respectively. Hence, the delay in the work zone is computed from Equation 14.25

as:

$$DELAYWZ_i = 713.3\left(\frac{1}{16} - \frac{1}{22.2}\right) = 12.45 \text{ sec per vehicle}$$

which, over a period of 10 minutes, translates to a delay cost of 2,500 10/60 12.45/3600 10 = \$14.41 for the 10 minutes analyzed.

Scenario 2

The delay calculations for a lane closure on two-lane undivided roads are different from the ones described under Scenario 1, because the traffic in each direction needs to alternate using the only remaining lane, while the opposing traffic lane is directed to stop by flagpeople (Figure 14.12b). Under this Scenario, there are two sources of delay: the one due to the reduced speed through the work zone and the other due to queuing while stopped. Assuming a constant speed through the work zone, the delay through the work zone can be calculated from Equation 14.25. The following describes how queuing delay is calculated.[36,62] The problem is equivalent to delay calculations through signalized intersections, where the cycle time *WZCYCLE* (sec) is the time required to complete a set of two alternating directional releases of traffic; inter-"green" time *IG* (sec) is the time required after a direction is switched to "red," until the opposite direction can be switched to green (i.e., the time it takes to clear from traffic the first direction); effective green time *EFFG* (sec) is the time actually available for movement in one direction; and the green-to-cycle-length ratio *GCRATIO* is expressed in effective green terms. These quantities are expressed as:

$$IG = \frac{WZLEN}{VELWZ} \tag{14.26}$$

$$EFFG = \frac{WZCYCLE}{2} + (IG + STLOSS + ENDLOSS) \tag{14.27}$$

$$GCRATIO = \frac{EFFG}{WZCYCLE} \tag{14.28}$$

where *STLOSS* is the lost time at the start of the cycle (sec) and *ENDLOSS* is the lost time at the end of the cycle (sec).

**Example 14.24**

Calculate the inter-green time, the effective green and the green-to-cycle ratio for a 500 m lane closure with a 600-second cycle on a two-lane undivided roadway with a posted speed through the work zone of 45 km/h (12.5 m/sec). Assume lost times at the beginning and end of the green interval of two seconds and one second, respectively.

**ANSWER**

Equations 14.26 to 14.28 give:

$$IG = \frac{500}{12.5} = 40 \text{ sec}$$

$$EFFG = \frac{600}{2} + (40 + 2 + 1) = 343 \text{ sec}$$

$$GCRATIO = \frac{343}{600} = 0.57$$

Defining $X$ as the arrival traffic volume over the capacity ratio, the unit delay caused by queuing through the work zone *DELAYQUE* (sec/pce) is calculated as the sum of the uniform delay *DELAYUN* (delay when all arrivals get processed during each green) plus the incremental delay *DELAYIN* (delay when not all arrivals in each direction get processed in each green cycle but need to wait the next green), expressed as:

$$DELAYQUE = DELAYUN + DELAYIN \tag{14.29}$$

The uniform delay is calculated by either:

$$DELAYUN = \frac{WZCYCLE - EFFG}{2} \tag{14.30a}$$

when $X > 1$, or:

$$DELAYUN = 0.38 \frac{WZCYCLE\ (1 - GCRATIO)^2}{1 - GCRATIO\ X} \tag{14.30b}$$

when $X \leq 1$.

The incremental delay is calculated as either:

$$DELAYIN = 0 \tag{14.31a}$$

when $X < Xo$, or:

$$DELAYIN = 15\ ANALINT\ X^{xn}\left[(X-1) + \sqrt{(X-1)^2 + \frac{720\ (X - Xo)}{ARRIVAL\ ANALINT}}\right] \quad (14.31b)$$

when $X \geq Xo$, with:

$$Xo = 0.67 + \frac{WZCAP}{3600} \quad (14.32)$$

and $xn$ is a model parameter, typically assumed to be equal to zero.
The average speed through the work zone $VELQUE$ is given by:

$$VELQUE = \left(\frac{V1}{2}\right)\left[1 + \sqrt{1 - \frac{CAPWZ}{CAPNORM}}\right] \quad (14.33)$$

where, $V1$ = the free-flow velocity (m/s), as defined in Figure 14.13 and $CAPNORM$ = the normal capacity of the work zone in pce/h.

**Example 14.25** Calculate the total hourly delay for the circumstance of the previous example, given that the directional volume is 900 pce/h and that the work zone capacity is 1,200 pce/h. What is the associated cost, assuming single-vehicle occupancy and a unit travel time cost of $10/hr?

**ANSWER**

The volume-over-capacity ratio is:

$$X = \frac{900}{1200} = 0.75$$

which, when substituted into Equation 14.30b gives:

$$DELAYUN = 0.38 \frac{600\ (1 - 0.57)^2}{1 - 0.57\ 0.75} = 73.6\ \text{s/pce}.$$

Equation 14.32 gives:

$$Xo = 0.67 + \frac{1{,}200}{3{,}600} = 1.0 > 0.75$$

which suggests that there no incremental delays. Hence, the total delay is 73.6 sec/pce which results in 73.6 / 3600 900 2 10 = $368/hour for both directions of traffic.

**OTHER NON-VEHICLE OPERATING COST COMPONENTS**

Other non-vehicle operating cost components include travel delays due to reduced speed caused by pavement roughness, cargo damage/packing, pavement-related occupational injuries, and pavement condition-related accidents.

High levels of pavement roughness may prevent vehicles from driving at the posted speed limit. The resulting increase in travel time can be quantified and included into pavement LCCA. A variety of empirical relationships are available in the international literature relating driving speed to pavement roughness (e.g., references 12, 34, 41, 45, 70). However, no consistent trends appear among them over the range of pavement roughness encountered in North America (Figure 9.23). As a result, this user cost component is typically not considered in state DOT-conducted pavement LCCA.

Freight insurance and packaging costs are sizeable compared to freight transportation revenues. Freight insurance alone comprises 2.2% of the gross revenue of intercity common carriers, as reported by the American Trucking Association. About half of this insurance cost is against damage during loading/unloading, while the remaining half is against in-transit damage. A considerable part of the latter can be attributed to pavement roughness. No relationships exist in the literature, however, relating directly freight insurance claims to pavement roughness. Furthermore, although considerable work has been done in quantifying freight packaging dynamics (e.g., references 2, 26, 49, 50, 56), little empirical evidence exists as to the packaging expenditure required solely for preventing freight damage due to pavement roughness. Hence, although the monetary implications of pavement roughness on freight insurance and packaging costs are recognized, they are typically not quantified in conducting pavement LCCA.

The literature contains an abundance of work on the effect of vibration on human physiology. In addition, considerable amount of work has been done in studying the occupational hazards of driving, particularly those related to heavy trucks (e.g., references 28, 64). However, no exclusive cause-effect relationship between roughness-induced vehicle vibrations and chronic medical problems has been established. As a result, including the cost of treating such medical problems into pavement LCCA is not yet possible.

Finally, the pavement condition attribute associated with vehicle safety is surface friction, which is related primarily to the texture of the pavement surface, as described in Chapter 9. Understandably, there is considerable difficulty in isolating pavement friction as the sole cause of vehicular accidents. Hence, associating the cost of an accident, which can be obtained from insurance claims, to pavement condition is tenuous at best. Thus, it is customary to use pavement surface friction as a trigger to pavement 4-R activity, rather than as a predictor of accident costs to be incorporated into pavement LCCA.

## References

1 AASHTO. (1990). *AASHTO Guidelines for Pavement Management Systems*, American Association of State Highway and Transportation Officials, Washington DC.

2 Antle, J. (1989). *Measurement of Lateral and Longitudinal Vibration in Commercial Truck Shipments*, School of Packaging, Michigan State University, Eastlansing, MI.

3 Bennett, C.R. (May 1998). *Predicting the HDM Volume of Wearable Rubber from Tyre Typology*, Highway and Traffic Consultants Ltd., Auckland, New Zealand.

4 Biggs, D.C. (1988). "ARFCOM—Models for Estimating Light to Heavy Vehicle Fuel Consumption," Research Report ARR 152, Australian Road Research Board, Nunawading.

5 Bonney, R. S. P., and N. F. Stevens (1967). "Vehicle Operating Costs on Bituminous, Gravel, and Earth Roads in East and Central Africa," Road Research Technical Paper, Number 76, London, England.

6 Brandenburg, R. K., and J. J. Lee (1991). *Fundamentals of Packaging Dynamics*, L. A. B, Skaneateles, NY.

[7] Brown, T. Julian, and J. J. Troon (1987). "Vehicle Operating Costs and Road Roughness Issues: A Southland Perspective," New Zealand Roading Symposium, Wellington, NZ.

[8] Butkunas, A. A. (1967). "Power Spectral Density and Ride Evaluation," SAE Paper No. 660138, Troy, MI.

[9] Carpenter, P., and P. Cenek (1998). Tyre Wear Model for HDM-4, Central Laboratories Report 98-529474, Opus International Consultants Limited, Wellington, New Zealand.

[10] Caruso, H., and W. Silver (1976). "Advances in Shipping Damage Prevention," *Shock Vibration Bulletin* 46 (4), pp. 41–47.

[11] Chaney, R. E., and D. L. Parks (1964). *Tracking Performance During Whole Body Vibration*, Boeing Company, Wichita, KS.

[12] Chesher, A., and Harrison, R. (1987). *Vehicle Operating Costs—Evidence from Developing Countries*, World Bank Publication, Johns Hopkins University Press, Baltimore, MD.

[13] Claffey, P. J. (1971). "Running Costs of Motor Vehicles as Affected by Road Design and Traffic," NCHRP Report 111, Washington, DC.

[14] Clouston, P. B. (1984). "The Effects of Road Roughness and Seal Extension Works on Vehicle Operating Costs," Institute of Professional Engineers New Zealand, Transportation and Traffic Engineering Group, Technical Session, Proceedings, Vol. 10, Issue 1.

[15] Coermann, R. R. (1960). "The Passive Mechanical Properties of the Human Thorax-Abdomen System and of the Whole-Body System," *Journal of Aerospace Medicine* 31, pp. 915–924.

[16] Cundhill M. (1995). "Recommended Relationships for Parts Modeling in HDM-4," Report to the International Study of Highway Development and Management Tools, University of Birmingham, UK.

[17] De Weille, J. (1966). *Quantification of Road User Savings*, Johns Hopkins Press, Baltimore, MD.

[18] Dieckmnann, D. (1955). The Effect of Mechanical Vibration upon Man—A Review and Summary of Research to Date, report of the Max-Planck Institut for Arbeitsphysiologie, Dortmund.

[19] du Plessis, H. W., and I. C. Schutte (1991). "Road Roughness Effects on Vehicle Operating Costs: Southern Africa Relations for Use in Economic Analyses and in Road Management Systems,"

Report Number PR 88/070/3, Division of Roads and Transport Technology, Council for Scientific and Industrial Research (CSIR), Pretoria, South Africa.

[20] du Plessis, H. W., and Meadows, J. F. (1990). "A Pilot Study to Determine the Operating Costs of Passenger Cars as Affected by Road Roughness," in *Road Roughness Effects on Vehicle Operating Costs*—Southern Africa Relations for Use in Economic Analyses and Road Management Systems, CSIR, Pretoria, South Africa.

[21] Duffner, L. R., L. H. Hamilton, M. A. Schmitz et al. (1962). "Effect of Whole-Body Vertical Vibration on Respiration in Human Subjects," *Journal of Applied Physiology* 33, pp. 914–916.

[22] Ernsting, J. (1960). "Respiratory Effects of Whole-Body Vibration," Air Ministry Flying Personnel Research Committee Report, Ministry of Defense, Air Force Department, London, UK.

[23] Fancher P. S., and C. B. Winkler (1984). "Retarders for Heavy Vehicles: Phase III Experimentation and Analysis; Performance, Brake Savings and Vehicle Stability," U.S. Department of Transportation, Report No. DOT HS 806672, Washington, DC.

[24] FHWA (1994). "Life-Cycle Cost Analysis," Summary of Proceedings, FHWA Life-Cycle Cost Symposium, *Searching for Solutions: A Policy Discussion Series*, No. 12, Federal Highway Administration, Washington, DC.

[25] FHWA (April 1994). Federal-Aid Policy Guide (FAPG), Section 500/Subpart B on Pavement Management Systems, FHWA, U.S. Department of Transportation, Transmittal 10, Washington, DC.

[26] Goff, J. W., and D. Twede (1983). *Shake and Break: Adventures in Package Dynamics*, available from the authors, Portland, MI.

[27] Goldman, D. E., and H. E. Von Gierke (1960). "The Effects of Shock and Vibration on Man," Lecture and Review Series No. 60-3, Naval Medical Research Institute, Bethesda, MD.

[28] Gruber, G. J. (1976). "Relationships between Whole-body Vibration and Morbidity Patterns among Interstate Truck Drivers," U. S. Department of Health, Education, and Welfare, Cincinnati, OH.

[29] Grenwood, I. D., C. R. Bennett, and A. Rahman (1995). "Effect of Pavement Maintenance on Road Users," Task 2070, the International Study of Highway Development and Management Tools, Birbingham, UK.

[30] Gruber, G. J., and H. H. Ziperman (1974). "Relationship between Whole-Body Vibration and Morbidity Patterns among Motor Coach Operators," U. S. Department of Health, Education, and Welfare, Cincinnati, OH.

[31] Guignard, J. C. (1964). A Note on the Heart Rate During Low-Frequency Whole-Body Vibration, Royal Air Force Institute of Aviation Medicine, Farnborough, UK.

[32] Guignard, J. C. (1971). " Human Sensitivity to Vibration, " *Journal of Sound Vibration* 15, pp. 11–16.

[33] Haas, R. C. G., R. W. Hudson, and J. Zaniewski (1994). *Modern Pavement Management*, Krieger Publishing Company, Malabar, FL.

[34] Haugodegard, T., J. Johansen, D. Bertelsen, and K. Gabestad (1994). *Norwegian Public Roads Administration: A Complete Pavement Management System in Operation*, Volume 2, Proceedings, Third International Conference on Managing Pavements, San Antonio, TX, May 22–26, Transportation Research Board, Washington, DC.

[35] Hide, H., S. W. Abaynayake, I. Sayer and R. J. Wyatt (1975). "The Kenya Road Transport Cost Study: Research on Vehicle Operating Costs," Transport and Road Research Laboratory Report, Number LR 672, Crowthorne, Great Britain.

[36] *Highway Capacity Manual* (1985). Special Report 209, Transportation Research Board, Washington DC.

[37] Hood, J. W. B., R. H. Murray, C. W. Urschel, J. A. Bowers, J. G. Clark, et al. (1966). "Cardiopulmonary Effects of Whole-Body Vibration in Man," *J. Applied Physiology* 21: 1725–1731.

[38] International Standards Organization (ISO) (1985). "Evaluation of Human Exposure to Whole-Body Vibration;" Part 1, *General Requirements*, Standard 2631, Geneva, Switzerland.

[39] Jones, A. J., and D. J. Saunders (1972). "Equal Comfort Contours for Whole-Body Vertical, Pulsed Sinusoidal Vibration," *Journal of Sound and Vibration* 23, pp. 1–4.

[40] Kent M. F. (1960). "Fuel and Time Consumption Rates for Trucks in Freight Service," Highway Research Board, National Academy of Sciences, Bulletin 276, Washington, DC.

[41] Karan, M. A. & Haas, R. & Kher, R. (1977). *Effects of Pavement Roughness on Vehicle Speeds*. Transportation Research Record 602, pp. 122–127.

[42] Kjrekostnadskomitéen, Transport-konomisk (1962). Urvalg; Handok for Beregning at Khreskostnader pa Veg. Norges-Naturvitenskapelige Forkningsrad, Oslo.

[43] Lamb, T. W., and S. M. Tenney (1966). "Nature of Vibration Hyperventilation," *Journal of Applied Physiology* 21, pp. 404–410.

[44] Lee, N. D. (1995). International Modeling Road User Effects in HDM-4, prepared for the Asian Development Bank as part of the International Study of Highway Development and Management Tools, Final Report.

[45] McFarland, W. F. (1972). *Benefit Analysis for Pavement Design Systems*, Texas Transportation Institute, Texas A&M University, College Station, TX.

[46] Memmott, J. L., and Dudek, C. L. (1982). "A Model to Calculate the Road User Costs at Work Zones," Research Report 292-1, Texas Transportation Institute, Texas A&M University College Station, TX.

[47] Moyer, R. A. (1939). "Motor Vehicle Operating Cost and Related Characteristics on Untreated Gravel and Portland Cement Concrete Road Surfaces," Highway Research Board, National Academy of Sciences, Washington DC, Proc. Vol. 19, pp. 68–98.

[48] Peterson, D. (December 1985). "Life-Cycle Cost Analysis of Pavement," Synthesis of Highway Practice No. 122, National Cooperative Highway Research Program, Transportation Research Board, Washington, DC.

[49] Pierce, C. D., S. P. Singh, and G. Burgess (1992). "A Comparison of Leaf-spring with Air-cushion Trailer Suspensions in the Transport Environment," *Packing Technology and Science* 5, pp. 11–15.

[50] Reinhall, P. G., R. Scheibe, M. A. Rauh, and S. V. Watkins ( 1995). The Effect of Vibration on the Shipment of Palletized Products, Proceedings of SAE Bus and Truce Meeting, Winston, NC.

[51] Rhodes, N. F. (1978). An Investigation of Vehicle Speed and Pavement Roughness Relationships for Texas Highways, Department of Civil Engineering, Texas A&M University, College Station, TX.

[52] Roberts, L. B., J. H. Dines, R. L. Hamlin, E. J. White (1969). Cardiovascular Effects of Vibration, Contract NGR 36-008-041 Columbus, OH. Department of Preventive Medicine, Ohio State University.

[53] Robinson, R., and P. A. Subramaniam (1976). Road Deterioration Parameters for Estimating Vehicle Operating Costs in Developing Countries, Overseas Unit, Transport and Road Research Laboratory (TRRL), Crowthorne, Berkshire, England.

[54] Sawhill, R.B., and J. C. Firey (1960). *Motor Transport Fuel Consumption Rates and Travel Time,* Highway Research Board, National Academy of Sciences, Bulletin 276 Washington, DC.

[55] Schmitz, M. A. (1959). "The Effect of Low-Frequency, High-Amplitude Whole-Body Vertical Vibration on Human Performance, Bostrom Research Laboratories," Milwaukee, WI.

[56] Sharpe, W. N., and T. J. Kusza (1973). *Preliminary Measurement of the Vibration Environment of Common Motor Carriers,* School of Packaging, Michigan State University, EastLansing, MI.

[57] Sime, M., and Ashmore, S. C. (January 2000). "WesTrack Roughness, Fuel Consumption, and Maintenance Costs," Federal Highway Administration Tech Brief, FHWA-RD-00-052, Washington, DC.

[58] Stephens, D. G. (1977). "Comparative Vibration Environments of Transportation Vehicles," Proceedings of the Design Engineering Technical Conference, Chicago IL, Sept. 26-28. pp. 59–72.

[59] ASTM (1996). Standard Test Method for Measuring Surface Macrotextrure Depth Using Volumetric Technique, ASTM Standard E 965, Volume 04.03, Sonshocken, PA.

[60] Stevens, H. (1961). "Line-Haul Trucking Costs in Relation to Vehicle Gross Weight," Highway Research Board, National Academy of Sciences, Bulletin 301 Washington, DC.

[61] Stikeleather, L. F., G. O. Hall, and A. O. Radke (January 1972). "A Study of Vehicle Vibration Spectra as Related to Seating Dynamics,"SAE Paper No. 720001, Tray, MI.

[62] Tate, F.N. (1993). "An Investigation into Traffic Delay at Road Surfacing Works," Appendix B in *Extended Guidelines for Road Surfacing Selection Based on an Analysis of Total Costs,* Research Report PR3-0051, Transit New Zealand.

[63] Transportation Efficiency Act for the 21st Century (TEA-21), H.R. 2400, enacted on June 9th 1998.

[64] Washington State Department of Labor and Industries, Occupational Hazards of Truck Driving, 1993.

[65] Washington State Department of Transportation, Standard Specifications for Roads Bridges and Municipal Construction, 2006 Edition. www.wsdot.wa.gov/fasc/EngineeringPublication/Manuals/2006SS.htm Accessed 07-2007.

[66] Watanatada, T., C. Harral, W. Paterson, A. Dhareshwar, A. Bhandari, and K. Tsunokawa (1987). *The Highway Design and Maintenance Standards Model*, Vol. 1, Description of the HDM-III Model, a World Bank Publication, Johns Hopkins University Press, Baltimore, MD.

[67] Watanatada, T., C. Harral, W. Paterson, A. Dhareshwar, A. Bhandari, and K. Tsunokawa (1987). *The Highway Design and Maintenance Standards Model*, Vol. 2, User's Manual for the HDM-III Model, a World Bank Publication, Johns Hopkins University Press, Baltimore, MD.

[68] White, J. A., K. E. Case, D. B. Pratt, and M. H. Agee, (1998). *Principles of Engineering Economic Analysis*, 4th ed., John Wiley & Sons, Ins., New York.

[69] Winfrey, R. (1969). *Economic Analysis for Highways*, International Textbook Company, Scranton, PA.

[70] Zaniewski, J. P. B. C. Butter, G. Cunningham, E. Elkins, M.S. Paggi and R. Machemehl (June 1982). ''Vehicle Operating Costs, Fuel Consumption, and Pavement Type and Condition Factors,'' Report No. FHWA/PL/82/001, Federal Highway Administration, Washington, DC.

## Problems

14.1 A company purchased a paving machine for $350,000 in 1992. What would the actual replacement cost for an equivalent machine have been in the year 2000; and, what would it be in, 2008, assuming an average inflation rate of 3.5%?

14.2 Explain why the incremental benefit-cost method is superior to the net method and the plain benefit-cost method in comparing economic alternatives.

14.3 Land was acquired for the right-of-way of a roadway in 1955 for $2 million. Its estimated value in 2001 was $36 million.

Calculate the inflation-free rate of return of this investment, given that the inflation runs at 2.5% average per annum. Carry out your calculations in terms of (a) actual dollars and (b) real (i.e., inflation-free) dollars.

14.4 Compare the agency costs for two flexible pavement treatments with the following characteristics:

- Capital cost of $60,000 per two-lane-km and an expected life of nine years
- Capital cost of $100,000 per two-lane-km and an expected life of 13 years

Assume a discount rate of 3.5%.

14.5 Compare the agency costs for a flexible and a rigid pavement with the following characteristics:

- Capital cost of $405,000 and an expected life of 15 years
- Capital cost of $675,000 and an expected life of 35 years

Assume a discount rate of 3.5%. Perform the calculations short-hand and through the built-in functions of a spreadsheet.

14.6 Compare the rolling resistance forces acting on a large passenger car operating on two different pavements with *IRI* roughnesses of 1 m/km and 6 m/km. Given:

- Vehicle speed of 110 km/h
- Radial tires
- *Tdsp* of 0.6 mm
- Vehicle mass $M$ of 1,800 kgr

Ignore the effect of pavement surface stiffness.

14.7 What is the difference in fuel consumption for the circumstance of the previous problem?

14.8 Compare the cost of vehicle maintenance parts and labor for a large passenger car that has an odometer reading of 100,000 km and operates on pavements with *IRI* of 6 m/km, versus operating on a pavement with an *IRI* of 1 m/km. Given:

- Purchase cost of the vehicle is $20,000
- Shop charges are $60 per hour labor

14.9 Calculate the delay and the associated cost for a two-lane lane closure of 250 m in length on a six-lane divided freeway. Given that:

- Arrival rate is 1,300 pce/hour in the 10-minute interval to be analyzed.
- No lower speed limit is posted through the work zone.
- All vehicles have double occupancy.
- Unit cost of travel time is $10/hour.

14.10 Calculate the total hourly delay for a 1,000 m lane closure on a two-lane undivided roadway with a directional volume of 800 pce/hour, a work zone cycle of 700 seconds, a work-zone capacity of 1,100 pce/hour, and a posted speed through the work zone of 45 km/hour (12.5 m/sec). Assume lost times at the beginning and end of the green interval of 2 seconds and 0 second respectively.

# Index